D0855552

Molecules and Life

MOLECULES AND LIFE

HISTORICAL ESSAYS ON THE INTERPLAY OF
CHEMISTRY AND BIOLOGY

JOSEPH S. FRUTON

Eugene Higgins Professor of Biochemistry
Yale University

WILEY-INTERSCIENCE, a Division of John Wiley & Sons, Inc.
New York · London · Sydney · Toronto

NMU LIBRARY

Copyright © 1972, by John Wiley & Sons, Inc.

All rights reserved. Published simultaneously in Canada.

No part of this book may be reproduced by any means, nor transmitted, nor translated into a machine language without the written permission of the publisher.

Library of Congress Cataloging in Publication Data

Fruton, Joseph Stewart, 1912–
Molecules and life.

Bibliography: p.
1. Biological chemistry—History.   I. Title.
QP511.F78   574.1'92   72–3095
ISBN 0–471–28448–3

Printed in the United States of America

10 9 8 7 6 5 4 3 2 1

This book is
dedicated to the memory of
my predecessors at Yale

RUSSELL HENRY CHITTENDEN

(1856-1943)

LAFAYETTE BENEDICT MENDEL

(1872-1935)

CYRIL NORMAN HUGH LONG

(1901-1970)

Preface

Recent achievements in the study of the chemistry of life have placed the biochemical sciences at the center of the scientific scene. One consequence of these successes has been the tendency to describe the present state of the interplay of chemistry and biology largely in terms of the events since 1950. The reasons are various; not the least of them is that the active scientific investigator usually values only the latest work bearing on his own research, the older results already having been woven into the conceptual fabric of his field. Moreover, some historical accounts have tended to nourish the illusion that the development of this interplay has been a succession of "breakthroughs" across a "frontier" that separates knowledge from ignorance. The euphoria that attends great advances, some of which have been labeled "revolutions," is understandable; a study of the historical record suggests, however, more continuity and complexity than are implied by such dramatic words. In these essays, I have tried to examine a portion of the record by tracing some of the efforts, during about 1800–1950, to study the chemistry of living organisms. I hope that students of the biochemical sciences will derive from this account of the past some illumination of the present, and that historians of science will find in it problems worthy of more intensive scrutiny.

The interplay of chemistry and biology since 1800 engaged the attention of investigators who followed different tracks, depending on their attitudes and on the tools and concepts they applied. To illustrate the interactions among these various kinds of scientists, I have selected five general topics for examination from a historical point of view. They are the nature of enzymes, the chemistry of proteins, the chemical basis of heredity, the role of oxygen in biological systems, and the chemical pathways of metabolism. Each of these interrelated topics presented problems that were considered important during the nineteenth century and that had been clearly defined by the middle of the twentieth.

In the preparation of this book, I have relied largely on the published description and discussion of scientific work shortly after it was done. The limitations inherent in later recollections of an event, especially when its significance has become evident, are well known to historians. This is no less true of the recollections of scientists, even though such reminiscences have a more human flavor than the usual austere scientific report. Although I have limited the account to the scientific developments, and have not dwelled on biographical details, I have attempted to bring the individual scientists closer to the reader by quoting, sometimes at length, from their scientific writings. The quotations are meant to be read as part of the running text, and are intended to provide a more accurate statement of a person's views than one based on hindsight. All the quotations are in English; except where otherwise noted, the translations from French and German are my own, and usually differ in detail from others that may be available. I am grateful to many authors and publishers for permission to quote or to translate copyrighted material; in each case, explicit indication of the source is given in the list of references.

I hope that the reader will not be unduly inconvenienced by the absence of a subject index. During the period 1800–1950 the same term was often used to denote different things, and what turned out to be the same thing was assigned different names. Given the difficult choice between an index likely to be misleading and no index at all, the second alternative seemed preferable.

The completion of this book was aided by a grant from the National Science Foundation (GS-30562). I am also indebted to fellow scientists who read the essays and offered valuable suggestions. My special gratitude goes to a former Yale colleague, Frederic L. Holmes, now Professor of the History of Medicine and Science at the University of Western Ontario; he read the entire manuscript in an early draft, and his cogent comments helped immeasurably in improving the text. Neither he nor any of the others should be held responsible for the shortcomings that will be apparent to the critical reader; for any errors of omission or commision, I alone am accountable. I reserve my final though no less heartfelt thanks for the devoted secretarial assistance of Mrs. Ida K. Nadel.

JOSEPH S. FRUTON

New Haven, Connecticut
July 1972

Contents

Molecules and Life

Introduction

These essays are an attempt to trace some of the threads in a pattern of scientific activity in which chemical explanations were sought for biological phenomena. We focus attention on the period after 1800, because this activity became a major preoccupation of leading scientists during the nineteenth century. The historical events we shall consider are reports of observations, experiments, and hypotheses, as well as the published discussions that attended these reports. What interests us about these particular events is that the people associated with them had similar objectives, but approached them with different attitudes and different methods. If, for convenience, we classify these various approaches into two types, the chemical and the biological, we also reflect the distinction made throughout the period 1800–1950 by the principal participants in the effort to understand the chemistry of life. As we trace the interplay of theory and experiment in this effort, however, it becomes evident that the pattern of its development was more complex than is suggested by a simple division of the participants into "chemists" and "biologists." Our historical problem, therefore, is to seek in the interactions of these various kinds of scientists some of the factors that were decisive in the development of the area of their common interest.

In searching for the influences that led to changes in direction or emphasis among a group of scientists working on similar problems, we must give first place to the discovery of new facts, the formulation of new theories, and the invention of new methods to test theories and to get facts. Consequently, although the history of science is more than a chronicle of the accumulation of scientific knowledge, no steps can be taken toward historical understanding without an accurate account of the relevant facts, theories, and methods. For this reason, I emphasize in these essays those observations, experiments, and hypotheses which, in my opinion, significantly influenced the course of the search for chemical explanations of biological phenomena.

1

Since science is a human activity, and not merely a catalogue of facts knitted by theory, such emphasis on the "internal" development of a scientific effort obscures the participants as individual human beings and as members of social groups. Many historians of science follow Carlyle's dictum that "history is the essence of innumerable biographies," and seek to discern from the life and work of a single prominent scientist the scientific temper of his time, the factors that influenced his development, and the impact of his work and thought on others. The historian is often obliged to demythologize men adulated by their contemporaries and depicted by biographers as heroes of science. Conversely, the biographical approach requires the historian to rescue from comparative oblivion men whose worth was appreciated only after many years of neglect. Such biographical studies are indispensable but run the risk of distorting the historical account by bringing an age to focus in the life of a single person.

In these essays, the work of some famous men will be given special attention because of their influence on the interplay of chemistry and biology, but there are also many men of lesser renown whose findings were important in their time; can we say that the latter deserve less attention as individuals? To accord each of these people a brief "potted" biography would hardly do justice to any of them, and would increase inordinately the size of this book. We must therefore leave to others the task of remedying the biographical deficiencies of these essays. At appropriate places in the text, biographies or obituary notices are cited; the reader is also urged to consult the valuable *Dictionary of Scientific Biography,* a multivolume work that began to be published in 1970.

There is a related aspect of our inquiry that cannot be evaded so cavalierly. The emergence of the interplay of chemistry and biology as an important scientific effort during the first decades of the nineteenth century was paralleled by a major change in the organization of scientific education; this was followed by the growing professionalization of science and the appearance of new social relationships among scientists. We must sketch briefly the course of this development, as it affected the chemists and biologists engaged in the study of the chemistry of life.

The changes in scientific education came initially in the universities of the German-speaking countries, and the chemists led the way; because of the numerous autonomous universities competing for talented teachers and students, and the growing interest in the applications of chemistry in agriculture and industry, the chemists succeeded in overcoming opposition to the support of their science. At the core of their programs was laboratory experimentation; relatively large numbers of students were trained, and by midcentury many of them had become professional scientists specializing in some area of chemistry. Other branches of science, such as mineralogy and physiology, also acquired an increasingly professional character; when William Whewell introduced the word "scientist" in 1840 (Ross, 1962), this process was already well advanced in the German universities.

The contrast to the teaching of chemistry in the eighteenth century was striking. In the universities, chemistry had been part of the medical curriculum, the other subjects usually being the theory and practice of medicine, "institutes of medicine" (later called physiology), anatomy, and botany. Hermann Boerhaave, the leading university teacher of chemistry during the first decades of the century, also taught medicine and botany at Leyden. Later, the position of leadership shifted to Edinburgh; in 1755 William Cullen became Professor of Chemistry and Physic there, and he was succeeded eleven years later by Joseph Black, his most celebrated pupil.

Eighteenth-century physicians did not need to do practical chemistry, however; this was the task of the pharmacists, who transmitted the art of chemical manipulation to their apprentices, and to affluent amateurs for whom chemistry was a diversion. A major center of such teaching was Paris, where pharmaceutical chemists like Guillaume François Rouelle, his younger brother Hilaire Martin, and Pierre Joseph Macquer taught chemistry to the men (among them Antoine Lavoisier) who fashioned what came to be called the Chemical Revolution. The subservience of chemistry to pharmacy, and its relation to botany, is evident from the fact that these teachers worked at the illustrious *Jardin du Roi*; it was renamed the *Jardin des Plantes* in 1790, and three years later it became the Museum of Natural History. This institution continued to be an important center of chemical research well into the nineteenth century, especially through the work of Michel Eugène Chevreul. As for the medical faculty of the university, before the Revolution Jean Bucquet taught pharmacy and chemistry there; his students included Antoine François Fourcroy, who succeeded Macquer at the royal garden in 1784. Among Fourcroy's laboratory apprentices was Louis Nicolas Vauquelin, his successor as Professor of Chemistry at the Faculty of Medicine and at the Museum. The tradition of practical chemistry was preserved during the Revolution, Directory, and First Empire; indeed, many of the principal participants in the interaction of chemistry and biology during the first half of the nineteenth century were French pharmacists (Berman, 1963). Chemistry did not develop rapidly in France, however; one consequence of the Napoleonic "reform" of French university education (in 1808), whereby it was placed under the domination of Paris, was the slow growth of science at the other universities. In this respect, the Germans were more fortunate; although they followed the French example in the encouragement of chemistry, and many of their leading chemists (including Justus Liebig) studied in Paris, they established numerous important university centers of chemical training. At first there were relatively modest efforts at Göttingen under Friedrich Stromeyer, and at Heidelberg under Leopold Gmelin. The decisive change came, however, after 1825, when Liebig set up his teaching laboratory at Giessen; this was soon followed by others, and by 1840 a constant stream of well-trained chemists was emerging in Germany. The chemistry carried forward in the previous century by practical chemists like Andreas Marggraf and Martin Heinrich Klaproth, with emphasis on its relation to pharmacy, metallurgy, and mining, had been in the tradition of the seventeenth-century adept Johann Rudolph Glauber; it now became a scholarly subject taught in universities to future highly specialized professionals. What happened in Germany was reflected on a smaller scale in Sweden, where pharmacy, metallurgy, and mining had provided much of the stimulus for the eighteenth-century work of Carl Wilhelm Scheele and Torbern Bergman; during most of the first half of the nineteenth century, the acknowledged leader of European chemistry was Jacob Berzelius, who taught at Uppsala. His most famous pupil was Friedrich Wöhler, whose teaching laboratory at Göttingen rivaled that of Liebig.

In the training of professional chemists, England lagged behind Germany and France. The amateur tradition of the Royal Society and the emphasis in Oxford and Cambridge on liberal, rather than professional, education were factors in the relative decline of British distinction in chemistry during the first half of the nineteenth century. The Royal Institution, founded in 1799 by Count Rumford, reflected the temper of English science; although equipped for the research of gifted men like Humphry Davy and, later, Michael Faraday, its principal educational function was to provide public enlightenment through popular lectures (Ironmonger, 1958). During the 1820s the leading English chemists included, in addition to Davy, William Hyde Wollaston and William Prout, both of whom were physicians who had private laboratories. By 1830 Davy and Wollaston had died, and the principal luminary was John Dalton. It is significant that he flourished in Manchester, a center of the Industrial Revolution, where middle-class virtues were more conducive to the development of professional science allied to technology. It was not until 1845 that the Royal College of Chemistry was opened in London, and headed by Liebig's former assistant, August Wilhelm Hofmann; eleven years later, a young student, William Henry Perkin, made the first of the aniline dyes (mauve). In 1851 Owens College, Manchester, was founded, and by the 1870s had the biggest school of chemistry in England. At the end of the century, chemistry was well established in the British universities, including Oxford and Cambridge. By that time, however, the chemical institutes of the German universities had turned out hundreds of professional chemists, most of whom were available to a burgeoning chemical industry that was exploiting with great efficiency discoveries such as that made by Perkin. In organic chemistry, the leading centers were those headed by Adolf von Baeyer (in Munich) and by his pupil Emil Fischer (in Berlin). Physical chemistry was emerging as a separate subject, and specialists were being trained in laboratories such as that of Wilhelm Ostwald (at Leipzig). The most talented of the chemical graduates remained in academic life and, as *Privatdozenten,* competed strenuously for professorial chairs. By that time also, the glory of French chemistry had begun to fade, and students no longer flocked to Paris from other countries. They went to Germany, and what they saw was copied at home. Before long, chemical research and instruction in American and Japanese universities were organized along the same lines as in the German-speaking institutions.

For our inquiry into the interplay of chemistry and biology it is significant that the professionalization of chemistry during the first decades of the nineteenth century was paralleled by a similar change in the university status of animal physiology. The initial steps were also taken in Germany, first by Carl Schultze at Freiburg and by Jan Evangelista Purkyně in Breslau, then (and most importantly) by Johannes Müller in Berlin. Although Müller's own outlook was largely that of a naturalist, he encouraged his students to use chemistry, physics, and microscopy in physiological research; the remarkable group (Theodor Schwann, Hermann Helmholtz, Rudolf Virchow, Emil duBois-Reymond, among others) trained in his laboratory included many of the leaders of nineteenth-century science. From 1865 onward, the most famous German institute of physiology was that of Carl Ludwig in Leipzig, but there were other important centers of professional training: Berlin (Emil duBois-Reymond), Bonn (Eduard Pflüger), Strassburg (Friedrich Leopold Goltz), and Vienna (Ernst Brücke) were only the most prominent ones. By that time, the German physiologists had largely freed themselves of the domination of the anatomists on the medical faculty; in many cases where physiology and anatomy were still combined, the professorships had been taken over by physiologists (see Eulner, 1970). Moreover, the scope of physiology was considered to range from the electrophysiology of duBois-Reymond to the physiological chemistry of Carl Voit (Munich) and Willy Kühne (Heidelberg), and to include microscopic anatomy. Thus Ludwig's institute had three departments: physiology, chemistry, and anatomy-histology. During the second half of the nineteenth century, the German physiologists clearly established their hegemony and, like the chemists, decisively influenced the development of physiology in other countries, including England, Russia, and the United States.

Although the main nineteenth-century centers of university research and instruction in experimental physiology were in Germany, the French physiologists also exerted a significant impact on the development of this field. François Magendie was the chief animator of the movement to replace anatomical observation by the use of vivisection and of chemical methods. It is noteworthy, however, that he gave private lectures and demonstrations in experimental physiology for about eighteen years before he was appointed Professor of Medicine at the *Collège de France* in 1831. His most famous pupil and successor, Claude Bernard, had many students after he began lecturing in 1853, and was later widely considered to be the greatest physiologist of the century. Bernard emphasized vivisection and chemical methods as the principal tools of experimental physiology, and tended to place less emphasis than his German colleagues on the use of physical instruments. This attitude may have been, in part, one of making a virtue of necessity, because of the limited support given physiological research (see Bernard, 1872, p. 207); no doubt it also reflected Bernard's own exceptional surgical skill. Subsequently, experimental surgery was extensively developed in other countries, notably in Russia, where Ivan Pavlov founded an important school of physiology. In England, the serious professional training of physiologists began around 1880, largely under the influence of Michael Foster, Professor of Physiology at Cambridge; before then, research in experimental physiology was conducted mostly by practicing physicians.

The transformation of animal physiology from a branch of human anatomy into a highly professionalized scientific discipline had its counterpart in the development of other subjects in the medical curriculum, for example pathology. Moreover, there was a parallel emergence during the nineteenth century of the professional training of biologists in university laboratories independent of the medical faculty. At the beginning of the century, part of what had been "natural history" came to be called "biology," a term introduced in 1800 by the German anatomist Karl Friedrich Burdach, and used in 1802 by the German physician and naturalist Gottfried Treviranus and the French naturalist Jean Baptiste Lamarck; they sought to emphasize the unique attributes and continuity of animal and plant life. The term was later popularized in the widely read books by Auguste Comte (*Cours de Philosophie Positive*) and by William Whewell (*Philosophy of the Inductive Sciences*). The clearest separation of biology from medicine came in botany. Thus, whereas in the eighteenth century the great Carl Linnaeus had been Professor of

Medicine and Botany at Uppsala, the leading botanists of the nineteenth century (for example, Carl von Nägeli) headed autonomous university institutes. Botany and zoology developed throughout the nineteenth century in university laboratories, in museums, and in biological stations such as the one opened in Naples during the 1860s. The English influence was strong, especially after Edwin Ray Lankester began to teach zoology at Oxford in 1870.

By 1900 the chemical and biological sciences had been organized in the universities in a pattern that was largely maintained throughout the first half of the twentieth century. New specialties continued to emerge, however; many among the growing scientific population were attracted to lines of research outside the main stream of established university subjects, and were seeking expanded opportunities for academic advancement. They demanded the right to train their own professionals and to be independent of their parent discipline. For the history of the interplay of chemistry and biology, the claims of the group that called themselves physiological chemists (later, biochemists) are of special interest.

Among the many fruits of the new chemistry developed by Lavoisier and his contemporaries was the rise of "animal chemistry." The impetus given this subject around 1800 by Fourcroy was later reinforced by Berzelius, and it was included in the programs of the new university chemical laboratories, notably that of Liebig. With the growing specialization in chemistry, animal chemistry came to be handled separately in the medical curriculum. An early example was at Tübingen, where Julius Eugen Schlossberger (who had studied with Liebig) in 1846 introduced practical instruction in medical chemistry; he had been preceded by Georg Sigwart who lectured on chemistry and botany from 1818 onward (Simmer, 1955). A parallel development was the emergence of agricultural chemistry as a separate subject, in response to increased official concern about human nutrition (Browne, 1944); the pioneers included Gerardus Mulder at Utrecht and Jean Baptiste Boussingault in Paris. By midcentury, chemistry was being applied to biological problems in three different kinds of university laboratories: those of chemistry (attached to the "philosophical" faculty), of medical chemistry (usually in the physiological or pathological institute of the medical faculty), and of agricultural chemistry (usually in technical schools).

During the second half of the nineteenth century, with the elaboration of the conceptual structure of organic chemistry as part of "pure" chemistry (*Experimentalchemie*), as well as its growing ties to the expanding chemical industry, the organic chemists paid less attention to biological problems. The main arena of the interplay of biology and chemistry was in laboratories of physiological (or pathological) chemistry of medical schools. This separation of the "pure" chemists from the "medical" chemists was accentuated by the fact that the style of research in medical chemistry was strongly influenced by the views of physiologists and pathologists, on whom the medical chemists depended for their academic posts. In many instances, it did not meet the experimental standards of the professional "pure" chemists; a widely quoted saying was *"Tierchemie ist Schmierchemie."* Although there were several leading physiological chemists (for example, Eugen Baumann at Freiburg) who did not merit this disdain, the quip was not entirely unjustified. Indeed, among the chemists working on biological problems, the agricultural chemists who studied plant products (especially Heinrich Ritthausen at Königsberg and Ernst Schulze at Zurich) came closer to the standards set by the professional organic chemists.

On the other hand, the nineteenth-century physiological chemists were exploring important biological problems, such as respiration or nutrition, and their labor produced important new knowledge. This was especially true of Felix Hoppe-Seyler and the large number of able men he trained (see Baumann and Kossel, 1895). He established the first independent institute of physiological chemistry at Tübingen in 1861; his students there included Friedrich Miescher. Hoppe-Seyler then moved in 1872 to Strassburg, where his laboratory became a leading center of professional training in the discipline; he was succeeded by Franz Hofmeister in 1896. For many decades, however, Tübingen and Strassburg remained the only German universities where physiological chemistry was an autonomous subject. In Berlin, for example, an independent institute of physiological chemistry was not set up until 1928; during the preceding fifty years outstanding biochemists (Ernst Salkowski, Albrecht Kossel, Leonor Michaelis, Hans Thierfelder, among others) held their posts in the physiological or pathological institute, or in one of the hospital clinics. A major part of the difficulty during this time lay in the attitude of leading physiologists. In 1877, when Hoppe-Seyler started the *Zeitschrift für physiologische Chemie,* he stated that "biochemistry . . . has grown to a science that has not only placed itself on a par with biophysics, but in activity and success competes with it for rank" (Hoppe-Seyler, 1877a, p. I). Pflüger, as editor of a journal that had published many papers by physiological chemists, countered with the statement that "the division of physiology into physiological physics and chemistry is philosophically inadmissible and impossible in practice" (Pflüger, 1877, p. 363). He concluded with an appeal to his colleagues "not to underestimate the danger . . . but to oppose strongly the divisive forces which now threaten the one great and glorious science of physiology" (*ibid.*, p. 365). This view was reflected in the statements of other German physiologists (for example, Ludimar Hermann in 1893), and partly explains the slow recognition of the claims made by their chemical associates for independent status. Nevertheless, around the turn of the century, the German physiological chemists exerted a major influence on the establishment of their discipline in other countries, for example Japan (Doke, 1969).

The importance of the attitude of influential physiologists to the aspirations of the physiological chemists was clearly evident in England, where Michael Foster (at Cambridge) and the neurophysiologist Charles Sherrington (at Liverpool) were largely responsible for establishing biochemistry as a separate university subject. In 1902, Benjamin Moore was made Professor of Biochemistry at Liverpool; four years later he founded the *Biochemical Journal*. In 1898 Foster invited Frederick Gowland Hopkins, a chemist at Guy's Hospital Medical School, London, to join his physiology department at Cambridge. It was not until 1914, however, that Hopkins was appointed Professor of Biochemistry (at the age of 53); after World War I his department became one of the leading centers of biochemical education in the world. In 1926 he described his view of the place of biochemistry in the university:

It does not seem to me enough that the biochemist, hitherto housed in an Institute of animal Physiology, and therefore mainly occupied with animal studies, should migrate to a separate building and retain his preoccupations. The greatest need of biochemistry at the moment in my opinion is equipment which shall make possible the study under one roof (of course, from its own special standpoint alone) of all living material. No full understanding of the dynamics of life as a whole, no broad and adequate views of metabolism, can be obtained save by studying with equal concentration the green plant and micro-organisms as well as the animal . . . . Such Institutes of General Biochemistry would have to be equipped for teaching as well as for research, for we have to prepare a future generation to bear burdens greater than our own. Modern physical chemistry, and modern organic chemistry, reacting as they already are—even the latter—to the newer concepts of Physics, are to provide entirely new concepts for biochemistry. There must be experts to understand and apply them; a task which will be as impossible for those who must continue to concern themselves with the rapid growth on other sides of physiology and biology, as it will be for the equally preoccupied pure chemist. I do not undervalue the difficulties of equipping and staffing an Institute of general biochemistry such as I am picturing, but I must not stop to discuss them. I will only state that we have, in not too ambitious fashion, attempted the task in Cambridge (Hopkins, 1926, pp. 35–36).

During the 1920s, research and training in the Biochemical Laboratory at Cambridge included microbial metabolism (Marjory Stephenson), chemical genetics (J. B. S. Haldane), and biological oxidations (Malcolm Dixon), in addition to Hopkins' own studies on a variety of topics. The biochemistry that developed after World War II was the general biochemistry that Hopkins had envisioned (see Fruton, 1951), and the people trained in his laboratory played a large role in its elaboration (Needham and Baldwin, 1949). Moreover, strong biochemistry departments had been established in other British universities, notably at Oxford under Rudolph Peters. When Jewish scientists left Germany during the 1930s to escape the Nazis many of them went to England; among them were young men, like Hans Krebs, who enriched British biochemistry further through their decisive contributions both to scientific knowledge and to the training of professional biochemists (Kornberg, 1968).

It was in the United States, however, that the most dramatic change occurred as a consequence of the emigration of German biochemists. Biochemistry was already firmly established in American universities, and had a strong tradition in such fields as nutrition, blood chemistry, and the chemistry of proteins and enzymes (Chittenden, 1930). In part, this biochemical strength arose from the interest in agricultural chemistry during the middle years of the nineteenth century; a leading figure in the early stages was Samuel W. Johnson of the newly established Sheffield Scientific School at Yale. One of his students was his son-in-law Thomas B. Osborne, whose work at the nearby Connecticut Agricultural Experiment Station greatly influenced the later development of protein chemistry. Another was Russell H. Chittenden, who studied during 1878–1879 with Kühne; in 1882 Chittenden became Professor of Physiological Chemistry at Yale. With his pupil and successor, Lafayette B. Mendel, he organized a leading center of professional training (see Rose, 1969). By 1910 several independent departments of physiological chemistry (or biological chemistry) had been established in American medical schools; Chittenden's students headed most of them. In view of the words of Hopkins, just quoted, it should be added that, in Chittenden's opinion,

. . . physiological chemistry should be considered as a biological science in the broadest sense of the term, as much so as zoölogy, comparative anatomy or physiology, ready however, to give its aid in any direction that might be called for but not be linked indissolubly with the science or art of medicine (Chittenden, 1930, p. 322).

These lines were written after Chittenden's unsuccessful opposition to the transfer in 1923 of the laboratory he had founded (then under Mendel) to the Yale School of Medicine.

During the period 1920–1940 the development of biochemistry in American universities occurred principally in medical school departments, but important centers of biochemical training had also been established in colleges of agriculture (especially the one at Wisconsin) and in some university departments of chemistry. Special mention must be made of the chemistry department at Illinois where an active division of biochemistry was directed from 1922 onward by William C. Rose, a student of Mendel. The biochemists trained there acquired an exceptionally thorough chemical preparation, and included many of the future leaders of their science. Another important center was the chemistry department at Columbia, where John M. Nelson worked on enzymes; the later work of his student John H. Northrop on the crystallization of enzymes marked a new stage in the growth of biochemistry. Most of the experimental papers published by American biochemists appeared in the *Journal of Biological Chemistry* (founded in 1905 by Christian A. Herter); by the 1930s its scope and quality made it one of the leading biochemical publications in the world (Clarke, 1955).

After 1932 the considerable strength of American biochemistry was enhanced by the addition of the organic chemistry of Emil Fischer's school, as well as the approach to metabolic problems developed by German physiological chemists. Among the numerous émigrés who exerted an effect on the further development of American biochemistry, Max Bergmann and Rudolf Schoenheimer were especially influential during the 1930s. With the decline of science in Germany, the United States and England became the principal centers of biochemical research and training. It should be added, however, that, during the Nazi rule of Germany, Otto Warburg in Berlin did work whose importance for the subsequent development of biochemical thought cannot be overestimated. Also, most of the leading organic chemists remained there, and several of them contributed significantly to the determination of the structure of biologically important substances.

During the first four decades of this century, the funds available to university biochemistry departments were fairly limited. For this reason, many of the significant advances were made in a few well-supported research institutes that were independent of the universities. Although they did not contribute to the education of predoctoral students, they offered exceptional opportunities for professional training at the post-doctoral level. Among the precursors of such institutions was the Pasteur Institute in Paris, opened in 1888 after funds had been collected by public subscription (Delaunay, 1962). Another was the Carlsberg Laboratory in Copenhagen, established in 1875 by the brewers Jacob Christian Jacobsen and his son Carl; it was an important center of research on proteins and enzymes under the successive leadership of Johan Kjeldahl, S. P. L. Sørensen, and Kaj Linderstrøm-Lang. Of special significance was the creation in 1901 of the Rockefeller Institute for Medical Research in New York (Corner, 1964); during the succeeding thirty years its members included such luminaries as Phoebus A. Levene, Jacques Loeb, Donald D. Van Slyke, Oswald T. Avery, Karl Landsteiner, Leonor Michaelis, and John H. Northrop, all of whom contributed decisively to the groundwork of modern biochemistry. The example of the Rockefeller Institute was imitated elsewhere; in 1911 the first of the Kaiser-Wilhelm Institutes was established in Germany (Fischer, 1924). During the 1920s those devoted to biochemistry represented major centers of research, with men such as Otto Warburg, Otto Meyerhof, and Max Bergmann in charge. In London, a counterpart of the Pasteur Institute was established in 1891; it later became the Lister Institute, where notable biochemical work was done by Arthur Harden, among others (Chick, Hume, and Macfarlane, 1971).

For the fruitful development of the interplay of chemistry and biology in the universities, few actions were so important as the decision of the Rockefeller Foundation in 1933 to encourage the application of physical and chemical techniques to the study of biological problems. The acumen of Warren Weaver, the Director of the Foundation's Division of Natural Sciences, led to major support of carefully chosen efforts such as those in genetics and structural chemistry at the California Institute of Technology, in X-ray crystallography at Cambridge, and in protein chemistry at Uppsala. Moreover, significant help was given to many young chemists and biologists, in particular those who had emigrated from Central Europe to the United States and England. The scientific advances after World War II made it clear that Weaver's choices during the 1930s reflected both remarkable foresight and sound judgment of the quality of men. Although operating with more limited funds, and restricted to Great Britain, the Medical Research Council (first under Walter Fletcher, then under Edward Mellanby) exerted a similar perceptive role in encouraging the efforts of selected biochemists and biologists (see Medawar, 1963). By the 1950s the seeds so nurtured in the earlier years were generously watered with government funds, in the wake of the expectatations generated by the success of nuclear physicists in constructing the atomic bomb and of chemists in producing penicillin; it seemed to many that a massive expenditure of money for biochemical research would produce a cure for cancer and other major diseases in a relatively short time (Strickland, 1971).

With the rapid growth of the scientific population after World War II, and the expanding support for research on the application of chemistry and physics to biological problems, new claims were made for the establishment of independent specialties. In 1945 biochemistry was still largely limited to the medical schools, but its successes during the preceding decade and the immediate postwar years led many biology departments to develop biochemical programs under such headings as "cell physiology." Similarly, some chemistry departments began to find the biological aspects of their subject less unattractive, and organized programs such as "biophysical chemistry." Moreover, the term "biophysics" changed its meaning. Since the nineteenth century it had been applied to the part of animal physiology that dealt with such phenomena as electrical conduction in nerve, muscular contraction, or the dynamics of blood circulation. A modern definition derived from this tradition is "the study of biological function, organization and structure by physical and physicochemical ideas and methods" (Hill, 1956, p. 1234). Hill emphasized that "the use of the latest physical methods and devices does not make one a physicist, and the employment of such things in a biological laboratory is not necessarily biophysics" (*ibid.*, p. 1235). Furthermore, he stressed the close interdependence of biophysics, as he defined it, and the biochemistry grounded in organic chemistry, since the full explanation of biophysical phenomena required a knowledge of the chemical dynamics in the underlying processes. From this point of view, biophysics and biochemistry are part of the "general physiology" promoted by Jacques Loeb, William Bayliss, and others during the early years of this century. After World War II, however, "biophysics" acquired other meanings, especially among physicists who turned to biological problems; around 1950, many of them considered it to be the application of nuclear physics to biology, as in the use of radioactive isotopes for metabolic studies or in research on the effect of ionizing radiations on biological materials. The importance of chemistry could not be long denied, however, and there began to appear such specialties as "molecular biophysics" (in many respects indistinguishable from biophysical chemistry).

By 1960 many of the adherents of the new biophysics, with some geneticists and cell physiologists, had shifted their allegiance to "molecular biology." The idea of the molecular basis of biological phenomena had reappeared frequently. To cite only two older examples, Henry

Bence-Jones (1867, p. 2) referred to "chemical or molecular disease," and Michael Foster (1877, p. 494) wrote: "The physiological function of any substance must depend ultimately on its molecular (including its chemical) nature." More recently, Weaver called attention to the fact that in his report for 1938 he had written:

> Among the studies to which the Foundation is giving support is a series in a relatively new field, which may be called molecular biology, in which delicate modern techniques are being used to investigate ever more minute details of certain life processes (Weaver, 1970, p. 582).

These studies included attempts to determine the structure of biological macromolecules by means of X-ray diffraction analysis. One of the leaders in this field during the 1930s was William Astbury; with characteristic exuberance he stated in 1945 that:

> We are at the dawn of a new era, the era of "molecular biology" as I like to call it, and there is an urgency about the need for more intensive application of physics and chemistry, and specially structural analysis, that is still not sufficiently appreciated (Astbury, 1947a, p. 326).

In recent years, there has been considerable discussion about the scope and origins of molecular biology [for example, see Stent (1968) and Hess (1970)]. Of special relevance are the comments of a British group, headed by John Kendrew:

> The fact is that the fashionable term "molecular biology" is unfortunate, on several grounds. Much of the research commonly held to be within this field (e.g., into the mechanism of protein synthesis and of DNA replication) is actually quite inseparable from biochemistry; in consequence, the term is resented by many biochemists who feel that in the eyes of the world they have no part in currently fashionable fields which in reality are their own territory and which in a sense they were the first to explore (Kendrew *et al.*, 1968, p. 2).

We leave it to future historians to sort out the multiple scientific, sociological, and personal factors in the competition during the 1960s among various scientific groups all engaged in the search for chemical explanations of biological phenomena. One of the symptoms of this competition was the proliferation, on an international scale, of new scientific periodicals and societies bearing distinctive names. Today, the participants include not only the numerous specialists who describe themselves as biochemists, biophysicists, or molecular biologists, but also members of a large variety of other professional groups in science, medicine, and technology. Indeed, the study of the chemistry of life has become the principal meeting-ground of present-day science.

This sketch of the changes, during 1800–1950, in the relationships among the scientists engaged in the interplay of chemistry and biology may perhaps suffice as a background for our consideration of the "internal" factors in its development. It will be evident that the systems of university organization, the relations of the participants to medicine, agriculture, and industry, the sources of financial support, and the competition among scientific groups for public attention were interrelated "external" determinants of our area of inquiry. As these have undergone change, in response to political and economic transformations in national societies, there was an impact on the climate for scientific work. Indeed, although we write of scientists largely in terms of their individual achievements in research, through insight, discovery, or invention, the professionalization of science has increased the relative importance of such qualities as the ability to organize team effort, to secure research funds, and to get favorable publicity. After 1900 the enormous growth in the scientific population and the mounting cost of research made these qualities even more important, and increased the significance of the social setting in which scientists work. In such circumstances, factors other than individual scientific genius assume a larger role, thus adding to the pitfalls in the biographical approach to the history of science.

In addition to the difficulties that face the historian of science in the evaluation of individual scientists through the study of their lives, or in the analysis of the impact of changing social relations on scientific effort, there are large questions about the extent to which the development of the interaction of chemistry and biology has been determined by prevailing philosophical ideas. Certainly, the problem of the relation of life to matter has been a compelling one throughout the history of recorded philosophical thought (Hall, 1969). Nor can there be doubt that, before 1800, the philosophical currents of the time were intimately interwoven with the study of biological phenomena (Roger, 1963). It is more difficult, however, to define the influence of nineteenth-century philosophical debates on the actual course of experimental work in chemistry and physiology. For example, there was much public discussion of the question whether the phenomena of life can be described without invoking "vital forces" inaccessible to study by the methods of chemistry and physics. Historical accounts of nineteenth-century science have tended to emphasize the significance of this debate as a factor in the development of the biological sciences, and we shall refer to it repeatedly in these essays. There is a strong likelihood, however, that such emphasis distorts the relative importance of the philosophical attitudes evident in the debate as compared with the increasing professionalization of chemists and physiologists, and the competition among them for preferment. It would seem that by midcentury, it was the particular scientific tradition and its methodology (*e.g.,* chemical analysis or vivisection) in which individual scientists had been trained that increasingly determined their choice of research problems and the mode of experimental attack. With the growing specialization in chemistry and physiology, the philosophical debates at the fringes of the scientific community had progressively less impact, and after World War I their effect on the course of the interplay of chemistry and biology appears to have been negligible.

Obviously, through their general education, modern scientists develop attitudes to philosophical questions that may determine whether they consider the search for chemical explanations of biological phenomena worthwhile. Once this commitment has been made, however, their work has usually followed a professional track, and the philosophical attitudes have receded into the background. If such attitudes were given prominence in a scientist's later account of a discovery that made him famous, his description of events often was more influenced by his philosophical views at the time of recollection than by any that he may have held when he did the work. Also, in addressing a wider audience, he may have wished to relate his work to the philosophical tradition in which his readers were educated. A notable nineteenth-century example is Claude Bernard's *Introduction to the Study of Experimental Medicine,* published in 1865; it became required reading for cultured Frenchmen as a description of the "scientific method." In his preface to the English translation of Bernard's book, Lawrence J. Henderson wrote that "we have here an honest and successful analysis of himself at work" (Bernard, 1927, pp. v–vi). Unfortunately for this judgment, the examination of Bernard's research notebooks and published articles around 1848, when he made the great discovery of the glycogenic function of the liver, reveals important inconsistencies with his later analysis (Grmek, 1968; Holmes, 1972). It would appear that this achievement was a consequence of assiduous following of trails started by others, and that Bernard's skill in experimental surgery may have been more important than the views he expressed later about the principles of the scientific method.

The debates over philosophical issues have continued to enliven public discourse about the relation of chemistry to biology, and the nineteenth-century battleground of vitalism versus mechanism has been replaced in the present century by organicism (or holism) versus reductionism. The historical record suggests, however, that these debates, and the catchwords associated with them, have reflected the acquisition of new knowledge more than they have inspired it. The impact of the findings of modern biology on the discussion of philosophical issues is undeniable; there is less evidence to indicate that the actual course of the search for chemical explanations of biological phenomena was directly influenced by these debates.

It must be added, however, that the nineteenth-century development of the concepts of matter and energy, with deep roots in philosophical thought, significantly affected the attitudes of the participants in the interplay of chemistry and biology. In seeking chemical explanations for biological phenomena, they attempted to describe the familiar attributes of "life"—reproduction, growth, development, nutrition, respiration, irritability, etc.—in terms of the unfamiliar units of which the chemical constituents of living organisms were thought to be composed. One of the fundamental questions was whether the specific structure and function of biological organisms reflect the properties of different kinds of units or differences in motion within a uniform matter. By 1860 these units were the molecules defined by the chemists as the smallest portion of a substance that retains its properties in chemical reactions, and by the physicists as the smallest portion of the substance that moves as a whole. The word "molecule" (Latin *moles*, mass) had been used in various ways. In 1749 the naturalist Buffon had suggested that living organisms are composed of *molécules organiques*, and forty years later the mineralogist Haüy proposed that crystals represent arrangements of *molécules intégrantes*; the analogy between the molecular structure of living matter and of crystals was drawn repeatedly during the period 1800–1950. After John Dalton's atomic hypothesis, Amedeo Avogadro wrote in 1811 of "integral molecules" and "elementary molecules" to denote what were later called the molecules of a chemical compound and the atoms of the elements of which it is composed. After 1860 the chemists had accepted Avogadro's principle as a basis for the calculation of the atomic weights of the elements, and were proceeding to the formulation of the molecular structure of organic compounds in terms of the ideas of valency and of stereochemistry. On the other hand, although the behavior of the molecules postulated in the kinetic theory of gases was consistent with the deductions drawn from the study of chemical combination, many nineteenth-century physicists considered the existence of molecules to be a fiction. Instead, emphasis was placed on energy relations in natural transformations, and a growing group of physical chemists stressed the importance of chemical thermodynamics. For several decades at the close of the nineteenth century, therefore, organic chemists were constructing an increasingly powerful conceptual edifice based on the detailed structure of molecules whose reality was questioned by many physicists and physical chemists. More specifically, around 1900, organic chemists like Emil Fischer considered proteins to be composed of molecules whose structural analysis was possible, whereas physical chemists like Wilhelm Ostwald and Svante Arrhenius emphasized their physical behavior in colloidal systems. These differences were not mutually exclusive, and found their meeting-ground during the 1920s, but the biologists who inclined toward the organic chemists frequently found themselves at odds with those who followed the physical chemists. As we trace the threads of other problems that occupied the attention of chemists, biochemists, biophysicists, and biologists, we will find repeated evidence of the differences in their attitude to fundamental questions regarding molecular structure and energy relations in biological systems.

A related source of divergence among investigators seeking chemical explanations of biological phenomena was the problem of the "purity" or "homogeneity" of the various materials obtained from living organisms. It was clearly recognized during the latter half of the eighteenth century that the classification of chemical substances and the development of chemical theory depend on one's ability to isolate pure substances with reproducible properties. The purest substances known at the start of the nineteenth century were crystalline materials, and the subsequent interplay of mineralogy (long a branch of natural history) with new chemical concepts and analytical methods laid much of the groundwork for the later development of structural chemistry (see Hooykaas, 1958). Many of the chemical materials considered to be associated with the phenomena of life (notably the albuminoid substances included among the "colloids") did not crystallize readily, however, and the chemists who studied them placed reliance on such criteria of purity as elemental analysis, color tests, and the like. The uncertainty of these criteria became evident during the nineteenth century, but so long as no better ones were available they continued to be used. Moreover, when the study of a physiological phenomenon suggested that it is associated with a discrete functional unit, the question arose whether this unit represents a chemical substance that can be isolated from the organism under study. Much of the history of such terms as "enzyme" (Kühne) or "organizer" (Spemann) reflects the difficulties encountered in attempts to isolate and to purify these presumed physiological units. Many entities of uncertain nature were given names whose adoption and frequent repetition conferred upon these units a reality later found to be illusory. So long as the available techniques did not permit the isolation of such units in the form of well-defined chemical substances, the possibility was considered that the biological process under study was caused by "forces" or "fields" associated with nonspecific physical phenomena (*e.g.*, adsorption on surfaces) or with biological "organization." Although the attitudes to such questions were not mutually exclusive, they led to separate experimental tracks and were often a source of dispute. In instances where the tracks merged, it was usually through the successful isolation and characterization of the physiological unit as a unique chemical substance whose molecular structure could be established. For this reason, an important place must be assigned in our account to technical advances in the methods for separating chemical substances. At each stage in the improvement of such techniques, however, questions remained regarding the chemical purity of isolated constituents of biological systems. For example, as recently as thirty years ago, N. W. Pirie (1940) and others called attention to the uncertainties in the definition of the homogeneity of large molecules such as the proteins and nucleic acids. Although some of these uncertainties have been removed since then, major difficulties remain, and continue to lead to disputes about the identification of a particular physiological effect with an individual chemical substance.

The various differences in attitude among chemists and biologists have made themselves evident through differences in their scientific language. As Hopkins stated in a lecture delivered in 1927:

> Biology and chemistry, though in their infancy both foster-children of medicine and passing their childhood in company, have long occupied domains which, though never really far apart, have sometimes seemed to be so. But their proper domains, while extending in all directions, have now definitely met at a frontier which is surely a region of supreme importance for both. At such a frontier, as at political frontiers, there is apt to be confusion of tongues and some mutual misunderstanding of aims and methods which may make difficult the proper cultivation of the soil. If inhabitants can be found for this borderland capable of learning the language and methods of both domains, they should certainly be sought and trained . . . . One reason for a certain lack of sympathy between the chemist and the biologist—fast disappearing doubtless, but still to be felt—as they meet at their borderland is a difference in language which has existed almost ever since either had much to tell the other. This is associated with a certain difference of mental attitude towards the facts of Na-

ture, which is again due to a difference in their special preoccupation. The biologist has been concerned with form and structure, and has tended to visualise Nature as a product of the eye itself—unaided or artificially aided. The evolution of form; the meaning of changes in form; the significance of what is visible in the phenomena of life; these, though not, of course, the concern of him alone, nor exclusive concern, have been the main preoccupations of the biologist. Hence arose what we may term biological language . . . . On the other hand, the pure chemist . . . long studied phenomena only after they had been first reduced by his own art and methods to their simplest form, and he dealt with systems made so far as possible homogeneous . . . . So far as he visualised, it was with his mind's eye; for the objective apparatus he used was of secondary importance and accidental. His thoughts were based on the properties of the invisible and indivisible atom.

It is making, I know, a great claim, but I believe that it will be the ultimate privilege of advancing biochemistry to tempt all biologists —including the physician—always to picture mentally—as a habit of mind—the molecular events which underlie the changes of form and visible appearances which interest them, and, on the other hand, to demonstrate to chemists that the molecular events they have studied so fully in systems more or less homogeneous, gain enormously in interest, in spite of the complications involved, when they are organised and co-ordinated in systems involving changing form and elaborate structure.

If it be admitted that the ultimate aim, or at least the constant endeavour, of biochemistry as an academic pursuit should be to describe the activities of living organisms in terms of physical and organic chemistry (and I for one believe that the description will in such terms be as complete as any that can be given in their own more specialised terms by the morphologist or by the physiologist when he deals with functions in terms of organs); this description will possess the merits due to the use of a more universal scientific language (Needham and Baldwin, 1949, pp. 179–180).

Indeed, during the forty-odd years since these words were spoken, the molecular interpretation of biological form and function has provided the rudiments of a language which, if not yet "universal," today makes communication among chemists and biologists less difficult than before.

Enough has perhaps been said to suggest that we shall find in the interplay of chemistry and biology during 1800–1950 an intricate pattern of scientific activity, in which a growing number of scientists have used an increasing variety of concepts and techniques in the search for chemical explanations of biological phenomena. As we trace and retrace the development of some of the problems attacked during this period, it will be seen that the accumulation of factual knowledge knitted by theory proceeded along separate tracks, some of which partly merged to produce new ones, whereas others lost importance for a time and were "rediscovered" later. All the tracks whose fusion contributed to the present-day knowledge of the chemistry of life are not included in these essays. This book does not constitute, therefore, a comprehensive history of physiological chemistry, or biochemistry, or molecular biology. One such work already exists in the valuable book by the Austrian biochemist Fritz Lieben (1935), who dealt with the development of physiological chemistry to about 1930.

To conclude this introduction, mention should be made of the difficulties facing a professional scientist who undertakes to describe the historical development of problems within his own specialty. Such efforts are not regarded highly by some present-day historians of science (for example, see Kuhn, 1971, p. 289), largely because a person bred in a particular scientific discipline readily falls victim to the fallacy of the "Whig interpretation of history," defined as

> . . . the tendency in many historians to write on the side of Protestants and Whigs, to praise revolutions provided they have been successful, to emphasise certain principles of progress in the past and to produce a story which is the ratification if not the glorification of the present (Butterfield, 1931, p. v).

In histories of science, this tendency usually finds expression in judgments whether a particular scientist exerted a favorable or unfavorable influence on the movement of a field toward its present state of development (more crudely, whether he was a "good guy" or a "bad guy"). Other manifestations are an excessive concern about matters of priority (who made the "breakthrough") and about anticipations of important advances (sometimes denoted "premature" discoveries). Professional historians are not immune to this tendency, but amateurs, such as the author of these essays, are peculiarly vulnerable. At issue is the process of selecting and weighting the evidence gathered in the study of a historical problem, and I must leave it to the informed reader and to later research to determine the extent to which my selection and emphasis require correction.

From Ferments to Enzymes

At the close of the nineteenth century, Eduard Buchner reported that he had prepared, from brewer's yeast, a cell-free press juice that caused carbon dioxide and ethyl alcohol to form in solutions of various sugars (sucrose, glucose, fructose, maltose). He concluded that

> . . . the initiation of the fermentation process does not require so complicated an apparatus as is represented by the yeast cell. The agent responsible for the fermenting action of the press juice is rather to be regarded as a dissolved substance, doubtless a protein; this will be denoted *zymase* (Buchner, 1897, pp. 119–120).

Although the preparation of Buchner's zymase presented difficulty, since some yeasts failed to give an active extract, and the first step of grinding pressed yeast with fine quartz sand and kieselguhr (a fine diatomaceous earth) was very tedious, several investigators (*e.g.*, Green, 1898; Macfadyen *et al.*, 1900) soon confirmed his report. The acceptance of Buchner's claim marked a decisive stage in a scientific controversy that agitated many minds throughout the nineteenth century. As in the development of other biochemical problems, the organismic view that a chemical process occurring within living cells is indissolubly linked to their life gave way to the conviction that it might be possible to dissect the intracellular apparatus involved in the process. In this chapter we shall attempt to examine why many of Buchner's contemporaries attached great importance to his discovery, and why one of them wrote: "The ancient conflict over the question 'What is fermentation?' had ended: fermentation is a chemical process" (Ahrens, 1902, p. 494). To set the stage for the controversy about fermentation during the nineteenth century, we begin with a brief sketch of the history of the problem to the time of Lavoisier.

22

The Problem of Fermentation to 1800

The process of fermentation had its origin in the kitchens of prehistory, probably with the rise of organized agriculture in Neolithic times. The long-known effect of leaven (a mass of yeast; Greek, *zymē*) on a cereal dough, with the evolution of gas and a change in the texture of the solid matter, had its counterpart in the effervescence when honey or sweet fruit juices were kept in a warm place. Before 2000 B.C., the Egyptians knew that when crushed fruits (*e.g.*, dates) were stored a material having a pleasantly intoxicating power was produced at first, but that, if the mixture was allowed to stand for a longer time it turned sour to yield vinegar, the strongest acid known to antiquity. This souring of wine was considered to be comparable to the souring of milk. By 1500 B.C., the use of germinated cereals (malt) for the preparation of beer from bread, and the formation of wines arising from changes in crushed grapes, were established technical arts in Mesopotamia, Palestine, and Egypt. The ancient artisans observed that, if the mixtures were allowed to stand a sufficiently long time, the production of beer, wine, or vinegar was followed by changes which liberated noxious odors; this slow putrefaction of plant material was compared to the more rapid decay of animal and human tissues. The practical arts of preserving animal foods —drying, smoking, curing, pickling in brine, treatment with dry granular salt—were well developed in the prehistoric Near East and Europe, and dehydration with natron (a mixture of salts, largely sodium carbonate) was a key operation in the mummification procedures in ancient Egypt (Forbes, 1954).

These ancient arts provided empirical data for the speculations of Greek natural philosophers about the nature of fermentation (Greek, *zymōsis*) and putrefaction (Greek, *sepsis*). Of particular importance in the development of Western thought about these phenomena were the writings attributed to Aristotle; the form in which the books of Aristotle exerted their effect was the text edited and published in 70 B.C. by Andronicos of Rhodes, the eleventh Scholarch of the Lyceum. As these writings passed through the hands of Greek, Syriac, Arabic, and Latin scholars during the succeeding thirteen centuries, the Aristotelian books became a unified and logically consistent system of philosophy and an encyclopedia of trusted knowledge about natural phenomena. By 1400 the Aristotle revered in the universities of western Europe was encrusted with the commentaries of Avicenna, Averroës, Maimonides, Albertus

Magnus, and Thomas Aquinas. This was the Aristotle whose description, comparison, and explanation of natural phenomena provided the basis of philosophical thought until the seventeenth century.

Aristotle considered the changes undergone by inanimate things to be analogous to those seen in the biological world. Thus grape juice is the infantile form of wine, and fermentation is a process of maturation; the further change to vinegar is the death of the wine. Furthermore, the juice matures through a "concoction" (*pepsis*) promoted by heat, as in the physiological transformation of nutrients through the innate heat of the animal body. In Aristotle's philosophy, such natural change, as distinct from the analogous artificial cooking of food, expresses the tendency of an object to function toward the attainment of a specific end; in a living thing, this property (*psyche*) is inherent in the organism as a whole and arises from the integrated functions of its parts (see *De Anima* 412^b, 415^a). In the centuries after Aristotle, the idea of *psyche* as a nonmaterial principle of biological organization became mingled with the idea of *pneuma,* the ethereal stuff postulated by the Stoics as a principle of cohesion and activity in all matter, both living and nonliving (Sambursky, 1959). Like *psyche, pneuma* is a word used in Greek literature for human breath, and the idea that *pneuma* is a subtle and volatile entity essential for life became popular among Greek physicians and natural philosophers before the time of Galen. Thus, in addition to the four fundamental elements—earth, water, fire, and air—there was thought to be a fifth essence (*quinta essentia*) which represents the vital principle that determines the specific nature and activity of all material things (Taylor, 1953). This *pneuma* became, in subsequent translation, the Latin *spiritus,* the French *esprit,* the German *Geist,* and the English *spirit.*

With the rise of alchemy during the first five centuries of the Christian era, the conviction grew that it should be possible to isolate the quintessences of things. Since *pneuma* was considered to be a volatile stuff, the early Greek alchemists distilled various materials, saw in the expulsion of fumes and vapors the liberation of the spirit characteristic of each material, and identified these fumes and vapors with powerful, divine agents that gave specific life to each thing, whether it was an egg or a metallic ore. In about the twelfth century the art of distillation had developed to the stage where highly volatile distillates could be collected by cooling the receiving flask, and the distillation of wine was found to yield an inflammable liquid (*aqua ardens*). By the thirteenth century the effects of drinking this "burning water" were well recognized, and it came to be called *aqua vitae,* whence all the familiar variants— akvavit, uisge beatha, usquebaugh, whisky, etc.—are derived. For the medieval alchemists, what had been isolated was the quintessence of wine, and it followed logically that in order to extract the proper quintessence from any other animal or vegetable material, one should allow it to ferment or putrefy, and then distill off a "water" that could be purified by redistillation. The persistence of this idea for about four centuries is indicated by the following extract from the popular chemical treatise of the French pharmacist Jean Beguin (Patterson, 1937). In this book, entitled *Tyrocinium Chymicum* (The Chemical Beginner), first published in Latin in 1610, Beguin wrote:

> The word quintessence refers to a substance that is ethereal, celestial, and extremely subtle . . . Some call it *Médecine par excellence,* others *Elixir,* because of the signal virtues it possesses in the protection of the human body from various ailments; others call it Heaven for two reasons. First of all, because heaven is composed, not of the four elements, but of a certain ethereal matter, or fifth element, and is entirely incorruptible . . . . Secondly, because just as heaven acts powerfully on sublunary things, giving life to all things and conserving their vitality, so the quintessence conserves the health of man, prolongs youth, retards old age, and drives away all sorts of diseases (Beguin, 1624, p. 414).

This passage may be compared to the lines in Ben Jonson's *The Alchemist* (first performed in 1610), in which the credulous Sir Epicure Mammon speaks of the elixir:

> 'Tis a secret
> Of nature, naturized 'gainst all infections,
> Cures all diseases, coming of all causes,
> A month's grief in a day; a year's in twelve:
> And, of what age soever, in a month.
> Past all the doses of your drugging doctors,
> I'll undertake, withal, to fright the plague
> Out o' the kingdom in three months   (Act II, Scene I).

Beguin's book contains detailed recipes for the preparation of the quintessences of various animal and plant materials (*e.g.,* the quintessence of blood—"a remedy of sovereign power"). In these recipes the first step in the operation was to imitate the process whereby crushed fruits are converted into wine. This process was termed *fermentatio;* when, as in the conversion of a cereal dough to bread, an agent such as yeast was required, the agent was termed *fermentum.* It should be

NMU LIBRARY

stressed, however, that medieval alchemists such as Albertus Magnus and Raymond Lull applied the word *fermentatio* to various kinds of natural change, in living and inanimate matter, and for them *fermentum* might be any reactive substance, or even the Philosopher's Stone. The historian of chemistry Hermann Kopp stated:

> I doubt that I can formulate a clear idea of the sense in which the alchemists used the terms *fermentatio* and *fermentum*. These expressions were applied in a very general manner by the alchemists of the 13th to the 15th century. No distinction was made between inorganic and organic substances; many thought that metals arose from a kind of seed; expressions which are now only applied to organic substances were also applied to inorganic materials. Thus, one frequently finds among the writers of that time the term putrefaction for the slow dissolution of an inorganic material . . . . (Kopp, 1847, Vol. 4, p. 286).

That this all-embracing and confused definition of *fermentatio* and *fermentum* lingered up to the seventeenth century is indicated by the following two extracts from a famous chemical treatise published in 1597:

> Fermentation is the exaltation of a substance in its essence through the admixture of a ferment which, by virtue of its spirit, penetrates the mass and transforms it into its own nature (Libavius, 1964, p. 103).

A small quantity of specially prepared "medicine" is thoroughly mixed with a substance (*e.g.*, a base metal) and brings it to life (*vivificatio*) or reawakens it (*resuscitatio*). "The ferment acts through its inner heat, but this must be activated by an external heat of not too high a degree, lest the spirit be driven off . . ." (*ibid.*, p. 104). The idea that fermentation was akin to cooking had pervaded Western thought since the time of Aristotle (see *Metereologica* 380[b]). As stated by Stahl a century after Libavius:

> From the origin of the word, *fermentatio* means the same as *fervimentatio* (heating) and *fermentum* the same as *fervimentum* (a thing that heats); in the use of the word, however, there is the difference that some saw the identity with heating in the effervescence, and others in the boiling or warming of the material (Stahl, 1748, p. 9).

Moreover, around 1700 the natural transformations seen in inanimate objects were often explained in terms of the familiar changes in living things; such biological explanations of chemical phenomena were consistent with the widely held view of "The Great Chain of Being" (Lovejoy, 1936).

During the seventeenth century these ideas about fermentation occupied a central place in the writings of a group of "chemical physicians" whose revolt against the medicine of Galen and the materia medica of the first-century herbalist Dioscorides had been launched in the sixteenth century by Theophrastus Bombastus of Hohenheim, who called himself Paracelsus. The Galenic doctrine, built upon the medical writings attributed to Hippocrates as well as the physiological ideas of Aristotle and the Stoics, assumed that human health depends on a balance of forces specifically associated with various fluids ("humors") of the body: blood, yellow bile, black bile, and phlegm (nasal outflow). These humors were related to each other by affinities or antagonisms defined by their presumed special relationship to the four elements of Empedocles: blood and fire (hot and dry), yellow bile and air (hot and wet), black bile and earth (cold and dry), phlegm and water (cold and wet). The objective of good medical practice, according to this doctrine, was to restore the balance of the humors by such treatment as bleeding or by purgation with plant extracts.

By the seventeenth century there was a ready ear among propertied classes for revolt against the inadequacies of Galenic medicine. The rapid and continuous development of Europe after the Crusades was marked by exploration, conquest, and exploitation of non-European territory, by the more efficient use of wind and water power, and by the adoption of numerous inventions (*e.g.*, windmill, compass, rudder) imported from Islam or Byzantium. At the end of the sixteenth century this development had produced great wealth, the growth of cities, and the emergence of a class of craftsmen skilled in the practical arts of shipbuilding, mining, and metallurgy, as well as in the preparation of materials (*e.g.*, leather, paper, ceramics, gunpowder) needed for trade and war. Because of the rise of these practical men, Europe began to export goods instead of bullion, and to assume the economic dominance that reached its peak during the nineteenth century. In this new prosperity, however, the medical problems arising from urbanization, long sea voyages, and armed conflict assumed larger social importance. The terror generated by diseases such as the bubonic plague (especially the Black Death of 1347–1348) and syphilis, and the inability of Galenic medicine to meet the challenge, made men receptive to calls for reform.

The revolt against Galenic medicine was also furthered, as were other uprisings of the sixteenth century, by the diffusion of knowledge through books written by scholars. Among these books those dealing with the pyrotechnical arts and with metallurgy played a large role in the rise of chemistry (Hirsch, 1950); such works as Brunschwig's *Liber de Arti distillandi* (1500), Agricola's *De Re Metallica* (1530), and Biringuccio's *Pirotechnia* (1540) were widely read and went through numerous editions. Not only did some of the scholars turn to craftsmen for knowledge of the practical arts as a way to understand the workings of nature, but they also did not scruple to soil their hands in the chemical laboratory. In thinking about natural phenomena, however, the sixteenth-century chemical philosophers were inspired not only by Aristotle, but also by the magic of Hermes Trismegistos and the Cabala, as popularized by Marsilio Ficino, Pico della Mirandola, and Cornelius Agrippa. Thus, in his attack on Galenic medicine, Paracelsus combined chemical technology and Hermetic mysticism to create a philosophical basis for medical practice. He assigned special importance to the use of mineral substances (*e.g.*, salts of antimony or mercury) in the treatment of disease, not only externally (a standard practice in medieval Arabic medicine), but for internal administration as well. He strongly advocated the medical values of vitriol and of various quintessences; one of the best-known of the elixirs was *aurum potabile* (first described in the fourteenth century), prepared by circulating the spirit of wine (he termed it *alcool vini*) with gold leaves for four weeks. This kind of "spagyric medicine" or "iatrochemistry" had its origins in fourteenth-century writings attributed to John of Rupescissa (Multhauf, 1954), but only achieved prominence in the sixteenth, and the gradual appearance of the chemical medicines in the materia medica of the early seventeenth century (as in the London pharmacopoeia of 1618) attests to their acceptance by many physicians (Debus, 1965). In France the Paracelsian medicine had a powerful ally in Duchesne, physician to King Henri IV, and in Germany Croll and Hartmann helped to spread the doctrine.

Much has been written about Paracelsus, who was one of the numerous controversial figures in a turbulent time, and many of the obscurities in his arcane writings have been interpreted by modern scholars, notably Sudhoff, Strunz, and Pagel (see Pagel, 1958, 1962a). Paracelsus proposed that the Galenic humors and their correspondence to the four "elements" be replaced by a triad of chemical properties: combustibility (termed "sulfur"), fluidity and changeability (termed "mercury"), solidity and permanence (termed "salt"). For him the interaction of these properties is responsible for health and disease in man, as well as for the generation of minerals and plants. Thus the task of the physician in treating a disease is to restore the interaction characteristic of health by using a medicine derived from an appropriate mineral or plant in which the desired property is predominant. In a sense, therefore, the medical doctrine of Paracelsus was a new humoralism, but it emphasized the use of specific medicines for specific diseases, and sought to achieve by means of chemical agents what, according to the Galenic doctrine, could be achieved within the human body only by treatments such as bleeding or purgation.

It is debatable whether Paracelsus' advocacy of chemical medicines greatly improved the medical practice of the sixteenth century. That he stirred up the medical profession is unquestioned; he is reported to have told an irate group of Basel physicians: "If you will not hear the mysteries of putrefactive fermentation, you are unworthy of the name of physicians" (Waite, 1888, p. 139). Nor can there be doubt that his ideas influenced seventeenth-century physicians to reformulate questions about the nature of living things in the language of the chemistry of their time, especially in relation to the process of fermentation. The most important of these followers of Paracelsus was Johannes Baptista Van Helmont.

In the writings of Van Helmont there is a large measure of Christian piety and of Hermetic mysticism (Pagel, 1944). Embedded in the mass of arcana, however, were reports of experimental observations that attracted the attention of physicians, apothecaries, and natural philosophers after the publication of his complete works [Latin, *Ortus Medicinae,* 1648; English, *Oriatrike,* 1662; French, *Œuvres,* 1671 (partial); German, *Schrifften,* 1683], and popularized in chemical textbooks such as that of LeFevre (in 1660). For example, Robert Boyle (1680, p. 112) found Van Helmont "more considerable for his experiments than many learned men are pleased to think him."

Fermentation occupied an important place in the natural philosophy of Van Helmont. He wrote that, if one burned 62 pounds of coal, and one pound of ashes was formed,

> . . . the 61 remaining pounds are the wild Spirit . . . . I call this Spirit, unknown hitherto, the new name of Gas, which can neither be constrained by Vessels, nor reduced into a visible body, unless the seed being first extinguished. But Bodies do contain this spirit . . . a Spirit grown together, coagulated after the manner of a body, and is stirred up by an attained ferment, as in Wine, the juyce of unripe Grapes, bread, hydromel, or water and Honey . . . . (Van Helmont, 1662, p. 106).

For Van Helmont, "gas" is different from an "air" because it is derived from a substance by the action of a ferment, and is different from "vapor" because, unlike water vapor, it is dry (Strunz, 1913; Pagel, 1962b). The "wild spirit" (*spiritus sylvester, gas silvestre*) also arose by the action of acids on "salt" (a word with many meanings at that time—in this case, a carbonate), and Van Helmont in noting (by taste) the acid character of gastric juice was led to explain digestion in the stomach as a process in which the action of a ferment was aided by acid (Multhauf, 1955; Pagel, 1955, 1956). He spoke of other kinds of fermentation to which food is subjected in passing through the body (the duodenum, the liver, the heart, etc.); later, terms such as chylosis, chymosis, hematosis, pneumatosis, and spermatosis were applied to these various kinds of fermentation. It is easy today to find in Van Helmont's ideas the outline of our present views about the role of enzymes in metabolism, and he has been called "the founder of Biochemistry" (Pagel, 1944, p. v); it should be noted, however, that for Van Helmont "gas" had an occult significance as a divine seed responsible for the specific form and function of a natural body, and "ferment" was an innate formative energy under the control of the *Archeus* (the Aristotelian principle, as modified by Gnostic philosophy, that represents the Idea of the body). It was left for later writers to strip away the mystical meanings of "gas" and "ferment," and to find in Van Helmont's empirical observations a basis for fruitful discussion of the problem of fermentation. Of special importance was the distinction he made between different gases. Thus, in addition to "gas sylvestre," there were an inflammable "gas pingue" (gas of dung) and a "gas sulfuris" formed on burning sulfur, among others. It was to be a long time before these three gases were to be named carbon dioxide, methane, and sulfur dioxide, but their description by Van Helmont represented the starting point of later investigation.

Among the contemporaries of Van Helmont were several men whose writings represented a bifurcation of the Paracelsian doctrine, with increased emphasis either on its mysticism or on its iatrochemistry. The mystics included Jakob Böhme who combined Paracelsus with the Bible to develop a pantheism that finds later reflection in the *Naturphilosophie* of Goethe (Gray, 1952, pp. 38–53) and of other German thinkers (Kant, Schelling, Hegel), as well as in the writings of William Blake. Another was the Oxford Rosicrucian Robert Fludd who emphasized the Hermetic and cabalistic aspects of Paracelsism. The iatrochemists who rejected much of the supernaturalism of Paracelsus included the influential Daniel Sennert whose interest in medical chemistry led him to adopt a corpuscular theory of matter, and to consider fermentation to be a process in which bodies are separated into their smallest indivisible parts, followed by the reunion of these atoms to form new bodies. By 1650 the ideas of the Greek atomists (Leucippos, Democritos, Epicuros) that natural phenomena are the expression of the size, shape, and motion of small indivisible particles of matter had become widely accepted, largely through the writings of Francis Bacon, Thomas Hobbes, René Descartes, and especially Pierre Gassendi (Kargon, 1966). This corpuscular philosophy was used by Walter Charleton and, in a more comprehensive way, by Robert Boyle to interpret chemical phenomena as the consequence of the mechanical interaction of atoms. The application of corpuscular philosophy to chemistry during the latter half of the seventeenth century represented not only the rejection of the mystical chemistry of Paracelsus and Van Helmont, but also marked the beginning of the separation of chemistry from its ties to medicine.

It was during this time that Van Helmont's ideas about fermentation became fused with the new corpuscular philosophy. In the writings of Sylvius (François Dubois or Franciscus de la Boë), the ferments of Van Helmont are no longer supernatural entities but become the most important chemical agents in human physiology; he attached special importance to the salivary ferment. In pace with the expanded knowledge about acids and alkalies, developed by such seventeenth-century chemists as Johann Rudolph Glauber, Sylvius combined his ideas about ferments with the view that the acidic or alkaline nature of biological constituents determine their physiological role, and that disease arises from an excess of acid or alkali in the body. The fermentative physiology of Van Helmont and Sylvius was translated into the language of contemporary mechanical philosophy by Thomas Willis. A highly regarded physician, and justly renowned as an anatomist for his dissection (with Richard Lower) of the human brain, Willis enthusiastically adopted chemical descriptions of physiological phenomena. In 1659 he defined fermentation as follows:

Fermentation is an intestine motion of Particles, or the Principles of every Body, either tending to the Perfection of the same Body, or because of its change into another. For the Elementary Particles being stirred up into motion, either of their own accord or Nature, or occasionally, do wonderfully move themselves, and are moved: do lay hold of, and obvolve one another: the subtil and more active, unfold themselves on every side, and endeavour to fly away; which notwithstanding being intangled, by others more thick, are detained in their flying away. Again, the more thick themselves, are very much brought together by the endeavour and Expansion of the more Subtil, and are attenuated, until each of them being brought to their height and exaltations, they either frame the due perfection in the subject, or compleat the Alterations and Mutations designed by Nature (Willis, 1684, p. 9).

And, in the same work he wrote that "We are not only born and nourished by the means of Ferments; but we also Dye; Every Disease acts its Tragedies by the strength of some Ferment" (*ibid.*, p. 14). It was such medical thought that caused Guy Patin, the famous French physician, to write in 1672 that "Descartes and the ignorant chemists strive to spoil everything, in philosophy as well as in good medicine" (Patin, 1846, Vol. III, p. 795).

By the end of the seventeenth century the reaction to iatrochemistry had gained renewed strength from the counsel of conservative physicians such as Thomas Sydenham, who encouraged more careful observation and clearer description of individual human diseases. Furthermore, whereas the serious biological research conducted before 1650 had been largely performed in Italy (Sanctorius, Fabricius, among others), except for the work of William Harvey, at the end of the century experimental biology was actively pursued by physicians in England, France, Holland, and Germany. The concern of physicians with chemistry was to continue well into the eighteenth century, but the growth of biological observation and experiment, without regard to chemical considerations, was an important consequence of the conflict over iatrochemistry during the seventeenth century. For the new biological physicians, the achievements of Harvey and Malpighi showed the way.

No less important was the emergence of chemical studies without regard to their medical applications. The pharmacists who served the physicians had inherited a massive, confused, and arcane body of experience about the properties of natural materials. It was the great service of Glauber, neither a physician nor an apothecary but a skilled, self-taught practical chemist, to develop reproducible methods for the preparation of many chemical substances, and thus to provide materials for later fruitful inquiry. In the middle of the seventeenth century it could be written that:

> Chymistry is not an art of its own kind, but meerly a preparation of medicaments, and therefore in proper speaking belongs to that part of Physicke called Pharmacie, and so ought not to be treated but in Pharmacie (Primerose, 1651, p. 34).

By the end of the century, however, the writings of Boyle and his followers had set a course that was to establish chemistry as an independent science a hundred years later.

The development of eighteenth-century chemistry was greatly influenced by Georg Ernst Stahl and Hermann Boerhaave, Professors of Medicine at Halle and Leyden, respectively. Both men opposed iatrochemistry in their teaching, and encouraged the development of chemistry as an independent subject; they differed, however, in their philosophical outlook. Stahl sought to develop a coherent theory of chemical composition and reaction, and followed in the footsteps of Johann Joachim Becher who assigned central importance to the chemical principle of inflammability (Greek, *phlogiston*). This property was elaborated by Stahl to form the core of his system of chemistry (see White, 1932; Partington and McKie, 1937–1939). Boerhaave did not adopt the phlogiston theory, and indeed was critical of the high-flown systems of chemistry advanced during the iatrochemical period (Jevons, 1962; Lindeboom, 1968), largely because of his preference for the iatrophysics that had emerged in Italy under the influence of Borelli in response to the successes of Galileo, Sanctorius, and Harvey. On the other hand, Stahl rejected both iatrochemistry and iatrophysics on the ground that the animal body is governed by the "sensitive soul" (*psyche, anima*); this biological philosophy was later termed "animism" to distinguish it from the kind of vitalism that did not identify the vital force with the soul. If Boerhaave's chemical writings were more highly valued during the eighteenth century as a guide to laboratory operations, Stahl's chemical philosophy offered a more satisfying theoretical basis for the study of the interactions of chemical substances (see Metzger, 1930). By the middle of the century, Stahl's ideas had been widely disseminated through books by disciples such as Johann Juncker and Caspar Neumann. In France, Guillaume François Rouelle and Pierre Joseph Macquer advocated modified versions of the phlogiston theory (Rappaport, 1961), and the Englishmen Joseph Priestley and Henry Cavendish were among its most famous adherents.

In the writings of Becher and Stahl, and especially in Stahl's treatise *Zymotechnia Fundamentalis* (1697; German translation, 1748), the problem of fermentation occupies an important place. Not long before (in 1677), a noted practical chemist had stated that "nobody who ever lived, now living, or to come, will understand correctly the nature of fermentation" (Kunckel, 1716, p. 697). Becher sought to clarify the use of the word *fermentatio* by distinguishing between three kinds of fermentation: that accompanied by evolution of gas, the alcoholic fermentation of sweet liquors (which he called *fermentatio proprie*), and the fermentation leading to the production of acid; to these Stahl added putrefaction, which he considered to be no more than the final stage of fermentation. Going beyond Willis, Stahl proposed that the action of a ferment is to communicate the motion of its particles to the particles (he called them *moleculae*) of the fermentable body so as to accelerate the decomposition of the latter; the separated particles then recombine to form more stable compounds, with the release as a spirit of "oily" ("sulfureous") particles which he identified with phlogiston. As we shall see later in this chapter, Stahl's theory, stripped of phlogiston, was to reappear in the nineteenth century as part of the background for Buchner's achievement.

Boerhaave's principal contribution to the problem of fermentation was to discuss carefully the available empirical knowledge, and the following extracts from his famous *Elements of Chemistry* (first authorized Latin edition, 1732) give some indication of his views:

> I say then, that in every Fermentation, there is an intestine motion of the whole Mass, and all the parts, so long as this physical action continues; and I call it an intestine one, because it chiefly depends upon the internal principles of the vegetable Substances that are fermenting . . . . But I add farther, that this intestine motion can be excited only in vegetable Substances . . . . I know very well, that some famous Authors make no scruple to assert the contrary; and therefore to distinguish here as nicely as possible, I define a true and perfect Fermentation by its proper effect, and that is, that always terminates in the production either of the Spirit, or Acid . . . . Putrefaction . . . . is quite different from every Fermentation; for I cannot allow any thing to come under this name which don't either generate inflammable Spirits, or an Acid. For the same reason therefore all the various kinds of effervescences . . . must be absolutely excluded likewise, though these properly come under the title of intestine Motions, and are often observed even in pure, vegetable Substances, as we see in very strong Vinegar, and a fixed alkaline salt . . . . A fermentable Body I shall call such a one as by the action described . . . may be so changed as to be capable of producing the Wine or Vinegar . . . . By the word Ferment, I shall mean any Substance, that being intimately mixed with the fermentable Vegetables . . . will excite, increase, and carry on the Fermentation describ'd . . . . Hence therefore it appears at one view, that such a Ferment must belong to the Class of Vegetables (Boerhaave, 1735, Vol. II, pp. 115–116).

Boerhaave's restrictive definition of fermentation does not appear to have been adopted by later eighteenth-century chemists, and around 1800 it was generally considered that three kinds of fermentation were possible both in plants and in animals: vinous or spirituous fermentation, acid or acetous fermentation, and putrid or putrefactive fermentation (Macquer, 1777); the last was usually associated with animal matter and, because of the release of ammonia, was also termed alkaline fermentation.

Boerhaave listed a number of materials as ferments. Among them were:

> The Yeast, or fresh Flowers of Malt Liquor, or Wine, which are thrown up to the top whilst they are in the action of Fermentation; for if this light, frothy Matter is mix'd with other fermentable Substances it

wonderfully promotes their Fermentation, provided these Flowers are fresh, and not fallen . . . . The same Matter, afterwards grown heavier, and subsided to the bottom, if it is not too old . . . . The acid, mealy, fermented Dough or Leaven of the Bakers. For if fresh, sweet, wheaten Flower is kept in a dry place, and secured from Insects, it may be preserved for years without Corruption; but if this be kneaded with Water into a soft, stiff, sweet Dough, and this is lightly covered in a warm place, it begins within the space of an Hour to grow lighter, puff up, and be full of Bladders, and lose its Smell, Taste, and Tenacity, and afterwards acquires both a sour Smell and Taste, which was then called *zyme, Fermentum,* a Ferment, and gave the first name to the whole Operation; for if this Leaven is mixed with fresh Dough not yet fermented, it will make it ferment much sooner, and more efficaciously than it wou'd do otherwise. Hence then we see, that a Ferment may be soon prepared from a Body in which no Ferment actually existed before (*ibid.,* pp. 119–120).

Boerhaave does not mention his compatriot Antony van Leeuwenhoek, who had reported in 1680 to the Royal Society that he had looked at "the Yeast" through his microscope, and had seen little globules; this observation was not to become widely appreciated until the nineteenth century.

It was only after 1750 that the fermentation problem emerged from the mysticism of Van Helmont and the mechanism of Willis and Stahl; the principal factor in this development was careful experimental work on the chemical properties of various "airs" (Macquer later reintroduced the term "gas"). The pioneers in this new "pneumatic" chemistry were Boyle, Robert Hooke, and especially John Mayow, whose discovery of the "nitro-aerial spirit" was a forerunner of the identification of oxygen a hundred years later (Partington, 1956). These men had new apparatus for such studies, notably the barometer (invented by Torricelli in 1643, published in 1663) and the air-pump (invented by von Guericke *ca.* 1654, published in 1672). The experimental studies of these English "virtuosi" provided much of the evidence for the chemical speculations offered in 1706 by the man acknowledged to be the greatest among them—Isaac Newton. In his book on *Opticks* he suggested that matter is composed of "solid, massy, hard, impenetrable, moveable Particles" (Newton, 1730, p. 375), and that chemical change is a consequence of their mutual attraction or repulsion:

Now the smallest Particles of Matter may cohere by the strongest Attractions, and compose bigger Particles of weaker Virtue; and many of these may cohere and compose bigger Particles whose Virtue is still weaker, and so on for divers Successions, until the Progression end in the biggest Particles on which the Operations in Chymistry, and the Colours of natural Bodies depend, and which by cohering compose Bodies of a sensible Magnitude . . . . The Particles when they are shaken off from Bodies by Heat or Fermentation, so soon as they are beyond the reach of the Attraction of the Body, receding from it, and also from one another with great Strength, and keeping at a distance, so as sometimes to take up above a Million of Times more space than they did before in the form of a dense body. Which vast Contraction and Expansion seems unintelligible, by feigning the Particles of Air to be springy and ramous, or rolled up like Hoops, or by any other means than a repulsive Power. The Particles of Fluids which do not cohere too strongly . . . are most easily separated and rarified into Vapour, and in the Language of the Chymists, they are volatile, rarifying with an easy Heat, and condensing with Cold. But those which are grosser, and so less susceptible of Agitation, or cohere by a stronger Attraction, are not separated without a stronger Heat, or perhaps not without Fermentation. And these last are the Bodies which Chymists call fix'd, and being rarified by Fermentation, become true permanent Air (*ibid.*, pp. 370–372).

Newton's chemical hypothesis was elaborated by John Keill (in 1708) and by John Freind (in 1712), whose writings encouraged the hope that chemical theory might develop on the lines of physical theory. Moreover, it stimulated the study and classification of the interactions between chemical substances in terms of the relative attraction of their particles; one of the early fruits of this interest was Geoffroy's table of chemical "*rapports*" (later, affinities), published in 1718. Although this table owed more to Stahl than to Newton, subsequent studies on chemical affinity (such as those of Bergman in 1775) reflected clearly the influence of the concept of Newtonian attraction.

The experimental "pneumatic" chemistry of Boyle and Mayow, and its Newtonian interpretation, were carried forward early in the eighteenth century by Stephen Hales, whose *Vegetable Staticks* (published in 1727) later influenced the work of Priestley and Lavoisier. It was not until 1755, however, that a new avenue was opened to the understanding of the nature of air, and of fermentation, by the studies of Joseph Black, one of the great teachers of chemistry, first at Glasgow, then in Edinburgh (Ramsay, 1918; Kent, 1950; Partington, 1960). In a memorable paper on *Magnesia alba, Quicklime, and some other Alcaline Substances*, Black showed that, when magnesia alba (basic magnesium carbonate) is heated, or when limestone (calcium carbonate) is treated with acid, a gas (which he called "fixed air") is liberated; this gas is readily absorbed by quicklime (calcium oxide) in water, with the formation of a milky precipitate (Guerlac, 1957). In 1757 Black used this lime-water test to demonstrate that "fixed air" is the gas evolved from a fermentation brew, and the one formed on burning charcoal in air; thus he had rediscovered the *gas silvestre* of Van Helmont. Black's results were confirmed in 1764 by Macbride, but were bitterly disputed by several German chemists (J. F. Meyer, Wiegleb); after decisive experiments by Jacquin, published in 1774, Black's conclusions were universally accepted.

After Black's work, came the quantitative studies of Henry Cavendish, whose first memoir (1766) on gases was entitled *Three Papers containing Experiments on facticious Air* ("in general any kind of air which is contained in other bodies in an unelastic state, and is produced from thence by art"). In these papers he described not only "inflammable air" (hydrogen) produced by the action of acids on metals, but also several studies on fixed air ("that species of factitious Air, which is produced from Alcaline Substances, by Solution in Acids or by Calcination"). In particular, Cavendish showed that, when brown sugar in water was treated with yeast, all the gas discharged during the fermentation was absorbed by "sope leys" (aqueous sodium hydroxide), and had the same density, solubility in water, and action on flame, as the fixed air derived from limestone. The identification of fixed air and inflammable air was followed by the work of Daniel Rutherford, who reported in 1772 that, after an animal had respired in a closed container, and then died, the residual *aer malignus* (after removal of fixed air with alkali) was common air highly charged with phlogiston. In the same year Carl Wilhelm Scheele established the separate identity of this "noxious air," and its inability to support life was indicated in the new French nomenclature of 1787 by the name *azote;* Jean Chaptal assigned the term "nitrogen" to it in 1790 to denote its association with nitric acid. Among the gases of the atmosphere, the center of the stage was of course held by the "dephlogisticated air" or "vital air" identified by Priestley and Scheele in 1772–1774, and to which Lavoisier later gave the name oxygen. In 1781 Cavendish exploded about two volumes of inflammable air with one volume of dephlogisticated air in a closed vessel and showed that the gases were completely converted into water. During the course of his studies, Cavendish greatly improved the technique (termed eudiometry) of measuring the volume of gases; the most accurate eudiometer described around 1780 was said to be that of Felice Fontana.

Along with other prominent adherents of the phlogiston theory, Cavendish did not accept Lavoisier's formulation of the composition of water as a compound of oxygen and hydrogen (Berry, 1960; McCormmach,

1969). Indeed, much of the empirical basis of the new antiphlogistic chemistry was provided by men who firmly held to the older doctrine. Among them was Andreas Marggraf, a pharmacist who pioneered in methods of chemical analysis; his isolation of pure sucrose from beets is an example of the best chemical art of his time (Lippmann, 1906). Another, even greater in his achievements, was Scheele, who developed new methods for the isolation of chemical substances from biological materials, and thus laid the groundwork for the organic chemistry of the nineteenth century (Zekert, 1931, 1963). Lavoisier overthrew the phlogiston theory, not primarily by discovering new chemical entities, but by thinking in a new way about the chemical composition of the substances carefully studied by many chemists, most of whom were adherents of the phlogistic doctrine.

Antoine Lavoisier, in initiating the Chemical Revolution (Berthelot, 1890; Meldrum, 1930; McKie, 1952; Guerlac, 1961), provided the chemical basis for all further studies on the nature of alcoholic fermentation. Starting from the principle that in every chemical operation there is an equal amount of matter at the beginning and at the end, Lavoisier introduced the idea of a chemical equation by writing "je puis dire que *mout de raisin = acide carbonique + alcool.*" Since grape-must scarcely qualifies as a pure substance, and since the idea of stoichiometry did not emerge clearly until after Lavoisier, this "equation" may be taken as (and was intended to be) nothing more than a statement of the principle of conservation of matter as applied to alcoholic fermentation. For Lavoisier the yeast was unchanged in the process, and he felt that he did not have to take account of it in the equation.

At the core of Lavoisier's system of chemistry was the idea that, in combustion, an inflammable principle (which he called caloric) resides in the "vital air" liberated on the calcination of calxes (metal oxides); he thus moved the principle from the inflammable body, where Stahl had placed phlogiston, to the gas that supports combustion. Furthermore, Lavoisier considered vital air to be composed of an acidifying principle (first, *principe oxigine;* later, *oxygène*) combined with caloric. When charcoal is burned in vital air, the air is decomposed with the absorption of the acidifying principle by the charcoal, and the release of caloric in the form of heat. The fixed air produced by this combustion was called "*acide crayeux*" (1777), or "*acide du charbon*" (1784), and finally "*acide carbonique.*" As in the case of other great scientific insights, it remained for later work to discard from Lavoisier's chemical philosophy those concepts, such as caloric or the acidifying principle, that were inconsistent with new experimental observations. Nevertheless, the results reported by Lavoisier, though offered in support of a theory later partially disproved, profoundly influenced the development of chemistry. His views led to the beginnings of the elemental analysis of organic compounds for carbon and hydrogen by burning them in oxygen, and by determining the amount of carbon dioxide and water produced; the analyses he performed (*e.g.,* for alcohol) were inaccurate, and it was left to nineteenth-century chemists to develop satisfactory analytical methods. Moreover, as we shall see in a later chapter, Lavoisier's experiments on animal respiration had a decisive impact on physiological thought and on the interplay of chemistry and biology during the nineteenth century.

Lavoisier's studies on alcoholic fermentation are described in Chapter XIII (*De la décomposition des oxydes végétaux par la fermentation vineuse*) of his remarkable book, *Traité Élémentaire de Chimie* (1789; English translation by Robert Kerr in several editions: 1790, 1793, 1796, 1799, 1801).

> We may lay it down as an incontestible axiom, that, in all the operations of art and nature, nothing is created; an equal quantity of matter exists both before and after the experiment; the quality and quantity of the elements remain precisely the same: and nothing takes place beyond changes and modifications in the combinations of these elements. Upon this principle, the whole art of performing chemical experiments depends: We must always suppose an exact equality between the elements of the body examined, and those of the products of its analysis.
>
> Hence, since from must of grapes we procure alkohol and carbonic acid, I have undoubted right to suppose that must consists of carbonic acid and of alkohol (Lavoisier, 1799, p. 187).

To the last sentence, Kerr found it necessary to append the footnote:

> In this assertion, the consequences do not strictly follow from the premises; because from the must of grapes we procure carbonic acid and alkohol, it is a necessary consequence that the original must contains the constituent elements of carbonic acid and of alkohol, but not that these products of fermentation are already formed (*ibid.,* p. 187).

In the description of his experiments, Lavoisier stated that

> . . . I did not make use of the compound juices of fruits, the rigorous analysis of which is perhaps impossible, but made choice of sugar, which is easily analyzed . . . . This substance is a true vegetable oxyd with two bases, composed of hydrogen and carbon, brought to the state of an oxyd, by means of a certain proportion of oxygen; and these

three elements are combined in such a way, that a very slight force is sufficient to destroy the equilibrium of the connection. By a long train of experiments, made in various ways, and often repeated, I ascertained that the proportion in which these ingredients exist in sugar, are nearly 8 parts of hydrogen, 64 parts of oxygen, and 28 parts of carbon, all by weight, forming 100 parts of sugar.

Sugar must be mixed with about four times its weight of water, to render it susceptible of fermentation; and even then the equilibrium of its elements would remain undisturbed, without the assistance of some substance to give a commencement to the fermentation. This is accomplished by means of a little yeast from beer; and, when the fermentation is once excited, it continues of itself until completed . . . I have usually employed 10 *libs.* of yeast, in the state of paste, for each 100 *libs.* of sugar, with as much water as is four times the weight of the sugar (*ibid.,* pp. 188–189).

Lavoisier then gave the amount (by weight) of the oxygen, hydrogen, and carbon in the water, sugar, and yeast; he also noted the content of azote (nitrogen) of the yeast. He conducted the fermentation with "a few pounds of sugar," and converted his results to the 100-pound scale mentioned in the excerpt above. From the 95.9 lb of sugar that had disappeared (corresponding to 26.8 lb of carbon, 7.7 lb of hydrogen, and 61.4 lb of oxygen), he reported the production of 57.7 lb of alcohol (obtained by distillation) corresponding to 16.7 lb of carbon, 9.6 lb of hydrogen, and 31.4 lb of oxygen, as well as 35.5 lb of carbonic acid (trapped in alkali) corresponding to 9.9 lb of carbon and 25.4 lb of oxygen. With the addition of a small amount (2.5 lb) of acetic acid produced in the fermentation, the total weight of the products was 95.5 lb, corresponding to 27.2 lb of carbon, 9.8 lb of hydrogen, and 58.5 lb of oxygen (all these figures are given to seven decimal places in Lavoisier's account). This agreement is all the more remarkable since all the analytical data must have been in error; the actual elemental compositions of sucrose and alcohol are very different from those given by Lavoisier. The alcohol was probably heavily contaminated with water, and he probably lost some of the carbon dioxide. Lavoisier concluded from his data that:

The effects of the vinous fermentation upon sugar is thus reduced to the mere separation of its elements into two portions; one part is oxygenated at the expense of the other, so as to form carbonic acid, while the other part, being disoxygenated in favor of the former, is converted into the combustible substance called alkohol; therefore, if it were possible to re-unite alkohol and carbonic acid together, we ought to form sugar (*ibid.*, p. 196).

In 1815 Joseph Louis Gay-Lussac reported that he had checked and corrected the figures given by Lavoisier and that, of 100 parts of sugar, 51.34 had been converted into alcohol and 48.66 into carbonic acid; this was interpreted to indicate that sugar had been converted into equal parts of the two products. This correction came from the development, by Gay-Lussac and Thenard, of the first successful general method for the analysis of compounds for carbon, hydrogen, oxygen, and nitrogen. It involved oxidation with potassium chlorate, and gave results that compare favorably with those obtained by means of the better analytical procedures developed later. In particular, they established the elementary composition of cane sugar; the term carbohydrate came from their finding that the proportion of hydrogen to oxygen in sugar (as well as in starch) was the same as in water.

As written in recent times, the "Gay-Lussac equation" is represented as

$$C_6H_{12}O_6 = 2\ CO_2 + 2\ C_2H_5OH$$

This is not the formulation given by Gay-Lussac but by organic chemists many years later. As noted by Dumas and Boullay in 1828, the calculations of Gay-Lussac only applied to sugars having a formula which they wrote as $C^{12}H^{12}O^{12}$, in keeping with the convention of their time which assigned a relative combining weight of 6 to carbon and of 8 to oxygen. For them, cane sugar had the formula $C^{12}H^{11}O^{11}$, and they concluded that its fermentation required the uptake of a molecule of water (then written as HO). This was followed by the finding of Augustin Pierre Dubrunfaut (1846) that before fermentation, cane sugar is cleaved, and later work showed the products to be the fermentable sugars glucose and fructose, each of which may be represented (in present-day terms) as $C_6H_{12}O_6$. Also the designation of alcohol as C_2H_5OH required many years of effort in the development of the concepts of organic chemistry; the recognition of the ethyl group came only during the 1830s. Most important of all was the acceptance, after 1860, of the ideas advanced in 1858 by Cannizzaro, who reformulated atomic and molecular weights on the basis of the molecular hypothesis proposed in 1811 by Avogadro and in 1814 by Ampère. This made possible the further development of organic chemistry in terms of molecular structure, and of the arrangement of atoms in space, by Kekulé, Baeyer, and van't Hoff (among others), and led to the work of Emil Fischer on the chemical structure of the sugars. Some facets of this efflorescence of organic chemistry during the nineteenth century are key elements in the history of the fermentation problem, and we shall return to them later.

To conclude this summary of the status of the problem at the end of the eighteenth century, we may note that in 1800 the Class of Physical and Mathematical Sciences of the French National Institute of Arts and Sciences (established in 1795 as a successor to the Academy of Sciences) offered a one-kilogram gold medal for the best answer to the question: "What are the characteristics which distinguish vegetable and animal substances that act as ferments from those that undergo fermentation?" The prize was never awarded, and withdrawn in 1804 because of lack of funds. The study of this question occupied the attention of leading scientists throughout the nineteenth century.

Microorganisms as Agents of Fermentation

In 1810 there appeared a book entitled *Le Livre de tous les Ménages ou l'Art de Conserver, pendant plusieurs Années, toutes les Substances Animales et Végétales,* by Nicolas Appert, a French manufacturer of confectionery, distilled spirits, and food products. In it he described methods for preserving foods by putting them into tightly closed vessels that were then heated in boiling water; his success won him wealth and recognition, and his work represents the beginning of the canning industry (Bitting, 1937). Gay-Lussac examined Appert's results and found that the air left in the closed heated vessels lacked oxygen. Since the fermentation of grape-must or the putrefaction of food products set in when the containers were opened to let in air, but not if the contents were protected from air under mercury, Gay-Lussac concluded that atmospheric oxygen was required for these processes. As he wrote in his report:

> These substances on contact with air, promptly acquire the tendency for putrefaction or fermentation, but when they are submitted to the temperature of boiling water in well-closed vessels, the absorbed oxygen produces a new combination which is no longer able to excite putrefaction or fermentation, or which becomes coagulated by the heat in the same manner as albumin (Gay-Lussac, 1810, p. 255).

The opinion that fermenting liquids deposit a nitrogenous material resembling coagulated albumin was based on the data of Louis Jacques Thenard (1803); this material was considered to be similar to the yeast deposited in the brewing of beer. Thenard also noted that, when brewer's yeast ferments pure sugar, it loses nitrogen. Earlier work by Giovanni Fabbroni (in 1787) had suggested that:

The material that decomposes the sugar is the vegeto-animal substance . . . . On crushing the grape one mixes this glutinous material with the sugar, as if one poured an acid into a carbonate in a container; as soon as the two substances are in contact, the effervescence or fermentation begins, as in every other chemical operation (Fabroni, 1799, pp. 301–302).

These various observations were interpreted as evidence for the theory that an essential component in the fermentation process is the decomposition of albuminlike matter in the fermenting body, and that oxygen promotes this process.

The idea that oxygen is necessary for fermentation was attractive because it was consistent with both the older identification of *fermentatio* with heating, and with the new explanation of combustion. Little attention was given, therefore, to the view that "the air is the vehicle of every kind of germs" and is the source of the ferment which ". . . lives and nourishes itself at the expense of the sugar, whereby there results a disruption of the equilibrium among the elementary units of the sugar" (Astier, 1813, p. 274). Nor was much account taken of the opinion of Christian Erxleben, expressed in his book *Uiber Güte und Stärke des Bieres und die Mittel, diese Eigenschaften richtig zu würdigen* (Prague, 1818), that fermentation is a chemical process associated with vegetation, or of the microscopic observations of Jean Desmazières in 1826, who described globules of yeast as living organisms (which he named *Mycoderma cerevisiae* Desmaz). Furthermore, in opposition to the conclusion of Gay-Lussac, Jean Colin (1825) reported that yeast can promote the fermentation of sugar in the absence of oxygen.

It is perhaps understandable that there should have been skepticism about a theory of fermentation that made living organisms the agents of the process. Such ideas had been advanced repeatedly since the seventeenth century in relation to the causation of contagious diseases, notably by Athanasius Kircher (in 1657) and by Plenciz (in 1762). In large part, these ideas were based on the observation that in all putrefying material there appeared "innumerable animalcules," widely believed to arise by spontaneous generation (*generatio equivoca*). Although Francesco Redi had already disputed this belief, it was greatly encouraged by the support given by Buffon to the claim made by Needham in 1745 that he had demonstrated the spontaneous generation of living things; the contrary evidence of Lazzaro Spallanzani, offered in 1765, swung some of the opinion the other way, but the issue was still in doubt at the end of the eighteenth century.

During the early years of the nineteenth century the most exciting new experimental approach to a chemical problem was to examine the effect of a current from Volta's electric "pile." Alessandro Volta described this "artificial electric organ" in 1800, and very large batteries were constructed for Gay-Lussac and Thenard in Paris, and for Humphry Davy in London. With this instrument, Davy founded the science of electrochemistry. Napoleon was greatly interested in the "pile," and the international competition that attended the construction of ever-larger Voltaic batteries may perhaps be compared to more recent competition in the construction of particle accelerators. It was inevitable, therefore, that some of the people who had access to a powerful Voltaic battery would test the effect of an electric current on alcoholic fermentation. Both Gay-Lussac and Colin did so and reported that the passage of the current excited the fermentation; later (in 1843), Hermann Helmholtz showed that electricity had no effect.

The demonstration that the agents of fermentations were living organisms came in 1837, when three investigators (Cagniard-Latour, Schwann, and Kützing) independently, and almost simultaneously, reported their microscopic observations and experimental results. As so often in the history of a scientific problem, this instance of multiple discovery (Merton, 1961) was an outgrowth of improvements in instrumentation, in this case in the construction of achromatic compound microscopes. The beginnings of microscopy in the seventeenth century (Stelluti, in 1625) led to the instruments used by Malpighi, Hooke, Grew, and, most importantly, Leeuwenhoek. Hooke observed pores (which he termed cells) in pieces of wood and cork, and Leeuwenhoeck saw spermatozoa, red blood corpuscles, and many kinds of protozoa and bacteria (which he called "little beasts"), as well as globules of yeast. The simple type of microscope that Leeuwenhoek used permitted magnifications of several hundred diameters, an achievement not to be duplicated with compound microscopes until after 1800. Before then, the limits of optical resolution were frequently exceeded by the imagination of the observers, and during the eighteenth century the use of the instrument fell into disrepute. By 1840, however, achromatic compound microscopes of various designs were widely used in several branches of science, including the new field of histochemistry, pioneered by François-Vincent Raspail (1830).

Let us first summarize the observations of Charles Cagniard-Latour; in 1837 he was a professor at the military school in Paris and a noted inventor. He presented his results initially in 1835–1836, then in a paper presented to the French Academy of Sciences on 12 June 1837 and published in 1838. In this paper he described the cells of brewer's yeast as spherical particles capable of multiplication by budding, and gave evidence for his view that alcoholic fermentation "résulte d'un phénomène de végétation." He also took special pains to exclude the possibility that the apparent multiplication of yeast during fermentation was a consequence of the precipitation of the albuminous material described by Thenard. He concluded, therefore, that

> . . . brewer's yeast, this widely-used ferment, is a mass of small globular bodies that can reproduce themselves . . . and not simply an organic or chemical substance, as had been supposed (Cagniard-Latour, 1838, p. 221).

Theodor Schwann was associated with the leading physiologist Johannes Müller in Berlin when he published his report on alcoholic fermentation in 1837. The year before, Schwann had announced the discovery of pepsin, and two years later he formulated his cell theory. In his 1837 paper he demonstrated that Gay-Lussac's view of the role of oxygen in fermentation was incorrect, since to prevent fermentation or putrefaction it was sufficient simply to heat the air before it came in contact with a previously heated infusion of plant or animal material. This experimental approach was an extension of the one described in 1836 by Franz Schulze, who showed that putrefaction of a previously boiled infusion could be prevented if the air in contact with the infusion had first been passed through concentrated sulfuric acid. Since these treatments (heat or acid) did not affect the atmospheric oxygen, Schwann concluded that the agents responsible for fermentation were brought in by the air. He then turned to the microscopic observation of yeast, saw the same budding globules as did Cagniard-Latour, and termed them *Zuckerpilz* ("sugar-fungus," later named *Saccharomyces*). For him the connection of the alcoholic fermentation and the growth of the yeast was established. He later wrote:

> That this fungus is the cause of the fermentation follows, in the first place, from its constant occurrence in fermentation, secondly because the fermentation ceases under all conditions which visibly kill the fungus, namely boiling, treatment with potassium arsenite etc., thirdly because the exciting principle in the fermentation process must be a material that is evoked and increased by the process itself, a phenomenon that applies only to living organisms (Schwann, 1839, p. 235).

Schwann concluded that the organism is a plant, rather than an animal, because it was resistant to the action of nux vomica. Of special interest was his view, later developed by Pasteur, that

> Alcoholic fermentation must therefore be regarded as the decomposition effected by the sugar-fungus, which extracts from the sugar and a nitrogenous substance the materials necessary for its own nutrition and growth, and whereby such elements of these substances (probably among others) as are not taken up by the plant preferentially unite to form alcohol (Schwann, 1837, p. 192).

Friedrich Kützing, first a pharmacist and after 1836 a science teacher at the *Realschule* in Nordhausen, published his microscopic observations in 1837 and reached the same conclusions regarding brewer's yeast as did Cagniard-Latour and Schwann, of whose work he knew when he wrote

> Now that three of us have made the same observations in regard to the truly organic nature of yeast, I am all the more happy that my findings were confirmed by other scientists. I therefore gladly renounce a claim to priority, since it does not matter for science who made the discovery first (Kützing, 1837, p. 386).

In Kützing's paper there are numerous provocative statements (including some of uncertain scientific merit); the following is of special interest:

> It is obvious that chemistry must now strike yeast off the list of chemical compounds, since it is not a chemical compound, but an organic body, an organism. Unfortunately, too many truly organized structures are still being included among the chemical compounds, where they do not belong (*ibid.*, p. 392).

Although he mentioned no chemists by name, it is clear that Kützing must have had in mind such statements as

> . . . the conversion of sugar into carbonic acid and alcohol, as it occurs in the process of fermentation cannot be explained by a double decomposition-like chemical reaction between a sugar and so-called ferment, as we name the insoluble substance under the influence of which the fermentation takes place. This substance may be replaced by fibrin, coagulated plant protein, cheese and similar materials, though the activities of these substances are at a lower level. However, of all the known reactions in the organic sphere, there is none to which the

reaction bears a more striking resemblance than the decomposition of hydrogen peroxide under the influence of platinum, silver, or fibrin, and it would be quite natural to suppose a similar action in the case of the ferment (Berzelius, 1836, p. 240).

The "similar action" was the new force of catalysis, a term introduced by Jacob Berzelius in 1836.

I do not consider this new force to be entirely independent of the electrochemical affinities of matter; I believe, on the contrary, that it is only a new manifestation of them, but so long as we cannot see their connection and mutual dependence, it will be more convenient to designate it by means of a separate name. I shall therefore term this force, catalytic force. I shall define catalysis as the decomposition of substances by this force, just as one defines analysis as the decomposition of substances by means of chemical affinity (*ibid.*, p. 243).

Furthermore, in his annual report on the progress of chemistry for 1839, Berzelius reviewed unfavorably the work of Cagniard-Latour, Schwann, and Kützing, and it was clear that the most influential chemist in Europe (Jorpes, 1966) was not willing to accept their evidence as proof of the thesis that alcoholic fermentation is effected by living organisms. As for Kützing, Berzelius wrote:

I regard it sufficient simply to indicate Kützing's work; it may have its worth as a microscopical study of various lower plant forms, and I pass over his philosophy regarding the organic and inorganic, which belongs to philosophical ideas that have long ceased to exert a harmful effect on the development of the sciences (Berzelius, 1839, pp. 402–403).

Other prominent chemists did not welcome the word "catalysis," and for many years preferred the earlier term, "contact" substances, suggested by Eilhard Mitscherlich to denote substances that accelerate chemical reactions without participating in them; among the reactions he listed as "decompositions or combinations by contact" were

. . . the breakdown of sugars to alcohol and carbonic acid, the oxidation of alcohol when it is converted to acetic acid, the reaction of urea and water to form carbonic acid and ammonia. As such, these substances undergo no change, but upon the addition of a small amount of ferment, which is the contact substance, and a definite temperature, these reactions take place at once (Mitscherlich, 1834, p. 281).

We shall return to these ideas of Berzelius and Mitscherlich in relation to the early history of work on the "soluble ferments."

One of the leading chemists who opposed Berzelius' introduction of the concept of catalysis was Justus Liebig, another of the many men who have been called "The Founder of Biochemistry." In a paper published in 1839 he explained fermentation on the basis of a general hypothesis that it was caused by

> . . . the ability of a substance in decomposition or combination, *i.e.* undergoing chemical reaction, to evoke in another substance with which it is in contact the same reaction, or to enable that substance to undergo the same changes that it undergoes itself (Liebig, 1839, p. 262).

Liebig also attached importance to the participation of atmospheric oxygen, thus combining Stahl's definition of fermentation with those of Lavoisier and of Gay-Lussac. In particular, Liebig suggested that the insoluble nitrogenous material that appeared during fermentation is derived from the soluble "gluten" by the action of oxygen; thus he considered yeast to be oxidized gluten in a state of putrefaction. Liebig's ideas were widely disseminated in his popular writings and were adopted by the French chemist Charles Gerhardt in his *Traité de Chimie Organique* (1853–1856):

> Berzelius, among others, explains [fermentation] by attributing to the ferment a particular virtue, a *catalytic force* which enables it to act solely by its presence, solely by its contact, and without the entry of its elements into the composition of the products of metamorphosis of the fermentable substance. This is obviously no explanation, because all substances, to react, must be in contact . . . . To call the phenomenon catalytic is not to explain it; it is nothing but the replacement of a common word by a Greek word (Gerhardt, 1856, Vol. IV, p. 538).

In writing about the nature of the ferment, Gerhardt stated that it is

> . . . any body that is in a state of decomposition and which by its contact with another substance promotes chemical changes in the latter. The same ferment, in passing through several stages of decomposition can react differently depending on its state of alteration (*ibid.*, p. 541).

And, in regard to the cellulose of plants,

> . . . freed of all foreign substances, it is perfectly stable in air; but when it is in plant tissues, where it is impregnated with nitrogenous

materials, it can undergo putrefaction upon contact with air and humidity, that is to say, to undergo a slow combustion (*ibid.*, p. 537).

These opinions may be compared with those of the Dutch chemist Gerardus Mulder:

All kinds of incorrect views have been formed regarding the nature of yeast. Recent experiments have convinced me that it is undoubtedly a cellular plant consisting of isolated cells. These little plants are vesicles of a substance similar to cellulose . . . [and] contain a protein which is insoluble in boiling alcohol, and is therefore not gluten. The substance forming the vesicle, which does not itself contribute to fermentation, is penetrated exosmotically by the protein . . . . The expelled protein, which is characterized by exceptional lability, soon undergoes decomposition to ammonia and a small amount of unidentified material . . . . The decomposition is transferred to the sugar, which is transformed into carbonic acid and alcohol (Mulder, 1844–1851, pp. 49–51).

It is clear from the foregoing that the chemists were divided about the nature of fermentation, with Berzelius and Mitscherlich advocating a catalytic or contact action, and with Liebig resurrecting the ideas of transmitted vibrations and oxidation. This disagreement, or the variations on the two themes, as in the writings of Gerhardt or Mulder, are less important, however, than their agreement that the cause of the fermentation is the decomposition of nitrogenous material. Mitscherlich accepted the idea that yeast is a living organism in 1841, and Berzelius admitted it in his annual report for 1848 (the year of his death), but they left open the question whether the fermentation of sugar was a process indissolubly linked with the life of the yeast cell. It should perhaps also be emphasized that the chemists involved in these arguments, and often stating ill-founded opinions about fermentation, were men whose researches during the period 1820–1850 laid the foundations of modern organic chemistry. In particular, the work of Mitscherlich on isomorphism, of Liebig (with Wöhler) on the benzoyl radical and on uric acid, and of Gerhardt on the systematization of organic structure paved the way for the advances made after 1860. The inadequacy of the views of these distinguished chemists regarding the problem of fermentation is only one instance of the limitations of chemical speculations about biological phenomena throughout the historical relationship of chemistry to biology and up to the present. Although the relevant chemical knowledge is essential for a fuller understanding of a biological process, a chemical model based on the limited knowledge available at any time has invariably turned out to be an oversimplification of the actual process. The recognition of the inadequacy of the model has in turn often led to new chemical discovery. This interplay between chemical speculation and biological experiment has been one of the most productive factors in the growth of biochemical knowledge.

It would be a distortion of the history of the fermentation problem, however, to suggest that, in the years immediately after 1837, the conclusions of Cagniard-Latour and of Schwann were not widely adopted and further developed. Because it is often believed that little happened until 1857, when Pasteur published his first paper on fermentation, at least a brief summary of the principal efforts during these two decades is essential.

In France, on behalf of a committee appointed by the Academy of Sciences to examine the validity of Cagniard-Latour's report, Pierre Turpin stated:

> Fermentation must be considered to be the cooperative action of water and living organisms which develop and nourish themselves by absorption of a structural element of the sugar, and separating from it alcohol and acetic acid; a purely physiological action, which begins and ends with the existence of the small plant or animal infusoria, whose life ceases upon the complete utilization of the saccharine nutriment, whereupon it is deposited as a slimy precipitate, or yeast, at the bottom of the container (Turpin, 1838, p. 402).

A comment of some historical interest was the following:

> To try to find out which of these three scientists (Cagniard-Latour, Schwann, Kützing) first discovered the organization and vegetation of yeast seems to be a matter of too small importance in itself to occupy us for a moment and, above all, at a time when so many investigators were working in every part of the world, and when, in view of the many means of publication, it was necessary to specialize in some branch of science. It is enough to know, and of this we are convinced, that these three investigators, without knowing of each other's work, arrived at the same result (*ibid.*, pp. 394–395).

A German abstract of Turpin's article appeared in Liebig's *Annalen* for 1839, and was followed by an anonymous satire (presumably written by Wöhler and Liebig) in which yeast was elaborately described as a tiny animal shaped like a distilling flask; under the microscope this organism could be seen to swallow sugar, digest it in its stomach, and excrete alcohol through its digestive tract and carbonic acid through its urinary bladder. This dubious humor was not the only example of its kind to appear in the *Annalen;* in 1840 a letter from "S. C. H. Windler" satirized the work of Dumas on substitution reactions of organic compounds, and was equally unjust.

In addition to Turpin, numerous investigators provided further evidence in support of the organismic theory of fermentation; among them were Quevenne, Schroeder and von Dusch, Ure, Thomson, as well as Hermann Helmholtz. In 1843, a year after completing work for his doctoral dissertation, Helmholtz reported that he had confirmed and refined Schwann's experiment on the effect of preheating the air in contact with an infusion of animal or plant material. In the same article, Helmholtz also described experiments similar to those described by Mitscherlich in 1841, showing that, if a yeast suspension is separated from a sugar solution by means of a parchment membrane, fermentation only occurs in contact with the yeast, and not on the other side of the membrane; in addition, Helmholtz found that boiled meat infusions putrified when separated in this manner from putrefying suspensions, and concluded that, in contrast to yeast fermentation, putrefaction did not necessarily depend on the participation of living organisms. Another noteworthy contribution during the twenty-year period after 1837 was that of Bouchardat, who in 1844 suggested that the globules of beer yeast need two kinds of nutrients for their life: sugar to provide heat, and nitrogenous material for their reproduction. Furthermore, in addition to the analyses performed by Mulder, Payen (in 1839) and Schlossberger (in 1844) reported dry yeast to contain nitrogenous matter (*ca.* 63%), cellulose (*ca.* 29%), fatty substances (*ca.* 2%), and mineral matter obtained by ashing (*ca.* 6%); such a complex composition was considered to be evidence of the biological nature of yeast.

An important factor in the climate of scientific opinion between 1837 and 1857 was that men concerned with yeast technology generally favored the organismic theory of fermentation. The brewing industry in England, Germany, and Bohemia had expanded considerably during the first half of the nineteenth century (Delbrück and Schrohe, 1904; Mathias, 1959; Teich, 1965a), and a number of academic scientists were closely connected with this development. One of these men was Karl Balling, whose influential treatise on fermentation (first edition, 1845–1847) advocated the organismic theory. Another was Friedrich Wilhelm Lüdersdorff, whose unsuccessful attempt to effect cell-free alcoholic fermentation was a forerunner of Buchner's work. The close connection between the development of ideas about fermentation and brewing technology later became evident in the work of practical men—Mayer, Delbrück, and Hansen, among others—who during the period 1870–1890 applied the findings of Pasteur (Grünhut, 1896).

By 1857 the idea that alcoholic fermentation is caused by a living organism was widely accepted, but the generalization of this idea to all the various processes considered to be fermentations was not. We have seen that the term fermentation was applied to all processes in which nitrogenous organic material (whether present as a living organism or as an "albuminoid" substance), acting in small amounts relative to the amount of matter that is fermented, was considered to cause a chemical breakdown of the fermented substance without supplying any of its elements to the products of fermentation. From sugar the products might be alcohol, or lactic acid (as in the souring of milk), or butyric acid (as in the souring of bread), but, as emphasized by Gerhardt in 1856, it had not been possible to demonstrate the presence of living organisms in processes other than alcoholic fermentation. Furthermore, the term fermentation was understood to include the transformation of gums, starch, dextrin, and mannitol to dextrose, as well as the cleavage of various glucosides (amygdalin, salicin, etc.). In animal physiology the term fermentation was applied to the formation of ammonia from urinary urea, and the transformation of proteins, starches, and fats by agents in the saliva, the gastric juice, and the pancreatic juice. As we shall see in the next section of this chapter, several of the agents causing glucose production (diastase, emulsin) or involved in digestive processes (pepsin) had been obtained in soluble form. Although produced by living organisms, these "soluble ferments" were considered to exert their action independently of the life of these organisms. For the organic chemist Marcelin Berthelot,

> . . . among the phenomena that are related to the transformations of matter in living things, whether during their life or their death, there are few that do not involve fermentations to a greater or lesser degree (Berthelot, 1860a, Vol II, p. 572).

Clearly, the term fermentation had been generalized to include many more kinds of chemical processes than had been envisaged a century before, but it was also being applied on a more rational basis than in the vague speculations of the alchemists and iatrochemists.

In Berthelot's view, one of the urgent scientific questions of the day was the relation of the soluble ferments to the living organisms that cause the fermentation of sugar, and he believed that processes such as alcoholic fermentation are the consequence of the production by the living organism of "insoluble" ferments similar to the known soluble ferments. His vision of the future development of the problem was that

> If a deeper study leads to the extension of the view that I propose here, and to its application with certainty to the insoluble ferments, as well as to the soluble ferments, all the fermentations would be brought back to the same general concept, and they could be definitively assimilated to the effects of acids provoked by contact, and of truly chemical reagents. That is an absolutely essential result. Indeed, in every fermentation, one must try to reproduce the same phenomena by chemical methods and to interpret them by exclusively mechanical considerations. To banish life from all explanations relative to organic chemistry, that is the aim of our studies (*ibid.*, pp. 655–656).

Louis Pasteur entered the fermentation problem with a point of view opposite that of Berthelot. After his remarkable work on the optical activity of the tartrates, reported in 1848, Pasteur studied in succeeding years the optical activity of asparagine, aspartic acid, malic acid, and amyl alcohol; these studies led him to the conviction that, if an organic substance possesses optical activity, it must have been formed by a physiological process. Pasteur's lecture in 1860 on what he termed the "molecular dissymmetry" of natural organic products is one of the great documents of nineteenth-century science; in discussing the difference between organic substances obtained from biological sources, and those prepared in the chemical laboratory, he concluded that "The artificial products do not have any molecular dissymmetry; and I could not indicate the existence of a more profound separation between the products born under the influence of life and all the others" (Pasteur, 1860a, p. 33).

As has been noted by his various biographers (Duclaux, 1896; Vallery-Radot, 1900; Dubos, 1950; Dagognet, 1967), Pasteur was a scientist of extraordinary experimental skill, and endowed with an exceptional ability to attack controversial problems by selecting for critical examination the weak points in a theory he intended to disprove. These qualities were evident in his first publication on the fermentation problem, which dealt with the formation of lactic acid from sugar, a process for which it had not been possible to show the participation of living organisms. Much was already known about this kind of fermentation. In 1780 Scheele had isolated lactic acid from soured milk, and the elemental composition of this acid had been established in 1833 by Gay-Lussac and Pelouze. It had been formed from sugar by means of "albuminoid" materials as the ferment, and in 1843 Pelouze and Gélis (following up earlier work of Boutron-Charlard and Frémy) demonstrated that the addition of chalk to the fermentation mixture markedly increased the amount of lactic acid produced. In his paper, Pasteur stated:

> I intend to establish in the first part of this work that, just as there is an alcoholic ferment, the yeast of beer, which is found everywhere where sugar is decomposed to alcohol and carbonic acid, so also there is a particular ferment, a lactic yeast, always present when sugar becomes lactic acid, and if all labile nitrogenous material can transform sugar into this acid, it is because it is a suitable nutrient for the development of this ferment (Pasteur, 1857, p. 914).

In this paper, and a fuller one published in the following year, Pasteur reported that he had in fact seen, under the microscope, "little globules . . . much smaller than those of beer yeast," and described their isolation "in a state of purity." He noted that:

> The purity of a ferment, its homogeneity, its free development, with the aid of nourishment well suited to its individual nature, these are among the essential conditions for good fermentations. Now, in this regard, one must know that the conditions of neutrality, alkalinity, acidity, or the chemical composition of the liquid play a large role in the preferential development of individual ferments, because their life does not accommodate itself to the same degree to various states of the media (Pasteur, 1858a, p. 414).

This passage provides a succinct statement of the reasons for Pasteur's successes in the study of fermentation; it also may be considered to mark the beginning of the science of microbiology.

During the period 1857–1859 Pasteur communicated, in rapid succession, a series of remarkable preliminary papers on alcoholic fermentation. In the light of the previous history of the problem, the most important of these reports was one showing that ammonia is not a normal product of the fermentation of sugar by yeast; on the contrary, if one added to a solution of pure sugar an ammonium salt (*e.g.*, ammonium tartrate), the mineral components of yeast (containing phosphate), and a very small amount of fresh yeast, the yeast developed and fermented glucose, while the ammonia disappeared.

> In other words, the ammonia is transformed into complex albuminoid material which enters the structure of the yeast, while at the same time the phosphate gives to the new globules their mineral principles.

As to the carbon, it is evidently furnished by the sugar (Pasteur, 1858b, p. 1012).

In other reports in this series, Pasteur denied that there is a chemical equation, such as that of Lavoisier and Gay-Lussac, for the fermentation of sugar to alcohol and carbonic acid, since he found succinic acid and glycerine to be normal products of alcoholic fermentation; he attributed the acidification observed in such fermentation to the succinic acid.

In 1860 Pasteur presented an extended and detailed report of his studies on alcoholic fermentation. He concluded from his analyses of the products that:

The variations in the proportions of succinic acid, of glycerine, and consequently of the other products of fermentation, should not be surprising in a phenomenon in which the conditions contributed by the ferment seem of necessity to be so changeable. What has surprised me, on the contrary, is the usual constancy of the results. The various analyses in this article provide us enough proof of this.

I am therefore much inclined to see in the act of alcoholic fermentation a phenomenon that is simple, unique, but very complex, as it can be for a phenomenon correlative with life, giving rise to multiple products, all of which are necessary.

The globules of yeast, true living cells, may be considered to have as the physiological function correlative with their life the transformation of sugar, somewhat like the cells of the mammary gland transform the elements of the blood into the various constituents of milk, correlatively with their life and the changes in their tissues.

My present and most fixed opinion regarding the nature of alcoholic fermentation is this: The chemical act of fermentation is essentially a phenomenon correlative with a vital act, beginning and ending with the latter. I believe that there is never any alcoholic fermentation without there being simultaneously the organization, development, multiplication of the globules, or the pursued, continued life of globules that are already formed. The totality of the results in this article seem to me to be in complete opposition to the opinions of MM. Liebig and Berzelius.

I profess the same views on the subject of lactic fermentation, butyric fermentation, the fermentation of tartaric acid, and many other fermentations properly designated as such [*fermentations proprement dites*] that I shall study successively.

Now, what does the chemical act of the cleavage of sugar represent for me, and what is its intimate cause? I confess that I am completely ignorant of it.

Will one say that the yeast nourishes itself with sugar so as to excrete it in the form of alcohol and carbonic acid? Will one say on the contrary that the yeast produces, during its development, a substance such as pepsin, which acts on the sugar and disappears when that is exhausted, since one finds no such substance in the liquids? I have no reply on the subject of these hypotheses. I do not accept them or reject them, and wish to constrain myself always not to go beyond the facts. And the facts only tell me that all the fermentations properly designated as such are correlative with physiological phenomena (Pasteur, 1860b, pp. 359–360).

Most of the remainder of this lengthy memoir is devoted to detailed data on the utilization of ammonium salts for the growth of yeast, and of a portion of the sugar for the formation of cellulose and fatty material. These data are presented after a historical review of earlier studies on the nature of yeast, with special emphasis on Liebig's views. It will be clear from the passages just quoted that Pasteur was not only a brilliant investigator, but also a skillful and forceful debater; this quality was to be evident in even greater measure in his replies to the counterattacks by Liebig in 1870, by Berthelot in 1878, as well as the various exchanges with others on the subject of the spontaneous generation of ferments. Pasteur disproved spontaneous generation once again, by means of experiments similar to those performed by Schroeder and von Dusch in 1854, in showing that the ferments came from the air.

In 1861 Pasteur described the ferment responsible for the transformation of sugar into butyric acid as an organism that not only lives without free oxygen, but one that is killed by oxygen; he later applied the term *"anaérobies"* to such organisms. He also found that, when the yeast of beer develops in the absence of oxygen, it is able to ferment sugar, but loses this ability when grown in air. From these and related observations, Pasteur concluded that fermentation is a consequence of anaerobic life.

In summary, besides all the hitherto known organisms which without exception (at least so it is believed) cannot respire and nourish themselves without assimilating free oxygen gas, there is a class of organisms whose respiration is sufficiently active so that they can live without the influence of air by taking the oxygen of certain compounds, which are thereby slowly and progressively decomposed. This second class of organisms is represented by the ferments, which are similar in all respects to the organisms of the first class, living as they do, assimilating carbon, nitrogen, and phosphates in the same way, and like them requiring oxygen, but differing in that they can, in the absence of free oxygen gas, respire with oxygen gas removed from relatively unstable compounds (Pasteur, 1861, pp. 1263–1264).

In later writings, Pasteur repeatedly spoke of this *vie sans air* as a process whereby an organism takes oxygen from a suitable compound, such as sugar, and he demonstrated experimentally the capacity of various microorganisms to live in the absence of oxygen gas, whereupon (according to his definition) they become ferments. In 1878 he summed up his position as follows:

> One agrees with me if: 1) it is accepted that the fermentations properly designated as such are absolutely dependent on the presence of microscopic organisms; 2) that these organisms are not of spontaneous origin; 3) that for every organism which can live in the absence of free oxygen, its life is concomitant with acts of fermentation, and that this is so for every cell which continues to produce chemical action (Pasteur, 1878, p. 1058).

Later work, during the twentieth century, modified and clarified the views expressed by Pasteur, but his discovery of anaerobic microbial life led to a fruitful sequence of researches by numerous investigators, notably Winogradsky (Waksman, 1953) and by the Delft school of microbiology (Beijerinck, Kluyver).

In their relation to the history of the fermentation problem, however, Pasteur's experimental findings and the conclusions he drew from them must be considered separately. There can be no doubt that his demonstration of the growth of yeast and other microorganisms in a medium devoid of albuminoid material demolished the theory popularized by Liebig who, in his counterattack of 1870, questioned the validity of Pasteur's claim. At the beginning of his lengthy critique, Liebig reiterated his position:

> I assumed that the breakdown of the fermentable substance to simpler compounds must be explained by a process of cleavage residing in the ferment . . . . The rearrangement of the sugar atoms in the sugar molecule is therefore a consequence of the decomposition or rearrangement of one or several constituents of the ferment, and occurs only when they are in contact (Liebig, 1879, p. 1).

The answer he received (Pasteur, 1871) was a terse and scornful challenge that the matter be submitted to impartial inquiry; Liebig never replied (he died in 1873), and the impact of Pasteur's work and words on Liebig appears to have been shattering (Volhard, 1909, Vol. II, pp. 88–103). Nor can there be any doubt that Pasteur's insistence on the necessity of working with pure cultures of microorganisms, of taking into account their age, and of controlling carefully the conditions of their growth, was an essential prerequisite for the development of a sound microbiology, or that the failure to consider these factors led many of his opponents into error. The successful application of these fundamental discoveries led to the fruitful study of infectious diseases by Pasteur (in collaboration with Chamberland and Roux), Robert Koch, and those who followed them.

In recognizing the greatness of Pasteur's experimental achievements, it is also necessary to note that he chose his scientific language in a manner best calculated to accord with his preconceived ideas. The use of the term *"fermentation proprement dite"* (which I have translated as: fermentation properly designated as such) at once narrows the ground of discourse to those fermentations that are caused by living organisms. To say, as Pasteur does, that such fermentations are "absolutely dependent on the presence of living organisms" is a skillful debating point, but is none the less a tautology. What about all the other phenomena termed "fermentations" by his contemporaries? Clearly, for Pasteur, they were not fermentations at all, at least in 1860. For example, in that year, Berthelot (1860b) reported that he had extracted from yeast a ferment that is soluble in water and that can change cane sugar into invert sugar. Pasteur had suggested that the inversion of cane sugar by yeast might be caused solely by the succinic acid produced upon fermentation. Berthelot showed this explanation to be incorrect, and proposed that there is no fundamental difference between alcoholic fermentation and the inversion of sugar. Pasteur's prompt rejoinder was that Berthelot

> . . . here calls *ferment* substances soluble in water, and able to invert sugar. Now everyone knows that there are very many [*une foule de*] substances that enjoy this property, for example all the acids.
>
> As for me, when it is a question of cane sugar and beer yeast, I only term ferment that which causes the fermentation of sugar, that is to say which produces alcohol, carbonic acid, etc. As to inversion, I have not occupied myself with it. In regard to its cause, I only proposed a doubt in passing, in a note in a memoir in which I summarized three years of study on alcoholic fermentation (Pasteur, 1860c, pp. 1083–1084).

This exchange had been preceded by the following:

As to the vitalist opinions adopted by M. Pasteur regarding the real causes of the chemical changes operative in alcoholic fermentation, I do not believe that the moment is appropriate to discuss them . . . . (Berthelot, 1859, p. 692).

I am completely in agreement with M. Berthelot on this point. Our writings and private conversations have shown us how much we differ in the interpretation of the facts . . . . Let us both retain our independence of opinion, and in awaiting the moment of discussion, let us follow the precept of Buffon: collect facts to have ideas (Pasteur, 1859, p. 740).

Aside from the personal rivalry between Pasteur and Berthelot, it is clear that the debate was being conducted with different scientific terminology and that, although Pasteur's convincing experimental results were widely accepted by his contemporaries, there was confusion about the generalizations he drew from his findings.

We may perhaps insert here a parenthetical note on the overtones of Franco-German antagonism evident in the debates over the nature of fermentation. For example, in a monograph by Gautier (1869), he speaks of the "French vitalist theory" and the "German theory" of Liebig. On the other hand, a German account of the history of the fermentation problem, by Ingenkamp (1886), complains of Pasteur's neglect of the contributions of Schwann. The feelings generated by the Franco-Prussian conflict appear to have been reflected in scientific controversy; they were to be exacerbated during World War I (see Duhem, 1915, 1916).

Although the nineteenth-century debate about the nature of fermentation is frequently considered to have culminated in the controversy between Pasteur and Liebig, with its overtones of nationalistic feeling, it would seem that at least equal importance should be given to the later confrontation between Pasteur and Berthelot over some posthumously published experiments on fermentation performed by the great Claude Bernard.

Bernard died on 10 February 1878; on 20 July of the same year there appeared in the *Revue Scientifique* an article submitted by Berthelot and containing notes regarding experiments on alcoholic fermentation conducted by Bernard during the fall of 1877. Although the notes were obviously not intended for publication in the form in which they appeared posthumously, it was clear that Bernard believed he had shown fermentation to occur in the juice of rotting fruit without the participation of living cells and that the conversion of sugar into alcohol could be effected by agents separable from living yeast. This sponsorship by

Berthelot of Bernard's work must not be taken to indicate that the two men held the same views regarding the organismic theory of fermentation. As we have seen, the chemist Berthelot wished "to banish life from all explanations relative to organic chemistry"; the physiologist Bernard, however, objected to the idea that a process of chemical degradation (as in the conversion of sugar into alcohol and carbonic acid) is correlative with life. During the 1840s Bernard had studied the action of pancreatic juice in causing the breakdown of fats, starch, and proteins ingested in the diet of animals. In continuing these studies, he found that, when he fed dogs meat (which he considered to be only a source of protein), and measured the amount of sugar in the blood leaving the liver, the liver put glucose into the blood under conditions when carbohydrate was absent from the diet. This observation led him in 1848 to his great discovery of the so-called glycogenic function of the liver, which he described more fully in a classic monograph published in 1853. Two years later he found that the liver contains a substance similar in its properties to starch, and which could be broken down to glucose by an extract of liver. He named this starch-like substance glycogen and explained the process of glucose formation by the liver of the dogs on a meat diet as follows: The protein that is fed is used by the liver to synthesize glycogen; this synthetic or "anabolic" process Bernard considered to be associated with the life of the organism. The glycogen was then degraded to glucose by what Bernard considered to be a purely chemical process, independent of intact cells, and involving a ferment of the kind already known in pancreatic juice.

The theme that recurs throughout Bernard's writings during the last twenty-five years of his life is this distinction between chemical synthesis as a phenomenon of life and chemical degradation as a process independent of life. In the last book he wrote, published ten days after his death, he summarized his views as follows:

> In my view, there are necessarily two orders of phenomena in the living organism: 1) The phenomena of *vital creation* or *organizing synthesis;* 2) The phenomena of death or *organic destruction* . . . . The first of these two orders of phenomena is alone without direct analogues; it is specific [*particulier, spécial*] to the living being: this evolving synthesis is what is truly vital.—I shall recall in this connection the formula I expressed a long time ago: *"Life is creation."* The second, namely vital destruction, is on the contrary of a physico-chemical order, most often the result of a combustion, of a fermentation, of a putrefaction, in a word of an action comparable to a large number of chemical decompositions or cleavages. These are the true phenomena of *death*, as ap-

plied to the organized being. And, it is worthy of note that we are here the victims of a habitual illusion, and when we wish to designate the phenomena of *life*, we in fact indicate the phenomena of *death* (Bernard, 1878–1879, Vol. I, pp. 39–40).

It is clear from this excerpt that Bernard could not accept Pasteur's designation of fermentation as a process correlative with life, and that he had set out to disprove it experimentally by showing that alcoholic fermentation was possible in the absence of living cells.

Pasteur's first reply to the notes published by Berthelot came only two days later, on 22 July. There then followed an acerbic but inconclusive debate before the Academy of Sciences; about Bernard's apparent belief that he had observed the cell-free formation of alcohol from sugar, Pasteur stated:

> I must add finally that it is always an enigma to me that one could believe that I would be disturbed by the discovery of soluble ferments in the fermentations properly designated as such, or by the formation of alcohol from sugar, independently of living cells. Certainly, I confess it without hesitation, and if one wishes, I am ready to explain myself on this point at greater length, I do not now see either the necessity for the existence of these ferments or the utility of their function in this kind of fermentation (Pasteur, 1878, p. 1057).

In the following year, Pasteur published a lengthy critique of Bernard's notes, based on new experiments showing that the fermentation of rotting fruit in air is in fact not observed if one carefully prevents the fruit from contact with microorganisms. It should be noted that in 1869 Lechartier and Bellamy had reported that various fruits can form alcohol from their sugar when kept in closed vessels; Pasteur confirmed this, and showed that, in the absence of oxygen and of microbial growth, cells of higher plants were able to form alcohol. In discussing Bernard's evidence for cell-free fermentation, Pasteur also questioned the reliability of Bernard's technique for determining the presence of alcohol, and indeed suggested that Bernard may have been so presbyopic that his visual observations were in serious error; thus Bernard's conclusion that he had observed cell-free fermentation was probably a consequence of his poor eyesight. And in characteristic style, Pasteur stated:

> As much as anyone, I attach importance to the substances that are called soluble ferments; I would not be at all surprised to see that yeast cells produce a soluble alcoholic ferment; I would understand

that every fermentation could be caused by a ferment of this kind; but
it is more difficult for me to imagine that such agents should be formed
by cells given over to organic destruction in a fruit or a cadaver that
is rotting (Pasteur, 1879, p. 54).

Once again, Pasteur held the field against his opponents, but there appears to have been some modification in his attitude toward the question
of a soluble alcoholic ferment, as compared with the one he evinced in
1860. Furthermore, many chemists continued to believe that it should be
possible to isolate such a ferment. We may note here the view expressed
in 1858 about the organismic theory of fermentation by Moritz Traube,
of whom we shall have more to say later:

> Even if all putrefactions depended on the presence of infusoria or
> fungi, a healthy science would not block the road to further research
> by means of such a hypothesis; it would simply conclude from these
> facts that the microscopic organisms contain certain substances which
> elicit the phenomena of decomposition. It would attempt to isolate
> these substances, and if they could not be isolated without changed
> properties, it would only conclude that all the separation methods had
> exerted a deleterious chemical effect on these substances (Traube,
> 1899, p. 74).

Indeed, before Buchner's work there had been several other reports of
attempts to prepare from yeast a soluble ferment that would convert
sugar into alcohol. In 1846 Lüdersdorff described experiments in which
he ground a small amount of wet yeast on a glass plate until no globules
were visible microscopically, and reported that the resulting material
failed to ferment sugar. In the following year, Carl Schmidt repeated
these experiments with longer trituration and explained the negative
results as a consequence of the destruction of the ferment, rather than
the failure to extract it; during the course of this work, Schmidt anticipated Pasteur's finding of succinic acid as a regular by-product in alcoholic fermentation. Of special interest was the report in 1872 of Marie
von Manassein who subjected yeast to heat or ground it extensively with
fine sand, and concluded that she had demonstrated cell-free fermentation. After the appearance of Buchner's first paper, she reiterated this
claim (Manassein, 1897), but the experimental deficiencies of her work,
especially as regards the possible presence of intact organisms, made
its validity uncertain (Buchner and Rapp, 1898). Moreover, Pasteur is
reported to have tried to extract a soluble alcoholic ferment, although
he "did not think it actually existed" (Roux, 1898), and Denys Cochin

(1880), working in Pasteur's laboratory, reported his unsuccessful attempts to do so.

To conclude this summary of the development of the idea that fermentations are caused by microorganisms, it should be added that the inferences drawn from Pasteur's studies on the fermentation of sugar to alcohol, lactic acid, and butyric acid were extended to other fermentations. Thus Pasteur explained the formation of acetic acid from alcohol as a process caused by an aerobic microorganism (which he termed *Mycoderma aceti*); part of his controversy with Liebig dealt with the question of the participation of such organisms in the conversion of wine into vinegar, since Liebig attached great importance to the fact that the oxidation had been readily effected by purely chemical means, notably by Döbereiner in 1831, with platinum black as a catalyst. Another kind of fermentation considered by Pasteur to belong to the "fermentations properly designated as such" was that of urea. He described the appearance of a microorganism in urine that had become ammoniacal, and stated:

I am led to believe that this production constitutes an organized ferment and that there is never any transformation of urea into ammonium carbonate without the presence and development of this small plant (Pasteur, 1862, p. 52).

Thus all these fermentations were considered to be associated with the life of specific microorganisms, termed "organized ferments." In 1876, however, Frédéric Musculus obtained from ammoniacal urine a "soluble ferment" (named urease in 1889 by Bourquelot) which readily converted urea into ammonium carbonate, and this finding was promptly confirmed by Pasteur (with Joubert). It is of interest to note Pasteur's conclusions:

Physiologists will no doubt notice that here is the first example of an autonomous organized ferment which can be grown in any media, provided only that they are suitable for its nutrition, and able to form, during its development, a soluble material that can cause the same fermentation as that effected by the microscopic organism. Diastase is not made by autonomous cells; it is the same for pepsin, synaptase [the more widely used term, emulsin, had been proposed by Liebig and Wöhler], the soluble ferments of the pancreas, etc. All are produced by cells that are part of higher organisms whose general life and functions are not concentrated in the secretion of these soluble ferments. The yeast of beer produces a soluble ferment which inverts cane sugar, but which is independent of the function of yeast, at least as it is ex-

erted on glucoses properly designated as such, where there is no need
for inversion. In other words, the function of the soluble inverting ferment of alcoholic yeast is not the same as the function of these yeasts.
It is not so with the soluble ferment of urea. The soluble ferment and
the organized ferment act in the same way on their fermentable material, *i.e.*, urea, because the soluble ferment presupposes the existence
of the organized being and, inversely, because the little plant necessarily
gives rise during its life to the soluble ferment (Pasteur, 1876a, pp.
7–8).

This remarkable distinction between soluble ferments such as pepsin and
urease gives further evidence of Pasteur's skill as a debater and his continued resistance to the views advanced by chemists such as Berthelot.
That the finding of a soluble ferment of urea fermentation did not fit
into his notions is indicated by his statement that this result

> . . . was not and could not have been foreseen. It is the first example
> of an autonomous organized ferment whose function merges with the
> function of one of its unorganized products. It is also a new example
> of a *diastase* produced during life and able to modify a substance by
> the fixation of water, in the same manner as for all the *diastases* (Pas-
> teur, 1876b, p. 10).

In Pasteur's discussion of the problem of urea fermentation we have an
example of a recurrent question in the history of biochemical thought:
To what extent is the presumed biological function of a cell constituent
a reliable basis for inference as to its nature and mode of action? We
shall see many instances in which the function assigned to a chemical
component of a biological system turned out to be incompatible with
new biological data, and we shall also see repeated evidence of the limitations imposed by the available scientific knowledge upon chemical
hypotheses based on presumed biological function. The two extreme
positions in regard to this question were epitomized in the one taken by
Berthelot, who wished "to banish life from all considerations relative to
organic chemistry," and that of Pasteur, who emphasized physiological
function as a guide to the chemical study of biological phenomena. In
1860 the state of chemical knowledge made Pasteur's approach the more
fruitful one in laying the groundwork for a sound chemical microbiology.
On the other hand, Berthelot's attitude did not, and could not, become
experimentally fruitful until the twentieth century, after the work of men
like Emil Fischer on the chemical structure of cell constituents and on
the nature and mode of action of the "soluble ferments." It is only in

retrospect, therefore, that we can applaud Traube who, in 1878, reiterated views he had expressed twenty years earlier:

1) The ferments are not, as Liebig assumed, substances in a state of decomposition, and which can transmit to ordinarily inert substances their chemical action, but are chemical substances related to the albuminoid bodies which, although not yet accessible in pure form, have like all other substances a definite chemical composition and evoke changes in other substances through definite chemical affinities. 2) Schwann's hypothesis (later adopted by Pasteur), according to which fermentations are to be regarded as the expressions of the vital forces of lower organisms is unsatisfactory . . . . The reverse of Schwann's hypothesis is correct: Ferments are the causes of the most important vital-chemical processes, and not only in lower organisms, but in higher organisms as well (Traube, 1878, p. 1984).

Traube's paper was in reply to one by Felix Hoppe-Seyler; they agreed that fermentation was a chemical process, but Traube emphasized the activation of molecular oxygen whereas Hoppe-Seyler stressed the role of the hydrogen derived from water.

During the 1880s, however, greater attention was paid to the views of the influential botanist Carl von Nägeli, who dismissed the "ferment theory" of Traube and Hoppe-Seyler. In Nägeli's view:

The agent of fermentation is inseparable from the substance of the living cell, *i.e.*, it is linked to the plasma [Nägeli's word for protoplasm]. Fermentation occurs only in immediate contact with plasma in so far as its molecular action extends. If the organism wishes to exert an effect on chemical processes in places or at distances where the molecular forces of living matter are without power, it excretes ferments. The latter are especially active in the cavities of the animal body, in the water in which molds live, and in the plasma-poor cells of plants. It is even doubtful whether the organism ever makes ferments that are intended to function within the plasma; since here it does not need them, because it has available to it in the molecular forces of living matter much more energetic means for chemical action (Nägeli, 1879, pp. 86–87).

Although Nägeli's theory was extensively discussed by German scientists, and was welcomed as a compromise between Liebig's views and those of Pasteur, it yielded little experimental fruit, and again emphasized the stubborn fact that the agent responsible for the process of alcoholic fermentation had not been obtained from yeast cells in the form of a soluble ferment. There can be little doubt that the attention given Nägeli's views heightened the impact of Buchner's report in 1897.

From Diastase to Zymase

In 1833 Anselme Payen and Jean Persoz reported that the addition of alcohol to an aqueous extract of germinating barley (malt) precipitated flocculent material which, when dried and redissolved in water, could liquefy starch paste and convert it into sugar. They gave the name diastase (Greek *diastasis,* making a breach) to this material, because they considered it to be responsible for bursting the outer envelope of the starch granules; they later identified it in germinating oats, wheat, corn, and rice. This work followed that of Parmentier and Fourcroy during 1780–1800 suggesting that acids can convert starch into sugar (recognized by taste), and especially that of Constantin Kirchhoff who, in 1811, reported that upon being boiled with dilute sulfuric acid, starch is converted into sugar, with no apparent change in the acid. Kirchhoff also found in 1814 that an aqueous extract of dry malt could effect this conversion and attributed it to some property of the gluten in the malt. These results were interpreted as indicating that "the gluten, in combining with the starch, appears only to accelerate a decomposition that the latter would have suffered in a longer time without this influence" (de Saussure, 1819, p. 407). A similar ability to convert starch into sugar was found in human saliva by Erhard Leuchs in 1831, and in 1845 Louis Mialhe precipitated an active material with alcohol from this source. This "salivary diastase" was later termed ptyalin. In the latter year, Bouchardat and Sandras also reported the diastase activity of pancreatic juice. Subsequently, diastase activity was found in a variety of other plant and animal materials; at the end of the nineteenth century, it was designated amylase (Latin *amylum,* starch).

As regards the identification of the products formed by the action of malt diastase on starch, a significant achievement was that of Augustin Pierre Dubrunfaut in 1847, when he succeeded in isolating a new crystalline sugar (maltose) which, upon treatment with acid, was converted into dextrose; at about the same time, he found that the "invert sugar" formed by the action of acids on cane sugar is a mixture of sugars, and he later showed that the mixture was composed of dextrose (glucose, grape-sugar) and levulose (fructose, fruit-sugar; isolated in 1847 by Apollinaire Bouchardat upon the cleavage of inulin). These identifications were based largely on measurements of optical activity (Lyle and

Lyle, 1964) and the reduction of cupric salts (later termed Fehling's test). Dubrunfaut's identification of maltose as a major product of the conversion of starch by diastase was not generally accepted until after 1872, when his work was confirmed and extended by Cornelius O'Sullivan; since maltose and glucose have different optical rotations and reduce cupric salts to a different degree, serious errors were made because of the neglect of Dubrunfaut's finding. O'Sullivan also extended earlier observations on the formation of a second, less well-defined product (dextrin) of the action of dilute acids or of diastase on starch; the recognition of the appearance of dextrin (a partially degraded starch) came from the use of iodine as a reagent for starch, a reaction introduced in 1814, shortly after the discovery of iodine by Bernard Courtois.

By 1833 a number of chemical phenomena had been reported whose similarity to the fermentation of sugar and to the cleavage of starch by acids and by diastase was considered to be striking. Among them were the observations of Louis Jacques Thenard on the decomposition of hydrogen peroxide by metals and by blood "fibrin," the work of Humphry Davy, Johann Wolfgang Döbereiner, and Michael Faraday on the "contact action" of platinum in promoting combustion reactions (notably the combination of hydrogen and oxygen), and, most importantly, the investigations of Eilhard Mitscherlich on the effect of sulfuric acid on the conversion of alcohol into ether. In 1834 Mitscherlich considered all these phenomena, whether caused by ferments or not, to be examples of a general process in which "decompositions and combinations by contact" are effected by substances whose components did not appear to take part in the reaction. And, in the following year, Berzelius introduced the term catalysis; in stressing the similarity between the action of the inorganic catalysts and of ferments, he suggested that, in living organisms

> . . . thousands of catalytic processes take place between the tissues and the liquids and result in the formation of the great number of different chemical compounds, for the production of which from the common raw material, plant juice or blood, no probable cause could be assigned. The cause will perhaps be discovered in the future in the catalytic power of the organic tissues of which the organs of the living body consist (Berzelius, 1836, p. 245).

Although the truth of Berzelius's suggestion was not demonstrated until the twentieth century, the preparation of diastase by Payen and Persoz was soon followed by the description of other "soluble ferments."

The role of ferments in the digestion of food by animals had been the subject of speculation since Van Helmont, and the eighteenth-century studies of Réaumur and of Spallanzani represent the beginnings of the scientific study of this process. During the early years of the nineteenth century the chemist Leopold Gmelin and the anatomist Friedrich Tiedemann collaborated in researches on digestion, and their treatise on this subject, published in 1826–1827, was widely read. The appearance in 1833 of William Beaumont's *Experiments and Observations on the Gastric Juice and the Physiology of Digestion* provided a powerful impetus to work in this field, and led Johann Eberle to show in the following year that an acid extract of gastric mucosa causes the dissolution of coagulated egg white. The high point in this development was the work of Theodor Schwann who identified the active principle as a soluble ferment that he named pepsin (Greek *pepsis*, cooking, digestion) and compared it to the agent in alcoholic fermentation; in the action of both there appeared to be "a spontaneous decomposition of organic materials, elicited by a substance acting (through contact?) in a minimal quantity" (Schwann, 1836, p. 110). He also described the relation between the hydrochloric acid (identified in 1824 as the acid of gastric juice by William Prout and by Prévost and LeRoyer) and the pepsin of the stomach. During the course of his studies, Schwann precipitated pepsin with lead acetate; treatment of the precipitate with hydrogen sulfide brought the active principle back into aqueous solution. In 1839, Wasmann described the precipitation of pepsin with alcohol.

The similarity in the action of acids and ferments led to the view, expressed by Sandras and Bouchardat in 1842, that the action of hydrochloric acid is sufficient to effect the gastric degradation of proteins such as fibrin or egg albumin. Also, Bernard and his chemical colleague Barreswil suggested in 1844 that gastric juice, pancreatic juice, and saliva contain the same active principle, the differential action on proteins or starch being largely determined by the acidity or alkalinity of the medium (like other French investigators, they accepted the conclusion of Leuret and Lassaigne, published in 1825, that lactic acid is the important acid of gastric juice). This was contradicted by Louis Mialhe, who stressed the idea of the specificity of the individual ferments.

Each of the ferments has an action appropriate to itself. One of them, salivary diastase, liquefies starch in less than a minute, and transforms it into dextrin and glucose; another, pepsin, which possesses no saccharifying action on starch, coagulates milk, fibrin, and gluten; then dissolves the coagulum, and subjects it to a very particular kind of molecular transformation. On the other hand, diastase exerts no action whatever on albuminoid fluids . . . . It is not possible to agree with Liebig, Bernard, Barreswil, and others, that the ferments are instantly produced and destroyed as soon as the need for their action is felt, or that these ferments are one and the same principle which exhibits different qualities depending on the medium in which it is placed, and depending on the substance to which it is exposed. For us, these materials are special and distinct, each one conserving its nature, its particular role, and its complete independence . . . . Up to the present, we know only two, diastase and pepsin, in animals, but there certainly exist others which also participate in the maintenance of life (Mialhe, 1856, pp. 35–36).

The discovery of soluble ferments that can effect the decomposition of animal nutrients was especially important during the period 1830–1850 in relation to the question whether a biological process could be explained in chemical terms. At the beginning of this period, a widely-held view was that

Even though the gastric fluid, as a result of its chemical composition, is the dissolving agent of both the simple and composite foods and its action on food is a chemical one, digestion is still a vital process conditioned by the life of animals (Tiedemann and Gmelin, 1827, Vol. I, p. 336).

As so often, both before and after this passage was written, the idea of a vital force was linked to emphasis on the inability of chemists to imitate, in the laboratory, the processes readily effected by living organisms. For example, in 1786 John Hunter had written that

. . . the action and production of actions, both in vegetable and animal bodies, have been hitherto considered so much under the prepossessions of chemical and mechanical philosophy, that physiologists have entirely lost sight of life . . . . No chemist on earth can make out of the earth a piece of sugar, but a vegetable can do it (Hunter, 1837, pp. 216–217).

Vital force (*Lebenskraft*) is a term that recurs in the writings of chemists and physiologists throughout the first half of the nineteenth century (Simmer, 1958; Teich, 1965b). Different meanings were attached to it by various writers; the sense in which many of the nineteenth-century chemists used this term was probably first formulated by Johann Christian Reil:

The characteristic nature of the matter of which animal bodies are composed provides the chief basis of the characteristic phenomena animals exhibit. The vital force, which we consider to be the cause of these phenomena, is not something different from the organic matter; rather, the matter itself, as such, is the cause of these phenomena. Most of the animal phenomena can largely be explained on the basis of the general forces of matter. We therefore do not need any vital force as a unique primary force to explain them; we only use the word to designate concisely the concept of the physical, chemical, and mechanical forces of organic matter, through whose individuality and cooperation the animal phenomena are effected (Reil, 1799, p. 424).

In the writings of many chemists there was considerable ambiguity in regard to the existence of a "unique primary force" in living organisms. Thus, in the 1827 edition of his famous textbook of chemistry, Berzelius defined organic chemistry as the chemistry of the substances that are formed under the influence of the vital force. Earlier, he had expressed the view that

. . . the cause of most of the phenomena within the Animal Body lies so deeply hidden from our view, that it certainly will never be found. We call this hidden cause *vital force;* and like many others, who before us have in vain directed their deluded attention to this point, we make use of a *word* to which we can affix no idea (Berzelius, 1813, p. 4).

Nor did Friedrich Wöhler's preparation (in 1828) of urea from ammonium cyanate sweep away the conviction that vital forces were operative in the chemical changes effected by living things (Walden, 1928; McKie, 1944, Lipman, 1964; Schiller, 1967a; Brooke, 1968). For the physiologist Müller,

. . . the mode in which the ultimate elements are combined in organic bodies, as well as the energies by which the combination is effected, are very peculiar; for although they may be reduced by analysis to their ultimate elements, they cannot be regenerated by any chemical process. Berard, Proust, Dobereiner, and Hatchett believe, indeed, that they have succeeded in producing organic compounds by artificial processes; but their results have not been sufficiently confirmed. Woehler's experiments afford the only trustworthy instances of the artificial formation of these substances; as in his procuring urea and oxalic acid artificially. Urea, however, can be scarcely considered as organic matter, being rather an excretion than a component of the animal body. In the mode of combination of its elements it has not perhaps the characteristic properties of organic products (Müller, 1843, p. 15).

At midcentury a distinction was drawn between those biological processes (such as digestion) that had been recognized to be chemical combinations and decompositions similar to those already observed in the chemical laboratory, and the biological phenomena in which the transformations of complex molecules lead to the formation of the liquids, solids, and tissues of living organisms. Although the latter phenomena were considered to be "the manifestation of an admirable power that could not yet be reproduced by chemical means" (Mialhe, 1856, p. 8), the successes achieved in the study of the soluble ferments involved in digestion were offered in evidence for the conviction that systematic chemical research would lead to chemical explanations of biological phenomena that had been ascribed to vital forces. Indeed, the recurrent theme during 1800–1950 that a particular physiological phenomenon was "linked to life" was intended by many investigators to mean that the underlying process had not yet been imitated with isolated chemical constituents of the biological system under study, rather than to indicate their adherence to a vitalist philosophy that denied the possibility of such chemical reconstruction of physiological processes.

Among the problems posed by the soluble ferments that caused the coagulation of proteins, and their subsequent dissolution, was the difficulty in defining their chemical action in a manner comparable to that worked out for diastase. For this reason, for example, it was not clear whether the "chymosin" (from calf stomach) described in 1840 by Jean Deschamps as the active principle that coagulated milk (*présure, Labferment,* rennet) was identical with Schwann's pepsin. Furthermore, the "pancreatin" identified by Tiedemann and Gmelin was not listed as a separate ferment until 1857, when Lucien Corvisart demonstrated the presence, in pancreatic juice, of a protein-degrading ferment active in alkaline media; it was later named trypsin by Willy Kühne. Subsequently, trypsin-like ferments were found in plant extracts; these "vegetable trypsins" included papain (isolated by Adolphe Wurtz in 1879 from the fruit of the pawpaw). During the 1850s the findings of Thomas Graham and others (*e.g.,* Dubrunfaut) on the diffusion of various substances through animal membranes were applied to the study of the action of ferments such as pepsin on the slowly diffusible proteins, and the more rapidly diffusible products were termed "albuminoses" (Mialhe) or peptones (Lehmann). Later work by Kühne involved the separation of the products by precipitation with salts, and only emphasized further the difficulty of the problem of defining the action of the protein-degrading ferments. The history of this problem is an integral part of the development of ideas about the nature of proteins, and throughout the latter half of the nineteenth century and the first half of the twentieth the study of protein structure and of the mode of action of the protein-degrading ferments went hand-in-hand.

Another of the soluble ferments to be identified during the 1830s came from the work of Liebig and Wöhler on the constitution of amygdalin (crystallized by Robiquet and Boutron-Charlard in 1830), obtained from bitter almonds. In 1837 Liebig and Wöhler found that an extractable "albuminoid material," present in both sweet and bitter almonds, decomposes amygdalin with the formation of sugar and hydrocyanic acid; they named this active principle emulsin and compared its action to that of yeast in alcoholic fermentation. The name was criticized in 1838 by Robiquet, who preferred to compare the process to that effected by diastase, and he termed the ferment synaptase (the name Pasteur used in 1878). Emulsin was the first soluble ferment to be described as having a chemical action on a well-defined crystalline compound whose structure was largely elucidated during the nineteenth century, and whose cleavage to benzaldehyde, HCN, and glucose was shown to involve the addition of the elements of water to the products (hydrolysis). In addition to amygdalin, some other naturally occurring derivatives of glucose (salicin, phlorizin, helicin, and arbutin) were also known to be hydrolyzed by preparations of emulsin. This knowledge served as a background for Emil Fischer's decisive experiments during 1894–1898, to which we will return shortly.

In view of the lively discussion around 1840 about the relation of ferments such as diastase and emulsin to those that caused alcoholic fermentation, it is of interest at this point to insert the comments of Boutron-Charlard and Frémy, who stated that

. . . we are persuaded that the discovery of diastase and emulsin has truly opened a new avenue to organic chemistry.

In this kind of research, it is much more useful to determine the action of various ferments on organic substances, to study the new materials they produce, to take account of the conditions that favor these changes, than to exhaust oneself in conjectures about the true cause that gives rise to them. We should use this force of fermentation in the manner that we use other forces that produce chemical combinations and decompositions, in noting the changes that result from it, while recognizing that we are still ignorant of the primary cause of their action (Boutron-Charlard and Frémy, 1841, p. 259).

Although the conjectures continued to proliferate, it was this empirical approach that was fruitful in laying the groundwork for later discovery.

We noted earlier, in connection with the Pasteur-Berthelot controversy, that in 1860 Berthelot had obtained from yeast a soluble ferment (*ferment glucosique*) that cleaves cane sugar (sucrose) to a mixture of dextrose and levulose. This mixture had long been known as "invert sugar," since polarimetric observation of the course of the process effected by dilute acids showed that the dextrorotatory sucrose had been converted into a levorotatory product; for this reason the ferment was named *ferment inversif* or invertine (in 1864 Béchamp called it zymase, but this name faded from view until it was revived by Buchner). As we shall see later in this chapter, the fact that the action of invertine (it was subsequently named invertase) on a well-defined crystalline compound could be followed with some precision by measurement of the change in optical rotation was important in the early studies on the kinetics of ferment action. In addition to invertase, another soluble ferment that was to play a significant role is urease which, as we noted earlier, had been identified by Fréderic Musculus in 1872; it was obtained from various bacteria by Pierre Miquel in 1890. During the nineteenth century, urease was considered to be a microbial ferment, but it was later found in plant seeds, notably jack beans, from which it was isolated in crystalline form by James Sumner in 1926.

In all, by the end of the nineteenth century, the list of soluble ferments had grown to about two dozen (Bourquelot, 1896; Green, 1899; Oppenheimer, 1900). It is noteworthy that, except for a few that catalyzed oxidation reactions (laccase, tyrosinase), all the known soluble ferments promoted the hydrolysis of their substrates (a term introduced by Émile Duclaux *ca.* 1880); in addition to those mentioned above, a ferment (lipase) that promoted the hydrolysis of fats had been identified in pancreatic juice (Bernard, in 1848) and in plant seeds. Also, specific ferments had been implicated in the breakdown of glycogen, inulin, pectin, and cellulose, and the hydrolysis of maltose to glucose was attributed to a widely distributed ferment (maltase) different from invertase. These soluble ferments were also termed "unorganized ferments" to distinguish them from the "organized ferments," *i.e.*, the microorganisms that caused the fermentation of sugar to alcohol, lactic acid, or butyric acid. Many of the French investigators, including Pasteur, used the term "diastase" to cover all the soluble ferments. In 1878 Willy Kühne suggested that the designations of organized and unorganized ferments

    . . . have not gained wide acceptance, in that on the one hand it was explained that chemical bodies, like ptyalin, pepsin, etc., could not be called ferments, since the name was already assigned to yeast cells and other organisms (Brücke), while on the other hand it was said that yeast cells could not be called ferment, because then all organisms, in-

cluding man, would have to be so designated (Hoppe-Seyler). Without
wishing to inquire further why the name has generated so much excitement from opposing sides, I have taken the liberty, because of this con-
tradiction, of giving the name enzymes to some of the better-known
substances, called by many "unorganized ferments." This is not intended
to imply any particular hypothesis, but it merely states that in zyme
[yeast] something occurs that exerts this or that activity, which is considered to belong to the class called fermentative. The name is not,
however, intended to be limited to the invertin of yeast, but it is intended to imply that more complex organisms, from which the enzymes
pepsin, trypsin, etc., can be obtained, are not so fundamentally different from the unicellular organisms as Hoppe-Seyler, for example, appears to think (Kühne, 1878, p. 293).

Although Kühne's proposal elicited the comment that "The new word
enzyme may be added to the large number of new names that Kühne has
proposed . . . for substances that are totally unknown" (Hoppe-Seyler,
1878, pp. 3–4), it was adopted fairly readily in Germany and England,
and less promptly in France; in 1903, Victor Henri still used the term
"diastase" rather than "enzyme." The adoption of "enzyme" by French
scientists was accompanied by a gradual shift during the twentieth century from its original feminine gender to the masculine (Plantefol, 1968);
according to *Le Monde* for 3 June 1970, the *Académie Française* had
ruled, on 5 February 1970, in favor of the feminine.

The problem of the chemical nature of the soluble ferments was of
continuing interest throughout the latter half of the nineteenth century. We
noted that Payen and Persoz obtained malt diastase by precipitation with
alcohol; they concluded that the water-soluble portion of the precipitate
was largely "albumin." The idea that the phenomenon of fermentation
was associated with albuminoid matter had been widely accepted on the
basis of the reports of Fabbroni and Thenard around 1800, so it was not
surprising that diastase and other soluble ferments should turn out to be
of an albuminoid nature. In 1838 the term "protein" was introduced by
Mulder to refer to a radical whose composition he gave in 1840 as
$C_{40}H_{62}N_{10}O_{12}$ ($C = 6$; $N = 7$; $O = 8$) and which, when modified by
variable amounts of sulfur and phosphorus, described the composition of
natural materials such as egg albumin. Because Mulder's designation of
"protein" as a fundamental radical proved to be at variance with later
analytical data, it was not generally used during the latter half of the
nineteenth century. More often a term such as "albuminoid bodies"
(French, *matières albuminoides;* German, *Eiweisskörper*) or "proteids"
was employed. If we refer to proteins in what follows, it is in the sense

in which the word has been used during the twentieth century, namely as a class of nitrogenous compounds of high molecular weight that includes such substances as egg albumin. The development of ideas about the nature of proteins, although closely related to the problem of the nature of enzymes, is a separate story which will be recounted in the next chapter.

Mulder's analyses of proteins, and the later ones performed in the laboratories of Liebig and Dumas around 1840, suggested that all protein materials (*e.g.*, egg albumin, casein, blood fibrin) had a nitrogen content of about 16 percent. When preparations of the soluble ferments were subjected to elemental analysis, a variety of values were obtained. For example, only about 8 percent of nitrogen was found in malt diastase, whereas emulsin appeared to contain nearly 19 percent. Indeed, as additional soluble ferments were added to the list, there arose doubts about the protein nature of enzymes. At the end of the nineteenth century, in writing about the nitrogen content of some of the better-known soluble ferments, Émile Bourquelot noted that:

> When it is higher, the elemental composition approaches that of albuminoid materials; in the cases where it is lower, the nitrogen content approaches zero. From this [have come] two diametrically opposed views regarding the possible results of a complete purification of the soluble ferments. For some people, this purification should lead to substances having the composition of albuminoids; for others, it should yield nitrogen-free compounds (Bourquelot, 1896, p. 77).

During the period 1850–1900 a variety of methods were described for the purification of enzymes such as diastase or pepsin; they all had in common the characteristic that the active principle was associated with "chemical precipitates of various kinds, particularly those in a finely divided state" (Mayer, 1882, p. 10). For example, in 1861 Ernst Brücke described a method for the purification of pepsin that involved its adsorption on a precipitate of calcium phosphate; the precipitate was then dissolved in dilute hydrochloric acid, and a solution of cholesterol in alcohol and ether was added. The pepsin was adsorbed to the cholesterol as it precipitated on contact with water and was brought back into aqueous solution by extraction of the cholesterol with ether. According to Brücke, this pepsin was highly active in the digestion of coagulated egg white, but it gave none of the reactions thought to be characteristic of proteins (*e.g.*, precipitation with tannic acid, a reaction considered to detect 0.0001 percent protein in aqueous solution). By 1900 a number of color reactions were used for the detection of proteins: among the oldest ones were the xanthoproteic test (the name given to the yellow color arising from the addition of nitric acid, followed by ammonia; described by Fourcroy and Vauquelin *ca.* 1800) and the reaction introduced by Nicolas Millon in 1849 (the red color formed with mercurous nitrate in acid solution). Solutions of Brücke's pepsin gave negative reactions with these reagents as well. Similar results were reported for several of the diastase and invertase preparations described before 1900. On the other hand, Wurtz found that the papain he had prepared contained 16.6 percent nitrogen and had all the properties then associated with proteins.

The situation was summarized as follows:

> The knowledge which we have as to the constitution or composition of enzymes is exceedingly scanty. This is owing in great measure to two causes. In the first place they are apparently very unstable bodies and undergo decomposition with great readiness. The methods of preparation which have been employed to isolate them show a considerable variety of detail, but they seem to agree in this, that they are all accompanied by a great loss of ferment power . . . . In the second place . . . we have no criterion of their purity, and it is consequently impossible to say whether any of the processes so far adopted has prepared a really isolated product (Green, 1898, p. 394).

Clearly, those who believed that enzymes are proteins offered in evidence the expected nitrogen content, the presence of sulfur, and positive reactions with various precipitants or reagents giving distinctive colors. Others considered as criteria of purity the absence of all these properties in a material that exhibited the enzymic activity to a high degree. In addition to the general protein reactions, several supposedly specific tests were offered. For example, in 1868, Christian Friedrich Schönbein reported that the addition of guaiac and hydrogen peroxide to diastase produced a blue color, and during the succeeding thirty years this test was repeatedly cited as characteristic for all ferments. Another test was based on the red color given upon heating a ferment preparation with orcinol and hydrochloric acid. By the end of the century, however, it was clear that these reactions were given by impurities in the preparations. In the face of such uncertainties, it is easy to understand the reasons for the verdict

> . . . that it is not yet possible to recognize the enzymes as having the characteristics of a reproducible chemical compound, and that in view of their precipitability and solubility they may be compared to substances of high molecular weight.

It is suggested by many observations, and widely assumed, that proteins as such or their close derivatives may be carriers of the enzymatic properties. This has by no means been proved. Indeed, the frequently-low nitrogen content of very concentrated enzymes does not make it likely (Mayer, 1882, p. 20).

This uncertainty regarding the nature of enzymes persisted for nearly fifty years after these opinions were expressed. For example, Marcellus Nencki stated in 1901:

> In recent years, when many investigators have worked on enzymes, the terms *enzyme, ferment* have been grossly misused. To the extent that we have come to know a few enzymes, we are justified in stating that they belong to the labile albuminoid bodies; it would be premature, however, to assert that all enzymes are albuminoid bodies (Nencki, 1904, Vol. II, p. 724).

As we shall see later, the view that enzymes are not proteins was widely held during the first three decades of the twentieth century and was powerfully reinforced by the evidence offered by the renowned chemist Richard Willstätter (1922).

It appears to be inevitable that, when there are ambiguities about the relation of an activity to well-defined chemical substances, hypotheses should be advanced in which the operation of "forces" is invoked. In the case of enzymes, Jager (1890) and Arthus (1896) proposed that their catalytic activity is merely a special property of a large variety of chemical substances, and that this property could even be exerted at a distance through a parchment membrane; this was disbelieved by William M. Bayliss (1908), who suggested that "there may have been minute holes." It is perhaps also not surprising that the newly discovered phenomenon of radioactivity should have been brought into the arena of enzymology; according to Barendrecht (1904), enzymes act as radioactive materials.

It is evident, therefore, that at the time of Buchner's successful preparation of the soluble ferment of alcoholic fermentation his opinion that it is "without doubt to be regarded as a protein" represented a point of view not shared by all his contemporaries. Among those who inclined to the view that enzymes are related to proteins was the leading organic chemist at the beginning of the twentieth century, Emil Fischer. Because of the central importance of Fischer's work during 1894–1898 on the specificity of enzyme action, his approach merits special attention.

Fischer brought to enzyme chemistry the power of the newly emerging synthetic organic chemistry in the preparation of substrates of known structure and configuration. The concepts of structural organic chemistry (developed by Kekulé and Baeyer, with both of whom Fischer had studied) and of the asymmetric carbon atom (proposed by van't Hoff and LeBel), formulated during 1860–1880, provided intellectual challenge to a generation of young chemists and yielded dramatic practical results in the form of new synthetic dyes and drugs. Indeed, Fischer's first researches in 1874 dealt with the synthesis of fluorescein and related triphenylmethane dyes. The following year he described phenylhydrazine, and in 1883 he discovered its reaction with aldehydes. The work of several chemists, notably Heinrich Kiliani, had already shown that the sugars glucose and galactose (identified by Pasteur as a cleavage product of the milk sugar lactose) are aldehydes, whose structure was written

$$CH_2(OH) \cdot CH(OH) \cdot CH(OH) \cdot CH(OH) \cdot CH(OH) \cdot COH$$

It was also known that fructose is a keto sugar of the structure

$$CH_2(OH) \cdot CH(OH) \cdot CH(OH) \cdot CH(OH) \cdot CO \cdot CH_2(OH)$$

Fischer used phenylhydrazine for the study of the chemical constitution of these and other sugars; in 1894, after a series of brilliant studies, he was able to describe the configuration about the asymmetric carbon atoms of the known 6-carbon and 5-carbon sugars (hexoses and pentoses) by means of projection formulas (Freudenberg, 1966). Thus he wrote the structures of the dextrorotatory glucose, mannose, galactose, and the levorotatory fructose as follows:

COH	COH	COH	CH$_2$OH
H—C—OH	HO—C—H	H—C—OH	CO
HO—C—H	HO—C—H	HO—C—H	HO—C—H
H—C—OH	H—C—OH	HO—C—H	H—C—OH
H—C—OH	H—C—OH	H—C—OH	H—C—OH
CH$_2$OH	CH$_2$OH	CH$_2$OH	CH$_2$OH
d–Glucose	*d*–Mannose	*d*–Galactose	*d*–Fructose

In the same year, he reported that these four sugars were readily fermented by a variety of pure strains of yeast, whereas several closely related hexoses (*l*-mannose, sorbose) were not. This finding led him to conclude:

Among the agents used by the living cell, the principal role is played by the various albuminoid substances. They are optically active, and since they are synthesized from the carbohydrates of plants, one may well assume that the geometrical structure of their molecules, as regards their asymmetry, is fairly similar to that of the natural hexoses. On the basis of this assumption, it would not be difficult to understand that the yeast cells, with their asymmetrically-constructed agent, can only attack and ferment those kinds of sugars whose geometry is not too different from that of grape sugar (Fischer and Thierfelder, 1894, p. 2037).

In another paper published in 1894, Fischer reported his experiments on the action of an aqueous yeast extract (which he called invertin) and of a preparation of emulsin on the isomeric glucosides he had prepared the year before by the reaction of a sugar with an alcohol in the presence of hydrogen chloride; he wrote the structure of these isomers as follows (later work led to a revision of the size of the oxygen-containing ring):

$$
\begin{array}{ll}
\text{H—C—O·R} & \text{R·O—C—H} \\
\quad\diagup \text{CHOH} & \quad\diagup \text{CHOH} \\
\text{O} \qquad\qquad & \text{O} \qquad\qquad \\
\quad\diagdown \text{CHOH} & \quad\diagdown \text{CHOH} \\
\qquad \text{CH} & \qquad \text{CH} \\
\qquad \text{CHOH} & \qquad \text{CHOH} \\
\qquad \text{CH}_2\text{OH} & \qquad \text{CH}_2\text{OH}
\end{array}
$$

Thus, with methyl alcohol and glucose, two methylglucosides were formed; the less soluble one was designated α-methylglucoside and the other was β-methylglucoside. Fischer found that α-methylglucoside was hydrolyzed by "invertin" but not by emulsin, whereas β-methylglucoside was cleaved by emulsin but not by "invertin." He concluded:

As is well known, invertin and emulsin have many similarities to the proteins and undoubtedly also possess an asymmetrically constructed molecule. Their restricted action on the glucosides may therefore be explained on the basis of the assumption that only with a similar geometrical structure can the molecules approach each other closely, and thus initiate the chemical reaction. To use a picture, I would say that the enzyme and the glucoside must fit each other like a lock and key, in order to effect a chemical action on each other .... The finding that the activity of enzymes is limited by molecular geometry to so marked a degree should be of some use for physiological research. Even more important for such research seems to me to be the demonstration that the difference frequently assumed in the past to exist between the chemical activity of living cells and of chemical reagents, in regard to molecular asymmetry, is nonexistent (Fischer, 1894, p. 2992).

In later papers he showed that maltose was cleaved by the yeast preparation that hydrolyzed α-methylglucoside, but not by emulsin, which was able to hydrolyze lactose. He summarized these results in a famous article entitled "The Significance of Stereochemistry for Physiology" (Fischer, 1898), and restated his "lock-and-key" hypothesis of enzyme action. Fischer's stereochemical ideas about the action of enzymes were taken up by many investigators; in particular, Armstrong and Armstrong (1904) offered imaginary diagrams of the catalytic sites of the enzymes that act on glucosides. Furthermore, Fischer's work paralleled that of Paul Ehrlich who proposed in 1897 a "receptor" theory of the action of bacterial toxins, a problem of great medical interest at that time. This theory was an outgrowth of Ehrlich's suggestion in 1885 that protoplasm is equipped with certain atomic groups ("side chains") whose function is to attach themselves selectively to food materials; in the later elaboration of the theory, the toxin was thought to unite specifically with particular side chains of cellular constituents by means of a "haptophore" and to exert its toxic action by virtue of its "toxophore." The appearance of this theory shortly after Fischer's lock-and-key analogy is of interest. Ehrlich's idea later reappeared in the "two-affinity" theory of enzyme action proposed in 1923 by Hans von Euler; he suggested that each enzyme has one group involved in the binding of the substrate and another responsible for the catalytic action. No doubt Henry E. Armstrong had Fischer's enzyme studies in mind when he stated:

> Whatever the nature of "protoplasmic" molecules, they must be of extreme complexity and, consequently, they must present very many active regions to which other groups may be attached and within which therefore circuits can be established: hence the marvellous power of protoplasm of conditioning a great variety of changes; and it is doubtless the marvellous complexity of the albuminoids which renders such diversity of type possible in both the animal and vegetable kingdoms (Armstrong, 1895, pp. 1138–1139).

It is difficult to escape the impression that Fischer's preference for the idea that enzymes are proteins, despite all the conflicting evidence on this question at the time, was prompted by his wish to explain the stereochemical specificity he had demonstrated so convincingly. His suggestion that the yeast protein discriminated in favor of glucose-like sugars because the protein was derived from glucose-containing carbohydrate is but another example of the limitations imposed on chemical speculation by the complexity of biological processes. On a more technical level, Fischer's use of the term "invertin" to designate the agent that cleaved α-methylglucoside and maltose is an early example of a problem that was to recur frequently during the twentieth century, as a consequence of the use of enzyme preparations whose purity was unknown. He was aware of this difficulty, but nevertheless was criticized by Bourquelot:

> Unfortunately, it has been found that the solution of invertin used by Fischer was not a solution of a single ferment . . . this solution contained invertin, maltase, and perhaps still other ferments . . . . The ingenious hypothesis he advances may perhaps correspond to the facts . . . but it will not be possible to study it until a means has been found to prepare a chemically-pure soluble ferment, something that it has not been possible to do until now (Bourquelot, 1896, pp. 133–134).

Although Bourquelot's criticism was not entirely unjustified, Fischer's work on the enzymic cleavage of the synthetic glucosides demonstrated clearly how synthetic organic chemistry could be applied fruitfully to the study of enzyme action, provided that individual enzymes could be obtained in a homogeneous state. During the twentieth century, as methods were developed for the laboratory synthesis of artificial substrates related to the proteins, the same problem arose in the study of the protein-degrading enzymes, and was not solved until after 1930, when John H. Northrop crystallized pepsin.

The belief that it should be possible to isolate pure enzymes was expressed much earlier, long before the methods that proved to be successful had been developed. In 1858, Moritz Traube stressed this point of view, although his theory of the nature of ferments was based on the idea that "Many, perhaps all, ferments have, at ordinary temperature, a strongly reducing power and possess it to a higher degree than all other organic compounds" (Traube, 1899, p. 110). Traube is rightly cited in later accounts (Sourkes, 1955) as a forerunner, with Berthelot, of the view that the chemical activity of living organisms is a consequence of the action of the individual ferments they contain, and that no distinction should be made between the organized and unorganized ferments so far as their chemical action is concerned; his generalization that all ferments act by oxidation and reduction is still another example of the limits imposed by the available empirical knowledge on speculations about biochemical processes. The same may be said of Hoppe-Seyler's proposal (in 1878) that emphasis should be placed, not on the reduction of oxygen, but on the oxidation of hydrogen (from water). In the writings of Traube and of Hoppe-Seyler, however, subsequent investigators found the beginnings of fruitful theories of intracellular oxidation-reduction reactions; we shall return to these developments in a later chapter.

We have now seen that, by the time of Buchner's preparation of zymase, a relatively large number of individual enzymes had been identified in terms of their action on particular substrates, and that Fischer's work brought stereochemical considerations to bear on the question of the mode of enzyme-substrate interaction. Although opinion was divided as to the protein nature of enzymes, and none of the available enzyme preparations was generally agreed to represent a pure substance, there could be no doubt regarding the significance of the new synthetic organic chemistry for the study of enzyme action. We must now add to these facets of the enzymology of the 1890s the important impetus that came to the study of enzyme action from the development, during the latter half of the nineteenth century, of physical chemistry, especially as related to chemical kinetics.

In 1850 Ludwig Ferdinand Wilhelmy described the first measurements of the velocity of a chemical reaction in a homogeneous medium; he determined polarimetrically the rate of inversion of cane sugar in the presence of various acids, and found that the amount of cane sugar converted (dZ) in an element of time (dT) is proportional to the amount of sugar (Z) at time T and the amount of acid (S). He integrated the differential equation $-dZ/dT = MZS$ (M was designated a "velocity coefficient") to give the "first-order" rate equation

$$\log Z_0 - \log Z = MST$$

where Z_0 is the value of Z for $T = 0$. Wilhelmy's work was followed by that of Berthelot and Saint-Gilles, in 1862, on the rate of the reaction alcohol + acid = ester + water; they derived the general rate equation for such a "second-order" process, as did Harcourt and Esson during 1864–1866. These studies served as a background to the proposal of Guldberg and Waage in 1864, usually denoted the law of mass action (Lund, 1965); this theory was developed by Jacobus Henricus van't Hoff in 1877 in terms of the concept that the velocity of a reaction is proportional to the "active masses" of the reacting species, the equilibrium constant being given by the ratio of the velocities of the forward and back reactions. Van't Hoff's development of an expression relating the equilibrium constant with temperature was followed, in 1889, by Svante Arrhenius' equation for the temperature dependence of the rate of a chemical reaction. By 1890, therefore, the theoretical groundwork of chemical kinetics was beginning to be firmly established, and it was reasonable for enzyme chemists to ask whether the rates of enzyme-catalyzed reactions could be described by means of the equations developed for ordinary chemical reactions.

During the last quarter of the nineteenth century a series of investigators (Barth, 1878; Duclaux, 1883; O'Sullivan and Tompson, 1890; Tammann, 1895) measured the rate of the conversion of sucrose by invertase, whose action could be followed with some precision by polarimetric measurements. They found that, in contrast to the acid catalysis of the inversion of sucrose, the data for the enzymic process did not accord with Wilhelmy's equation for a first-order reaction. Indeed, Duclaux (1898) concluded that enzyme action cannot be described by means of equations based on the law of mass action; this view appears to have been widely held, together with the opinion that enzymes only catalyze "exothermic" reactions (Oppenheimer, 1900, p. 14). Thus the fundamental ideas of chemical kinetics and thermodynamics were considered by some writers to be inapplicable to enzymic catalysts. The view that, like inorganic catalysts, enzymes promote the rate of reversible reactions in both directions was affirmed by leading chemists (van't Hoff, 1898); however, the report by Arthur Croft Hill (1898) that maltase promotes the synthesis of maltose from glucose was not accepted by more physiologically oriented critics (Oppenheimer, 1900, p. 23). At the time of Buchner's preparation of a cell-free system for alcoholic fermentation, therefore, there was sharp division of opinion, not only regarding the nature of enzymes, but also about the nature of enzymic catalysis.

Wilhelm Ostwald, who did much at the end of the nineteenth century to propagate the new physical chemistry, differentiated between the action of catalysts that act only by "contact" (in the sense of Mitscherlich and Berzelius) and those that combine with the substances undergoing change to form reactive intermediates. The first type of catalysis would be expected to show adherence to the law of mass action, but in the latter case more complex mathematical relationships would be required. An example of the formation of an intermediate compound in catalysis had been provided by the work of Desormes and Clément in 1806; they found that nitric acid promotes the oxidation of sulfurous acid by molecular oxygen, and attributed the accelerating effect to brown oxides of nitrogen (formed by the reaction of HNO_3 with O_2), which then react with the sulfurous acid to form sulfuric acid (Lemay, 1949). Presumably, Berzelius omitted this example from his list of known catalytic phenomena because the reaction was explained on the basis of "chemical affinity" rather than solely as a consequence of the "presence" of the catalyst.

During the first decade of the present century the action of invertase on sucrose was explained by several investigators, notably Victor Henri (1903), in terms of the intermediate formation of an enzyme-substrate complex, whose "active mass" determined the rate of the cleavage of the substrate. It should be recalled, however, that the idea that an enzyme acts on a substrate by first combining with it in a specific manner was clearly stated by Emil Fischer in 1898, and strongly influenced the physical-chemical approach to the study of the kinetics of enzyme action (Segal, 1959). Although this idea was foreshadowed by Adolphe Wurtz, who in 1880 observed the formation of a precipitate upon the addition of papain to a protein substrate, it was not generally accepted until after 1900. The further development of this approach to the study of enzyme kinetics involved the quantitative formulation of the concept of "pH" by Søren P. L. Sørensen (1909), and the extensions of Henri's mathematical treatment by Leonor Michaelis and Maud L. Menten (1913) and by George E. Briggs and J. B. S. Haldane (1925).

In the light of the subsequent history of the fermentation problem, it is clear that the years around 1900 were important in the advance of knowledge regarding the mode of action of the soluble ferments, and Eduard Buchner's achievement in preparing a soluble agent of alcoholic fermentation brought the study of this process into the stream of enzyme research. Buchner recognized that zymase differed in many respects from enzymes such as invertase and pepsin, most notably in its great lability, but he considered that he was dealing with a single catalytic agent similar to the known enzymes. As an organic chemist, he based his views on the chemical reactions of glucose and its derivatives, and likened the action of zymase to those effected by heat and dehydrating agents. It was natural that, as a former student of Adolf von Baeyer, Buchner should give weight to his teacher's theory, published in 1870, that the conversion of glucose into alcohol and carbon dioxide might be formulated as a series of chemical reactions for which there were known analogies. Although Baeyer's theory may be considered to be an additional item in the large catalogue of chemical speculation about biological phenomena, it is noteworthy that the theory predicted correctly that the two central carbon atoms of the hexose should appear in the carbon dioxide, and the other four carbon atoms in the two molecules of ethyl alcohol (Baeyer, 1870). This prediction had only indirect impact, however, on the developments that led to the elucidation of the chemical pathway of this process; these depended on the empirical researches of many biochemists, notably Harden, Neuberg, Embden, Meyerhof, and Warburg, among others. What Buchner thought to be a reaction catalyzed by a single enzyme turned out to be a complex series of reactions effected by a dozen separate enzymes; it was not until the 1930s, however, when the protein nature of enzymes had been firmly established and methods for their crystallization had been developed, that the isolation and purification of the individual enzymes of alcoholic fermentation was possible. It may be noted that Richard Neumeister (1897) suggested that the fermentation effected by Buchner's yeast juice is not caused by a single substance, but by a more complex system of cell constituents, but this was rejected by Buchner:

> So long as no experimental data whatever can be offered in favor of this complicated hypothesis, it is provisionally expedient to adhere to the simpler assumption of a homogeneous zymase as the agent of fermentation. Neumeister's view appears to have arisen from the need to explain more easily the "complicated function" of zymase, whose action as a single substance seemed difficult to understand (Buchner, 1903, p. 38).

In concluding this chapter it is necessary to emphasize that Eduard Buchner's achievement, so significant in relation to the debates that preceded it and the experimental advances that followed, was essentially a matter of improved technique rather than new theoretical insight. The preparation of the cell-free fermentation juice was an outgrowth of work he, as a chemist, and his brother Hans (a bacteriologist) were doing during the 1890s to develop more effective means of breaking microbial cells so as to prepare extracts that might be of medical value (Buchner, 1903, p. 20). Hans Buchner wished to obtain cell-free protein preparations from microorganisms because he espoused the view that pathogenic bacteria produce not only the toxins that cause disease, but also the antitoxins present in the sera of immunized animals (see Kohler, 1971). After 1890 there was intense interest in the preparation of antitoxins for the treatment of infectious diseases, largely as a consequence of the dramatic report by Behring and Kitasato (in that year) that they had successfully protected animals against diphtheria or tetanus by the prior injection of sera from animals that had been immunized to the disease. If Hans Buchner's idea about the origin of the antitoxins was correct, it should have been possible to replace such serum therapy by the administration of the protoplasmic protein (he termed it "plasmine") obtained by the mechanical disintegration of microbial cells. In their efforts to prepare such cell-free extracts, the two brothers were aided by Martin Hahn, who developed a method of grinding cells with quartz sand and adding kieselguhr to give sufficient consistency to the resulting paste so that it could be safely subjected to the high pressure of a hydraulic press. It was the latter technical improvement that proved to be decisive; mere grinding with sand had been used extensively during the previous two decades (*e.g.*, Mayer, 1882, p. 14) to break up microbial cells. When Hahn applied his method to yeast, the resulting press juice underwent change rapidly, and various substances were added as possible preservatives. Among the substances tested was sucrose (in high concentration, as in the preservation of fruits), and its fermentation by the cell-free juice was observed. Because the controversy about the nature of alcoholic fermentation occupied an important place in the thinking of chemists and biologists during the latter half of the nineteenth century, the significance of this observation was readily apparent. An indication of the value that Eduard Buchner's chemical contemporaries attached to his contribution (see Harries, 1917) was his designation as Nobel Laureate in Chemistry for 1907. Among some noted physiologists, however, there was a lingering attachment to the organismic view of fermentation. Many of Buchner's papers, including his last (in 1914), were defenses of his position against such as that of Max Rubner (1913, p. 55) who emphasized the fact that the cell-free extract was much less effective in fermenting sugar than was intact yeast, and who maintained that, although a small part of yeast fermentation might be caused by Buchner's zymase, the major role was played by the "living proteins" in the protoplasmic structure. Furthermore, Buchner was obliged to prove experimentally that the cell-free extract did not contain "bits of protoplasm," as suggested by critics who accepted Nägeli's theory of fermentation. The debate concerning the significance of Buchner's achievement was to continue for several decades (see Nord, 1940); thus in 1937 Willstätter expressed the view that the action of Buchner's yeast juice on sugar is different from that in living yeast. In his autobiography, Willstätter recalled that, after the appearance of Buchner's first paper on zymase, their teacher Baeyer said "This will bring him fame, even though he has no chemical talent" (Willstätter, 1949, p. 63).

Despite these doubts and criticisms, it was the study of the properties of Buchner's cell-free extract that led to new insights into the nature of enzyme action and of metabolic pathways in living organisms; we shall return to these developments in later chapters. Moreover, as Jacques Loeb expressed it,

Through the discovery of Buchner, Biology was relieved of another fragment of mysticism. The splitting up of sugar into CO_2 and alcohol is no more the effect of a "vital principle" than the splitting up of cane sugar by invertase. The history of this problem is instructive, as it warns us against considering problems as beyond our reach because they have not yet found their solution (Loeb, 1906, p. 22).

The Nature of Proteins

We have seen that many of the nineteenth-century students of the fermentation problem attached great importance to what were usually termed *Eiweisskörper, matières albuminoides,* or albuminoid substances. At the beginning of the century these materials were identified largely by means of such familiar phenomena as the coagulation of egg white by heat, the curdling of milk by acids, or the clotting of blood. Some of the names given to such coagulable materials had a long heritage; for example, *album ovi* had been termed *albumen* by Pliny the Elder. Antoine François Fourcroy (Smeaton, 1962) wrote about blood serum as follows:

> The serum, when subjected to heat, coagulates and hardens like egg white. This property is one of its striking characteristics; it is attributed to a particular substance which is thereby readily recognizable, and which is named *albumine,* because it is the one present in egg white, termed *albumen* (Fourcroy, 1801, Vol. V, p. 117).

The heat-coagulable "albumin" in egg white and blood serum thus became a substance whose presence could be recognized in other biological fluids, or extracts of animal tissues, by virtue of this property. Other characteristics assigned to this substance by 1800 included the presence of sulfur (Scheele showed that, when an alkaline solution of egg white is acidified, a "hepatic smell" arises and blackens silver), the appearance, on treatment with nitric acid, of a yellow color that turns orange on the addition of ammonia (later termed the xanthoproteic reaction), and the appearance of a purple color in concentrated hydrochloric acid. In addition, it was known that albumin is precipitated by salts of lead and mercury as well as by tanning agents.

The separation of the serum from clotted blood gave a red, water-insoluble material; when washed thoroughly, the resulting product (named *fibrine* by Fourcroy) lost its red color. For a time during the eighteenth century the coagulable fibrous principle of blood was termed "gluten" (Latin *gluten,* glue) or gelatin, because of its resemblance to the jelly formed upon extraction of many animal tissues (skin, ligaments, cartilage, etc.) with boiling water, followed by cooling. A similar gluten was described in 1747 by Iacopo Beccari, who obtained it by kneading wheat flour with water to remove the starch; the tendency of this plant gluten to undergo putrefaction suggested to Beccari a similarity to animal materials (Beach, 1961). The gluten of wheat, as well as of other plant materials used for food, was extensively studied by Antoine Augustin Parmentier during the 1770s.

The red "coloring matter" of the blood had been shown by Antony van Leeuwenhoek to be associated with "globules" (Heinemann, 1939); his microscopic observations were followed, during the eighteenth century, by those of Giovanni della Torre, whose description of their appearance was disbelieved. After the reports of William Hewson, who described them as flat "vesicles," the particulate nature of the coloring matter was generally accepted, and it was also believed that the color was due to iron, whose presence in blood had been reported by Vincenzo Menghini in 1747 and confirmed by Hilaire Martin Rouelle in 1773. In addition to blood, an animal fluid subjected to considerable chemical scrutiny was milk, from which a curdy precipitate was obtained on acidification; this acid-insoluble material (*caseum*) acquired the name "casein" early in the nineteenth century.

The rise of the new chemistry that emerged from the work of Black, Cavendish, Priestley, Scheele, and Lavoisier led to the discovery in 1785 by Claude Louis Berthollet that, when animal materials are treated with nitric acid, they release azote (nitrogen) in relatively large amounts; he concluded that azote is a characteristic principle of animal organisms, and explained the formation of ammonia during putrefaction as the combination of the nitrogen with hydrogen derived from the decomposition of "oils" or of water. In confirming this observation, Fourcroy urged

. . . the savants who occupy themselves with animal physics to continue research on this important point, and above all to determine whence this principle comes, and how and in what organ it is fixed in animals (Fourcroy, 1789a, p. 46).

He also called attention, however, to the presence in plants of a nitrogenous material that resembled the albumin of egg white and serum (Fourcroy, 1789b), and later work by Nicolas Deyeux and Louis Nicolas Vauquelin (in 1795) extended the list of plant extracts that gave products similar to those described earlier by Beccari, Parmentier, and Fourcroy. Because of these findings, the terms albumin, fibrin, and casein were soon applied to nitrogenous plant materials that resembled the corresponding animal products in their solubility properties. Thus, when some clarified vegetable juices (*e.g.*, from cauliflower, asparagus, turnips) were boiled, the coagulum that separated was indistinguishable from heat-coagulated egg white or blood serum, and was therefore vegetable albumin. When in 1817 Joseph Proust found that the "gluten" of barley was different from that of wheat, he named it "hordein," and considered it to be a vegetable albumin. Similarly, in 1827 Henri Braconnot obtained from the seeds of leguminous plants a nitrogenous material which he called *legumine;* because of its precipitability by acid it was classified as a vegetable casein. Furthermore, the water-insoluble material deposited when vegetable juices were allowed to stand was termed fibrin. In addition to nitrogen, all these products contained sulfur, and gave positive reactions with color tests considered to be characteristic of animal albumin, casein, and fibrin.

During the early years of the nineteenth century, much impetus was given to the study of these animal and plant products by the revival of interest in medical chemistry and the emergence of agricultural chemistry. The disrepute into which iatrochemistry had fallen during the eighteenth century gave way to the conviction among many physicians that chemistry could give new insights into animal physiology, and that it was necessary to apply the fruits of the Chemical Revolution to the careful study of the characteristic constituents (*principes immédiats*, proximate principles) of animal fluids and tissues. Within a few years, during 1800–1810, there appeared a number of books that collected the available knowledge on "animal chemistry"; foremost among them were Fourcroy's *Système des Connaissances Chimiques* (1801), Johnson's *History of the Progress and Present State of Animal Chemistry* (1803), and Berzelius's *Föreläsningar i Djurkemien* (1806–1808). In his autobiography, Berzelius noted that, in writing his book

> . . . it was soon apparent that neither correct data nor correct interpretations were to be found in the works on animal chemistry which already existed. I must therefore undertake an extensive analytical investigation on animal substances, such as blood, bile, milk, the fluids and membranes of the eye, bones, bone-marrow, membranes in general and nerves . . . (Berzelius, 1934, p. 62).

No doubt the anatomist Bichat had in mind some of the work criticized by Berzelius when he wrote:

One analyzes urine, saliva, bile, etc. taken haphazardly from this or that subject, and from their study emerges animal chemistry: so be it, but that is not physiological chemistry, it is, if I may say so, the post-mortem anatomy of fluids (Bichat, 1805, pp. 81–82).

In 1808 a "Society for the improvement of Animal Chemistry" was established as part of the Royal Society of London. The founding members resolved

> That the branch of the science of chemistry which comprehends the Analysis and examination of Animal Substances, and which may properly be called Animal Chemistry, has not hitherto been sufficiently extended; and that the advances already made in it clearly demonstrate its great importance, and hold forth well-grounded hopes, that if proper pains are hereafter taken to advance it, the most important discoveries will be the result.
>
> That this most useful pursuit cannot be advantageously carried on, unless the united talents of persons well versed in Chemistry, in Anatomy and in Physiology, are combined in the advancement of it (Coley, 1967, pp. 174–175).

The impact of the new chemistry on agriculture was furthered by several men active during the early years of the nineteenth century. In France, Jean Chaptal emphasized agricultural chemistry in his *Élémens de Chymie* (3rd. ed., 1796); he later wrote the important *Chimie appliqué à l'agriculture* (1823; 2nd ed., 1829). In Germany, Sigismund Friedrich Hermbstädt founded the first journal devoted to agricultural chemistry (*Archiv der Agriculturchemie*, Vol. 1, 1804). In England, Humphry Davy delivered a course of lectures during 1802–1812, and these were published in 1813 under the title *Elements of Agricultural Chemistry*. The famous "sprig of mint" experiment of Joseph Priestley, followed by the work of Jan Ingen-Housz and of Jean Senebier, stimulated the research of Théodore de Saussure on the role of oxygen and carbon dioxide in the growth of plants; his book, *Recherches chimiques sur la Végétation* (1804), is a landmark in the history of agricultural chemistry.

The search for better knowledge of the chemical constituents of plants and animals used for human food was given encouragement in England and France by the recurrent food riots both before and after 1789 (Rudé, 1966, p. 226). In eighteenth-century England the enclosure of farm lands was followed by higher prices and shortages of flour and bread in the towns, many of which were acquiring a growing population of factory workers. In prerevolutionary France the impact of taxation and the application of *laissez-faire* capitalism to agriculture had a similar effect; during the years of the French Revolution the food situation worsened and contributed to the rise of Napoleon. Later, his policy of closing the European continent to British commerce, as well as the British blockade of French ports, stimulated research on new methods in agriculture (as for the manufacture of sugar from beets) as well as in industry (*e.g.*, the dyeing of cloth). In these efforts the leading French chemists played a key role; for example, during Napoleon's Hundred Days, Chaptal was appointed Minister of Agriculture, Commerce, and Industry (Crosland, 1967, p. 116). In England the government was less involved in the promotion of science, but Davy delivered his lectures to the members of the Board of Agriculture as part of a policy to convert farming "from a mere art of blind processes into a rational system of science" (Treneer, 1963, p. 90).

After 1800, therefore, a major preoccupation of many chemists became the study of the "immediate principles" of animals and plants. Biological tissues were subjected to extraction with acids, alkalis, and alcohol. The isolation of pure substances from such extracts, or from natural biological fluids, was sought by means of various methods of precipitation. Some chemists thought that, when all the immediate principles had been identified by such "analysis," and after they had been studied as regards their elementary composition and their properties, it would be possible to explain many biological phenomena, including those ascribed to "vital forces." At midcentury an example offered as evidence of success in this effort was the explanation of digestion in terms of the action of the acid in the stomach and of the salivary and gastric ferments.

What was meant by the term "immediate principles" may be seen from the following quotations from the writings of Michel Eugène Chevreul:

> In employing the least energetic methods of analysis, one reduces plants and animals to principles that are called *immediate*, because having been separated in the state in which they existed before the chemical operations, one is justified in attributing to them properties of the plant or animal to which they belonged, and to consider them as their essential or immediate constituents." (Chevreul, 1847, p. 577.)

> Some savants think that the expression immediate principles is faulty because it is repugnant to reason to apply the word principle to compound bodies; I do not share this opinion, and here is why. If one considers in general the composition of a salt, as established by Lavoisier, it is evidently formed by the union of an acid with an alkali, rather than the union of the elements of the acid and those of the alkali; since, if you conceive these elements to be united in proportions other than

those that constitute an acid substance or an alkaline substance, these elements will no longer give you the idea of a salt. Consequently, it seems logical to say that the acid and the alkali are the *two immediate principles of salts*. It is the same for sugar, gum, starch, lignin, etc., in relation to a plant; of fibrin, albumin, cellular tissue, etc., in relation to an animal; one should consider these substances as immediate principles, and characteristic of the plant or animal to which they belong, whereas oxygen, azote, carbon, and hydrogen are their ultimate or elementary principles (Chevreul, 1824, pp. 22–23).

For Chevreul, whose work during 1811–1823 on the nature of animal fats represented the first significant success in the chemical analysis of a complex organic material (Costa, 1962), it was important to include among the immediate principles only those compounds in which the elements are present in definite proportions. Thus, because he found that sheep fat, on treatment with alkali (saponification), gave rise to the fatty acids stearic acid, "margaric acid" (later shown to be a mixture of stearic and palmitic acids), and oleic acid, in addition to glycerine, Chevreul concluded that the fats are composed of two or more immediate principles, and he asks, in regard to the sheep stearin, "May we not believe that, in a state of purity, it would only yield glycerine and stearic acid?" (Chevreul, 1824, p. 129). In applying this definition to gluten, Chevreul noted that it had been separated into an alcohol-soluble component (named gliadin by Gioacchino Taddei) and an insoluble residue; his reference to Taddei does not include the prior observations of Einhof in 1805–1806 on the solubility of a portion of gluten in alcohol:

> Although I do not believe at all that these substances have been obtained in a state of purity . . . the experiments of M. Taddey seem to show clearly that gluten should no longer be listed among the immediate principles (*ibid.*, p. 159).

The problem of the purification of the chemical constituents of biological systems was to plague investigators throughout the nineteenth and twentieth centuries, up to the present, and we shall return to it repeatedly. In the early 1800s the techniques of extraction (*e.g.*, with alcohol) and of selective precipitation with acids, alkalis, or metallic salts often led to the isolation of crystalline materials; these methods had been used with great art by Carl Wilhelm Scheele, and it was Chevreul's success in crystallizing the individual acids formed on saponification of the fats that enabled him to draw fruitful conclusions about their constitution.

One of the important criteria of the purity of an immediate principle was, for Chevreul and his contemporaries, an elementary composition in accordance with the laws of definite and multiple proportions that had emerged from the theoretical considerations of Richter and Proust. As we noted on p. 40, the values obtained by Lavoisier for the elementary composition of cane sugar were inaccurate, and it was not until after 1810, when Gay-Lussac and Thenard described their analytical procedure, that the quantitative analysis of organic compounds began to give reproducible and reliable data. By 1820, owing to the influence of Berzelius, the atomic theory was generally accepted among chemists as a basis for the conversion of the percentage composition (by weight) of each element in a compound (as found by oxidation to carbon dioxide and water) into a formula that indicated the relative proportion of the elements in that compound. Although the question of the existence of atoms was to be hotly debated throughout the nineteenth century (Knight, 1967), the value of such formulas for the designation of the chemical composition of organic substances was accepted. The conversion of the analytical data into the so-called empirical formulas required knowledge of the relative combining weights of the elements; before 1860 the values used were in the ratio $H = 1$, $O = 8$, $C = 6$, $N = 7$ (or 14), $S = 16$, and for a time the molecular weights were calculated on the basis of an atomic weight for oxygen of 100, in accordance with the proposal of Berzelius. There was much confusion about the assignment of atomic weights to the elements, and the same compound might be denoted by different empirical formulas. It was agreed, however, that, whatever formula was used, its validity depended on the homogeneity of the isolated compound, and the precision of the methods employed to determine its elementary composition. An example of what a skilled experimenter could do before 1820 is provided by William Prout, who isolated a sample of urea that he found to contain 19.99% carbon, 6.66% hydrogen, 46.66% nitrogen, and 26.66% oxygen; since the theoretical values for $CO(NH_2)_2$ are 19.99% carbon, 6.71% hydrogen, 26.64% oxygen, and 46.66% nitrogen, it is safe to say that no present-day analyst could have done better.

By 1835 the improvements introduced by Berzelius, Gay-Lussac, and Liebig had transformed the original analytical procedure of Gay-Lussac and Thenard into a reliable method for the determination of carbon and hydrogen; the substance was now burned in a long tube, copper oxide was used in place of potassium chlorate as the oxidant, and the carbon dioxide was collected in an alkali trap. Jean Baptiste Dumas had developed a combustion method for the determination of the nitrogen content of organic substances; it involved the measurement of the volume of nitrogen gas after absorption of the water and carbon dioxide. At mid-century the Dumas method was largely replaced by that of Will and

Varrentrapp (described in 1841), which involved decomposition of a substance with alkali, collection of the liberated ammonia in acid, and precipitation of the ammonia as its chloroplatinate for gravimetric determination. This procedure was in turn supplanted in 1883 by the Kjeldahl method (Vickery, 1946). The importance of these technical advances for the development of organic chemistry during the nineteenth century cannot be exaggerated, since the concepts inherited by men like Emil Fischer about the structure and reactions of organic compounds depended on the accurate determination of the elementary composition of carefully purified compounds and of products of their chemical transformation. By 1835 it was clear that the new analytical methods had yielded exciting knowledge of the nature of such constituents of biological systems as the fats, the sugars, urea, uric acid, hippuric acid, and amygdalin. It was natural, therefore, for someone to apply the improved methods of elementary analysis to the study of albumin, casein, and fibrin. Although some data on the composition of albuminoid substances had been reported during the period 1810–1836, notably by Jean Baptiste Boussingault, the first systematic attack on the problem was that of Gerardus Mulder during the 1830s; soon after he began, Dumas and Liebig independently embarked on this effort.

An important stimulus to Mulder's work was the growing recognition of the importance of the albuminoid substances as components of the diet of animals. The studies of François Magendie, published in 1816, had shown that dogs fed only sugar and fat would not survive, and Prout's classification (in 1827) of foodstuffs into the three categories of "saccharinous," "oleaginous," and "albuminous" was widely accepted. Furthermore, throughout the rest of the nineteenth century, the conviction grew that the albuminoid substances occupied a central place in the metabolism of animals and plants. After 1840, with the rise of the cell theory, the idea that a jellylike material (the sarcode of Dujardin, the protoplasm of Purkyně, Cohn, Schultze, and Nägeli) constituted the intracellular stuff associated with the phenomena of life was coupled to the observation that this material exhibited many of the properties of albumin. Thus Carl von Nägeli defined protoplasm as ". . . the semi-liquid mucilaginous content of the plant cell, which consists of variable amounts of insoluble and soluble albuminates" (Nägeli, 1879, p. 86), and, in a famous lecture delivered in 1868, Thomas Henry Huxley popularized the idea of protoplasm—a "proteinaceous" semifluid substance—as the "physical basis of life." By the end of the century this opinion found its most extreme expression in the hypotheses of Eduard Pflüger and Max Verworn, who wrote of "living albuminoid substances" as the distinctive components of biological systems. Indeed, the idea that "living proteins" were different from the proteins isolated from organisms carried through into the twentieth century; an example is the conclusion of Claudio Fermi (in 1910) that the resistance of living organisms to pepsin or trypsin, in contrast to the rapid digestion of dead organisms, is a consequence of a difference in the nature of "living" and "dead" proteins.

The interest of Mulder, Dumas, and of Liebig in the albuminoid substances reflected the attitude of chemists who felt that the new chemistry could make decisive contributions to the problems of physiology (Holmes, 1963; Lipman, 1967). From their work emerged valuable practical knowledge of the elementary composition of foodstuffs, as reflected in such influential books as that of Ernst von Bibra, *Die Getreidearten und das Brot,* published in 1860. There also emerged the word "protein," numerous speculations about the role and fate of the albuminoid substances in plants and animals, and disappointingly little lasting insight into the chemical nature of these substances. Despite this meager yield, their work merits our attention, because it had considerable influence on the subsequent development of what came to be called "physiological chemistry," and because their failure was to be reflected in a succession of attempts, during the succeeding 100 years, to solve the problem of protein structure.

The Organic Chemistry of Proteins 1840—1920

In plants as well as in animals there is present a substance that is produced in the former, constitutes part of the food of the latter, and plays an important role in both. It is one of the very complex compounds, which very easily alter their composition under various circumstances, and serves especially in the animal organism for the maintenance of chemical metabolism [*Stoffwechsel*], which cannot be imagined without it; it is without doubt the most important of all the known substances of the organic kingdom, and without it life on our planet would probably not exist.

It is found in all parts of plants, in the roots, stems, leaves, fruits, and juices, as well as in very dissimilar parts of the animal body. In plants it occurs in three different forms, as water-soluble, water-insoluble, or alcohol-soluble; in animals it occurs in a large variety of forms, being sometimes soluble, sometimes insoluble in water, and in its insoluble form its structure is variable. It combines with sulfur or phosphorus, or both, and thereby exhibits differences in its appearance and its physical properties. The substance has been named *protein,* because it is the origin of very different substances and therefore may be regarded as a primary compound (Mulder, 1844–1851, pp. 300–301).

Although the word "protein" first appeared in the chemical literature in a paper by Mulder (1838), it was suggested to him by Berzelius, in a letter dated 10 July 1838 (see Vickery, 1950):

> I consider it to be sufficiently well established that the immediate organic substances either are oxides of compound radicals or are combinations of two or even several oxides of this kind. It is necessary first to look for this radical. This is easy when it contains only two elements. The addition of nitrogen complicates matters a bit, but in general the difficulty is not great . . . . Now I presume that the organic oxide which is the base of fibrin and albumin (and to which it is necessary to give a particular name, *e.g., protein*) is composed of a ternary radical combined with oxygen . . . . The word protein that I propose to you for the organic oxide of fibrin and albumin, I would wish to derive from *proteios,* because it appears to be the primitive or principal substance of animal nutrition that plants prepare for the herbivores, and which the latter then furnish to the carnivores (Berzelius, 1916, pp. 104–109).

During the years around 1840 these ideas about the chemical nature of "protein," and its place in the order of nature, were widely accepted. Although they were abandoned soon thereafter, the impetus that Mulder gave to the study of materials such as fibrin and albumin was a lasting one, and the name he introduced remained long after the chemical and physiological ideas that it was intended to connote had been discarded. These ideas were based on the results of elementary analysis.

From the results of his analytical determinations, Mulder concluded that egg albumin, serum albumin, and fibrin had the same content of carbon, hydrogen, nitrogen, and oxygen, corresponding to an empirical formula which he wrote $C_{40}H_{62}N_{10}O_{12}$, and that this unit (termed "protein") was combined with an atom of sulfur and an atom of phosphorus to form fibrin and egg albumin, and with two sulfurs and one phosphorus to form serum albumin. Mulder also reported that, by the treatment of the albumins and fibrin with dilute alkali, it was possible to remove the sulfur and phosphorus completely and to isolate the fundamental unit common to all three albuminoid materials. Furthermore, an analysis of plant gluten that had been treated with alkali gave the same analytical values that had been found for the protein derived from animal materials.

It appears, therefore, that animals draw their most important proximate principles from the plant kingdom . . . . The herbivorous animals are, from this point of view, no different from the carnivores. Both are nourished by the same organic substance, protein, which plays a major role in their economy (Mulder, 1839, p. 140).

These conclusions were soon adopted by Liebig (1841) on the basis of analyses performed by his associates Johann Joseph Scherer and Henry Bence-Jones; on 28 June 1841 Liebig wrote to his friend Wöhler:

I have been working on the legumin of leguminous plants and have obtained the remarkable result that it is casein in all its properties and its composition. We have therefore a complete analogy, we have plant albumin, plant fibrin, and plant casein, all three identical with each other and with the animal proteins that bear their names (Hofmann, 1888, Vol. I, p. 185).

These all-embracing principles of the identity of the fundamental protein units, and of their passage from plants to animals, were attractive in their simplicity and gained much attention when they were popularized by Liebig in his book *Animal Chemistry, or Organic Chemistry in its Application to Physiology and Pathology* (first published in 1842). It was this book that elicited from Berzelius the comment that:

This easy kind of physiological chemistry is created at the writing desk, and is the more dangerous, the more genius goes into its execution, because most readers will not be able to distinguish what is true from mere possibilities and probabilities, and will be misled into accepting as truths probabilities that will require great effort to eradicate after they have become imbedded in physiological chemistry. To the extent that it is easy in this manner to do physiological chemistry, which has many chemical facts with great possibility of combination, it is likely that the aspiration to be the first to bring this *probability-physiology* to market will produce conflict over priority, to the detriment of science (Berzelius, 1843, pp. 535–536).

Despite this negative verdict, Liebig's views were widely acclaimed. Furthermore, as noted by Berzelius, and as was to happen so often in the later history of biochemistry, the euphoria generated by an attractive generalization was accompanied by claims for priority in having enunciated it. Thus the great German chemist Liebig was opposed by the great French chemist Dumas who published (in 1842, with Cahours) a long paper in which he disputed the identity of legumin with casein, found a higher nitrogen content in fibrin than in egg albumin, but nonetheless concluded that ". . . the animal receives and assimilates almost intact the neutral nitrogenous substances which it finds fully formed in the animals and plants that form its food" (Dumas, 1844, p. 41).

There can be little doubt that Mulder's results elicited a generally favorable initial response among chemists, as evidenced by the approval they received in the 1842 edition of Liebig's *Animal Chemistry*, the paper of Dumas and Cahours, and books such as Thomson's *Chemistry of Animal Bodies,* published in 1843 (he listed "protein" among "animal amides") or Simon's *Animal Chemistry with Reference to the Physiology and Pathology of Man,* whose English translation appeared in 1845.

In 1845, however, Liebig wrote to Wöhler that "After so much has been prattled and written about protein and protein oxide, it is a source of despair to have to see that there is no such thing as protein" (Hofmann, 1888, Vol. I, p. 263). This verdict came from extensive experimental work in Liebig's laboratory, largely by Nicholas Laskowski, who concluded:

1) that the empirical formulas proposed by Mr. Mulder for the albuminoid substances are unacceptable, since they do not agree with the analytical results;

2) that the material prepared as protein by Mr. Mulder contains sulfur, *i.e.,* it is something other than what was designated by that name, and so far as our experience goes, [protein] does not exist at all in the isolated state;

3) that since the acceptance of the substance described by Mr. Mulder as protein was based solely on the belief that it has been isolated free of sulfur,—the substance isolated by Mr. Mulder contains sulfur, and the one described by him cannot be isolated,—there is no basis left for the assumption that protein is a hypothetical fundamental substance (Laskowski, 1846, p. 165).

Laskowski's long paper had been preceded by a short note, several months before, in which Liebig stated these conclusions and concluded with the request "Would Mr. Mulder please describe his procedure in full detail?" (Liebig, 1846a, p. 133). This immediately drew from Mulder an indignant reply in the form of a pamphlet entitled *Liebig's Question to Mulder Tested by Morality and Science,* and a revised theory that albuminoid substances are combinations of a hypothetical protein that cannot be isolated (its empirical formula became $C_{36}H_{54}N_8O_{12}$) and linked to various amounts of "sulfamide" or "phosphamide." In return, Liebig wrote in 1847:

A theoretical view in natural science is never absolutely true, it is only true for the period during which it prevails; it is the nearest and most exact expression of the knowledge and the observations of that period. In proportion as our knowledge is extended and changed, this expression of it is also extended and changed, and it ceases to be true for a later period, inasmuch as a number of newly acquired facts can no longer be included in it. But the case is very different with the so-called proteine theory, which cannot be regarded as one of the theoretical views just mentioned, since, being supported by observations both erroneous in themselves and misinterpreted as to their significance, it had no foundation in itself, and was never regarded, by those intimately acquainted with its chemical groundwork, as an expression of the knowledge of a given period (Liebig, 1847a, pp. 18–19).

At midcentury the various forms of Mulder's theory were largely discounted:

It must rather excite our surprise that chemists should have hazarded any theory of their composition, than that nothing positive should as yet have been ascertained regarding their composition and mutual relations. Although we have the most accurate analyses of the protein-compounds, it is impossible to form any decisive conclusion regarding their internal constitution (Lehmann, 1855, Vol. I, p. 290).

Although the names of Mulder, Liebig, and Dumas are the ones usually mentioned in connection with the excitement caused by the idea that a single fundamental unit characterizes the albuminoid substances, others of lesser renown contributed evidence to support this view. Among them was Apollinaire Bouchardat, who reported in 1842 that the fundamental unit (which he chose to call "albuminose") has the same optical activity no matter what its source (egg albumin, casein, gluten, etc.), but this claim was later disproved by Antoine Béchamp. After the decline of the Mulder theory, some suggested that the difficulty lay, not in the theory, but in the purification of the albuminoid substances. As expressed by Gerhardt,

The composition of the albuminoid substances is extremely complex; it appears to be the same for all of them . . . . If one considers that these substances behave in an identical manner under the influence of agents that transform them, one is led to attribute to impurities the small differences encountered in the results of analyses (Gerhardt, 1856, Vol. IV, p. 432).

That the constitution of the albuminoid substances was indeed very complex became evident by the 1850s from studies on their chemical cleavage. This approach, used successfully by Chevreul in his work on fats, gave a bewildering variety of products when it was applied to the protein compounds. Thus in 1846, after Liebig had turned from speculations about their elementary composition to the isolation of products of their chemical breakdown, he found that treatment of casein with alkali gave crystals of two different products—one of them was a substance he named tyrosine (Greek, *tyros*, cheese), and the other was leucine, a material that had been obtained from albuminoid substances by Proust and Braconnot a quarter-century earlier. By 1860 tyrosine and leucine had been found as well-defined products of the cleavage of many proteins by means of acids (sulfuric acid, hydrochloric acid) or alkalis (potassium hydroxide, barium hydroxide); during most of the remainder of the nineteenth century these two crystalline products were widely considered to be the most important constituents of the proteins. In addition to acids and alkalis, various oxidizing agents were used to good advantage by the chemists of the 1840s for the transformation of organic compounds; when casein, albumin, or fibrin was subjected to such treatment in Liebig's laboratory, a large variety of aldehydes and acids was formed. In characteristic fashion, Liebig added to his condemnation of Mulder's work the following statement:

> The study of the products, which caseine yields when acted on by concentrated hydrochloric acid, of which, as Bopp has found, Tyrosine and Leucine constitute the chief part, and the accurate determination of the products which the blood constituents, caseine, and gelatine, yield when oxidized, among which the most remarkable are oil of bitter almonds [benzaldehyde], butyric acid, aldehyde, butyric aldehyde, valerianic acid, valeronitrile, and valeroacetonitrile, have opened up a new and fertile field of research into numberless relations of the food to the digestive process, and into the action of remedies in morbid conditions; discoveries of the most wonderful kind, which no one could have even imagined a few years ago (Liebig, 1847a, p. 27).

During the 1850s, oxidation with potassium permanganate gave benzoic acid, among other products, and the succeeding decades of the century were dotted with reports that added further complexity.

The idea that the albuminoid substances might be related to the substances named "amides" had been mentioned repeatedly since Dumas proposed in 1830 that the oxamide he discovered (by treatment of ethyl oxalate with ammonia) might be a structural unit of the nitrogenous constituents of animals, and such amides as urea and uric acid were considered to be derived from the metabolic breakdown of albuminoid substances in the animal body. Furthermore, in 1833 Ferdinand Rose observed that albumins give a blue-violet color with an alkaline solution of copper sulfate, and in 1848 Gustav Wiedemann showed that this color is given by the substance he called biuret, formed by prolonged heating of urea at high temperatures. This "biuret test" for proteins was to figure largely in later speculations on the chemical nature of proteins.

The Rise of Organic Chemistry. The emergence, during the middle of the nineteenth century, of new concepts regarding the structure of organic compounds was decisive in providing insights into the nature of proteins; a brief summary of the major developments is therefore essential. We have already noted the importance of the methods of elementary analysis introduced by Gay-Lussac and Thenard, and improved by Berzelius, Liebig, and Dumas, among others. The results provided by these methods made possible the formulation, in the 1860s, of a theory of organic chemical structure based on valency. In the early stages of this development, during the period 1830–1840, a key figure was Dumas. Like Berzelius and other leading chemists of the time, he accepted John Dalton's atomic theory because it provided a satisfactory basis for explaining the chemical properties of substances in terms of the laws of definite and multiple proportions, and because it had led to successful predictions, such as Eilhard Mitscherlich's law of isomorphism:

> The same number of atoms combined in the same manner produce the same crystalline form; the crystalline form is independent of the chemical nature of the atoms, and is determined solely by their number and their relative positions (Mitscherlich, 1821, p. 419).

Dumas made a notable contribution to chemical thought in asserting that chemical compounds should be considered unitary structures, rather than binary combinations of units of opposite electrical charge, as proposed in the electrochemical theory of Berzelius. Dumas also emphasized the importance of the relative positions of the atoms of a substance in relation to the recently discovered phenomenon of isomerism, and attempted to develop classificatory schemes based on experiments in which certain elements or groups of elements were substituted by others during the course of chemical reactions; among his important empirical results was the conversion (in 1838) of acetic acid into trichloroacetic acid by treatment with chlorine in sunlight.

Special mention must be made of Auguste Laurent, whose bold imagination brought new insights to the problem of the classification of the growing number of organic compounds (Kapoor, 1969). In 1836 he advanced the hypothesis that every organic compound is derived from a hydrocarbon, constituting a fundamental unit in which hydrogen can be replaced by other elements or groups of elements to form derived units. Laurent's ideas were not well received by leading chemists, such as Berzelius and Liebig, partly because of his penchant for bizarre chemical nomenclature. He also antagonized Dumas in claiming priority for the substitution theory. Laurent said of the theory "if it falls, I will have been its author; if it succeeds, another will have done it" (Laurent, 1837, p. 326), and Dumas stated in 1838 that he did not accept Laurent's extension of his theory, and that Laurent's analyses were inaccurate. A further source of difficulty for Laurent was his association with Charles Gerhardt, both in chemistry and in the 1848 revolution; Gerhardt had also incurred the wrath of Dumas in connection with a priority claim in the use of the word "homology" to describe regular differences among a series of carbon compounds (see Kahane, 1968). Although Laurent and Gerhardt were ostracized by most of the established French chemists, with the notable exception of the aged Jean Baptiste Biot and the younger Adolphe Wurtz, and were denied professorships in Paris, their writings had a large effect on young chemists elsewhere, especially in England. An incidental item worthy of mention is that Laurent appears to have stimulated Pasteur to undertake his famous work on the optical activity of the tartrates (Dubos, 1950, p. 91).

From the collective, though often competitive, efforts of Dumas, Laurent, and Gerhardt there emerged by 1850 a classificatory theory of organic compounds according to "types," in each of which one or more of the hydrogen atoms were replaced by "compound radicals." The first of the types to be established experimentally was the "ammonia" type, in 1849, by Wurtz and Hofmann through their studies on aliphatic amines and aniline derivatives. In the succeeding year, Alexander Williamson showed the alcohols and ethers to be of the "water" type, and by 1858 three other types ("hydrogen," "hydrochloric acid," and "marsh gas") had appeared. Thus there was added to the "radical" theory that had emerged from the work of Liebig and Wöhler the idea of substitution in a number of "types." That the type theory had predictive value was shown when Gerhardt converted acetic acid, considered to belong to the water type $\genfrac{}{}{0pt}{}{H}{H}\Big\}O$ and therefore written $\genfrac{}{}{0pt}{}{C_2H_3O}{H}\Big\}O$, into an acid anhydride $\genfrac{}{}{0pt}{}{C_2H_3O}{C_2H_3O}\Big\}O$, as expected from the theory. Its limitations became evident when it was realized that many compounds could not be denoted as belonging to one of the established types, and in 1857 August Kekulé wrote of "mixed types"; the grotesque formulas that resulted is exemplified by the one for glycine (empirical formula $C_2H_5NO_2$; $C = 12$, $N = 14$, $O = 16$):

$$\left.\begin{array}{c} H \\ H \end{array}\right\} N$$
$$\left.\begin{array}{c} C_2H_2O \\ H \end{array}\right\} O$$

to indicate that it belonged both to the ammonia and water types (Kekulé, 1861, Vol. I, p. 758). It should be emphasized that, for the adherents of the type theory, such designations were not structural formulas, but rather shorthand representations of the reactions that a compound could undergo. The suggestion that formulas might be used to represent the arrangement of the constituent atoms in a compound, made tentatively by Williamson in 1852, was fully developed some 6 years later by the Scotsman Archibald Scott Couper and the Russian Alexander Butlerov, both working in Wurtz's laboratory in Paris, and by the German Kekulé who had associated with Gerhardt and Wurtz in Paris and with Williamson and Odling in London. William Odling was responsible for the English translation (1855) of Laurent's *Méthode de Chimie*, published posthumously in 1854. This group of young men, stimulated by the ideas of Laurent and Gerhardt, charted a new course for organic chemistry.

It was Couper who first expressed most forcefully the idea that the type theory should be abandoned in favor of the "combining power" of individual atoms, and who stressed the necessity of writing structural formulas, which he considered to represent physical reality. Shortly after the appearance of his paper in 1858, Couper lost his reason and lived in retirement until his death, but his formulas became widely known, especially through Lothar Meyer's *Modernen Theorien der Chemie*, published in 1864. In assuming the combining power of carbon to be four, and the linking of carbon atoms with one another, Couper had independently developed ideas with which Kekulé's name is most prominently identified. In a famous paper, published in 1858, Kekulé wrote of carbon:

When the simplest compounds of this element are considered (marsh gas, methyl chloride, chloride of carbon, chloroform, carbonic acid, phosgene, sulfide of carbon, hydrocyanic acid, etc.) it is seen that the quantity of carbon which chemists have recognized as the smallest possible, that is, as an atom, always unites with 4 atoms of a monatomic or with 2 atoms of a diatomic element; that in general the sum of the chemical units of the elements united with one atom of carbon is 4. This leads us to the view that carbon is tetratomic or tetrabasic . . . .

In the cases of substances which contain several atoms of carbon, it must be assumed that at least some of the atoms are held in the same way in the compound by the affinity of carbon, and that the carbon atoms attach themselves to one another, whereby a part of the affinity of the one is naturally engaged with an equal part of the affinity of the other (Kekulé, 1858, p. 153).

Although in these ideas of "combining power" or "basicity" Couper and Kekulé had been partly anticipated by Edward Frankland, whose work on organometallic compounds (*e.g.*, zinc methyl) led him to a similar view in 1852, it was principally Kekulé's influence that placed the concept of valency at the center of chemical thought (Anschütz, 1929). The term *Valenz* was introduced in 1868 by Wichelhaus to denote what had also been termed "saturating capacity" or "atomicity."

Closely related to the rise of the theory of atomicity was the question whether atoms exist; during the 1860s the debate was lively, especially in England, and Kekulé felt obliged to rise to the challenge:

All chemists, or at least the greater number of them, are at present adherents of the so-called theory of atomicity . . . . Some regard atomicity as a fundamental property of matter, and therefore fixed and unalterable, like the weight of atoms, whereas others define it as the maximum of saturating capacity; others again—and this last-mentioned view appears at present to have the greatest number of adherents— assume a varying atomicity, regarding the same element as monatomic, biatomic, triatomic, or even pentatomic etc., according to circumstances . . . . We cannot say with certainty whether the theory of atomicity has gained in clearness and scientific value by these developments; but we must confess that to us the contrary appears to be the case, and that the contempt with which mathematicians and physicists regard the present direction of theoretical chemistry may, perhaps, be attributed to these lawless extensions of the atomic hypothesis . . . .

The question whether atoms exist or not has but little significance in a chemical point of view: its discussion belongs rather to metaphysics. In chemistry we have only to decide whether the assumption of atoms is an hypothesis adapted to the explanation of chemical phenomena. More especially have we to consider the question, whether a further development of the atomic hypothesis promises to advance our knowledge of the mechanism of chemical phenomena.

I have no hesitation in saying that, from a philosophical point of view,

I do not believe in the actual existence of atoms, taking the word in its literal signification of indivisible particles of matter. I rather expect that we shall some day find, for what we call atoms, a mathematico-mechanical explanation, which will render an account of atomic weight, of atomicity, and of numerous other properties of so-called atoms. As a chemist, however, I regard the assumption of atoms, not only as advisable, but as absolutely necessary in chemistry. I will go even further, and declare my belief that *chemical atoms exist,* provided that term be understood to denote those particles of matter which undergo no further division in chemical metamorphoses. Should the progress of science lead to a theory of the constitution of chemical atoms— important as such a knowledge would be for the general philosophy of matter—it would make but little alteration in chemistry itself. The chemical atom will always remain the chemical unit . . . . (Kekulé, 1867, pp. 303–304).

The issue whether chemical atoms exist was a crucial one for many reasons, not the least of which was the necessity for agreement on the atomic weights to be assigned to the constituent elements. We have already seen that this vexing problem faced the chemists of the 1840s when they converted the results of elementary analysis into empirical formulas. The story of the confusion over the precise meaning of the terms "atom," "molecule," and "equivalent" is too lengthy to be recounted here, except to note that Laurent and Gerhardt persistently drew attention to the fact that atoms and chemical equivalents were not necessarily identical. The "new" atomic weights proposed by Gerhardt (*e.g.*, $C = 12$, $O = 16$, $S = 32$), which were based on the reduction of the formulas of all volatile organic compounds to equal volumes, were adopted in the 1850s by some of the younger chemists (Odling, Williamson, Couper, Kekulé), but many influential men (*e.g.*, Kolbe in Germany, Berthelot in France) considered them to be unacceptable. It was the intervention of Stanislao Cannizzaro at a meeting of chemists in 1860 at Karlsruhe that gave decisive support to the adherents of Gerhardt's system; he reminded the chemical world of the hypothesis advanced by Avogadro and by Ampère that, at constant temperature and pressure, the number of molecules in all gases is always the same for equal volumes. When nine years later the new atomic weights appeared in Mendeleev's Periodic Table, only a few stubborn organic chemists still refused to accept them in writing formulas.

We saw before that in 1858 Couper had written formulas that he considered to represent the actual arrangement of the atoms in an organic compound. It should be noted here that Kekulé wrote in his textbook:

It must be kept in mind that the rational formulas . . . are nothing more than expressions for the transformations of compounds, and for the comparison of various substances with each other; but that they are in no way intended to express the constitution, *i.e.*, the arrangement of the atoms in the compound in question (Kekulé, 1861, Vol. I, p. 157).

Nonetheless, Kekulé drew pictures that suggested the arrangement of atoms, Butlerov in 1861 introduced the term "chemical structure," and in 1864 Alexander Crum Brown and Lothar Meyer drew graphical formulas similar to those in use a century later. After 1865, when Kekulé made his famous proposal regarding the hexagonal structure of benzene, structural formulas appeared with increasing frequency in chemical articles. The most persistent critic of the new structural chemistry was Hermann Kolbe; as editor of the *Journal für praktische Chemie*, he filled its pages with vitriolic attacks on Kekulé and his followers, Baeyer and Fischer (Phillips, 1966).

The capstone of the new structural chemistry was the independent discovery in 1874, by Jacobus Henricus van't Hoff and Joseph Achille LeBel, of the asymmetric carbon atom as an explanation of the optical activity of organic compounds; although there had been anticipations of this discovery (Larder, 1967), its impact was felt only after the formulation of a theory of valence and structure. Their description of isomerism in terms of differences in the arrangement of atoms in space (stereo-isomerism) provided a convincing argument for considering structural formulas as representations of reality. Van't Hoff also made it clear that the repeated objections to the structural formulas on the basis of the kinetic theory of matter could be countered by assuming that the atoms oscillated rapidly about their mean position in the molecule. When in 1877 van't Hoff's book *La Chimie dans l'Espace* appeared in a German translation, with a favorable introduction by the respected Johannes Wislicenus, it drew from Hermann Kolbe the following famous remark:

A certain Dr. J. H. van't Hoff, employed at the Veterinary School at Utrecht has, it seems, no taste for exact chemical investigation. He has deemed it more convenient to mount Pegasus (evidently hired at the veterinary stables) and to proclaim in his "La Chimie dans l'Espace" how, during his bold flight to the top of the chemical Parnassus, the atoms appeared to him to have grouped themselves throughout universal space (Kolbe, 1877, p. 473).

Van't Hoff's reply came the following year, on assuming his professorship in Amsterdam; his inaugural lecture was appropriately entitled *Imagination in Science* (Cohen, 1912).

Before returning to the consideration of nineteenth-century studies on the organic chemistry of proteins, it should be noted that the rise of structural organic chemistry coincided with a lessened interest among organic chemists in the nature of the complex constituents of living systems and their metabolic transformations. Whereas during the 1840s renowned chemists such as Liebig and Dumas were active participants in the study of albuminoid substances, the younger men who followed Laurent and Gerhardt sharply separated such research from the mainstream of organic chemistry. The situation around 1860, so far as leading organic chemists were concerned, appeared to be that albuminoid substances represent amides of an extremely complex constitution, and because they were noncrystalline materials whose purification was very difficult it was well to leave them alone. An extreme view was:

> The albuminoid substances . . . do not constitute, properly speaking, chemical species; they are organs or debris of organs whose history should belong to biology rather than to chemistry (Naquet, 1867, p. 612).

The most authoritative statement of the separation of the new organic chemistry from chemical physiology was that of Kekulé:

> In particular it must be emphasized that organic chemistry has nothing to do with the study of the chemical processes in the organs of plants and animals. This study forms the object of physiological chemistry. This deals therefore with the chemical changes that occur within the living organism and is divided, depending on whether it treats the chemical part of life processes of plants or of animals, into plant chemistry or phytochemistry and animal chemistry or zoochemistry (Kekulé, 1861, Vol. I, p. 11).

As we shall see in our further discussion, however, the new organic chemistry, based on the concepts of structure, valency, and stereoisomerism, provided indispensable fundamental knowledge that contributed to the elucidation of the chemical nature of proteins, of other complex chemical constituents of biological systems, and indeed of their transformation during the course of metabolic processes. This was envisaged by Felix Hoppe-Seyler in the foreword to the first volume of his new *Zeitschrift für physiologische Chemie*:

> The upsurge experienced by organic chemistry during the past few decades enables it not only to analyze biological problems in the manner attempted before, but also to conduct searching experiments on

the chemical processes in the living organism. The synthetic results, in providing insights, of which the recent past may be proud, into the structure of chemical substances and their transformations through chemical processes, have provided the means and directions to investigate, with hitherto unexpected assurance, the causes of vital phenomena in the structure and relationship of the substances that are active in biological organisms.

Biochemistry has thereby grown to a science, from its natural and necessary analytical beginnings . . . . (Hoppe-Seyler, 1877a, p. I).

A somewhat different attitude was expressed by an English physiological chemist:

. . . [O]f late years, there has been introduced into chemical teaching the so-called "structure" hypothesis of carbon-compounds, under which these latter are represented as structures comprised of constituent atoms arranged graphically. This graphical arrangement is arrived at from a knowledge of the way in which substances decompose when subjected to particular processes. That is to say, if they yield by some process a particular substance, this is considered to be sufficient evidence that they contained in their structure a particular group of atoms, and so far so good. But many chemists go further than this, and say not only that such and such a group is present, but also that it is present in a certain position. It is this last-named assumption that is so extremely unprofitable and unmeaning, particularly for physiological chemistry (Kingzett, 1878, pp. 27–28).

Similar reservations about the new organic chemistry were expressed by other physiological chemists, and contributed to their separation from the fruitful lines of research pursued by organic chemists such as Emil Fischer.

Amino Acids and Peptides. In 1860 it was generally agreed that the compounds leucine and tyrosine are characteristic cleavage products of the albuminoid substances; also, it was known that gelatin, a material known to be related to these substances, yields on acid hydrolysis the substance glycocoll (glycine) and known to be aminoacetic acid. Glycine had been obtained by Henri Braconnot in 1820, and because of its sweet taste he considered it to be a sugar (*sucre de gélatine*). In 1845, Victor Dessaignes showed that acid hydrolysis of hippuric acid (excreted in the urine when benzoic acid is fed) gives glycine, and the first statement of its correct empirical formula was provided in the following year by

Gerhardt. The elucidation of its structure as aminoacetic acid only came in 1858.

The chemical relationship of glycine to leucine was indicated by Laurent and Gerhardt in 1848; they considered the two compounds to be members of a homologous series having the composition $C_nH_{2n+1}NO_2$, and later successively termed "amido acids" or "amino acids." Two years later Adolf Strecker provided, by chemical synthesis, a compound he named alanine and which he recognized to be aminopropionic acid, another member of this series; in 1888 Theodor Weyl isolated alanine as one of the products of the acid hydrolysis of silk, a source from which Emil Cramer had obtained in 1865 the substance serine. Neither alanine nor serine (later proved to be α-amino-β-hydroxypropionic acid), however, was regarded as a widely distributed constituent of albuminoid substances until after 1900. The same was true of the leucine-like substance obtained from tissue extracts in 1856 by von Gorup-Besanez; by 1878 it was established that this substance (valine) is aminoisovaleric acid, and 13 years later leucine was proved to be aminoisocaproic acid. In 1881, Ernst Schulze and Johann Barbieri identified phenylalanine as a constituent of plant proteins; in the following year it was synthesized by the method Strecker had used to make alanine, and its structural relation to tyrosine (p-hydroxyphenylalanine) became clearly evident. The recognition of tyrosine as a phenol identified this protein component as being responsible for the Millon and xanthoproteic reactions. Also, phenylalanine was recognized to be the source of the benzoic acid formed on oxidation of albuminoid substances with reagents such as potassium permanganate.

Of special interest in the history of protein chemistry was the discovery by Heinrich Ritthausen that the long-known aspartic acid is formed on acid hydrolysis of plant proteins. This acidic compound had been the object of extensive chemical study, having been obtained in 1827 (by Plisson) from the substance (asparagine) isolated from asparagus juice by Vauquelin and Robiquet in 1806. By 1838 work by Pelouze and by Liebig had established the elementary composition of these two compounds, and during the 1840s it was recognized that aspartic acid is converted into malic acid upon treatment with nitrous acid (Piria) and could be made by heating the ammonium salts of malic or fumaric acids (Dessaignes). The latter observation was of considerable interest to Pasteur, as both the malic acid and the aspartic acid derived from biological sources were optically active, whereas the aspartic acid prepared by Dessaignes turned out to be optically inactive; Pasteur's famous study during the early 1850s on the properties of the optically active and inactive (racemic) malic and aspartic acids laid the groundwork for the understanding of the stereoisomerism of the amino acids, after the papers of van't Hoff and LeBel. It may also be noted in regard to the work of Ritthausen that the method he used for the isolation of aspartic acid was, in principle, the one employed by Scheele in the previous century for the isolation of acids (lactic acid, tartaric acid, malic acid) from biological fluids, namely, the precipitation with alcohol of their barium, calcium, or zinc salts. The structure of asparagine as the β-amide of aspartic acid was established by synthesis in 1887.

Ritthausen's finding of aspartic acid came in 1868; two years earlier he had obtained from the acid hydrolysate of plant proteins the acidic substance "glutaminic acid" (glutamic acid) soon recognized to be homologous with aspartic acid. By 1890 the structure of aspartic acid (aminosuccinic acid) and of glutamic acid (aminoglutaric acid) had been established by synthesis, and their widespread occurrence as products of the acid hydrolysis of plant and animal proteins was generally acknowledged.

During the period 1870–1890, as the number of the known cleavage products of the albuminoid substances increased, the chemists working in this field hoped to develop hydrolytic methods that would yield these products in amounts that would account for the weight of the material subjected to hydrolysis. Thus Hlasiwetz and Habermann (1873) believed that hydrolysis of casein with hydrochloric acid in the presence of tin gave leucine, tyrosine, aspartic acid, glutamic acid, ammonia, and little else. Others, however, could account for only a fraction of the starting material in terms of these components, and it was realized that "in the cleavage of the albuminoid substances by means of concentrated hydrochloric acid, there arise other hitherto unidentified products" (Drechsel, 1889, p. 426). By 1891, through the use of the reagent phosphotungstic acid (employed for the precipitation of alkaloids) and laborious fractionation, Edmund Drechsel and his associates were able to isolate, from an acid hydrolysate of casein, the hydrochloride of a basic substance (lysine), whose structure was established in 1899 to be α,ϵ-diaminocaproic acid. By 1896 two other basic amino acids (arginine and histidine) had been found to be protein constituents; the story of their discovery is recounted in detail by Vickery and Schmidt (1931). For a time after its discovery in 1877 (by Max Jaffé), ornithine (α,δ-aminovaleric acid) was thought to be a protein constituent, but its occasional appearance was later explained by the finding of the enzyme arginase (by Kossel and Dakin in 1904) which cleaves arginine to ornithine and urea.

Before the end of the nineteenth century still another amino acid was added to the list of protein constituents. It had been found in 1810 as a component of a new type of kidney stone by William Wollaston, one of the most accomplished chemists of his time; he termed the substance cystic oxide, and Berzelius renamed it cystine. In 1837 it was shown to contain sulfur, but the correct elementary composition was not obtained until 1884 (Külz, Baumann). Eugen Baumann also demonstrated that cystine can be reduced to a substance he named cysteine, and he recognized the relation of the two compounds to be that of a disulfide (cystine) to the corresponding mercaptan (cysteine). His structural formula for cystine was incorrect, however, and its chemical structure was not established definitely until 1903, when Ernst Friedmann converted cystine into taurine (known to be amino ethyl sulfonic acid) and into cysteic acid. When in 1899 Karl Mörner reported that he had isolated cystine from an acid hydrolysate of horn, and a year later he and Gustav Embden obtained it from various other protein materials, it was concluded that this constituent provided the alkali-labile protein sulfur, whose presence had been detected by Scheele over a century earlier. Some years before, Krüger (in 1888) and Suter (in 1895) had noted that a portion of the sulfur present in proteins was not released as sulfide by alkali; the correct explanation was not to come until 1922, however, when the amino acid methionine (having a thioether group) was identified as a protein constituent.

Other amino acids were to be discovered as protein constituents before the list was brought to its present state in 1935, but it is necessary to look back into the nineteenth century and to note that the cleavage methods employed frequently yielded a variety of products of uncertain identity and purity, and to which names were given that have since disappeared from the chemical vocabulary. Examples are "lysatine" (which turned out to be lysine mixed with arginine), "tyroleucine" (which may have been phenylalanine), and "glycoleucine" (which turned out to be racemized leucine). It required many years of painstaking work to sort out the genuine products, and the new synthetic organic chemistry, in which men like Strecker were pioneers, frequently decided the issue. As we shall see, however, the twentieth century was not free of such problems.

Aside from those products of acid hydrolysis whose identity and purity were uncertain, other methods of degradation yielded results of even greater complexity. Around 1875 much attention was given to the extensive work of Paul Schützenberger on hydrolysis with barium hydroxide, and to his idea that ureido groups ($-NH-CO-NH-$) are significant structural features of proteins. Also, some of the methods found useful in the study of simpler organic compounds (fusion with alkali, or oxidation with potassium permanganate, chromic acid, hydrogen peroxide, or ozone) were applied to proteins, notably by Béchamp, Maly, von Fürth, Schulz, and Harries. These violent reagents gave a variety of products, many of which were ill-defined, and whose names soon lapsed into obscurity. Despite the limitations of such methods, some investigators (for example, Niels Trønsegaard) pursued this approach as late as the 1920s. In addition, many studies were conducted toward the end of the nineteenth century on the chemical modification of proteins by treatment with chlorine, bromine, and iodine. All these various approaches yielded results that compounded the confusion, but by 1900 it was evident that the hydrolytic methods were likely to be the most fruitful ones. It was also clear, however, that the products of the hydrolytic cleavage of proteins were so numerous, as compared with those obtained from such materials as the fats or starch, that elementary analysis could contribute little to the understanding of the intimate structure of proteins. As noted by Kossel and Kutscher in 1900, it was necessary to consider the problem in terms of the arrangement of the amino acid units rather than on the basis of individual atoms.

In 1899 Emil Fischer entered the field of protein chemistry, and during the succeeding two decades decisively influenced the course of its subsequent development (Forster, 1920; Hoesch, 1921). We have mentioned some highlights of his work on the sugars in its relation to the problem of enzyme action. He had also already achieved great success in the synthesis of purines, and we shall have occasion to mention this in considering the early history of the nucleic acids. The Nobel Prize in Chemistry that Fischer received in 1902 was in recognition of his work on carbohydrates and purines. In summarizing the beginnings of his work on proteins, Fischer wrote:

Since the proteins participate in one way or another in all chemical processes in the living organism, one may expect highly significant information for biological chemistry from the elucidation of their structure and their transformations. It is therefore no surprise that the study of these substances, from which chemists have largely withdrawn for more than a generation, because they found more worthwhile work in the development of synthetic methods or in the study of simpler natural compounds, has been cultivated by the physiologists in ever-increasing degree and with unmistakable success. Nevertheless, there was never any doubt that organic chemistry, whose cradle stood at the proteins, would eventually return to them. There has been, and still continues to be difference of opinion only regarding the date when the collaboration of biology and chemistry will be successful.

Whereas cautious colleagues fear that a rational study of this class of substances will encounter insuperable difficulties, because of their highly inconvenient physical properties, other optimistically-inclined observers, among whom I number myself, believe that one should at least attempt to besiege the virgin fortress with all the present-day resources; since only through daring can the limits of the potentialities of our methods be determined (Fischer, 1906, p. 530).

When Fischer began his work on proteins, about a dozen amino acids (glycine, alanine, leucine, phenylalanine, tyrosine, serine, aspartic acid, glutamic acid, lysine, arginine, histidine, and cystine) were known to be cleavage products. By 1906, when he presented a summary report on the progress he and his associates had made, he had added to the list valine, proline, hydroxyproline, and diaminotrioxydodecanic acid (he withdrew the last in 1913). In 1901 Frederick Gowland Hopkins and Sydney Cole had found the indole-containing amino acid tryptophane (now tryptophan), of which more later, and Felix Ehrlich had found isoleucine in 1903. Fischer's discovery that valine, proline, and hydroxy-proline are cleavage products of proteins came from the application of a general method he introduced in 1901 for the separation of the amino acids present in an acid hydrolysate of a protein by the distillation of their esters. Earlier work in 1883–1888 by Theodor Curtius had shown that the methyl or ethyl esters of various amino acids (*e.g.,* glycine ethyl ester, $NH_2CH_2CO—OC_2H_5$) can be distilled under reduced pressure without decomposition, and Fischer took advantage of this property to subject the mixture of amino acid esters derived from a protein hydroly-sate to fractional distillation. In addition to the amino acids known to be protein constitutents, proline (pyrrolidine carboxylic acid, synthesized by Richard Willstätter in 1900) was obtained from a casein hydrolysate in 1901, and in the following year the closely related oxyproline (now termed hydroxyproline) was obtained by the ester method from a hy-drolysate of gelatin.

In parallel with his development of the ester method for the separation of the amino acids present in protein hydrolysates, Fischer embarked on a systematic program to synthesize the individual amino acids. However, as he noted in his first paper on the subject:

> Whereas the hydrolytic cleavage of proteins usually produces op-tically-active amino acids, the artificial products are racemic. The com-plete synthesis of the natural compounds is only achieved when one can cleave the racemic mixture into the optically active components (Fischer, 1899, p. 2451).

In addition to improving older synthetic methods for making racemic amino acids and developing new ones, Fischer also devised procedures (based on principles stated by Pasteur) for the resolution of racemic products into their optically active components. Since all the amino acids Fischer handled had one asymmetric carbon, two optically active isomers were obtained, and comparison of the sign of rotation with that of the natural amino acid identified the desired isomer.

It is clear from Fischer's early papers in the protein field that he had set himself a bold objective, namely, the synthesis of proteinlike substances. In his 1906 lecture he stated that

> . . . six years ago, when I decided to devote myself to the study of the proteins, I began with the amino acids, in order to gain new viewpoints and methods for their complex derivatives. The success has not disappointed me. First it was possible to find a new separation method for the monoamino acids through the use of esters . . . . More significant seem to me to be the similarly found methods for the conversion of the amino acids into their amide-like anhydrides, for which I have chosen the collective name "polypeptides." The higher members of this group of synthetic materials are so similar to the natural peptones in their external properties, certain color reactions, and behavior toward acids, alkalis and ferments that one must consider them to be very close relatives, and that I may describe their preparation as the beginning of the synthesis of the natural peptones and albumoses (Fischer, 1906, p. 531).

In our consideration of the "soluble ferments" that act on albuminoid substances we noted that nineteenth-century investigators had found, among the products, materials they termed albuminoses (or albumoses) and peptones. The study of these products was intensively pursued by Willy Kühne after 1875 in collaboration with Russell H. Chittenden and Richard Neumeister. They followed up the observations of Georg Meissner published around 1860 that pepsin digests proteins only incompletely, and that pepsin-resistant products are formed. Kühne proceeded to examine the nature of the products formed by the action of pepsin and of trypsin through laborious fractionation of the resulting mixtures by means of precipitation with salts (sodium chloride, ammonium sulfate, magnesium sulfate) under various conditions of temperature and acidity. From our present vantage point in the development of protein chemistry, it is easy to see why this approach yielded little fruitful knowledge; the separation methods available to Kühne and his contemporaries were not equal to the task of separating the complex mixture of the very similar products present in an enzymic digest of a protein. In their view the action of pepsin on albumins first gave "syntonin," which was then cleaved to "protoalbumose" (soluble in water) and "heteroalbumose" (soluble only in salt solutions); these "primary albumoses" were thought to be cleaved further to secondary albumoses ("deuteroalbumoses"). Analogous products from other proteins were designated "globuloses," "caseoses," etc. Tryptic digestion was considered to give approximately equal amounts of "hemipeptones" which were subjected to further cleavage to amino acids (including leucine and tyrosine), and "antipeptones" which were resistant to further enzymic attack. By 1887 Neumeister had elaborated this idea into a scheme of considerable complexity, and, since none of the products (except the amino acids) could be characterized in a manner acceptable to the chemists of his time, it was of interest only to physiologists; many other notable figures (Maly, Kossel, Pekelharing) contributed to a massive literary output characterized by an efflorescence of new names, such as the "kyrines" (basic peptones) of Siegfried. For the physiology of the late nineteenth century, peptones were important because they diffused through membranes, and were therefore considered to represent the form in which the products of the digestion of proteins were transported across the intestinal wall. This idea was developed most extensively by Franz Hofmeister and his associates (notably Pick and Zunz) during the period 1880–1900, and was not laid to rest until after Otto Cohnheim had shown in 1901 that the intestinal mucosa elaborates an enzyme ("erepsin") which cleaves peptones to amino acids.

Although the chemical studies of Kühne, Hofmeister, and their followers provided an inadequate basis for the physiological conclusions that were drawn, the clear demonstration that amino acids were formed upon the enzymic cleavage of proteins was of considerable chemical importance. The conditions under which these digestions were conducted were so much milder as regards temperature and acidity than those used by the protein chemists that the appearance of amino acids after extensive enzymic digestion strengthened the conviction that these products were not simply artifacts of acid hydrolysis but were present in proteins in the form of "amide-like anhydrides." Such results also silenced some adherents of the idea of "living" protein molecules; for example, Loew (1896, p. 22) had stated that "we can conclude with a high degree of probability that the leucine radical does not exist as such in the proteids" and had compared the formation of amino acids upon acid hydrolysis of proteins to the effect of such treatment on sugars to yield various decomposition products. In his work, Fischer used digestion with pepsin and trypsin (pancreatin) to show that proline is a genuine constituent of proteins, and this method was used for other amino acids as well. But, as to the problem of the albumoses and the peptones, Fischer was clear:

Despite all the acknowledgment that I gladly pay to the researches of Kühne, Hofmeister, Neumeister, *et al.*, I cannot rid myself of the conviction that the precipitation methods they used are not able to yield pure products with such complex mixtures as arise upon protein breakdown, and that therefore the various kinds of albumoses and peptones with which the physiologists deal are for the chemist only inextricable mixtures. The next aim of research in this field must be directed to the isolation from them of chemically definable homogeneous substances. In all probability, however, this will not succeed without the discovery of new and more effective separation methods. My own experience in this field is still very limited (Fischer, 1906, pp. 605–606).

Neither Fischer nor the protein chemists who followed him could solve this problem satisfactorily before the introduction of chromatographic separation methods into protein chemistry after World War II. Some of the simplest polypeptides (dipeptides) had been isolated in the form of their cyclic derivatives (diketopiperazines) during the nineteenth century, and in 1902 Fischer and Bergell isolated glycyl-alanine from a partial hydrolysate of silk fibroin (Fischer introduced the terms "glycyl," "alanyl," etc., to denote an amino acid unit acting as an acyl group). Although numerous other dipeptides were obtained in this manner before 1940, the isolation of longer polypeptides in a homogeneous state was extremely difficult (Synge, 1943).

The peptones therefore were, for Fischer, complex and unresolvable mixtures of polypeptides, and as a chemist he chose to make peptonelike polypeptides by synthesis. Efforts in this direction had been reported during the nineteenth century; for example, in 1871 Eugen Schaal had heated asparagine to 180° to yield a material that was studied further by various workers, notably Hugo Schiff, in 1897–1899; although Schiff recognized the product to be a polymer, he did not formulate it as a polyamide. This product gave the biuret reaction, as did a material prepared by Paul Schützenberger in 1891 by heating a mixture of various amino acids with urea. As Fischer stated in his first paper on the synthesis of polypeptides:

All the products described by them are amorphous substances that are difficult to characterize, however, and one can say as little about their structure as about the extent of their relationship to natural protein compounds. If one wishes to obtain secure results in this difficult field, one must first find a method that permits one to combine, in anhydride-like linkage, successively and with definable intermediates the molecules of various amino acids (Fischer and Fourneau, 1901, p. 2868).

Theodor Curtius had already made important contributions to this end, and during 1900–1910 he was to add others of lasting value to the synthesis of polypeptides. The magnitude of Fischer's attack on the problem was so much greater, however, that the immediate impact of his results tended to obscure the work of others working in this field. In his 1901 paper he described the preparation of the simplest possible peptide, glycyl-glycine (NH_2CH_2CO—$NHCH_2COOH$), and by 1907 he had prepared a polypeptide chain composed of 18 amino acid units (3 leucines, 15 glycines) linked to each other by CO—NH bonds ("peptide bonds"). During the period 1901–1919 he described synthetic procedures for making a host of polypeptides containing some of the amino acid units known to be present in proteins, and he took special pains to use optically active amino acids. The details of Fischer's synthetic methods need not be elaborated here; it should be noted, however, that, although the strategy of his synthetic approach has guided all further research in the field of peptide synthesis, the tactical methods he introduced have largely been supplanted by newer procedures. An especially important contribution was that of Fischer's student Max Bergmann who, with his associate Leonidas Zervas, introduced the so-called "carbobenzoxy" method in 1932; it broadened the scope of peptide synthesis to include amino acids that Fischer's methods could not handle.

When Fischer reached the stage of the 14-amino acid compound, and found that it gave a strong biuret reaction and was precipitated by protein reagents, he stated (in January 1907) that

> . . . one cannot avoid the impression that this tetradecapeptide represents a product quite closely related to the proteins, and I believe that with the continuation of the synthesis to the eicosapeptide one will come within the protein group (Fischer, 1923, p. 13).

Later in the same year, when he had reached the octadecapeptide (molecular weight, 1213), he wrote:

> If one imagines the replacement of the many glycine residues by other amino acids, such as phenylalanine, tyrosine, cystine, glutamic acid, etc., one would soon attain 2–3 times the molecular weight, hence values assumed for some natural proteins. For other natural proteins the estimates are much higher, up to 12,000–15,000. But in my opinion these numbers are based on very insecure assumptions, since we do not have the slightest guarantee that the natural proteins are homogeneous substances.

From the experience thus far, I do not doubt that the synthesis can

be continued by means of the same methods beyond the octadecapeptide. I must however provisionally waive such experiments, which are
not only very laborious but also very expensive, and I will attempt to
construct combinations with a larger number of different amino acids,
if possible where similar units are not located next to each other in the
chain. I believe that these completely mixed forms are preferred in
Nature (Fischer, 1907a, pp. 1757–1758).

Many of the polypeptides that Fischer and his associates prepared
were hydrolyzed by pancreatin. He concluded that

> . . . the attack of the pancreatic juice depends partly on the nature of
> the amino acids, partly on their sequence, also on the chain length, and
> finally and most especially on the configuration of the molecule
> Fischer, 1906, p. 580).

None of his synthetic polypeptides was cleaved by pepsin, however; this
fact was to be raised during the 1920s in connection with the discussion
of the structure of proteins.

In 1902 Franz Hofmeister wrote an important review of the status of
the problem of protein structure and, after considering the various proposals that had been made regarding the mode of linkage of the amino
acid units, he concluded that proteins ". . . are formed by the condensation of α-amino acids bound through the regularly recurrent
—CO—NH—CH= group" (Hofmeister, 1902, p. 792). Among the hypotheses that Hofmeister judged to be less well-founded was Schützenberger's view (in 1875) that the amino groups of polyamino acid chains
are linked by CO groups to form ureido linkages (—NH—CO—NH—),
and that of Albrecht Kossel who in 1900 considered complex proteins to
be built around a central nucleus that resembled the protamines, relatively simple basic materials discovered in 1874 by Miescher. Kossel's
idea was based on the observation that the protamines did not give the
usual color tests for proteins, except for the biuret reaction; he proposed
that this reaction is specific for the protamine nucleus of proteins, but this
view had to be abandoned when Schiff showed that the biuret reaction is
given by a wide variety of organic compounds having two amide groups
in a straight-chain molecule. In connection with Hofmeister's espousal of
the "peptide hypothesis" of protein structure, it is of interest that in 1878
he confirmed the conclusion of Arthur Henninger that peptones (whose
chemical relation to the proteins was still a matter of dispute) are products
of protein hydrolysis; they offered in evidence the finding that, if peptone

was treated with a dehydrating agent (acetic anhydride), a nondialyzable product was formed.

Hofmeister is usually credited, with Fischer, for having formulated clearly the peptide theory of protein structure. It should be stressed, however, that, in Fischer's view,

> . . . simple amide formation is not the only possible mode of linkage in the protein molecule. On the contrary, I consider it to be quite probable that on the one hand it contains piperazine rings, whose facile cleavage by alkali and reformation from the dipeptides or their esters I have observed so frequently with the artificial products, and on the other hand the numerous hydroxyls of the oxyamino acids are by no means inert groups in the protein molecule. The latter could be transformed by anhydride formation to ester or ether groups, and the variety would increase further if poly-oxyamino acids are assumed to be probable protein constituents (Fischer, 1906, pp. 607–608).

These speculations were to be revived after 1920, when the status of the peptide theory of protein structure became uncertain. Although the new organic chemistry had made possible the identification of the structure of many amino acid units derived from proteins, and the synthesis of polypeptides that had some of the properties of proteins, the problem of the chemical structure of proteins was far from being settled. Nevertheless, "Beginning with the announcements of the polypeptide hypothesis, the study of proteins became once more an occupation in which a self-respecting organic chemist might take a part" (Vickery and Osborne, 1928, p. 418).

Before concluding this section, a postscript must be added about the amino acid tryptophan. At intervals during the nineteenth century there were reports (*e.g.*, by Tiedemann and Gmelin in 1826, by Bernard in 1856) of the appearance of a red color when chlorine water was added to albuminoid matter that had undergone decomposition. By 1875 it was known that indole (a product of animal putrefaction) gives this color reaction, and in 1890 Neumeister concluded that the chromogenic material arises in all processes where proteins undergo extensive degradation; he named the substance "tryptophane." During the succeeding decade several investigators attempted to isolate this substance, but without success. Another color reaction observed with albuminoid bodies was found by Albert Adamkiewicz in 1874; a violet color appeared when sulfuric acid was added to a solution of egg albumin in glacial acetic acid. Hopkins and Cole found in 1901 that the reaction was caused by the presence of glyoxylic acid in the acetic acid. They then used this color test to concentrate the chromogen and isolated a material that gave both the chlorine and the glyoxylic reactions; in 1907 Alexander Ellinger synthesized the amino acid and showed it to be β-indolylalanine. Because tryptophan, like other indole derivatives, is decomposed by acids, Hopkins and Cole isolated it (in 1902) from a digest of casein by trypsin.

Tryptophan was identified as a protein constituent because of a color reaction; before the catalogue of the amino acids present in proteins was to be completed in 1935, three other amino acids were to be added (methionine, threonine, and hydroxylysine). The discovery of the first two came from the careful study of the nutritional factors required for the growth of a bacterium and the rat. We shall mention them again later in considering the history of the organic chemistry of proteins after 1920. Other amino acids added to the list of protein constituents during the twentieth century, but later withdrawn, included (beside the diaminotrioxydodecanic acid of Fischer) α-aminobutyric acid, norleucine (α-aminocaproic acid), "prolysine," oxytryptophan, and β-hydroxyglutamic acid.

The Multiplicity and Individuality of Proteins

After 1850, when some of the facile generalizations based on elementary analyses had been abandoned, the recognition grew that there were many albuminoid substances differing in their properties. The impetus to this development came principally from the introduction of a new method for their fractional precipitation by means of neutral salts ("salting out"); this method was applied to the separation of the components present in human blood, in other animal tissues (*e.g.*, muscle), and in plants of importance in human nutrition. Indeed, one of the first hints that egg albumin and serum albumin might be different came from the observation of Peter Ludwig Panum (in 1852) that, whereas the addition of solid sodium chloride to blood serum gave a precipitate (that redissolved in water), no precipitate was formed with egg white. Shortly thereafter Robin and Verdeil noted the precipitation of both egg albumin and serum albumin with magnesium sulfate, and in 1854 Rudolf Virchow made extensive observations on the precipitation of albuminoid substances with a variety of neutral salts; he interpreted the failure of a salt to precipitate all of the protein in a biological fluid to indicate that the protein was present in solution in different states. Although other reagents, especially those found to be useful for the precipitation of plant alkaloids (phosphotungstic acid, tannic acid, trichloroacetic acid, picric acid, etc.) continued to be used to precipitate proteins, it was recognized that, whereas the products obtained in this manner could not be redissolved in water, the protein precipitates yielded by the salting-out method were water-soluble.

The systematic use of the salting-out method by Prosper-Sylvain Denis, who published his results in two monographs in 1856 and 1859, provided the first clear indications of the multiplicity of the serum proteins. From the work of Denis and of those who followed him came the separation of a material soluble in water (serum albumin, which Denis termed "serine") from a material soluble in dilute salt solutions but insoluble in water; the latter subsequently came to be termed "globulin" (Weyl, 1877), a designation that originated with Berzelius to denote the material he believed to be associated with the iron-containing component of the "red coloring matter" (hematoglobulin) of red blood corpuscles. Denis also obtained from plasma a fraction ("soluble fibrin") which had the properties of a soluble precursor of fibrin (Virchow named it fibrinogen). The continued study of the process of blood coagulation later came to a focus in the work of Alexander Schmidt and of Olof Hammarsten; by 1880 it was clear that the formation of the insoluble fibrin involves the action of a "fibrin ferment" (now called thrombin) on fibrinogen. This advance, made possible by the work of Denis and others on the salting out of proteins, put the physiological phenomenon of blood coagulation within the scope of chemical investigation, and the ability of the circulating blood to remain fluid was no longer considered to be the expression of a vital force, as believed by John Hunter in the previous century. During the twentieth century the fractionation of plasma proteins became a matter of military importance in connection with blood transfusion, and was notably furthered by Edwin Cohn during World War II.

In 1859 Willy Kühne reported the extraction from frog muscles of a coagulable globulin-like substance which he called myosin; this material could be precipitated from a solution in 10% sodium chloride by the addition of solid salt. He also found that, if myosin is dissolved in dilute acid, it is transformed to a material ("syntonin") that is insoluble in salt solution; Kühne considered this product to be identical with the "muscle fibrin" of Liebig. Another indication of the rapid adoption of the method developed by Denis was Hoppe-Seyler's work on the albuminoid component of the yolk of hen's eggs. After extraction of the yolk with ether, Hoppe-Seyler obtained a material (he named it vitellin) soluble in 10% sodium chloride, and which could be precipitated from salt solutions by the addition of water. Substances similar to vitellin had been found in the eggs of other animals, notably fish and amphibia, and Hoppe-Seyler grouped them in the class of globulins that included the materials identified in blood plasma and in muscle. By 1870 it was becoming clear that these various globulins might be different, and new names appeared to denote the proteins obtained from a particular tissue of an individual animal species; thus the vitellin-like protein of fish eggs was named "ichthulin."

The salting-out technique of Denis was applied to the isolation of many other protein preparations; an important addition to the salts he used was ammonium sulfate, introduced by Camille Méhu in 1878, whose high solubility in water made it the protein precipitant of choice. Its use came into general favor after 1885, when Heynsius and Kühne described the fractionation of mixtures of proteins, albumoses, and peptones by means of ammonium sulfate precipitation.

Protein Crystals. Denis also demonstrated that albuminoid substances present in plant seeds are extractable with 10% sodium chloride, but the decisive studies on these proteins were conducted later in the nineteenth century by two agricultural chemists: Heinrich Ritthausen and Thomas B. Osborne. In a report of the work he had done since 1860 on the albuminoid constituents of a large variety of plant seeds, Ritthausen concluded that, despite the similar elementary composition of plant and animal proteins, there are significant differences between them, especially as regards their content in tyrosine, leucine, and aspartic acid, and that the metabolic transformation of plant proteins into animal proteins is more complex than had been assumed.

> Does not perhaps the circumstance that the plant proteins casein, albumin are distinguished from the animal materials by designating them plant casein, plant albumin conceal the idea that, despite their similar composition and properties, there are differences among them, and that it is necessary to consider the substances as distinct and not merely to regard them as identical? (Ritthausen, 1872, p. 236).

Because Ritthausen's protein preparations were largely obtained by extraction with acid or alkali, Hoppe-Seyler considered them to correspond

> . . . not to pure unchanged albuminoid substances, but to more or less decomposed and insufficiently purified substances whose properties and composition tell nothing about the ones from which they were derived (Hoppe-Seyler, 1877b, Vol. I, p. 76).

Hoppe-Seyler's judgment was based on the results obtained by his associate, Theodor Weyl (1877), who applied the methods of Denis to the extraction of plant seeds. According to Weyl, there were two globulin-like types of protein in seeds: one of them was termed plant vitellin because like the material from egg yolk it was not salted out by excess sodium chloride, and the other was termed plant myosin because it was insoluble in saturated sodium chloride. As for the material that Ritthausen termed plant casein, Weyl considered such alkali-soluble preparations (called albuminates) to be transformation products of the plant globulins present in the cell. In particular, Weyl based his conclusions on the properties of a crystalline seed globulin prepared by another of Hoppe-Seyler's associates, Oswald Schmiedeberg. In rebuttal, Ritthausen published during 1880–1884 a series of papers to show that many of his preparations had retained their solubility in saline solutions, and that their elementary composition was the same as that of the corresponding crystalline globulins. Furthermore, during the course of this work he prepared crystalline preparations of globulins from the castor bean, hemp seed, and sesame seed in the manner that Edmund Drechsel (in 1879) and Georg Grübler (in 1881) had obtained crystals from Brazil nut and squash seed. The procedure was simply to allow a warm sodium chloride solution saturated with the protein to cool slowly, whereupon well-developed crystals appeared.

The doubts expressed by the followers of Hoppe-Seyler about the validity of Ritthausen's results were settled by the work of Thomas B. Osborne. In an important paper published in 1892, he wrote of the crystalline plant proteins:

> The fact that these proteid substances can be artificially crystallised is not only interesting in itself, but is important as presumably furnishing a means for making preparations of undoubted purity which will afford a sure basis for further study of their properties. The contradictory statements made by the various investigators, not only in regard to properties and composition of these bodies but also in respect to the value of the methods of solution and separation which have been employed hitherto, render an exact knowledge of all the facts relating to these substances a matter of the highest scientific and practical importance (Osborne, 1892, p. 663).

During the succeeding years, Osborne provided such "exact knowledge," which he collected in his book *The Vegetable Proteins*, first published in 1909 (second edition, 1924). Thus, in the case of the Brazil-nut globulin, Osborne's careful analyses of the crystalline material gave values close to those reported by Ritthausen and showed that Weyl's sulfur determination was incorrect. In this and many other instances, Osborne's work established the importance of Ritthausen's contributions and counteracted the disfavor into which they had fallen among physiologists owing to the verdict of Hoppe-Seyler.

The significance of Osborne's work in clearly demonstrating the multiplicity and individuality of closely related proteins cannot be overestimated (Vickery, 1931). During the ten-year period 1890–1900 the proteins of over 30 different species of seeds were examined, and as the work progressed it became clear that the globulins previously grouped under the term "vitellin" were sufficiently different in one or another chemical respect that the term had to be abandoned. Osborne was obliged to assign new names to the individual globulins; thus the one from Brazil nut became excelsin, the one from hemp seed edestin, etc. The alcohol-soluble proteins (which he termed prolamines) gliadin (from wheat), hordein (from barley), and zein (from corn) were also included in Osborne's study. In particular, he attempted to compare the various proteins on the basis of the amount of the individual amino acids formed upon hydrolysis; after Fischer's ester method became available, it was applied with considerable skill, but this procedure, together with others (such as that of Kossel and Kutscher for lysine, arginine, and histidine), could account for only a fraction (*ca.* 50 percent) of the starting material. Despite this deficiency, it was clear by 1909 that proteins which resembled each other closely in elementary composition and in solubility properties differed significantly in the amounts of individual amino acids formed upon acid hydrolysis. As we noted before, the problem of the quantitative analysis of the amino acid composition of protein hydrolysates was not solved until after 1945. Nonetheless, Osborne's analytical data provided knowledge of great importance to the later nutritional studies he conducted in association with Lafayette B. Mendel. By 1914 they were able to state:

> . . . [T]he relative values of the different proteins in nutrition are based upon the content of those special amino acids which cannot be synthesized in the animal body and which are indispensable for certain, as yet not clearly defined, processes which we express as maintenance or repair (Osborne and Mendel, 1914, p. 340).

This fruitful collaboration, which depended so heavily on Osborne's painstaking characterization of the plant proteins, laid the groundwork for much further work in the field of protein metabolism and in the study of vitamins.

For our story, however, the most dramatic evidence of the individuality of the seed proteins came from Osborne's experiments (in collaboration with H. Gideon Wells) on their so-called anaphylactic reactions. During the period 1900–1910 much attention was given to the fatal effect of a second administration to guinea pigs, after a delay of some days, of various protein-containing materials. In particular, Rosenau and Anderson demonstrated in 1906 that an animal injected with horse serum was not hypersensitive to the serum of other animals. Clearly, there was evidence of specificity with respect to the nature of the constituents of the serum. The availability of the purified crystalline seed proteins permitted Osborne and Wells to examine whether the anaphylactic reaction would be produced by injecting a particular protein into a guinea pig that had been sensitized by another protein. With only a few exceptions, the seed proteins tested proved to be specific, and there was clear evidence that, despite many apparent similarities, the various crystalline proteins represented different chemical individuals. Subsequent research on the properties of proteins as antigens, and their specificity in serological reactions (Landsteiner, 1945), has brought the study of immunity even closer to the problem of the intimate structure of individual proteins.

We saw earlier that some of the plant globulins (*e.g.*, that of Brazil nut) had been obtained in crystalline form by 1877. These crystals had been observed under the microscope as "aleurone" grains in plant seeds by Theodor Hartig (in 1850), and their albuminoid nature was demonstrated by Ludwig Radlkofer in 1859. The crystalline nature of these structures was questioned in 1862 by the influential botanist Carl von Nägeli, whose microscopic observations led him to designate them as "crystalloids," and Ritthausen used this term for his preparations of seed globulins. Among the various hypotheses advanced by Nägeli during his scientific career (we have already noted his contribution to the problem of fermentation), one of the earliest (*ca.* 1860) was the idea that organic substances are arranged in the plant cell into a mosaic of submicroscopic particles (micelles); in many plant structures, as in the cell wall or starch granules, the micelles were thought to have a crystalline appearance. The micellar hypothesis of Nägeli was adopted by leading plant physiologists during the latter half of the century, notably by Wilhelm Pfeffer, whose *Pflanzenphysiologie* (published in 1881) was widely respected. The beautiful crystals obtained in bulk by Osborne made it clear that the doubts of Nägeli and Pfeffer were not justified in this case; on the other hand, the apparently crystalline "yolk-platelets" seen by Frémy and Valenciennes (in 1854) in the eggs of some fish and amphibia did not turn out to correspond to well-defined proteins.

Even before the first recorded microscopic observation of the crystalline plant proteins, it had been shown that the "red coloring matter" of earthworm blood could be brought to crystallization. In 1840, Friedrich Ludwig Hünefeld wrote:

I have occasionally seen in almost dried blood, placed between glass plates in a desiccator, rectangular crystalline structures which under the microscope had sharp edges and were bright red (Hünefeld, 1840, p. 160).

Although there were other isolated reports between 1840 and 1851 of the appearance of "blood crystals" in partly dried blood from various animal species, it was only in the latter year that Otto Funke described the phenomenon carefully, and laid the groundwork for further study. The red coloring matter had been called hematosine (Lecanu) and hematoglobulin (Berzelius); in 1864 Hoppe-Seyler shortened the latter designation to hemoglobin. By 1871, when Wilhelm Preyer collected the available knowledge about crystalline hemoglobin in his book *Die Blutkrystalle,* the blood from over 40 animals had been examined, and it was clear that there were wide differences in the forms of the crystals, depending on their biological source. It was also known, from the work of Hoppe-Seyler and of Stokes during 1862–1864, that hemoglobin solutions take up molecular oxygen to form oxyhemoglobin, and Kühne had shown that oxyhemoglobin is less soluble in water and crystallizes more easily than does deoxygenated ("reduced") hemoglobin. A decisive factor in the recognition of the reversible interconversion of the two forms of hemoglobin and, in the later elucidation of the phenomenon, was the use of the spectroscope, introduced into chemistry in 1860 by Kirchhoff and Bunsen (McGucken, 1969). The importance of this instrument for the subsequent development of biochemistry cannot be overestimated, and it underlines the determinant role of new experimental tools in the emergence of new knowledge. We shall have frequent occasion to refer to spectroscopic observations, especially in connection with the study of intracellular respiration.

By 1870 it was also known that treatment of hemoglobin with acid gives a colorless albuminoid constituent (globin) and a red iron-containing material named hematin. To denote nonalbuminoid constituents such as hematin, Hoppe-Seyler introduced the term "prosthetic group," and he designated the conjugate with the albuminoid substance a "proteid." (The latter term was subsequently used, however, for all proteins, thus contributing to the already considerable confusion in the nomenclature.) Ludwig Teichmann had shown in 1853 that treatment of dried blood with sodium chloride and hot glacial acetic acid readily gives crystals of "hemin"; by 1857 this reaction had been introduced into forensic chemistry as a test for blood. Between 1880 and 1900 various methods for the preparation of hemin gave products that differed in their elementary composition; in 1901, however, William Küster showed that there was a single product of the composition $C_{34}H_{33}O_4N_4ClFe$. The elucidation of the chemical structure of hemin, and of its relation to the pigment of hemoglobin, is one of the many great achievements of the new organic chemistry as applied to the study of biological materials. The chief figure was Hans Fischer who in 1929 established the structure of the tetrapyrrole unit (porphyrin) of hemin by synthesis; before Fischer's work, important contributions had been made by Nencki, Willstätter, and others. On the basis of the knowledge available in 1913, Küster suggested that the four pyrrole rings of a porphyrin are linked by four single carbon atoms to form a 16-membered ring system. Such large ring compounds were unknown at that time, however, and both Willstätter and Fischer preferred alternative formulas. In the face of this opposition, Küster withdrew his proposal in 1923, but Fischer's synthesis proved that Küster had been right. Fischer's work on hemin also set the stage for the elucidation of the structure of the chlorophylls; it had been known from earlier studies (notably by Leon Marchlewski) that chlorophyll is a porphyrin derivative.

The fact that hemoglobin contained a prosthetic group led to some discussion (see Wurtz, 1880, p. 297) whether it should be classified as an albuminoid substance, but by the end of the century it was agreed that, in analogy to other known conjugated proteins, hemoglobin should be designated a "chromoprotein." At that time the most extensively studied group of conjugated proteins were the "glycoproteins"; the application of color tests for carbohydrates to egg albumin had shown this protein to belong to this group, and the studies of Oswald Schmiedeberg on chondroitin sulfate (from cartilage) and of Scherer and Hammarsten on mucins (from ovarian fluid, etc.) had indicated the widespread distribution of such conjugates of proteins with carbohydrate material. Another class of conjugated proteins discovered during the latter half of the nineteenth century were the nucleins or nucleoproteins; we shall consider them in the next chapter. It may be added that among the chromoproteins were included the hemocyanins, copper-containing respiratory pigments present in the blood of invertebrates; although studied early in the nineteenth century, their relation to the hemoglobins was clearly recognized only in 1878 (by Léon Fredericq), and crystalline hemocyanin (from octopus blood) was first described by Martin Henze in 1901.

The emergence of this variety of conjugated proteins led to the suggestion that the albumins and globulins isolated during the period 1860–1880 did not represent true constituents of protoplasm. For example, Olof Hammarsten stated:

So far as I can conclude from my own observations and investigations, our present ideas on the albumin-like constituents of protoplasm can scarcely be correct. The chief mass of free cells and of organs rich in cells—like glands—consists mainly, according to my experience, not of genuine protein materials in the usual sense, that is not of globulins and albumins, but of far more complex proteid substances which commonly contain, besides nitrogen and sulfur, also phosphorus and iron. These proteid substances represent the true protoplasmic protein and the albumins and globulins should rather be considered partly as nutrient material of the cell, and partly as breakdown products in the chemical transformation of protoplasm (Hammarsten, 1885, p. 449).

As we shall see in succeeding chapters, this idea recurred repeatedly during the period 1890–1910 in connection with speculations about the chemical basis of heredity and about intracellular respiration.

To return to the crystalline hemoglobins and the evidence they provided for the multiplicity and individuality of proteins, special mention must be made of the painstaking researches of Edward T. Reichert and Amos P. Brown. In a volume summarizing their results, they concluded:

The oxyhemoglobin obtained from the same blood crystallizes in the same form, with the same axial ratio, though often with different habit, when obtained by different methods of preparation . . . The crystals obtained from different species of a genus are characteristic of that species, but differ from those of other species of the genus in angles or axial ratio, in optical characters, and especially in those characters comprised under the general term of *crystal habit,* so that one species can usually be distinguished from another by its hemoglobin crystals (Reichert and Brown, 1909, pp. 326–327).

This crystallographic evidence for the individuality of the hemoglobins was supplemented in 1923 by the immunological studies of Karl Landsteiner and Michael Heidelberger.

As we have seen, crystalline preparations of albuminoid substances were known throughout the latter half of the nineteenth century. It is of interest, therefore, to find the following in a lecture given by Pasteur on 22 December 1883:

You know that the most complex molecules of plant chemistry are the albumins. You also know that these immediate principles have never been obtained in a crystalline state. May one not add that apparently they cannot crystallize . . . . If there are some that crystallize, such as hemoglobin, one may suppose that these substances are not asymmetric, or that they contain only two asymmetric groups, not three, not four, etc. (Pasteur, 1886, p. 36).

In 1889, however, Hofmeister crystallized egg albumin from a half-saturated solution of ammonium sulfate; his procedure was improved by Hopkins and Pinkus ten years later. In 1894, August Gürber crystallized horse serum albumin, and in 1899 Arthur Wichmann obtained a crystalline albumin from milk. During the twentieth century, proteins were to be crystallized in ever-increasing number, especially after the crystallization of jack bean urease (Sumner, 1926) and of swine pepsin (Northrop, 1930).

Although it was generally agreed that the crystallization of proteins, and especially their repeated recrystallization, are valuable means of purification (as in the case of simpler organic compounds), many investigators expressed doubts regarding the homogeneity of the crystalline proteins. For example, it was remarked that "There is scarcely a crystalline substance which takes up dissolved substances, like a sponge, to such a high degree as does albumin" (Wichmann, 1899, p. 584). Also, in speaking of the crystals of oxyhemoglobin, Emil Fischer stated that ". . . the existence of crystals does not in itself guarantee chemical individuality, since isomorphous mixtures may be involved, as is frequently evident in mineralogy for the silicates" (Fischer, 1913, p. 3288). Nevertheless, the value of crystallization as a means of purification has remained unchallenged, even though many of the crystalline protein preparations have been shown to be inhomogeneous. In part, the appeal of crystallization arises from the fact that "the beauty of crystals lies in the planeness of their faces" (Tutton, 1924, p. 5). More importantly, the availability of well-formed crystals of such proteins as oxyhemoglobin permitted the study of their intimate structure by the technique of X-ray diffraction.

For many years after the discovery of X-rays by Wilhelm Röntgen in 1895, there was intensive but inconclusive experimentation to determine whether they were waves or particles. In 1912 Max von Laue suggested to his associates Paul Knipping and Walther Friedrich that, since the presumed wavelength of X-rays was shorter than the distance between the regularly spaced atoms in a lattice of a crystal, diffraction effects should be produced when these rays are passed through crystals. The striking symmetrical patterns of diffraction spots that were indeed found provided convincing proof of the wave nature of X-rays, and of their interaction with crystal structures. Furthermore, the diffraction patterns accorded with the geometrical theory of crystal structure developed during the nineteenth century by Auguste Bravais and Evgraf Fedorov, among others, after René Just Haüy had transformed the molecular theory of crystalline matter into a mathematical science of crystallography (Metzger, 1918; Burke, 1966). In a letter dated 2 October 1912, Fedorov wrote of von Laue's discovery:

> For us crystallographers this discovery is of prime importance because now, for the first time, we can have a clear picture of that on which we have but theoretically placed the structure of crystals and on which the analysis of crystals is based (Ewald, 1962, p. 347).

The idea that X-rays might be used to determine the structure of a crystal by working back from the angles of reflection and the intensities of the diffracted rays was developed immediately after von Laue's discovery by William H. Bragg and his son W. Lawrence Bragg. By 1913 they had achieved striking success with several inorganic crystals, notably sodium chloride; to the consternation of many chemists, it turned out to be a regularly spaced array of sodium ions and chloride ions. Many years later, Henry E. Armstrong commented:

> . . . Prof. W. L. Bragg asserts that "In sodium chloride there appear to be no molecules represented by NaCl. The equality in number of sodium and chlorine atoms is arrived at by a chess-board pattern of these atoms; it is a result of geometry and not of a pairing off of the atoms."
>
> This statement is more than "repugnant to common sense." It is absurd to the n . . . th degree, not chemical cricket. Chemistry is neither chess nor geometry, whatever X-ray physics may be. Such unjustified aspersion of the molecular character of our most necessary condiment must not be allowed to pass unchallenged . . . . It were time that chemists took charge of chemistry once more and protected neophytes against the worship of false gods: at least taught them to ask for something more than chess-board evidence (Armstrong, 1927, p. 478).

After their study of sodium chloride, the Braggs determined the structure of diamond in 1914, and gave striking evidence for the tetrahedral carbon atom proposed by van't Hoff and LeBel (Bragg, 1967).

During the 1920s W. H. Bragg and his associates began work on the crystals of organic compounds; in 1929 Kathleen Lonsdale solved the structure of hexamethylbenzene. Although exact solutions of more complex structures were not yet possible, by 1932 X-ray diffraction could play an important role in deciding between alternative structures, as was shown by John Desmond Bernal in the case of the carbon skeleton of the sterol nucleus. Subsequently, new approaches to the handling of X-ray data (*e.g.*, the vector method of Patterson, the Beevers and Lipson strips) and the introduction of techniques such as isomorphous substitution with heavy atoms in 1937 led to the solution of the structure of the porphyrin-like compound phthalocyanine (see Robertson, 1972). During World War II the growing power of crystal structure determination was made clearly evident by Dorothy Crowfoot's analysis of the structure of benzylpenicillin.

After 1945 the development of high-speed electronic digital computers enormously increased the ability of X-ray crystallographers to solve complex organic structures (Hamilton, 1970); an especially dramatic demonstration was provided by the announcement, in 1957, of the complete solution of the structure of vitamin B_{12}, a molecule having the elementary composition $C_{63}H_{88}N_{14}O_{14}PCo$ (Hodgkin, 1965). Max Perutz and John Kendrew undertook the study of horse oxyhemoglobin and of whale myoglobin (a hemoglobin-like protein of muscle); by 1960 the three-dimensional structure of the simpler myoglobin molecule had been worked out. We shall return to the X-ray studies on proteins later in this chapter.

From Colloids to Macromolecules

In 1861 Thomas Graham described an instrument he called a *dialyser*, consisting of a bell jar closed at the large end with parchment paper and immersed in a reservoir of water; with this instrument he studied the diffusion across the membrane of substances in an aqueous solution within the bell jar. He found that albuminoid substances dialyzed very slowly, whereas crystalline compounds such as cane sugar diffused rapidly (Mokrushin, 1962). In writing of the slowly diffusible materials, Graham stated:

Among the latter are hydrated silicic acid, hydrated alumina, and other metallic peroxides of the aluminous class, when they exist in the soluble form; and starch, dextrin and the gums, caramel, tannin, albumen, gelatine, vegetable and animal extractive matters. Low diffusibility is not the only property which bodies last enumerated possess in common. They are distinguished by the gelatinous character of their hydrates. Although often soluble in water, they are held in solution by a most feeble force. They appear singularly inert in the capacity of acids and bases, and in all the ordinary chemical reactions. But, on the other hand, their peculiar physical aggregation with the chemical indifference referred to, appears to be required in substances that can intervene in the organic processes of life. The plastic elements of the animal body are found in this class. As gelatine appears to be its type, it is proposed to designate substances of the class as *colloids* [*colla*, glue], and to speak of their peculiar form of aggregation as the *colloidal state of matter*. Opposed to the colloidal is the crystalline condition. Substances affecting the latter form will be classed as *crystalloids*. The distinction is no doubt one of intimate molecular constitution (Graham, 1876, p. 553).

Graham's *dialyser* was nothing more than the "endosmometer" (minus the long tube at the narrow opening of the bell jar) used by Henri Dutrochet in his studies on osmosis. The movement of liquids through natural membranes had been the subject of several studies since its first clear description in 1748 by Nollet, and throughout the nineteenth century the phenomenon of osmosis was intensively investigated by plant physiologists. Furthermore, the interest of animal physiologists (*e.g.*, Karl Ludwig) in the rates of diffusion across membranes led to the important general equations developed by Adolf Fick in 1855 (Tyrrell, 1964). For Graham the movement of water in osmosis was

> . . . an affair of hydration and of dehydration in the substance of the membrane or other colloid septum, and that the diffusion of the saline solution placed within the osmometer has little or nothing to do with the osmotic result, otherwise than as it affects the state of hydration of the septum . . . . The degree of hydration of any gelatinous body is much affected by the liquid medium in which it is placed. Placed in pure water, such colloids are hydrated to a higher degree than they are in neutral saline solutions. Hence the equilibrium of hydration is different on the two sides of the membrane of an osmometer . . . . It is not attempted to explain this varying hydration of colloids with the osmotic effects thence arising. Such phenomena belong to colloidal chemistry, where the prevailing changes in composition appear to be of the kind vaguely described as catalytic. To the future investigation of catalytic affinity, therefore, must we look for the further elucidation of osmose (Graham, 1876, pp. 599–600).

With the development of artificial semipermeable membranes (*e.g.*, copper ferricyanide) by Moritz Traube in 1867, quantitative studies on osmotic pressure followed, notably by the plant physiologist Wilhelm Pfeffer, whose data were collected in his book *Osmotische Untersuchungen*, published in 1877. Pfeffer's results provided the basis for van't Hoff's formulation in 1885 of the osmotic pressure equation, and the development of a theory of solutions that connected osmotic pressure, freezing-point depression, and the lowering of vapor pressure as thermodynamic properties. This followed the work of François Raoult (published in 1882) on the freezing points of solutions, and the use of this property for the determination of the molecular weights of dissolved substances; soon thereafter Sabaneev and Alexandrov (in 1891) reported a molecular weight of 14,000 for egg albumin, based on measurements of freezing-point depression, and Reid (in 1905) gave a value of 48,000 for hemoglobin (osmotic pressure measurements). In addition, Picton and Linder examined the ability of solutions of crystalline hemoglobin to scatter light (Tyndall phenomenon) and concluded that ". . . there is no hard and fast line between colloidal and crystalloidal solution" (Picton and Linder, 1892, p. 159). Thus, although proteins were considered to be colloids, it was deemed appropriate to apply to their study the theory of solutions developed for "crystalloids," and the molecular weights so obtained suggested that proteins were substances of appreciable size. This was further indicated by the "ultrafiltration" experiments of Martin in 1896; he described a filter (gelatin or silicic acid) that held back dissolved proteins but let salts and sugars through. This method was further developed by Bechhold in 1907.

The Molecular Size of Proteins. In discussing the size of the colloids, Graham used the word "equivalent" to denote molecular weight:

> The equivalent of a colloid appears to be always high, although the ratio between the elements of the substance may be simple. Gummic acid, for instance, may be represented by $C_{12}H_{11}O_{11}$, but judging from the small proportions of potash and lime which suffice to neutralize this acid, the true numbers of its formula must be several times greater . . . . The inquiry suggests itself whether the colloid molecule may not be constituted by the grouping together of a number of smaller crystalloid molecules, and whether the basis of colloidality may not really be the composite character of the molecule . . . . A departure from its normal condition appears to be presented by a colloid holding so high a place in its class as albumen. In the so-called blood crystals of Funke, a soft and gelatinous albuminoid body is seen to assume a crystalline contour. Can any fact more strikingly illustrate the maxim that in nature there are no abrupt transitions, and that distinctions of class are never absolute? (Graham, 1876, pp. 596–598).

Although Graham's work was often quoted during the succeeding years, and his dialysis method was used by Kühne and others for the fractionation of peptones from proteins, the colloidal nature of the albuminoid substances was not studied intensively before about 1890, with the rise of the new physical chemistry promoted by Wilhelm Ostwald.

The problem of the size of proteins was of continuing interest, however, throughout the latter half of the nineteenth century, and there were repeated statements that proteins, like polysaccharides, are polymeric substances (*e.g.*, Hlasiwetz and Habermann, 1871; Grimaux, 1886). In 1903, when Friedrich Schulz collected the available data on this problem, he wrote:

> There is no doubt that the protein molecule is relatively large, much larger than most objects of chemical investigation. The general properties of the proteins are influenced to a high degree by this circumstance; the biological significance of the proteins is based largely on properties determined by molecular size; the multiplicity of the proteins is an expression of the magnitude of the molecular weight; the difficulties in the investigation of the chemical properties and our deficient knowledge of the proteins find reasonable explanation in the magnitude of the molecular weight (Schulz, 1903, p. 1).

For the chemist an accurate elementary analysis can give an estimate of the minimal molecular weight required to account for one atom of an element present in the compound in smallest amount. As early as 1872, Ludwig Thudichum stated that analytical data for the amount of iron in crystalline oxyhemoglobin (he called it hematocrystalline) had indicated

> . . . an atomic weight of about 13,000. Persons who have not studied this branch of chemistry and . . . do not read of atomic weights rising above 500, may wonder at the high atomic weight here assigned to hematocrystalline. But this body can now be obtained pure in quantity, and the analyses of crystals have always shown them to contain four tenths per cent. of iron (Thudichum, 1872, p. 27).

Subsequent iron determinations gave values ranging from 0.335 (Zinoffsky, 1886) to 0.47 percent (Hüfner, 1884), corresponding to a range of minimal molecular weight of 16,700 to 12,000. From the capacity of hemoglobin to bind oxygen or carbon monoxide, a molecular weight of about 16,700 was estimated on the assumption that there was a 1:1 combination. Since it was known that the iron did not belong to the protein part of oxyhemoglobin, and the sulfur did, the sulfur analysis might have been a more reliable guide; the ratio of sulfur to iron was close to 2:1, and Oscar Zinoffsky suggested that two globin molecules are linked to one hematin, making the minimal molecular weight of globin about 7000. From sulfur analyses for other proteins, as well as the chemical substitution of proteins (*e.g.*, iodination of crystalline serum albumin), values in the range 2000 to 6000 were obtained. Attempts were made to determine quantitatively some of the amino acids (tyrosine, cystine) formed on acid hydrolysis, but the results of different investigators were in wide disagreement, as the analyses depended on the isolation of these amino acids, and this could not be done quantitatively.

After summarizing the evidence, Schulz concluded:

> On the whole the picture is not heartening. We are far from being able to state the molecular size of proteins with any degree of confidence. In this opinion, I am in disagreement with the current widespread view which regards it as almost firmly established that the molecular weight of oxyhemoglobin is about 15000, that of crystalline egg albumin is 5–6000, etc. (Schulz, 1903, p. 101).

Similar doubts were expressed by Emil Fischer in 1907; in writing of the large number of amino acids that had been identified as products of the hydrolysis of proteins, he stated:

> If they were really all constituents of the same molecule, this must be a frighteningly large complex, and in fact older estimates of the molecular weight of several proteins gave a value of 12–15,000, which would exceed that of the fats by 15–21 fold.
>
> I am however of the opinion that these calculations rest on a very insecure basis, principally because we do not have the slightest guarantee for the chemical homogeneity of the natural proteins; moreover, I believe that they are mixtures of substances whose composition is in fact much simpler than has been inferred hitherto from the results of elementary analysis and hydrolysis (Fischer, 1923, p. 8).

Fischer reiterated his doubts about the molecular weights of the proteins in 1916:

> . . . Hofmeister based his considerations on a protein molecule of about 125 units (amino acids), as is thought to be the case for hemoglobin.
>
> In my opinion, however, the methods applied to the determination of the molecular weight of the hemoglobins are less certain than had been assumed previously. Although they crystallize beautifully, no guarantee of homogeneity is given, and even if one concedes this and accepts the validity of a molecular weight of 15000–17000 for several hemoglobins, it should always be remembered that the hematin, from all that we know of its structure, can bind several globin units . . . .
>
> On the other hand, I gladly concur in the view of Hofmeister and

many other physiologists that proteins of molecular weight 4000–5000 are not rare. If one assumes an average molecular weight of 142 for the amino acids, this would correspond to a content of 30–40 amino acids (*ibid.*, p. 40).

Three years before, Fischer reported that he and Karl Freudenberg had prepared, during the course of their studies in tannin chemistry, a hepta(tribenzoylgalloyl)*p*-iodophenyl maltosazone which, though not crystalline, was considered to be pure.

The structure follows from the synthesis, and the calculated molecular weight for $C_{220}H_{142}O_{58}N_4I_2$ is 4021. This structure greatly exceeds in molecular size all hitherto-obtained synthetic products of well-defined individuality and known structure (Fischer and Freudenberg, 1913, p. 1118).

In a lecture given in the same year, Fischer said that this value

. . . exceeds that of the largest product of polypeptide synthesis, *l*-leucyl-triglycyl-*l*-leucyl-triglycyl-*l*-leucyl-octaglycyl-glycine, by a factor of 3, and I believe that it also exceeds that of most natural proteins (Fischer, 1913, p. 3288).

It was clear, therefore, that in 1916 the verdict of the leading organic chemist working on proteins was that values above 5000 for their molecular weights were not acceptable. In the following year, however, there appeared a series of papers entitled *Studies on Proteins* by Søren P. L. Sørensen, whose work profoundly influenced the development of protein chemistry (Linderstrøm-Lang, 1939). In the last paper of the 1917 series, Sørensen reported that measurements, under carefully controlled conditions, of the osmotic pressure of solutions of highly purified crystalline egg albumin indicated a molecular weight of 34,000 for this protein; in 1925 the osmotic pressure method gave him a value of 45,000 for serum albumin. That these enormous values (by Fischer's standards) were in fact too low was demonstrated by Gilbert Adair, in whose hands this method reached new experimental and theoretical refinement. In 1925 Adair startled protein chemists by reporting that the molecular weight of oxyhemoglobin is 66,800, or about four times the minimal value calculated from its iron content. Furthermore, recalculation of Sørensen's data for egg albumin and serum albumin by Adair's method gave molecular weights of 45,000 and 74,000, respectively. In a paper published in the same year, Edwin Cohn (with Hendry and Prentiss) collected the available data on the minimal molecular weights of several proteins; among them were egg albumin, serum albumin, and hemoglobin, for which their preferred values were 33,800, 45,000, and 50,000, respectively.

Whatever doubts may have been entertained regarding the validity of Adair's values were soon stilled by the results obtained with the ultracentrifuge, described by Theodor Svedberg in 1925. In this instrument, whose design was later extensively modified and improved, protein solutions could be subjected to high centrifugal forces, and the rate of the movement of the protein outward from the axis of rotation could be measured by means of ingenious optical techniques. Not only were the rates of sedimentation in accord with the molecular weights determined by Adair, but the homogeneity of many proteins was evident from the single sharp boundary between the part of the solution containing the sedimenting protein and the solvent above it. Within about ten years a large number of purified proteins had been examined in the ultracentrifuge, and a range of molecular weights from 17,200 (for myoglobin) to 6,680,000 (for snail hemocyanin) was found. During the course of the studies on some of the hemocyanins, it was found that the species of highest apparent molecular weight was in equilibrium with smaller units which sedimented more slowly, and whose molecular weights were approximately one-eighth and one-sixteenth of the largest value. The data available to Svedberg in 1937 led him to suggest that all proteins, regardless of apparent size, are aggregates of subunits having a molecular weight of about 17,600:

> Not only the molecular weights of the haemocyanins but also the mass of most protein molecules—even those belonging to chemically different substances—show a similar relationship. This remarkable regularity points to a common plan for the building up of the protein molecules. Certain amino acids may be exchanged for others, and this may cause slight deviations from the rule of simple multiples, but on the whole only a very limited number of masses seems to be possible. Probably the protein molecule is built up by successive aggregation of definite units, but that only a few aggregates are stable. The higher the molecular weight the fewer are the possibilities of stable aggregation. The steps between the existing molecules therefore become larger and larger as the weight increases (Svedberg, 1937, p. 1061).

This attractive generalization thus assigned a value of 17,600 to a fundamental protein unit, and the finding in 1938 that, under certain conditions, oxyhemoglobin (molecular weight 68,000) could dissociate into "half-molecules" appeared to offer support to Svedberg's hypothesis. Later work was to show, however, that, although many well-defined proteins are indeed composed of subunits (*e.g.*, hemoglobin has four subunits), the idea that all subunits have the same fundamental molecular weight was not correct. As we shall see in the next section of this chapter, "the hypnotic power of numerology" (Chibnall, 1942, p. 159) was at work during the 1930s in other laboratories engaged in the study of proteins.

Whatever the actual molecular weights of individual proteins might be under various experimental conditions, Svedberg's results provided convincing evidence for the view that proteins were very large molecules ("macromolecules"), as compared with the largest synthetic polypeptide known to Fischer. Furthermore, the sedimentation behavior of many highly purified proteins indicated them to be well-defined physical entities, rather than mixtures of particles whose size could vary between wide limits. There were some knotty questions, however, arising from the designation of the proteins as macromolecules. Were the physical entities studied in the ultracentrifuge chemical molecules in the traditional sense, namely, the smallest unit held together by covalent bonds, as well as the smallest subdivision of a sample of a substance that retains all the chemical properties of that substance? For the organic chemist, the concept of the molecule, since the time of Cannizzaro and Kekulé, included not only Avogadro's principle but also a structural formula eventually based on the synthesis of the compound in question; during the 1930s, however, this criterion was not applicable to the proteins. Also, if the natural proteins were composed of fundamental units of 17,600, as suggested by Svedberg, were all the subunits of a given protein the same kind of chemical molecule? If not, comparison of the minimal molecular weight, as calculated from the analytical value for an element or an amino acid, with the value obtained in the ultracentrifuge was not justified, because the element (or amino acid) chosen might have been present in one kind of subunit and not in another. The experimental means for seeking answers to these questions were not available during the 1930s, and there was some discussion whether the term "molecular weight" should be applied to the values estimated from measurements in the ultracentrifuge or by the osmotic pressure method. Indeed, it was stated that ". . . it is certain that, in almost all these cases, what is measured is the size of a *micelle* and not that of a *molecule*" (Roche, 1935, p. 740). It was also proposed "to use a symbol for one-sixteenth of the weight of an oxygen atom and to refer to particle weights in terms of this unit" (Pirie, 1940, p. 380). It is now customary to denote the particle weights of macromolecules in terms of daltons (1 dalton = mass of one hydrogen atom = 1.67×10^{-24} gram).

The molecules studied in the ultracentrifuge were of the kind defined in the molecular-kinetic theory developed during the nineteenth century by Joule, Clausius, Maxwell, and Boltzmann. Aside from the problem of the relation of these molecules to those considered by organic chemists to be involved in chemical reactions, there was continual debate whether molecules existed at all. Around 1900 the debate became particularly animated, largely because of Wilhelm Ostwald's campaign to banish materialism from science, and to base all physical and chemical theory on his new *Weltanschauung*, energetics. The philosophical aspects of this controversy enlivened German and Austrian intellectual life at the turn of the century, and Ostwald, as an adherent of the positivist philosophy of Ernst Mach, propagated the idea that atoms and molecules were hypothetical constructs and were inappropriate for "true" explanations of natural phenomena. Instead, he urged reliance on thermodynamic theory, whose power to treat chemical problems had been notably enlarged by J. Willard Gibbs; during the twentieth century, Gibbs's concept of the chemical potential was to have lasting and far-reaching impact on the development of physical chemistry. Thanks to Ostwald's influence, his great achievements were becoming appreciated in Europe (Wheeler, 1951).

By 1910, however, the debate (at least among scientists) was over, and even Ostwald agreed that molecules existed. The events that decided the issue were the formulation, by Albert Einstein and Marian von Smoluchowski in 1905 and 1906, of equations for Brownian motion (Teske, 1969), and their experimental verification, during 1907–1909, by Svedberg and by Jean Perrin. In 1828 the botanist Robert Brown reported that he had observed, under the microscope, an agitated movement of a suspension of particles derived from pollen grains; he attributed the motion to "active molecules" (Layton, 1965). During succeeding years, the observation was confirmed repeatedly by others, and the movement was usually attributed to convection currents. Ludwig Wiener recognized in 1863, however, that the agitation was caused by characteristic motions within the fluid, and subsequent work (notably by Bodaszewski in 1881 and by Gouy in 1888) called attention to the possibility that the Brownian motion is a consequence of the kind of molecular movement postulated in the kinetic theory of heat. With the invention of the ultramicroscope in 1903, by Henry Siedentopf and Richard Zsigmondy, it became possible to determine quantitatively the fluctuation in the number of colloidal particles in a given volume; the results were those predicted on the basis of molecular-kinetic theory, and gave a value for Avogadro's number of 6×10^{23}. It was clear that colloidal suspensions obeyed the laws of dilute solutions, which linked the properties of osmotic pressure, freezing-point depression, and lowering of the vapor pressure, as all being determined only by the number of chemical entities (ions, molecules, or particles) in the solution.

Proteins as Electrolytes. In 1910 it was not yet clear, however, whether proteins should be regarded as colloidal aggregates of indefinite composition, or whether they represent well-defined molecular entities amenable to chemical study in terms of their stoichiometric interactions. Leading colloid chemists such as Wolfgang Ostwald and Herbert Freundlich advocated the former view, but the work of protein chemists during the succeeding two decades showed the latter to be correct. This transition in attitude, during the period 1910–1930, was associated with experimental and theoretical achievements of lasting importance, and marked a decline in the popularity of colloid chemistry, as compared to its position during 1890–1920.

Two properties of colloidal solutions came to the fore during the latter half of the nineteenth century: one was coagulation (in Graham's terminology, the conversion of a "sol" to a "gel"), and the other was the adsorption of other substances by colloids. Among biologists, interest in protein coagulation was a continuing one; Nägeli's hypothesis regarding the nature of protoplasm (Wilkie, 1961) assigned great importance to the idea that gels were aggregates of molecules (micelles), and a succession of workers saw in the aggregation of proteins (*e.g.,* the formation of fibrous material on shaking a solution of egg albumin) phenomena analogous to the deposition of microscopically discernible intracellular structures. During the period 1860–1880, leading physiological chemists (Kühne, Hoppe-Seyler, Halliburton) attempted to characterize individual proteins by measurement of the critical temperature for heat coagulation. From their studies it became evident that the presence of neutral salts, or the addition of acid or alkali, markedly altered this property, in analogy to the effect of such agents on inorganic colloids. With the increasing use of the salting-out method for the precipitation of proteins, much work was done to determine the efficiency of various salts; this effort culminated in the conclusion of Hofmeister in 1888 that, whatever the nature of the protein, the salts fell into the same order in the relative concentrations required to cause incipient precipitation. This "Hofmeister series" was soon extended to the effect of salts on the ability of protein gels to swell by the imbibition of water.

Near the end of the century, a common feature of the explanations offered for the coagulation or precipitation of proteins (and other sols) was the emphasis on adsorption, the name given to the binding of substances at the surface of various materials. Adsorption phenomena had been described by Tobias Lowitz during the eighteenth century; by 1812 the use of animal charcoal to decolorize sugar solutions was a well-known practical application. Experiments on the adsorption of various substances by filter paper were reported by Schönbein in 1861, but the fruitful exploitation of this method was not to come until some eighty years later. We have mentioned the use of adsorption techniques for the purification of enzymes; some colloid chemists (*e.g.*, Bayliss, 1908) considered the interaction of an enzyme with its substrate to be an adsorption phenomenon. For Wilhelm Ostwald, in 1885, phenomena of this type were consequences of mechanical affinity, rather than of chemical interaction, and his opinion was widely accepted. Thus, when William Bate Hardy examined the effect of acidity or alkalinity on the direction of movement of protein particles in an electric field, he interpreted his results from this point of view.

At the time Hardy began his work, nearly all chemists (with the notable exception of H. E. Armstrong) had accepted the theory of Svante Arrhenius, proposed in 1884, that the conductivity of electrolytes is an expression of their dissociation into ions. Also, studies on the effect of ions on the migration of colloidal particles in an electric field (electrophoresis) had been correlated with precipitation phenomena; the work of Picton and Linder on colloidal arsenic sulfide during 1892–1895 was of special importance in this regard. In 1899 Hardy subjected a suspension of heat-coagulated egg albumin to an electric field under various conditions of acidity or alkalinity:

I have shown that the heat-modified proteid is remarkable in that its direction of movement is determined by the reaction acid or alkaline, of the fluid in which it is suspended. An immeasurably minute amount of free alkali causes the proteid particles to move against the stream while in the presence of an equally minute amount of free acid the particles move with the stream. In the one case therefore the particles are electro-negative, in the other they are electro-positive.

Since one can take a hydrosol in which the particles are electro-negative and, by the addition of free acid, decrease their negativity, and ultimately make them electro-positive it is clear that there exists some point at which the particles and the fluid in which they are immersed are iso-electric.

The iso-electric point is found to be one of great importance. As it is neared, the stability of the hydrosol diminishes until, at the iso-electric point, it vanishes, and coagulation or precipitation occurs, the one or the other according to whether the concentration of the proteid is high or low, and whether the iso-electric point is reached slowly or quickly, and without or with mechanical agitation (Hardy, 1900, p. 110).

A few years later, Hardy wrote of the properties of globulins in solution:

> They are not embraced by the theorem of definite and multiple proportions. Therefore they are conditioned by purely chemical forces only in a subsidiary way. A precipitate of globulin is to be conceived not as composed of molecular aggregates but of particles of gel. I have shown elsewhere that gelation and precipitation of colloidal solutions are continuous processes. These particles of gel when suspended in a fluid containing ions are penetrated by those ions. Let the fundamental assumption be that the higher the specific velocity of an ion the more readily it will become entangled within the colloidal particle. Then as H and OH ions have by far the highest specific velocity the colloidal particle will entangle an excess of H ions in acid and thereby acquire a + charge and of OH ions in alkali and thereby acquire a − charge. These charges will decrease the surface energy of the particle and thereby lead to changes in their average size (Hardy, 1903, p. xxix).

In a popular article, Hardy stated:

> Proteids unquestionably are the material basis of life, but when isolated after the death of the cell they are not living. They are chemically stable bodies. They show no signs of the characteristic chemical flux. It is therefore conjectured on experimental grounds that the living molecule is built up of proteid molecules, that it is so complex, so huge, as to include as units of its structure even such large molecules as these. But when such very large molecules enter into chemical combination with one another, whether by reason of the great magnitude of the masses of matter in each in relation to the magnitude of the directive forces, or because the molecules themselves, owing to their great size, to a certain extent cease to be molecules at all in the physical sense, and possess the properties of matter in mass, it is any rate certain that in their chemical combinations they cease to follow the law of definite combining weights which is the basis of chemistry (Hardy, 1906, pp. 197–198).

This denial of stoichiometric chemical interaction was not accepted by other investigators. The ability of amino acids and of proteins such as the albumins and globulins to act both as acids and as bases had been clearly established by 1900, and Georg Bredig had applied the term "amphoteric electrolyte" to such substances; later, T. Brailsford Robertson called them "ampholytes." During the period 1902–1904 there appeared several articles, including one by Jacques Loeb, in which the behavior of proteins in an electric field was explained in terms of their ability to act as amphoteric electrolytes; shortly afterward, Hardy also embraced this viewpoint.

In 1904 Hans Friedenthal recommended that the acidity or alkalinity of an aqueous solution be denoted by the numerical value of the hydrogen-ion concentration (Szabadvary, 1964), and five years later Sørensen proposed that, for convenience, the negative logarithm of this value (pH) be used; according to this convention, a solution whose hydrogen ion concentration is 2×10^{-6} normal ($10^{-5.7} N$) has a pH of 5.7. Sørensen (1909) introduced the pH concept in connection with experiments on enzymes; he showed how important it was to control the pH of enzymic reactions, and described the use of buffer solutions for this purpose. The English word buffer and the German *Puffer* are translations of the French *tampon*, introduced by Fernbach and Hubert in 1900. The importance of the pH concept was quickly recognized by Leonor Michaelis, who by 1910 had determined the pH values for the isoelectric points of several proteins. After World War I, the invaluable book of William Mansfield Clark, entitled *The Determination of Hydrogen Ions* (first edition, 1920; third edition, 1928), exerted a lasting influence on the subsequent development of biochemistry.

Among the many consequences of the recognition of the importance of controlling the pH of protein solutions was the convincing evidence presented by Jacques Loeb against the view, still held by some colloid chemists during the 1920s, that proteins did not conform to the laws of classical chemistry. His book *Proteins and the Theory of Colloidal Behavior* (first edition, 1922; second edition, 1924), in which he summarized his results, had a considerable effect on a generation of young men who identified themselves as general physiologists. It should be added that his work on proteins came at the close of a remarkable scientific career (Osterhout, 1930) in which such diverse problems as animal tropisms, artificial parthenogenesis, and the diffusion of ions through biological membranes were intensively studied in relation to an underlying philosophical outlook expounded in his *The Mechanistic Conception of Life* (published in 1912).

In their impact on the further development of the physical chemistry of proteins two advances, both made in 1923, were of greater importance in the rejection of the colloidal approach. One of these was the theory of Peter Debye and Erich Hückel, which described the behavior of an ion in relation to its charge and radius, to the other ions in the solution, and the dielectric properties of the medium. The other was the demonstration by Niels Bjerrum that amino acids in their isoelectric state are not uncharged molecules (*e.g.*, NH_2CH_2COOH) but are *Zwitterionen* (dipolar ions) of the structure $^+NH_3CH_2COO^-$. By the same token, the isoelectric forms of proteins were not, as had been thought by Loeb and his predecessors, uncharged molecules; they were instead highly charged, with a net charge of zero. These two important insights laid the ground-work for the systematic study of proteins as electrolytes, a field in which Edwin Cohn (1925) and his associates played a leading role during the succeeding three decades (Edsall, 1955). In particular, it became clear that, if the Fischer-Hofmeister theory of protein structure is correct, and the individual amino acid units are joined in a chain-like array by peptide bonds linking the α-amino group of one amino acid to the α-carboxyl group of another, the behavior of proteins as electrolytes is a consequence of the presence of ionized groups such as those derived from the unsub-stituted carboxyl group of aspartic acid or glutamic acid, and the basic groups of arginine, lysine, and histidine. Extensive studies were conducted to determine the ionization constants of such groups, and correlations were sought with the quantitative data on the acid-base behavior of pro-teins. Reliable data on the complete amino acid composition of proteins were not available before 1940, but there could be no doubt that proteins were macromolecular polyelectrolytes. Furthermore, the development by Arne Tiselius (in 1937) of moving-boundary electrophoresis, and the subsequent improvement of this technique by Lewis Longsworth, made possible more precise measurements. One of the first important successes of the electrophoretic method was Tiselius's separation of the proteins of blood plasma, and his demonstration that the globulin fraction contains multiple components, some of which (the so-called gamma-globulins) represent the antibodies.

Protein Denaturation. Although many proteins had indeed been shown to be well-defined chemical entities whose behavior as acids and bases obeyed the laws of stoichiometry, the property of coagulation, long con-sidered to be their chief characteristic, remained obscure. The biological importance of this property was repeatedly emphasized, not only by biologists, but also by chemists. Thus, Samuel Schryver in discussing the proteins (he terms them albumens) stated that

> . . . *the molecule of the albumen, which performs the chemical func-tions necessary for the maintenance of life is, compared with ordinary molecules with which we are in the habit of dealing in the chemical laboratory, of enormous size and very unstable;* it loses many of its properties on the application of a very moderate amount of heat (Schryver, 1906, p. 168).

Others were more explicit in writing about the nature of the transformation of water-soluble proteins to insoluble materials by the application of heat:

> We have in chemistry a whole series of analogous phenomena with substances of known molecular structure and this process, as with the aldehydes and cyano compounds, can only be conceived as a consequence of the rearrangement of atoms in the molecule . . . . If we wish to approach the phenomena associated with the word "life," research on the chemistry of the albuminoid bodies must take a new direction. As Pflüger rightly stated over 10 years ago, the protein of the living cell must have an entirely different molecular structure from that of dead tissues, and the evidence that the protein of living cells has a labile aldehyde structure increases daily (Nencki, 1885, p. 343).

The idea that the heat coagulation of proteins was associated with the presence of a "labile aldehyde structure" was later propagated by Loew (1896), who suggested the presence, in "living albumen," of amino-aldehyde structures, and offered in evidence the fact that substances (*e.g.*, cyanide or hydroxylamine) known to react with aldehydes are toxic to living organisms.

Similar chemical speculation abounded at the turn of the century; by 1910, however, the coagulation process began to be studied carefully with purified proteins. Of special importance was the work of Harriette Chick and Charles Martin who showed that the heat coagulation of "native" egg albumin involves an initial process, termed "denaturation," followed by the precipitation of the coagulated protein. Their work, together with that of Michaelis and of Sørensen at about the same time, demonstrated that

> Under otherwise equal conditions, the rate of denaturation will be higher, the higher the concentration of hydrogen ions in the solution, but the flocculation of the denatured protein only takes place at iso-electric reaction, or near it. The addition of salt extends the limits for complete flocculation . . . . (Sørensen and Sørensen, 1925, p. 26).

With the recognition that the precipitation of the denatured protein does not require heat, and depends only on the pH and salt content of the solution, the problem of the cause of coagulation became the problem of protein denaturation. During the period 1910–1935 the effect of many chemical agents (acids, alkalis, alcohol, etc.) or physical treatments (heat, shaking, ultraviolet irradiation, etc.) were studied; special mention must be made of the finding that, at high concentrations, urea is an extremely effective denaturing agent (Hopkins, 1930a). It was also noted that, upon denaturation, some proteins (*e.g.*, egg albumin) acquire the ability to react with nitroprusside, a reagent for sulfhydryl (SH) groups; since such groups were associated with the amino acid cysteine, it was evident that the process of denaturation had altered the protein in a manner that made the SH group of the cysteine reactive. Moreover, the important studies of Mortimer Anson and Alfred Mirsky during the 1920s on the denaturation of hemoglobin demonstrated that denatured globin reacts with the iron-porphyrin to produce a substance that is unable to bind oxygen, differs from native hemoglobin spectroscopically, and cannot be crystallized. Later (in 1931) they showed that the denaturation of hemoglobin was, under suitable circumstances, a reversible process, and drew the important conclusion that the native and denatured forms were in equilibrium with each other. With the isolation of enzymes in crystalline form, it was possible to offer as evidence for their protein nature the fact that the denaturation (as measured by solubility) closely paralleled the loss of catalytic activity, and that under conditions of reversible denaturation both properties were restored together.

A remarkable feature of protein denaturation was found to be the temperature coefficient of the process; whereas an increase of 10° in temperature usually causes a 2–3 fold increase in the rates of chemical reactions, the increase in the first-order rate of protein denaturation was as high as 600 fold. The nature of the chemical reaction (or reactions) characterized by this high temperature coefficient was not clear in 1935, since it had been established that the apparent molecular weight of denatured egg albumin was the same as that of the native protein and no covalent bonds appeared to have been cleaved. Chick and Martin had noted:

> What precisely happens when denaturation of albumen in solution occurs, we are not yet able to say. In cases where agglutination is absent and the solution may remain completely limpid, its viscosity will be found to have considerably increased (Chick and Martin, 1912, p. 261).

Later work showed that this increase in viscosity was consistent with the transformation of a rather rounded particle into an elongated structure. These various characteristics of the process of protein denaturation led Hsien Wu to propose, in 1931, that the essential feature of the process is the unfolding of tightly coiled peptide chains of the native protein, leading to the disorganization of its internal structure. Although this hypothesis was widely adopted, and soon received additional experimental support, the question of the chemical forces that held the native protein in its native conformation (the word "configuration" was used at the time) was unclear. An important contribution to the discussion of this problem was the proposal, made by Alfred Mirsky and Linus Pauling and by Maurice Huggins, that "hydrogen bonds" (formed by the tendency of a hydrogen atom bound to one atom to share the electrons of another atom, as in —N—H . . . O=C—) were responsible for the unique structure of native proteins:

> Our conception of a native protein molecule (showing specific properties) is the following: The molecule consists of one polypeptide chain which continues without interruption throughout the molecule (or in certain cases, of two or more such chains); this chain is folded into a uniquely defined configuration, in which it is held by hydrogen bonds between the peptide nitrogen and oxygen atoms and also between the free amino and carboxyl groups of the diamino and dicarboxyl amino acid residues . . . . The characteristic specific properties of native proteins we attribute to their uniquely defined configuration.
>
> *The denatured protein molecule we consider to be characterized by the absence of a uniquely defined configuration* (Mirsky and Pauling, 1936, p. 442–443).

In this closely reasoned and persuasive paper, Mirsky and Pauling attributed the denaturing action of urea or alcohol to their ability to form hydrogen bonds, and thus to disrupt the internal hydrogen bonds of the proteins. They also suggested that the association of protein subunits (as in hemocyanin) may involve hydrogen bond formation. Another hypothesis, offered by Henry Eyring and Allen Stearn (in 1939), suggested that denaturation involves the breaking of "salt linkages" (formed by the mutual attraction of oppositely charged ions) and of covalent bonds (possibly the disulfide bonds of cystine units); this proposal was not received with the same approval accorded the Mirsky-Pauling theory. Subsequent work, especially on the mode of action of urea, forced a revision of the generalization that hydrogen bonding is the principal means whereby the internal structure of a globular protein is established. In particular, it became clear after 1950 that the kind of interaction dear to the colloid chemists—the affinity of hydrophobic groups for each other when they are present in an aqueous environment—played a major role in determining the structure and the interactions of protein molecules; such hydrophobic interactions were extensively studied around 1920 by William Harkins, in his measurements of the surface tension of monomolecular

films of organic substances. In large part, this revision came from the
knowledge gained after World War II about the organic chemistry of
proteins, and the use of this knowledge in the study of protein structure
by means of X-ray crystallography. Before we turn to the background of
these advances, during the period 1920–1950, it is necessary to reempha-
size the importance of the Mirsky-Pauling paper in its influence on further
work on the problem of the nature of proteins.

The Arrangement of Amino Acids in Proteins

Near the close of the period we shall now consider, Frederick Sanger
wrote in a review article with the title above:

> As an initial working hypothesis it will be assumed that the peptide
> theory is valid, in other words, that a protein molecule is built up only
> of chains of α-amino (and α-imino) acids bound together by peptide
> bonds between their α-amino and α-carboxyl groups. While this peptide
> theory is almost certainly valid . . . it should be remembered that it
> is still a hypothesis and has not been definitely proved. Probably the
> best evidence in support of it is that since its enunciation in 1902 no
> facts have been found to contradict it (Sanger, 1952, p. 3).

It was indeed Sanger's work on the structure of the protein hormone
insulin that contributed decisively to the proof of the peptide theory,
whose vicissitudes during 1920–1940 deserve to be recalled. In retracing
the stumbling steps of protein chemists during that period, we shall see
that the confusion was compounded of a variety of factors; foremost
among them was the fact that the available experimental methods were
not adequate to resolve doubts or shake adherence to inherited belief.

By 1920 it was agreed that crystalline proteins such as egg albumin,
oxyhemoglobin, or the plant globulins yielded, on hydrolysis with acids
or alkalis, a variety of amino acids. They included glycine, alanine, valine,
leucine, isoleucine, serine, cystine, proline, hydroxyproline (then termed
oxyproline), phenylalanine, tyrosine, tryptophane (the final *e* was later
dropped), aspartic acid, glutamic acid (then termed glutaminic acid),
oxyglutaminic acid (later withdrawn as a protein constituent), arginine,
lysine, and histidine. To this list of 18 protein amino acids, to which the
German workers applied the word *Bausteine* (whence the journalistic
term "building blocks"), was added methionine in 1922. The studies of
J. Howard Mueller on the nutritional requirements of some bacteria
(*e.g.*, Streptococci) that require protein hydrolysates for growth led him

to discover, in casein, a new sulfur-containing amino acid. In 1923 he described the properties of a pure sample, and by 1928 the structure of methionine was established by George Barger and Frederick Coyne. Then in 1935 William C. Rose demonstrated that hydrolysates of fibrin contain a hitherto-unknown amino acid essential for the growth of the immature rat; it proved to be a β-hydroxyamino acid related to serine, and to the four-carbon sugar threose (whence its name threonine). The last amino acid to be added to the list of protein amino acids was hydroxylysine, discovered in 1938 by Donald D. Van Slyke, who found it as a minor component in hydrolysates of gelatin. Hydroxylysine and hydroxyproline were recognized to be limited in their occurrence to proteins such as the collagen of connective tissues (or the gelatin derived from it); another amino acid already known to be associated with a unique protein was the iodine-containing hormone thyroxine. The studies of Eugen Baumann, who made important contributions to several aspects of physiological chemistry, showed in 1895 that an iodine-containing organic compound in the thyroid gland is the active agent in the treatment of goiter. In 1900 Adolf Oswald prepared, from the thyroid, an active globulin (thyroglobulin), and in 1915 Edward C. Kendall isolated thyroxine, whose structure was definitely established by Charles Harington and George Barger in 1927.

Even with the omission of the amino acids obtained only from unique proteins (hydroxyproline, hydroxylysine, thyroxine), the list of 18 widely distributed amino acids was a formidable one to an organic chemist interested in protein structure. By 1940 this list had to be expanded to 20, since it had been established that some of the aspartic acid and glutamic acid residues of proteins are present in the form of their β-amides (asparagine and glutamine, respectively). As we noted before, asparagine had been obtained in the free state as early as 1806, and free glutamine had been isolated from plants in 1877 by Ernst Schulze, who clearly established their importance in plant metabolism (Chibnall, 1939). There had long been considerable evidence for the view that the ammonia formed upon acid hydrolysis of proteins was largely derived from the cleavage of the amide group of bound asparagine and glutamine, but the definitive proof of their membership in the list of protein amino acids only came around 1930, with their isolation from enzymic digests; under these conditions, the β- (or γ-) amide group remains intact. We can only note here in passing that the extensive work, during 1870–1920, on the source of ammonia found on the hydrolysis of proteins by acids and alkalis was frequently indecisive, because of the partial (or total) decomposition of certain amino acids (*e.g.*, serine, arginine) during hydrolysis.

In the accompanying table, the chemical structures of the 20 widely distributed protein amino acids are given. It should be added that, by 1940, it was recognized that some proteins may contain cysteine units (formed by conversion of the disulfide group of cystine to two sulfhydryl groups); the cysteine-cystine pair may be considered together. Cysteine had also been found to be a constituent of a naturally occurring tripeptide, glutathione (of which more will be said in a later chapter).

Except in glycine, the carbon atom linking the α-amino and α-carboxyl groups of the protein amino acids is an asymmetric carbon atom in the sense of van't Hoff and LeBel. We have mentioned that Fischer had studied the optically active forms of the protein amino acids he knew about, and subsequent work led to the recognition by 1940 that all 20 are characterized by the same type of arrangement of the four groups about the α-carbon, *i.e.*, they all have the same configuration (now usually designated with the prefix L-). During the 1930s the isomeric (D) amino acids were often termed the "unnatural" antipodes, but this practice disappeared after 1945, as many naturally occurring polypeptides (*e.g.*, antibiotics produced by molds) were found to contain D-amino acid units. In 1939 there was a brief flurry of excitement when Fritz Kögl reported that the proteins of tumor tissues contain D-glutamic acid units, but the careful studies of Charles Chibnall and others showed this claim to be invalid.

For the protein chemist of the 1920s, therefore, it was already clear that he was dealing with a type of structure constituted of at least 16 different kinds of units, and it was also evident that, if progress was to be made in the elucidation of the structure, it was necessary to devise analytical methods that would permit him to determine quantitatively the amount of each amino acid produced upon hydrolysis of a highly purified protein preparation. Conceptually, this procedure is exactly equivalent to the quantitative analysis of a simpler organic compound in order to establish an empirical formula, as developed by Berzelius and his contemporaries; the only difference is that the units of structure are not the elements C, H, N, O, S, etc., but the amino acids. In 1900, Albrecht Kossel and Friedrich Kutscher had emphasized the need for the quantitative accounting of a protein in terms of the amino acids produced on hydrolysis, and efforts in this direction were made during 1910–1920 by estimation of the distribution of nitrogen in the form of amino-N, amide-N, and basic amino acid-N. This approach, initiated by Walther Hausmann in 1900, was developed by Thomas B. Osborne and Donald D. Van Slyke. Although valuable general analytical techniques became available for the determination of amino-nitrogen (the Sørensen "formol" titration method, the Van Slyke gasometric and manometric

NH$_2$CH$_2$COOH

Glycine (Gly)

Three–letter abbreviations denote amino acid
residues (−NHCH(R)CO−), not free amino acids

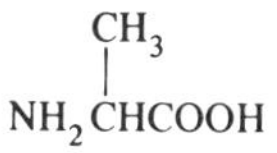

CH$_3$
|
NH$_2$CHCOOH

Alanine (Ala)

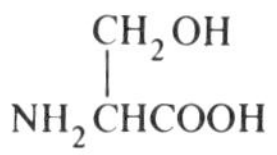

CH$_2$OH
|
NH$_2$CHCOOH

Serine (Ser)

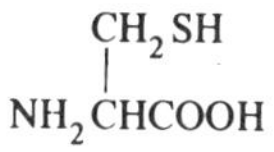

CH$_2$SH
|
NH$_2$CHCOOH

Cysteine (Cys)

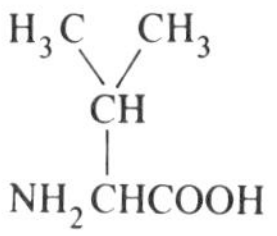

H$_3$C   CH$_3$
\ /
CH
|
NH$_2$CHCOOH

Valine (Val)

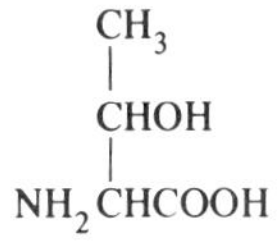

CH$_3$
|
CHOH
|
NH$_2$CHCOOH

Threonine (Thr)

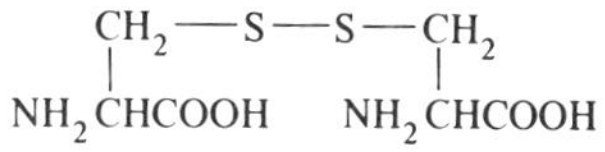

CH$_2$—S—S—CH$_2$
| |
NH$_2$CHCOOH   NH$_2$CHCOOH

Cystine (Cys   Cys)

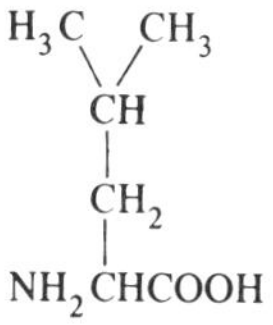

H$_3$C   CH$_3$
\ /
CH
|
CH$_2$
|
NH$_2$CHCOOH

Leucine (Leu)

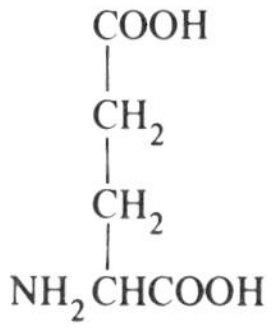

COOH
|
CH$_2$
|
CH$_2$
|
NH$_2$CHCOOH

Glutamic acid (Glu)

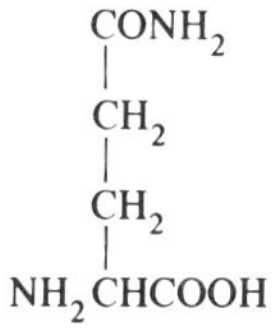

CONH$_2$
|
CH$_2$
|
CH$_2$
|
NH$_2$CHCOOH

Glutamine (Gln)

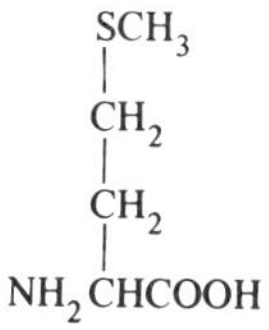

SCH$_3$
|
CH$_2$
|
CH$_2$
|
NH$_2$CHCOOH

Methionine (Met)

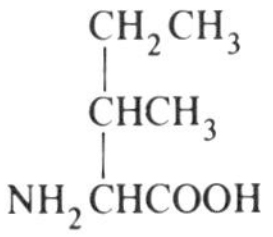

CH$_2$CH$_3$
|
CHCH$_3$
|
NH$_2$CHCOOH

Isoleucine (Ile)

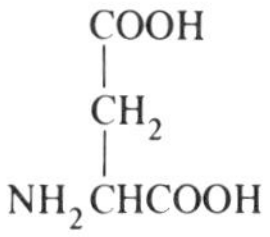

COOH
|
CH$_2$
|
NH$_2$CHCOOH

Aspartic acid (Asp)

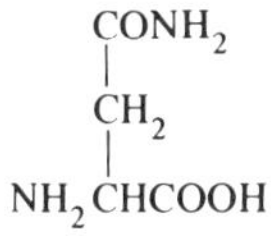

CONH$_2$
|
CH$_2$
|
NH$_2$CHCOOH

Asparagine (Asn)

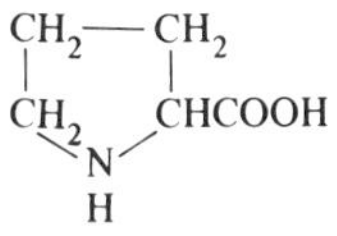

CH$_2$—CH$_2$
| |
CH$_2$ CHCOOH
      \ N /
         H

Proline (Pro)

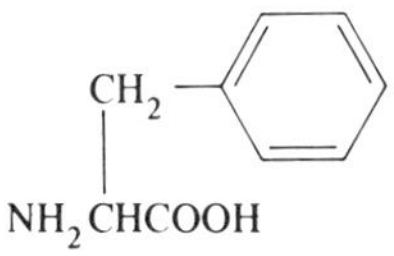

CH$_2$—⟨phenyl ring⟩
|
NH$_2$CHCOOH

Phenylalanine (Phe)

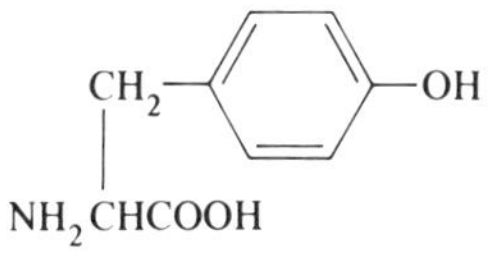

CH$_2$—⟨phenyl ring⟩—OH
|
NH$_2$CHCOOH

Tyrosine (Tyr)

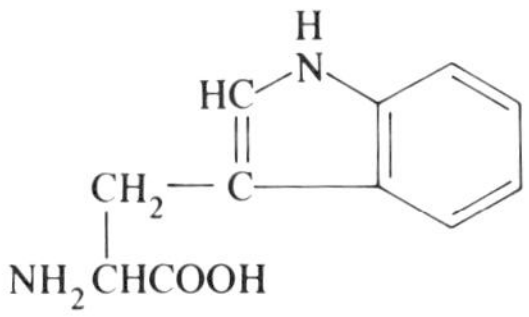

CH$_2$—⟨indole ring⟩
|
NH$_2$CHCOOH

Tryptophan (Trp)

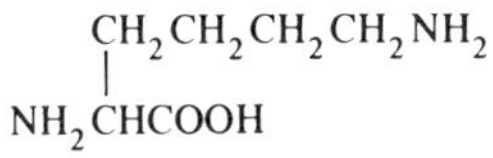

CH$_2$CH$_2$CH$_2$CH$_2$NH$_2$
|
NH$_2$CHCOOH

Lysine (Lys)

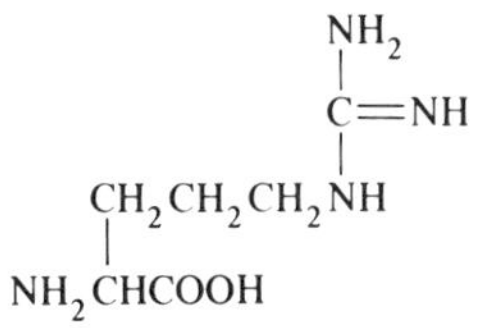

NH$_2$
|
C=NH
|
CH$_2$CH$_2$CH$_2$NH
|
NH$_2$CHCOOH

Arginine (Arg)

CH$_2$—⟨imidazole ring⟩
|
NH$_2$CHCOOH

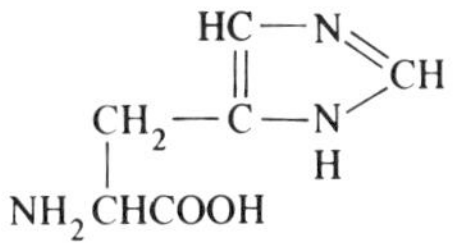

Histidine (His)

methods), the problem of the individual amino acids remained intractable. Hubert B. Vickery and Charles S. Leavenworth, in 1928, markedly improved the procedures for the determination of arginine, histidine, and lysine which, because of their basic character, could be selectively precipitated; they also called attention to the use of organic sulfonic acids as precipitants for histidine and arginine. During the 1930s Max Bergmann, in association with William H. Stein and Stanford Moore, attempted to extend the use of such sulfonic acids for the determination of other amino acids, but it was soon evident that the limitations of this procedure were considerable. Despite many isolated contributions, during the period 1920–1940, to the techniques of amino acid analysis by selective precipitation, the situation was not a heartening one. A few amino acids could be isolated or estimated fairly quantitatively, approximate estimates could be made of the others, and statements could be made that certain proteins were relatively rich in arginine (*e.g.,* the protamines) or in glycine and alanine (*e.g.,* silk fibroin), but no single method, such as Henry Dakin's modification of Fischer's ester method, or combination of methods, had solved the problem. It is necessary to emphasize this point, as the difficulties encountered in taking the first necessary step in the study of proteins as organic substances created a situation in which many fanciful speculations could not be subjected to critical experimental test. As we shall see later, the introduction of partition chromatography by Archer Martin and Richard Synge (in 1941) resulted, within ten years, in the solution of the problem of the quantitative amino acid analysis of protein hydrolysates, by Moore and Stein, with dramatic consequences for the study of protein structure.

The Peptide Theory. During 1920–1940 doubt about the validity of the peptide theory was usually based on the reiteration of the idea that "living" proteins represent labile structures which, upon denaturation, are converted into polypeptide-like structures. For example, Emil Abderhalden, for many years associated with Fischer in the latter's work on proteins, wrote in 1924:

> Every idea regarding the specific structural relations in proteins, and especially of those that play a significant role in cellular life, must take into consideration their entire structure, their ready transformation from the native to the denatured state, *i.e.,* their greater or lesser lability. It is certain that protoplasmic proteins have properties of which we at present ignorant. We study proteins almost invariably in a more or less altered state. On the one hand, we have proteins which in a certain sense are denatured in the organism, and that undergo

further changes outside of the organism. Thus, for example, we see the threads of the silkworm, the web of the spider . . . solidify. A labile form has been transformed into a very stable structure. On the other hand, cellular functions are only consistent with the delicately adjusted, undoubtedly more reactive structure of the proteins. When death ensues, the cell proteins coagulate and profoundly alter their state (Abderhalden and Komm, 1924, pp. 190–191).

We noted that in 1906 Fischer had considered it to be "quite probable" that proteins contain diketopiperazine rings (formed by the cyclization of a dipeptide), and in 1924 Abderhalden advanced this idea to the status of a theory of protein structure. Diketopiperazines had been isolated repeatedly from protein hydrolysates since 1849, when Friedrich Bopp found "leucinimide" (the cyclization product of leucyl-leucine), and Abderhalden proposed that a protein molecule consists of a number of diketopiperazine-containing complexes that are held together by the kind of partial valency (*Nebenvalenz*) postulated by Johannes Thiele in 1899 for the interaction of unsaturated organic compounds. A variety of chemical evidence was offered, but to Vickery and Osborne (1928, p. 426) it did "not seem that the enormous labors of Abderhalden and his associates have really furthered his fundamental view of the structure of the protein molecule." On the other hand, a leading colloid chemist expressed the belief ". . . that the evidence which has been presented by Abderhalden for the presence of diketopiperazine rings in proteins amounts almost to proof" (Gortner, 1929, p. 344). In 1924–1925, Max Bergmann proposed a theory that also assumed the association of diketopiperazine-like structures, with the difference that his presumed units were unsaturated diketopiperazine derivatives of the sort he had prepared from dipeptides containing serine or cysteine. Furthermore, Bergmann's model compounds were well-characterized substances and, although his chemical work on their reactions did not greatly advance the understanding of protein structure, it represented a valuable contribution to peptide chemistry. The attraction of the idea that diketopiperazines formed structural units of proteins also drew Paul Karrer during 1923–1925.

The reasons for the upsurge in popularity of the diketopiperazine structure of proteins are not difficult to surmise. There was in the first place the inherited tradition of Fischer's views on their probable occurrence in proteins, and his disbelief in molecular weights above 5000. A second, and more immediate, reason was the report by Rudolf Brill (in 1923) that silk fibroin gave an X-ray diffraction pattern that was consistent with either a long polypeptide chain consisting of glycine and alanine units, or the association of glycyl-alanine diketopiperazine units. The X-ray examination of natural fibers, such as cellulose or silk fibroin, was begun by Nishikawa and Ono in 1913, but was not systematically pursued until Reginald Herzog's work immediately after World War I. This work showed that natural fibers, though seemingly noncrystalline, are ordered structures, with one- or two-dimensional repeating packings. Among Herzog's collaborators was Michael Polanyi, who examined the X-ray diffraction pattern of cellulose in 1921. He later recalled that

> . . . I evaluated the elementary cell of cellulose and drew the conclu-sion that the structure of cellulose was either a straight giant molecule composed of a single file of linked hexoses, or else an aggregate of hexobiose-anhydrids; both structures were compatible with the sym-metry and size of the elementary cell—but unfortunately I lacked the chemical sense for eliminating the second alternative (Ewald, 1962, p. 631).

Independent work by Olenus Lee Sponsler led him to favor the first alter-native, which had indeed been suggested in 1913 by Richard Willstätter and Laszlo Zechmeister. Preference was given to the second alternative, however, because Bergmann, Kurt Hess, and Karrer had reported molec-ular-weight data that indicated cellulose and starch to be aggregates of small molecules, and it was not until the subsequent work (in 1928–1929) by Kurt Meyer and Hermann Mark, together with the organic chemical studies of Haworth, that the long-chain character of the poly-saccharides was established. Thus, in the beginnings of X-ray studies on natural products, there was a necessary interdependence between the knowledge provided by the techniques of organic chemistry and the interpretation of the diffraction patterns in terms of structure. During 1920–1923 most of the leading organic chemists were not yet ready to accept the evidence presented by Hermann Staudinger for the high-molecular character of linear polymers (he introduced the word "macro-molecule" in 1922), although such products had been prepared repeatedly since Agostinho Lourenço had condensed succinic acid and ethylene glycol in 1863. The turning point in attitude appears to have come about 1925, when Staudinger presented impressive evidence for the macro-molecular nature of rubber; his results disproved the idea that rubber was an aggregate of relatively small isoprene derivatives held together by secondary valences, and that linear polymeric structures of molecular weights up to 70,000 were possible (Staudinger, 1961). Also, Wallace Carothers (1931) demonstrated clearly the long-chain nature of condensation polymers (*e.g.*, polyamides of the kind described in the 1880s by Grimaux). The further development of the field of polymer chemistry (Flory, 1953), especially in the relatively recent discovery of stereo-specific polymerization (Natta, 1965), is so large a subject, and so important in its impact on technology, that no brief summary is possible. Apart from the many practical consequences of his work, Staudinger's emphasis on the continuum embracing both large and small molecules had considerable influence on chemical thought. He wrote, during the course of a discussion with Meyer (1928), that the latter

> . . . speaks of micelles and micellar forces, and seeks thereby to establish an inappropriate connection with the ordinary concepts of colloid chemistry; he thus destroys, however, the connection with the simple organic compounds. I speak of large molecules and of intermolecular forces and thereby assume that the macromolecular substances have the same structural principle as those of low molecular weight (Staudinger, 1929, pp. 72–73).

The rise of polymer chemistry during the late 1920s thus marked a change in the attitude of organic chemists toward very long chains in which the structural units are joined to each other by normal valence bonds. Protein chemists were influenced by this change, and after 1930 there was little effort to propagate the ideas advanced by Abderhalden, Bergmann, and Karrer during the previous decade.

There were some difficulties, however. To quote from the review by Vickery and Osborne,

> Taken together this body of evidence is as nearly conclusive proof that the peptide bond occurs to a very considerable extent as could be wished. There are, however, certain facts which remain unexplained upon the view that the protein is essentially a single large polypeptide.
> In the first place the nature of the bonds that are attacked by pepsin is unknown . . . . [P]epsin has never been found to have an effect upon any synthetic model substance whether it contained a peptide bond or not (Vickery and Osborne, 1928, p. 417).

Similarly, Ross A. Gortner noted that:

> Fischer concluded that the chains of the polypeptides were not long enough for pepsin to act upon them, but it seems more probable that

pepsin attacks some linkage other than the linkage in the peptide group (Gortner, 1929, p. 319).

These views were expressed before the crystallization of pepsin by John H. Northrop, and the developments in the methods of the synthesis of polypeptides initiated by Max Bergmann and Leonidas Zervas (1932).

The crystallization of pepsin represented a turning point for both enzyme chemistry and protein chemistry. We noted before that the problem of the protein nature of enzymes had been debated extensively and inconclusively during the nineteenth century. The purification of enzymes was plagued by many difficulties; for example, in 1901 Nencki had purified pepsin, and believed it to be a labile and complex material containing nucleoprotein, lecithin, and iron. He reflected the attitude of many of his contemporaries in his discussion of this result:

> The necessity for the presence of certain apparently foreign substances so that enzymes can exert their specific action brings us to the analogy with living organisms. In every living cell there are present, besides the albuminoid bodies, the fats, lecithins, and other organic and inorganic substances essential for life. With the extremely complex protein molecules that we find in cells, it is difficult to distinguish what is only mechanical admixture from what is a real constituent of the molecule (Nencki, 1904, Vol. II, pp. 724–725).

With the rise of colloid chemistry after 1900, and increased emphasis on adsorption phenomena, the idea that an enzyme is a catalyst of low molecular weight which is readily adsorbed and stabilized by a colloidal carrier, such as a protein, became even more attractive than before. Thus, according to Bayliss:

> The chemical nature of enzymes is probably very various. There is direct evidence that some are not proteins, and it is doubtful whether any are. Some appear to be complex systems of colloids with inorganic components, or other simple compounds (Bayliss, 1924, p. 331).

Upon the entry of the renowned Richard Willstätter into the enzyme field, after World War I, a persuasive voice was raised in favor of this view. With a group of outstanding associates, notably Richard Kuhn, Willstätter proceeded to set new standards for the assay of the catalytic activity of enzymes, and to develop valuable new adsorption techniques (*e.g.*, the use of aluminum hydroxide gels) for their purification. The issue of the protein nature of enzymes came to a focus about 1925 over the saccharase (invertase) preparations obtained by Hans von Euler and by Willstätter. In 1927 Willstätter stated:

A further subject of controversy was the protein nature of saccharase and other enzymes. E. Fischer, in his Faraday Lecture (1907), not without a certain plausibility concluded from the observations available at that time that the ferments are produced from proteins and that many of them possess protein-like character. But saccharase even in our earliest investigations seemed to have been obtained free from protein as well as from carbohydrates. Euler, however, on the basis of his analyses, adhered to the assumption that the enzyme is a protein. The enzyme-protein was, according to Euler and Josephson, supposed to possess a specific structure, characterized by a remarkably high proportion of tryptophane . . . . When now this paper of Euler's appeared, I sought out a preparation of my very first research on saccharase, which had been looked upon as protein free. It was of course far from being of so high a saccharase-value as the later preparations. But in fact the old preparations did not give the tryptophane reaction at all. On the other hand the much purer preparations . . . really contained tryptophane, and even more than the preparation of Euler . . . But in a few single cases we even succeeded, without loss of activity, in quantitatively removing the tryptophane from preparations of a very high degree of purity. Although we have indeed no systematic method for freeing the enzyme from this protein derivative, in my opinion it is sufficient to have freed a few times an enzyme preparation or a hormone without changing its activity, from a substance accompanying it in order to prove the superfluousness of that accompanying substance.

The protein therefore is no part of the enzyme . . . . An enzyme seems to consist of a specific active group and a colloidal carrier. With the latter, other substances of high molecular weight are linked in a varying manner (Willstätter, 1927, pp. 48–52).

These statements were made during a lecture at Cornell University where James Sumner had

. . . recently reported that he had obtained urease in the form of pure crystals which he identifies as those of a globulin. It would perhaps be premature to judge whether the globulin crystals actually are the pure enzyme or whether they only contain the latter in an adsorbed state (*ibid.*, p. 53).

In Sumner's own report, he wrote:

> After work . . . that extends over a period of a little less than 9 years, I discovered on the 29th of April a means of obtaining from the jack bean a new protein which crystallizes beautifully and whose solutions possess to an extraordinary degree the ability to decompose urea into ammonium carbonate (Sumner, 1926, p. 435).

Several years later, a disciple of Willstätter interpreted work on catalase and peroxidase (both iron-porphyrin enzymes, which we shall mention again later) to indicate that, as in hemoglobin, "their active groups are non-protein in character" (Waldschmidt-Leitz, 1933, p. 189) and regarded the preparations of Sumner and Northrop

> . . . as adsorption compounds of the true enzymatic component plus crystalline protein to which they have special affinity. The finding of crystalline protein-enzyme compounds may lead to the concept that enzymes are merely proteins, and thus cause investigators to discard enzyme specificity which can only be explained by the existence of highly specialized active groups (*ibid.*, p. 189).

Sumner replied that he thought

> . . . there is little danger of this. The enzyme as I consider it, is in some cases a simple protein, in others a conjugated protein where the properties are to be ascribed to the molecule as a whole. But whether the specific active groups are in the protein part or in the side chain, the enzyme is a protein, as I demonstrated in 1926 (Sumner, 1933, p. 335).

This ambiguity about the location of the "active groups" of the enzymes isolated as conjugated proteins was to continue for over a decade, until it was realized that in all cases the protein has structural units responsible for catalysis, and that the prosthetic group may be a substrate molecule more or less tightly bound to the catalytic protein.

Whereas Sumner's initial evidence for the protein nature of urease was insufficient to sway opinion from Willstätter's view, the data presented by Northrop made the conclusion inescapable that pepsin is a protein. In 1933 Dyckerhoff and Waldschmidt-Leitz reported that crystalline non-enzymic proteins such as edestin can adsorb pepsin, and concluded that the edestin removes the active group from the pepsin protein. Within a few months, Northrop countered with the clear demonstration that edestin and pepsin combine to form a complex from which the pepsin protein can be isolated. Between 1930 and 1935 the results of a variety of physical and chemical studies left no doubt regarding the protein nature of pepsin, and Northrop noted:

> Numerous experiments have been reported in the literature in which solutions of pepsin and other enzymes have been found to give negative protein tests although they are active. These experiments are also inconclusive, since the activity test is far more delicate than the chemical test for protein (Northrop, 1935, p. 280).

By 1935 a pepsin preparation was available, therefore, of unquestioned identity as a protein, although Northrop's first judgments as to its purity required later revision.

It was my good fortune to have the opportunity, while working in Bergmann's laboratory at the Rockefeller Institute, to find the first model peptides that were hydrolyzed by pepsin; they were made by the new synthetic method he and Zervas had described in 1932. The Bergmann-Zervas procedure vastly enlarged the scope of peptide synthesis in that it made possible the inclusion of complex amino acids; the methods developed by Fischer, while yielding long polypeptides, were limited to relatively few amino acids such as glycine or leucine. The first synthetic substrate for pepsin (Fruton and Bergmann, 1938) was cleaved very slowly, but later more sensitive peptide substrates were prepared, and it was clear that pepsin could catalyze the hydrolysis of peptide bonds.

During the 1930s Northrop's colleague Moses Kunitz had crystallized, from extracts of beef pancreas, two protein-splitting enzymes which he named trypsin and chymotrypsin (both are components of the "trypsin" of Kühne), and synthetic peptide substrates for these two enzymes were also found in Bergmann's laboratory during 1936–1938. It seemed clear, therefore, that there was no need to assume the existence of linkages other than peptide bonds in the union of the amino acid units of proteins. Indeed, this belief guided our work, and we attached greater significance to the remarkable difference in the specificity of closely related enzymes such as trypsin and chymotrypsin. We found that, although these two protein-splitting enzymes both catalyze the hydrolysis of peptide bonds in the interior of synthetic polypeptides, trypsin acts preferentially at bonds involving an arginine or lysine residue, whereas chymotrypsin prefers a bond involving an aromatic amino acid (*e.g.*, phenylalanine). It was evident that this catalytic selectivity reflected a specific interaction of the side-chain groups of the peptide substrate with the enzyme, along the lines of Emil Fischer's "lock-and-key" hypothesis.

Moreover, since Northrop and Kunitz had established the protein nature of pepsin, trypsin, and chymotrypsin, the specific enzyme-substrate interactions in their catalytic effect were examples of the type of chemical "recognition" to be expected in the interactions of proteins with each other. During the 1920s such protein-protein interactions were considered to involve largely the mutual attraction of oppositely charged side-chain groups of amino acids (*e.g.*, arginine and glutamic acid units), and in the 1930s Mirsky and Pauling added the hydrogen bond as a probable mode of interaction between the polypeptide chains of proteins. The preference of chymotrypsin for uncharged side chains, such as that of phenylalanine, indicated still other modes of specific chemical recognition. It may be noted in passing that one of the factors in the finding of specific synthetic substrates for pepsin and trypsin was the difficulty I found in accepting the widely held view that pepsin acts preferentially on positively charged substrates and trypsin acts on negatively charged substrates. This idea, which was based on the fact that pepsin acts optimally at acid pH values, whereas trypsin prefers an alkaline pH, was inconsistent with the available data on the relation of the amino acid composition of protein substrates to the rates at which they were cleaved by these two enzymes. It seemed that, if proteins were long polypeptides, pepsin should act preferentially on acidic substrates, especially those containing tyrosine or phenylalanine units, and trypsin should act best on basic substrates, with lysine or arginine as structural units. The reasoning was naïve, and later work necessitated revision of the idea, but the outcome of the experiments which followed from it was gratifying.

Although it appeared that the finding of synthetic peptide substrates for pepsin had removed one of the objections to the peptide theory of protein structure, there still remained the problem of denaturation. Indeed, in 1938 Kaj Linderstrøm-Lang noted:

> It is well known that genuine proteins (at pH 7) are attacked by crystalline trypsin which, on the other hand, is able to split synthetic peptides. This has been taken as support for the view that these proteins contain peptide bonds in their molecules. We wish, however, to point out that in the light of the following consideration, this support loses a great deal, if not all, its importance. If according to Anson and Mirsky, denaturation is reversible, then in a solution of a given globular protein there is an equilibrium between genuine and denatured protein, $G \rightleftharpoons D$. Hence it is sufficient that D and only D should contain peptide bonds open to fission by trypsin, because by removal of D by hydrolysis this process is forced in the direction from left to right and

G will gradually disappear as well (Linderstrøm-Lang *et al.*, 1938, p. 996).

Experiments on the temperature coefficients of the cleavage of a native and denatured protein substrate were consistent with this consideration, and led to the suggestion that they

> . . . provide sufficient basis for giving a warning against the conclusion that genuine proteins contain peptide bonds because they are split by proteinases like trypsin. They give a certain indication that peptide bonds are formed or "appear" (like SH-groups) upon denaturation, but they are not conclusive enough to decide whether or not some hydrolysable peptide bonds are pre-formed in the molecules of the genuine globular proteins (*ibid.*, p. 997).

This cautious doubt was expressed at a time when other voices were raised in favor of the recurrent idea that linear polypeptides are merely the denatured forms of proteins, whereas native proteins are characterized by the presence of covalent linkages, other than the peptide bond, which are decisive in holding the structure together. We saw earlier that Mirsky and Pauling had proposed in 1936 that hydrogen bonds (which are weaker than the usual covalent linkages) were responsible for the folding of the polypeptide chain to form globular proteins. Considerable attention was also given, however, to several proposals that assumed the transformation of the polypeptide chain to form ring structures involving covalent bonds. Thus in 1930, from X-ray diffraction data on fibrous proteins (wool keratin, tendon collagen), William Astbury (Bernal, 1963) inferred the existence of hexagonal folds, and one of the suggested structures (Frank, 1936) was developed by Dorothy Wrinch (in 1936), who offered a "cyclol" theory of protein structure. In the Wrinch hypothesis, which was based largely on topological considerations, it was proposed that a native globular protein exists as a honeycomb-like structure constructed of six-membered rings formed by the covalent union of the NH of one amino acid unit with the CO group of another amino acid to give $=N-C(OH)=$ bonds. Among the protagonists of the cyclol theory was the physicist Irving Langmuir, who wrote in 1939:

> We may sum up the present position with regard to the structure of proteins as follows: A vast amount of data relating to protein structure have been collected by workers in a dozen different fields. No reasonable doubt remains as to the chemical composition of proteins. The

original idea of native proteins as long chain polymers of amino-acid residues, while consistent with the facts relating to the chemical composition of proteins in general, was not a necessary deduction from these facts. Moreover it is incompatible with the facts of protein crystallography, both classical and modern, with the phenomena of denaturation, with Svedberg's results which show that the native proteins have definite molecular weights, and with the high specificity of proteins discovered in studies in immunochemistry and enzyme chemistry. All these facts seem to demand a highly organized structure for the native proteins, and the assumption that the residues function as two-armed units leading to long-chain structures must be discarded. The cyclol hypothesis introduced the single assumption that the residues function as four-armed units, and its development during the last few years has shown that this single postulate leads by straight mathematical deductions to the idea of a characteristic protein fabric which in itself explains the striking uniformities of skeleton and configuration of all the amino-acid molecules obtained by the degradation of proteins. The geometry of the cyclol fabric is such that it can fold round polyhedrally to form closed cage-like structures. These cage molecules explain in one single scheme the existence of megamolecules of definite molecular weights capable of highly specific reactions, of crystallizing, and of forming monolayers of very great insolubility. The agreement between the properties of the globular proteins and the cyclol structures proposed for them is indeed so striking that it gives an adequate justification for the cyclol theory, especially in view of the fact that this great variety of independent facts are on this theory seen to be logical consequences of one simple postulate (Langmuir, 1939, p. 611).

In the same year, however, Linus Pauling and Carl Niemann wrote:

We have carefully examined the X-ray arguments and other arguments which have been advanced in support of the cyclol hypothesis, and have reached the conclusion that there exists no evidence whatever in support of this hypothesis and that instead strong evidence can be advanced in support of the contention that bonds of the cyclol type do not occur at all in any protein (Pauling and Niemann, 1939, p. 1860).

They also stated:

It is our opinion that the polypeptide chain structure of proteins, with hydrogen bonds and other interatomic forces (weaker than those corresponding to covalent bond formation) acting between peptide chains, parts of chains and side chains, is compatible not only with the chemical and physical properties of proteins, but also with the detailed information about molecular structure in general which has been provided by the experimental and theoretical researches of the last decade (*ibid.,* p. 1860).

One of the arguments offered in support of the cyclol theory was that it explained the Svedberg unit of 35,000 in terms of a cyclol containing 288 amino acid residues, a number independently suggested by Bergmann and Niemann on the basis of their estimates of the amino acid composition of several proteins. According to Pauling and Niemann:

> Considerable evidence has been accumulated suggesting strongly that the stoichiometry of the polypeptide framework of protein molecules can be interpreted in terms of a simple basic principle. This principle states that the number of each individual amino acid residue and the total number of all the amino acid residues contained in a protein molecule can be expressed as the powers of the integers two and three (*ibid.,* p. 1867).

The principle referred to was deduced by Bergmann and Niemann during 1936–1938 from their analyses for some amino acids in acid hydrolysates of several protein preparations (hemoglobin, fibrin, silk fibroin, gelatin), and from data reported by others for egg albumin. The total number of amino acid residues per "molecule" of protein appeared to fall into a series of multiples of 288; thus fibrin had 576 ($2^6 \times 3^2$) and silk fibroin 2592 ($2^5 \times 3^4$) residues, respectively, corresponding to molecular weights of 69,300 and 217,000. Furthermore, not only was the content of an amino acid denoted by the formula $2^n \times 3^m$ (where n and m are integers), but also Bergmann and Niemann proposed, as a law of protein structure, that every amino acid residue occurs at a regularly periodic interval in the polypeptide chain of a protein. Thus the polypeptide chain of silk fibroin, which has a preponderance of glycine (G) and alanine (A) residues (the others being denoted X), was considered to be represented by the sequence

$$\text{-G-A-G-X-G-A-G-X-G-A-G-X-G-A-G-X-}$$

Periodicity hypotheses of this kind had been offered before. Albrecht Kossel concluded, from analytical data he obtained with Henry Dakin in 1904, that arginine, which represents the predominant amino acid unit of the protamine clupeine, recurs regularly in this protein in the repeating triplet -Arg-Arg-X-. In 1934 Astbury inferred from the available data for the amino acid composition of gelatin that every third residue could be a glycine unit and every ninth a hydroxyproline unit. The Bergmann-Niemann hypothesis, therefore, was a generalization to all proteins of these earlier specific suggestions.

Opinions regarding the validity of the periodicity hypothesis, and the numerical rule on which it was based, were divided, even in Bergmann's laboratory. Some geneticists welcomed the hypothesis (see Goldschmidt, 1938), since it appeared to suggest the operation of a mathematical principle that might link Mendel's laws to the structure of proteins, at that time considered to be the most probable chromosomal constituents responsible for the transmission of hereditary characters; we shall return to this problem in the next chapter. Among protein chemists who had worked for many years to improve the methods for the analysis of protein hydrolysates, however, the attitude toward the Bergmann-Niemann hypothesis, when expressed publicly, was one of skepticism tempered with the respect due to a chemist of Bergmann's distinction. The leading British protein chemist, who later succeeded Hopkins as Professor of Biochemistry at Cambridge (Chibnall, 1966), had the following to say:

> To those of us interested in the subject it seemed like the dawn of a new era, and as Bergmann (1938 [p. 427]) himself so appropriately remarked: "Everyone who is familiar with the history of protein chemistry may feel somewhat amazed on being confronted with a simple stoichiometry of the protein molecule." Some of us, nevertheless, were not prepared to give this attractive generalization our immediate and unqualified support, for we questioned the reliability and completeness of the amino acid analyses on which it was based (Chibnall, 1942, p. 138).

Indeed, by 1939 Bergmann was obliged to abandon the periodicity hypothesis, after his associate William H. Stein had produced incontrovertible analytical data that could not be accommodated by the $2^n \times 3^m$ rule. Furthermore, during the period 1939–1941 there was a succession of adverse judgments by highly regarded protein chemists (Neuberger, Pirie, Bull, among others). Chibnall's searching critique in 1942 made it clear, therefore, that the Bergmann-Niemann hypothesis had suffered the fate of other mathematical theories of protein structure insecurely founded on experiment. By 1943 even X-ray evidence was cited as being inconsistent with the Bergmann-Niemann rule (Astbury, 1943).

One of the articles critical of the Bergmann-Niemann hypothesis was a 1941 paper in which the authors attempted to test its validity by analyzing the mixture of small polypeptides produced by the partial acid hydrolysis of proteins. They used various analytical methods then available for individual amino acids, and an electrodialysis method for separating the basic peptides; although their work was not sufficiently conclusive

> . . . to form a rigid refutation of the Bergmann-Niemann hypothesis, it seems to suggest for proteins a considerably more complicated structure. The hypothesis of random structure, with which it is scarcely incompatible, may be rejected on physical and biological grounds (Gordon *et al.,* 1941, p. 1385).

It was evident that better methods were needed for the quantitative separation and analysis of the hydrolytic products derived from proteins, if more definitive experimental results were to be obtained.

In an adjoining article, two of the authors, Archer Martin and Richard Synge, described an approach that set the stage for the solution of this problem, through the use of the technique known as chromatography, a term introduced by Michael Tswett in connection with his studies on plant pigments, notably chlorophyll. Because the impact of this technique has been so great, a brief digression on its development before 1941 is necessary.

Chromatography. In 1906 Tswett reported:

> If a petroleum ether solution of chlorophyll is filtered through a column of an adsorbent (I use mainly calcium carbonate which is stamped firmly into a narrow glass tube), then the pigments, according to their adsorption sequence, are resolved from top to bottom into various colored zones, since the more strongly adsorbed pigments displace the more weakly adsorbed ones and force them further downwards. This separation becomes practically complete if, after the pigment solution has flowed through, one passes a stream of solvent through the adsorbent column. Like light rays in the spectrum, so the different components of a pigment mixture are resolved on the calcium carbonate column according to a law and can be estimated on it qualitatively and also quantitatively. Such a preparation I term a chromatogram and the corresponding method, the chromatographic method.
>
> It is self-evident that the adsorption phenomena described are not restricted to the chlorophyll pigments, and one must assume that all kinds of colored and colorless chemical compounds are subject to the same laws (Tswett, 1906, p. 322).

We saw earlier that, during the first decade of this century, the attention of leading physical chemists was directed to the study of the colloidal state of matter, and the adsorption phenomena associated with it. In a book (in Russian) published in 1910, Tswett discussed various theoretical aspects of the chromatographic method from the point of view of then-new adsorption theory of Herbert Freundlich (Synge, 1962). Tswett's work was not favorably received, largely because Leon Marchlewski, the leading expert on chlorophyll before Willstätter, did not value the chromatographic method highly as a means of purification, and severely criticized Tswett's chemical work (Robinson, 1960). As it turned out, many of Marchlewski's criticisms were unjustified, but the credit for some of Tswett's findings (*e.g.*, the existence of two forms of chlorophyll in leaves) went to Willstätter, whose remarkable chemical studies during 1906–1915 overshadowed all that had been done on the subject previously. Willstätter made little use of Tswett's technique, although he knew his 1910 book; when Willstätter's manuscript copy of a German translation passed into the hands of Richard Kuhn, during the 1920s, the latter appreciated its significance and used adsorption chromatography extensively in his studies on the carotenoid pigments, as did Leroy Palmer before him. By the 1930s, because of the work of Kuhn, Edgar Lederer, and Laszlo Zechmeister, the method described by Tswett was widely used in laboratories of organic chemistry.

The chromatographic method developed by Martin and Synge in 1941 differed from adsorption chromatography in that it depended on the establishment of an equilibrium between two liquid phases (*e.g.*, chloroform and water), one of which (water) is immobilized by being held by a solid support (*e.g.*, silica gel), as the other (chloroform) flows through the column. This method thus involves the distribution of the molecules of a dissolved substance between the two phases, and was later termed "partition chromatography." To apply this method to the amino acid analysis of protein hydrolysates, it was necessary to convert the amino acids (which are insoluble in chloroform) to acetylamino acids (*e.g.*, $CH_3CO—NHCH_2COOH$); these are partitioned sufficiently differently between chloroform and water, depending on the nature of the amino acid, to permit the nearly quantitative separation of closely related acetylamino acids, such as those derived from valine and isoleucine. A disadvantage of the method was the necessity for the acetylation of the amino acids, since one could not be certain in the case of a protein hydrolysate that the acetylation had been complete. It may be added that the choice of acetylamino acids came from work Synge had done in 1938 in an attempt to extend the methods available for the isolation of amino acids from protein hydrolysates, and Martin had previously used a "countercurrent liquid-liquid" extraction apparatus for the separation of closely related substances through systematic consecutive distributions of the substances between two immiscible solvents. Earlier in 1941 Martin and Synge reported the application of this procedure to the separation of acetylamino acids; the results were not satisfactory, and they decided to modify the extraction method by holding one of the solvents stationary on a solid support (Martin, 1948). The power of the countercurrent distribution method was greatly increased through the later work of Lyman Craig.

The appearance of the 1941 paper of Martin and Synge on partition chromatography coincided with efforts in several laboratories to develop new methods for the quantitative analysis of the amino acids present in a protein hydrolysate. Aside from the impetus given this search by the Bergmann-Niemann hypothesis, other factors also encouraged these efforts. During World War II, greater urgency was given to the need for accurate data on the amino acid composition of food proteins, and there was much interest in antibacterial agents (*e.g.*, gramicidin, penicillin) which had turned out to be related to polypeptides. The use of fastidious microorganisms which, like the one whose study led to the discovery of methionine, required individual amino acids for growth, was extensively explored (Snell, 1945). In addition to the studies of Arne Tiselius on adsorption chromatography of amino acids (*e.g.*, on charcoal), their separation on the basis of ionic interaction with solid supports having charged groups ("ion-exchange chromatography") began to be studied systematically. Furthermore, the possibilities of the electrophoretic separation of amino acids were examined, and various "ionophoretic" procedures were described. Older chemical methods for the estimation of individual amino acids were subjected to renewed scrutiny and, in some cases, notably improved. Indeed, in 1945 Erwin Brand, through the use of a combination of chemical and microbiological techniques, was able to account for 99.6 percent of the amino acid residues of β-lactoglobulin, and the minimum molecular weight he calculated (42,000) was identical with the value obtained with the ultracentrifuge. In addition, he and his associates stated:

> It can be concluded that the constituent amino acids are primarily linked by typical peptide bonds; other linkages, such as atypical peptide bonds (as in glutathione), esters, anhydrides, or imides, if they are present at all, can be present only in small numbers (Brand *et al.*, 1945, p. 1531).

This considerable achievement required much time and effort, however, and the possibility was not excluded that the complete accounting was the result of the compensation of errors.

It was clear that a general method for the amino acid analysis of protein hydrolysates was needed. In 1945 Martin and Synge reviewed the status of this problem, and expressed the opinion that

> The time has nearly arrived . . . when it will be possible to determine with equal accuracy *all* of the amino acids in a protein hydrolysate. The main types of method which will make this possible can already be outlined . . . the ionophoretic and chromatographic methods have inherent advantages . . . . In combination, they are likely to prove very powerful analytical weapons. They have the additional advantage of being "comprehensive" analytical procedures, more likely to reveal unsuspected amino acids in a mixture than methods directed to the determination of particular amino acids (Martin and Synge, 1945, pp. 12–14).

We mentioned that the partition chromatography of acetylamino acids had deficiencies as an analytical tool; similar procedures were therefore sought for the separation of the free amino acids. In 1943 Gordon, Martin, and Synge attempted to use silica gel for this purpose, without success, but obtained good separations by means of cellulose in the form of strips of filter paper. This method was developed by Consden, Gordon, and Martin, and the publication of their article in 1944 brought to the attention of biochemists the technique of "paper chromatography," whose immediate and widespread adoption had a profound effect on many branches of biochemistry. They noted that their method

> . . . is rather similar to the "capillary analysis" method of Schönbein and Goppelsroeder . . . except that the separation depends on the differences in partition coefficient between the mobile phase and the water-saturated cellulose, instead of differences in adsorption by the cellulose (Consden, Gordon, and Martin, 1944, p. 225).

After the work of Runge, published during the 1850s, on the adsorption of chemicals by paper and other porous materials, Schönbein (1861) had noted the selective adsorption of the components of a mixture by observing that they rose to different heights on a strip of filter paper dipped into the solution; the analogy to the long-known rise of liquids into capillary tubes was obvious. This work on "capillary analysis" was continued by his student Friedrich Goppelsroeder (1901). At that time the geologist David T. Day was using such procedures for the fractionation of petroleum. Although not unknown to organic chemists, the method does not appear to have been highly regarded, probably because of their understandable preference for crystallization and fractional distillation as techniques for the separation and isolation of compounds. For biochemists who were often obliged to work with very small amounts of a natural product, however, paper chromatography provided the solution to many intractable problems. During 1945–1955 innumerable variations of the method were described, and repeated efforts to devise a quantitative procedure for the analysis of protein hydrolysates were reported. It became clear, however, that, although paper chromatography had considerable value as an analytical tool, its accuracy was too low for the reliable analysis of the amino acid composition of proteins.

The next step toward the solution of the problem was the use of columns, with starch as the support for the stationary aqueous phase; because the rate of flow through such columns is very slow, an analysis of a protein hydrolysate might require several days, and an automatic collecting device was necessary. By 1946 William H. Stein and Stanford Moore had developed this technique, and in 1949 they reported the complete amino acid analysis of β-lactoglobulin, as determined by partition chromatography on starch columns with the aid of an automatic fraction collector; although some of the values reported in 1945 by Brand were confirmed, others were found to be incorrect, and Brand's complete accounting was recognized to have been the result of a compensation of errors. The disadvantages of the starch columns led Stein and Moore to develop, by 1951, a procedure in which partition chromatography was replaced by the use of an ion-exchange resin (sulfonated cross-linked polystyrene, a polymer first described by Staudinger). The remarkable ability of this resin, which acts both by ion-exchange and adsorption, not only to separate free amino acids, but also closely related polypeptides, made the new Stein-Moore method the procedure of choice. An important factor in the ready acceptance of the method was the meticulous care with which it was described. Within a few years Stein and Moore, together with Darrel Spackman, replaced the collection of individual fractions, as they emerged from the column, by an automatic recording assembly. This instrument measured and recorded the intensity of color produced by the emergent amino acids upon reaction with a nonspecific reagent (ninhydrin, described by Siegfried Ruhemann in 1910). By 1958, when the details of the Spackman-Stein-Moore apparatus were fully described, it was evident that the problem of the complete amino acid analysis of protein hydrolysates had been solved.

The development of the Spackman-Stein-Moore apparatus transformed the complete amino acid analysis of proteins from a laborious operation, requiring much material, time, and skill, into a reliable and relatively rapid method, which has since been further accelerated and automated. As in every analytical procedure, there are precautions to be taken, but the clearest evidence of the utility and reliability of the method was its immediate and continued acceptance. As we have already noted, however, the information provided by the complete amino acid analysis of a protein is only the equivalent of that given by the elementary analysis of an organic compound of low molecular weight. Just as for the organic chemist, the next step has been the determination of the arrangement of the atoms in his compound through the study of its chemical transformations (he now has powerful physical methods, such as infrared spectroscopy, nuclear magnetic resonance, X-ray crystallography, and mass spectrometry, which often give direct information regarding the structure), the challenge to the protein chemist was to determine the arrangement of the amino acid units in the protein under study.

The Structure of Proteins. The challenge was taken up by Frederick Sanger, even before the general methods of amino acid analysis had been developed; within a ten-year period, 1945–1955, he had established the sequence of the amino acid units in the polypeptide chains of the protein hormone insulin. Although the experimental methods Sanger employed were new, for he depended heavily on the use of chromatographic and ionophoretic techniques and of crystalline proteinases as reagents, the basic approach was the time-honored one of the nineteenth-century organic chemists, namely, the partial cleavage of the molecule, identification of the structure of the fragments, and the use of this information to formulate the structure of the intact molecule. It should also be added that, in view of the various speculations during the period 1920–1940, Sanger's decision to embark on a long and difficult task clearly reflected the conviction that proteins are well-defined molecules in which the amino acid units are linked by peptide bonds to form long polypeptide chains. His success, not only in showing this to be true, but also in giving others confidence to attack the sequences of other proteins, marked the end of a century-long collective effort in which many eminent chemists and biologists made important discoveries but also took wrong turnings.

Sanger's choice of insulin was both wise and fortunate. In 1921 this hormone had been extracted from pancreatic tissue by Frederick Banting and his associates, and many investigators thereupon sought to isolate it in pure form. The hormones known at that time were all relatively small molecules. The first to be crystallized, the epinephrine (adrenalin) of the adrenal medulla, was obtained by Takamine in 1901, and its structure was established by 1904; it turned out to be a compound of low molecular weight related to tyrosine. Secretin, the gastric hormone described in 1904 by William M. Bayliss and Ernest H. Starling (who introduced the word hormone; Greek *hormon,* arousing, exciting) was recognized to be a dialyzable substance; though its structure was unknown, it was also considered to be a small molecule. In 1915 Kendall had identified thyroxine as an amino acid. In 1925 John Murlin reported that he had prepared a "biuret-free insulin," and concluded that "insulin *per se* is not a true protein in the sense of being composed of nothing but amino acids" (Allen and Murlin, 1925, p. 138). There was considerable dismay, therefore, when in the following year John Jacob Abel reported that he had purified insulin in the form of a crystalline protein. As we have seen, wide credence was given at that time to Willstätter's view of enzymes as small reactive molecules that are readily adsorbed and stabilized by colloidal carriers; Sumner's announcement of the isolation of urease as a crystalline protein, a few months after Abel's report, was therefore received with skepticism. In the opinion of many physiologists, enzymes and hormones (along with vitamins) were related substances, so that it was reasonable for them to believe that Abel's crystals were simply another instance of the adsorption of a biologically active substance on a colloidal protein carrier (Murnaghan and Talalay, 1967). In 1934 a leading colloid chemist could still write:

A comparative recent epoch ascribed to the proteins the central position in the life process. Today, however, they appear rather as passive carriers of the life phenomena, determining the important chemical and physico-chemical properties of the medium and the mechanical properties of the tissue or forming the base for certain attached active groups. The true directive participation in the maze of vital chemical reactions on the other hand devolves on scattered specific substances which, in general, belong to entirely different classes of bodies . . . . (Pauli, 1934, p. 111).

It required over five years for the general acceptance of Abel's view that insulin is a protein; aside from tests of the biological potency of his crystalline preparation, extensive chemical studies by Hans Jensen, Vincent duVigneaud, and Oscar Wintersteiner contributed much to the resolution of a prolonged controversy among German, Dutch, British, and American workers. The fact that the discussion about the protein nature of insulin overlapped the dispute about the nature of Sumner's urease and Northrop's pepsin may be taken to indicate the extent to which many leading chemists and physiologists had accepted the idea that biologically active substances such as enzymes, hormones, and vitamins are small reactive molecules that are readily adsorbed by colloidal substances such as the proteins.

By 1940 it could fairly be said that insulin had been studied more thoroughly from a chemical point of view than any other well-defined protein. Seventeen different amino acids had been identified in hydrolysates of cattle insulin and, in particular, the protein was found to contain an unusually large proportion of cystine residues, the integrity of whose disulfide (—S—S—) bonds was essential for the hormonal activity. Chibnall selected insulin as a test protein for his critical examination, during 1940–1945, of the reliability of the methods available for the determination of individual amino acids in a protein hydrolysate. By 1945 his analyses accounted for 96.4 percent of the protein, and he concluded from his data that its minimal molecular weight was about 12,000. Earlier data, obtained by Dorothy Crowfoot by means of X-ray crystallography, had suggested a molecular weight of 36,000, and studies in the ultracentrifuge had given values as high as 46,000; it seemed, therefore, that, like the hemocyanins, insulin subunits of 12,000 could aggregate to form larger particles. Consequently, when Sanger began work on cattle insulin in Chibnall's laboratory at Cambridge, he assumed that he was dealing with a chemical species whose molecular weight was near 12,000.

If proteins were indeed long-chain polypeptides, then at one end of the chain there should be a free amino group, and at the other end there should be a free carboxyl group, as in NH_2CH_2CO—$(NHCH_2CO)_n$—$NHCH_2COOH$. Before 1945, however, no satisfactory methods were available for the quantitative determination of these "end groups" (Fox, 1945); such end-group analyses had been important around 1930 in the demonstration that polysaccharides such as cellulose and starch are macromolecular substances. The unavailability of reliable methods for the end-group analysis of proteins made it difficult to test hypotheses, based on the yield of products formed upon complete hydrolysis, about linkages other than ordinary peptide bonds; for example, in 1942 Chibnall revived the idea that carboxyl groups in separate chains might be linked by NH groups to form a —CO—NH—CO— (imide) bridge, but no direct means of testing this hypothesis was at hand.

Sanger's first step toward the elucidation of the structure of insulin was to develop a method for the determination of the nature and amount of the amino-terminal amino acid units in proteins. In a paper published in 1945, he described the use of a reagent (dinitrofluorobenzene) for this purpose and, with the aid of partition chromatography on silica gel, he was able to establish that, per unit weight of 12,000, insulin contains two amino-terminal glycine residues and two amino-terminal phenylalanine residues (the qualitative identification of the amino-terminal phenylalanine had been reported in 1935 by Jensen and Evans). Sanger concluded that

The presence of four α-amino groups suggests that the submolecule is built up of four open polypeptide chains bound together by cross-linkages, presumably chiefly —S—S— linkages. It is, of course, possible that other chains may be present in the form of a ring structure with no free amino groups (Sanger, 1945, p. 514).

It may be noted that the cross-linking of separate peptide chains by disulfide linkages had been considered by Karl Freudenberg in 1935. As regards the possible occurrence of ring structures, they had been found in some of the peptide antibiotics (*e.g.*, gramicidin), and their appearance as parts of proteins was not excluded. Indeed, when Sanger's end-group method was later applied to some proteins, and no evidence of an amino-terminal amino acid residue was found, it was inferred that a cyclic structure was present; further work showed, however, that the amino-terminal residue was blocked (for example, by an acetyl group).

The development of Sanger's method and the widespread adoption of paper chromatography for the detection of small amounts of amino acids were followed by the introduction of other procedures for the determination of amino-terminal amino acid residues in proteins. An especially important contribution was that of Pehr Edman who described, in 1950, a method for the successive chemical removal of individual amino acids from the amino-terminus of a peptide chain; the chemical reaction on which Edman's degradation method was based had been known since the work of Paul Schlack and Walther Kumpf in 1926. Also, chromatographic analysis made it possible to use, for the determination of carboxyl-terminal amino acid units, cleavage by the enzyme carboxypeptidase; this enzyme, which is restricted in its catalytic action to attack at carboxyl-terminal peptide bonds, had been identified in 1926 by Ernst Waldschmidt-Leitz as a component of Kühne's "trypsin" and had been crystallized in 1937 by Mortimer Anson. Before the use of paper chromatography, no methods were known for the unequivocal demonstration of the release of a single amino acid from the end of a long polypeptide chain.

By 1949 Sanger was able to report a method for the separation of the four polypeptide chains assumed to be present in the insulin molecule. The procedure involved the oxidative cleavage of the disulfide bridges, and separation of the resulting mixture into two fractions, one of which (Fraction A) had only glycine-terminal chains, the other (Fraction B) having only phenylalanine-terminal chains. Further work indicated that each of the two fractions contained a single polypeptide, and Sanger (with Hans Tuppy) embarked on the determination of the amino acid residues in the B-chain. The procedure was to subject the polypeptide to partial hydrolysis with acid or various enzymes (pepsin, trypsin, chymotrypsin), to separate the peptide fragments formed, and to determine the structure of these fragments. Their remarkable skill in the use of paper chromatography for the isolation of pure samples of the many fragments that were formed, and for the determination of the amino acid composition of each purified fragment, permitted Sanger and Tuppy to report in 1951 the amino acid sequence of the 30-member polypeptide constituting the B-chain of insulin. Two years later Sanger and E. O. P. Thompson reported the sequence of the A-chain, which turned out to have 21 amino acid residues. A subsequent study established the location of the glutamic acid and glutamine units; all the aspartic acid formed on acid hydrolysis of insulin was shown to be derived from asparagine units.

By 1952 it was known from the critical analytical studies of Craig that the minimal molecular weight of insulin is not 12,000, but one-half this value, and it was clear that the intact protein contains two (not four) polypeptide chains held together by disulfide bridges. In 1955 Sanger (with Ryle, Smith, and Kitai) described an extended series of experiments which established the positions of the disulfide cross-linkages, and completed the elucidation of the arrangement of the amino acid units of cattle insulin. The structure is shown in the accompanying diagram. A remarkable feature of this structure is that, of the three disulfide bonds present in the molecule, two are involved in holding the A- and B-chains together, while the third forms a ring structure in the A-chain. This ring is the same size as that found by Vincent duVigneaud in 1953 in the peptide hormones oxytocin and vasopressin (from the posterior pituitary). In that year, duVigneaud and his colleagues established the structure of these hormones (composed of 8 amino acid units) by unequivocal chemical synthesis, and by the demonstration of the identity of the synthetic products with the natural materials.

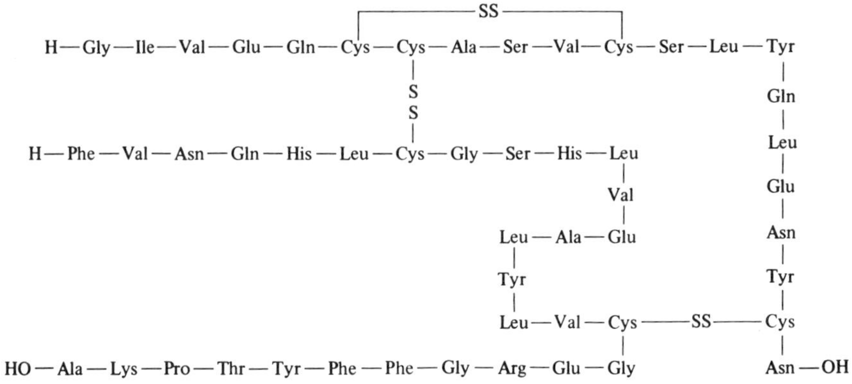

Amino Acid Sequence of Cattle Insulin

We mentioned earlier the decisive contributions of Fischer, Curtius, and Bergmann to the art of peptide synthesis. After World War II, interest in the possibility of making peptide hormones and antibiotics by chemical synthesis provided a powerful spur to the elaboration of new methods. Thus, after Paul Bell and his associates had reported in 1954 the complete 39-amino acid sequence of the adrenocorticotropic hormone (ACTH) of the anterior pituitary, not only had the entire chain been synthesized within ten years, but many structural analogues had also been made. It was not surprising, therefore, that, by 1965, chemists in three laboratories (in the United States, Germany, and China) had announced the preparation of synthetic insulin; although the synthetic procedures were extremely laborious, and the yields were poor, biologically active material was obtained. Much remained to be done to develop the art of peptide synthesis; nevertheless, the laboratory synthesis of a biologically active protein provided convincing proof of the value of the classical route for the study of the structure of an organic substance of biological interest.

Sanger's achievement was soon followed by another—the elucidation of the sequence of the 124-amino acid chain of the enzyme ribonuclease by Moore and Stein in association with Hirs and Smyth. We shall refer to this enzyme more fully in the next chapter. Although the basic strategy was similar to that employed by Sanger, the greater chain length presented special problems and required a greater variety of quantitative methods; the availability of the powerful new chromatographic technique developed by Spackman, Stein, and Moore yielded a solution within a much shorter time, however. Since the work of Sanger on insulin and of Moore and Stein on ribonuclease, the complete amino acid sequence of many proteins has been determined, and new chemical methods (*e.g.,* automatic Edman degradation) or physical tools (*e.g.,* mass spectrometry) have made the task easier. The important knowledge that has flowed from these sequence studies may be found in current textbooks, and need not be discussed here; one of the early achievements requires special mention, however.

The role of the detailed amino acid sequence in determining the properties of a protein was strikingly brought out by studies on the hemoglobin from normal adults (hemoglobin A) and from human subjects suffering from sickle-cell anemia (hemoglobin S). In 1949 Harvey Itano and Linus Pauling showed that the two hemoglobins differ in their electrophoretic behavior, and in 1958 Vernon Ingram demonstrated that trypsin digests of the two proteins, when examined by paper chromatography and electrophoresis ("fingerprint" technique), differ only with respect to the nature of a single peptide. The subsequent elucidation of the complete amino acid sequence of hemoglobin A showed that a single glutamic acid unit (which is negatively charged) is replaced in hemoglobin S by an uncharged valine unit in one of the two kinds of polypeptide chains in the hemoglobins. Since sickle-cell anemia was long known to be a hereditary trait, these findings were important in providing evidence for the genetic determination of protein structure.

Lest the reader who is not familiar with the present state of protein chemistry be led to believe that the problem of the determination of the amino acid sequence of polypeptide chains has been fully solved, a cautionary note must be appended. That the methods now used require improvement is amply attested by the numerous instances in which a sequence proposed by one research group has required revision after it had been subjected to experimental scrutiny by another; a notable exception, it must be added, is Sanger's sequence for insulin.

We have seen that during the 1930s there was much speculation about the manner in which the polypeptide chains of proteins might be folded in globular proteins. Some of the speculations were based on X-ray data obtained with silk fibroin or wool keratin fibers. It seemed more likely, however, that significant information about the three-dimensional structure of globular proteins would be obtained from X-ray patterns of crystals, and the first reasonable patterns of this kind were reported in 1934 by John Desmond Bernal and Dorothy Crowfoot, who used single crystals of pepsin immersed in the liquid in which they had been crystallized. After giving an estimate of the molecular weight based on the dimensions of the unit cell of the crystal, they stated:

> Not only do these measurements confirm such large molecular weights but they also give considerable information as to the nature of the protein molecules and will certainly give more when the analysis is pushed further. From the intensity of the spots near the centre, we can infer that the protein molecules are relatively dense globular bodies, perhaps joined together by valency bridges, but in any event separated by relatively large spaces which contain water. From the intensity of the more distant spots, it can be inferred that the arrangement of atoms inside the protein molecule is also of a perfectly definite kind, although without the periodicities characterising the fibrous proteins . . . . Peptide chains in the ordinary sense may exist only in the more highly condensed or fibrous proteins, while molecules of the primary soluble proteins may have their constituent parts grouped more symmetrically around a prosthetic nucleus (Bernal and Crowfoot, 1934, p. 794).

During the succeeding ten years, a number of crystalline proteins (insulin, β-lactoglobulin, chymotrypsin, etc.) were examined (Hodgkin and

Riley, 1968); among them was hemoglobin, data for which were reported in a 1938 paper by Bernal, Isidor Fankuchen, and Max Perutz. Although the X-ray photographs obtained in these studies were of excellent quality, the methods that had proved to be applicable to the study of small organic molecules were inadequate in the face of the very large number of diffraction spots given by protein crystals. The key to successful protein structure analysis was provided in 1954 by Perutz, who had continued work on the X-ray crystallography of hemoglobin; by attaching heavy atoms (*e.g.*, mercury) at specific sites on the surface of the protein molecules ("multiple isomorphous replacement") he was able to solve one of the most troublesome problems. By means of this technique, and computer analysis of the data, John Kendrew (a colleague of Perutz at Cambridge) succeeded by 1958 in establishing the general structure of whale myoglobin; this hemoglobin-like protein was known to have a molecular weight of about 17,500, or one-quarter that of hemoglobin. Subsequent high-resolution analysis permitted Kendrew and his associates (1960) to construct a three-dimensional model that not only revealed the convolutions of the single peptide chain of the protein and the position of the iron-porphyrin, but also showed some of the amino acid residues with a distinctive structure (*e.g.*, tyrosine). It was clear, however, that at the level of resolution in this work the uncertainties in the X-ray pattern required a prior knowledge of the amino acid sequence for a complete model to be constructed. In one of the memorable papers reporting their results, they noted:

> Perhaps the more remarkable features of the molecule are its complexity and its lack of symmetry. The arrangement seems to be almost totally lacking in the kind of regularities which one instinctively anticipates, and it is more complicated than has been predicted by any theory of protein structure (Kendrew *et al.*, 1958, p. 665).

In writing this passage, the authors may have had in mind the following, written eight years earlier:

> There appears to be a real simplicity of chain structure in myoglobin, which will perhaps be shown by other favourably built proteins, and which makes it particularly suitable for intensive X-ray investigation (Bragg *et al.*, 1950, p. 356).

Some of the most striking features of the three-dimensional structure of myoglobin were the tight packing of the polypeptide chain, and the predominance of the ionizable groups (amino, carboxyl) on the surface and of the "nonpolar" amino acid units (*e.g.*, leucine) in the interior (for a vague anticipation in 1909 of this discovery, see Chibnall, 1967). Of particular interest was the fact that large sections of the polypeptide chain were twisted into a type of coil with 3.7 amino acid residues per turn (α-helix) that Linus Pauling and Robert Corey had proposed in 1950 as a general structural feature of proteins. They based this proposal on data obtained by Corey for the bond lengths and angles in small peptides, and the assumption that the CO and NH groups of long polypeptide chains tend to form either internal hydrogen bonds to produce a helix, or to form hydrogen bonds with adjacent polypeptide chains. Kendrew's model for myoglobin thus provided the first direct evidence for the occurrence of the α-helix in a globular protein. It may be added that a helical structure had previously been proposed for the long polyglucose chains ("amylose") of starch by Charles Hanes (in 1937) and by Rundle and French (in 1943).

Just as Sanger's determination of the amino acid sequence of insulin marks the dividing line in the study of the linear arrangement of the amino acid units in proteins, so does Kendrew's model for myoglobin represent the transition from geometrical or chemical speculation about the three-dimensional structure of proteins to its direct experimental study. Since Kendrew's report, David Phillips has produced a three-dimensional model of an enzyme (lysozyme), and subsequently models of several other enzymes (ribonuclease, chymotrypsin, papain, carboxypeptidase, etc.) have been derived from X-ray data. By 1969 the three-dimensional structure of the insulin molecule, which presented special difficulty, had been worked out by Dorothy Crowfoot Hodgkin, who had taken the first good X-ray photographs of a protein crystal in 1934. In some of these proteins the proportion of the polypeptide chain that is coiled into an α-helix is very small, but some of the hydrogen bonding postulated by Pauling and Corey for adjacent parts of the folded chain is evident. As noted by Kendrew for myoglobin, the three-dimensional structure of proteins is much more complicated than had been imagined. Nevertheless, although much remains to be done, especially in learning about the intimate chemical mechanism of the catalytic action of enzyme proteins, and of the specific interactions of proteins with each other and with other molecules, the "living albuminoid bodies" of the nineteenth century or the colloidal aggregates of 1900–1930 are now considered to be well-defined molecules accessible to direct chemical study.

The impact of this development on biological thought has been far-reaching. As we shall see in the succeeding chapters, the emergence of modern protein chemistry is historically related to the recognition that the processes whereby matter and energy are transformed in living things depend on the integrated action of enzymes, that the biosynthesis of these catalytic proteins is under genetic control, and that the transmission of hereditary characters from parent to progeny links biological evolution to chemical events in individual organisms. In this chapter we have traced the tortuous history of the organic chemistry of proteins, from the satisfying simplicity of Mulder's protein radical to the growing confusion arising from the variety of cleavage products, a confusion only partly dispelled by the achievements of Emil Fischer. We have seen something of the essentially independent but equally uncertain course followed by the study of the colloidal properties of proteins. The strands that link these efforts during 1850–1920 are the development of methods for the purification and characterization of individual proteins, and the discovery and identification of the constituent amino acids; they were intimately interwoven with the advances made in constructing the theoretical and experimental fabric of chemistry. By the early 1920s the problems of protein structure had been defined, but there was doubt regarding the basic questions of the macromolecular nature of proteins and the validity of the peptide theory. Twenty years later these doubts had been set aside, largely as a consequence of new methods for the study of macromolecules and of the proof that enzymes are proteins. The successful application of the techniques of chromatography and of X-ray crystallography then made possible, after World War II, the detailed study of proteins as chemical individuals. The structures that emerged from such studies have required chemists to consider the role of multiple cooperative interactions in the specific coiling of polypeptide chains to form functional protein molecules, in the self-assembly of protein subunits, and in the interaction of proteins with other cell constituents. For each generation of investigators, from Mulder's time to the present, there were experimental advances and setbacks, as well as an efflorescence of theoretical insights. It is easy to choose, in retrospect, those contributions that were decisive; in the face of the uncertainties that beset the problem, such judgments were more difficult at the time these contributions were made. There can be little question, however, that the recognition of the biological importance of proteins provided a constant stimulus to their chemical study.

From Nuclein to the Double Helix

The emergence of organic chemistry during the first decades of the nineteenth century, with its emphasis on the isolation and chemical characterization of the "immediate principles" of living things, was paralleled by the rise of animal physiology, and by the growing conviction that bodily functions should be studied experimentally with the aid of vivisection, physical apparatus, and chemical methods (Temkin, 1946). The principal animator of this movement was François Magendie, whose most famous pupil, Claude Bernard, provided at midcentury striking evidence of the value of chemistry for the study of physiological function. In Germany the fruitful collaboration of Friedrich Tiedemann with the chemist Leopold Gmelin, during the 1820s (Mani, 1956), was followed two decades later by the impetus given by Johannes Müller to the chemical, microscopic, and physical researches of a remarkable group of younger physiologists; his students included Helmholtz, duBois-Reymond, Virchow, and Schwann. During the third quarter of the nineteenth century the German counterpart of Bernard was Karl Ludwig (Schröer, 1967); in his famous *Lehrbuch der Physiologie des Menschen* (first edition, 1852 and 1856), Ludwig stated:

> After organic chemistry has recognized the weighable mass of the animal body as a collection of molecular [*atomistischer*] entities, it is the task of physiology to determine what functions are assumed in the animal organism by each of the more or less complex molecules. This task will have been completed when one knows the arrangement of the elements within the complex molecule, the amount of its latent heat, and the manner in which each individual molecule expresses its relationship to all the others present in the animal organism under given conditions (Ludwig, 1852, Vol. I, p. 15),

and in a letter (11 August 1851) to Emil duBois-Reymond, he wrote:

180

We are completely blocked by the slow progress in our chemical knowledge of the albuminoid substances; in my opinion, a new day will only dawn in physiology when one has crystalline albumin in one's hands, and at last can study with precision the many remarkable transformations and decompositions that it undergoes (duBois-Reymond, 1927, pp. 102–103).

Another influential protagonist of the chemical approach to physiological problems was Rudolf Virchow, whose view that the cell constitutes the fundamental unit of life was widely accepted in German-speaking countries.

The physiological chemistry that developed during 1860–1900, principally in Germany, was largely led by men educated in the experimental tradition carried forward by the pupils of Magendie and Müller, rather than in the speculative physiology of the chemists Dumas and Liebig. In a sense, these chemical physiologists and their biophysical colleagues represented a mid-nineteenth-century version of "molecular biology"; in 1848 one of them envisioned the dissolution of physiology into "organic physics and chemistry" (duBois-Reymond, 1912, Vol. I, p. 21), and they rejected Liebig's view that

Another fundamental error entertained by some physiologists is that physical and chemical forces alone or in combination with anatomy can suffice to explain vital phenomena (Liebig, 1846b, p. 225).

Moreover, the physiologists of Magendie's school shared the opinion of Bernard that

. . . in order to avoid error and to render all the services of which it is capable, chemistry must never venture alone in the study of animal functions; I think that in many cases it alone can resolve the difficulties that block physiology, but cannot anticipate it, and I think finally that in no case can chemistry consider itself authorized to restrict the resources of nature, which we do not know, to the limits of the facts or processes which constitute our laboratory knowledge (Bernard, 1848, p. 317).

This attitude reflected the view expressed earlier by Auguste Comte (1838, Vol. III, p. 385) that "only biologists are competent to apply physical theories successfully to the rational solution of physiological problems."

Thus, before embarking on their independent careers, Willy Kühne worked with duBois-Reymond and with Bernard, and Felix Hoppe-Seyler worked with Virchow. When Kühne and Hoppe-Seyler emerged in the 1860s as dominant figures in German physiological chemistry, the study of physiological function, both at the level of the organism and of the cell, included the identification, isolation, and characterization of chemical substances whose properties appeared to be related to the function under study. For example, Kühne's study of myosin (in 1864) was an approach to the problem of muscular contraction, and his isolation ten years later of a photosensitive pigment (rhodopsin) from retinas was a contribution to the study of the visual process. Furthermore, the emergence of a distinctive group of physiological chemists was encouraged by the rejection of their objectives by organic chemists like Kekulé who were seeking to develop the conceptual framework of their subject in terms of structure and valence, rather than physiological function. A minority voice was that of Emil Erlenmeyer:

. . . [I]f a man like Ludwig identifies the progress of physiology with the progress of chemistry, and finds in chemistry, because of its ability to explain the most delicate and complex processes, part of the salvation of physiology, then this must be a stimulus to chemists toward physiological investigations (Erlenmeyer and Schöffer, 1859, p. 316).

As we have seen from the early history of protein chemistry, the physiological chemists encountered many difficulties, not the least of which was the uncertainty whether a homogeneous compound was being studied. To cite only one of many examples, during the 1890s Frederick Pavy suggested that all proteins are glucosides, because the products of hydrolysis included material that reacted with phenylhydrazine. This idea, though welcomed by some physiologists who saw in tissue proteins the source of tissue carbohydrates, was soon rendered untenable when hydrolysates of more carefully prepared protein preparations, from which carbohydrate impurities had been removed, failed to give the phenylhydrazine test.

Similar problems recurred repeatedly in the work on other materials isolated during the course of the search for chemical explanations of physiological phenomena. Because the criteria of purity and identity were limited, there were many controversies about the nature of newly discovered substances throughout the latter half of the nineteenth century. For example, the work of Ludwig Thudichum on the isolation of new chemical constituents (*e.g.*, sphingosine) from brain tissue was not accepted by leading physiological chemists (Drabkin, 1958), and Thudichum (1881a) questioned the validity of Kühne's conclusions about rhodopsin. It was a time in which names given to new products soon lapsed into obscurity, together with the names of their creators; sometimes, as with Thudichum, later work with improved methods (in this case by Levene, Rosenheim, and Thierfelder) established the validity of a portion of the disputed work. In many instances, however, contemporary disbelief was justified by subsequent research. What usually decided the issue, if it was decided at all, was the application of better methods for the isolation, purification, and chemical characterization of the substances under discussion. When such methods were inadequate, as in the case of the proteins, the uncertainty, controversy, and fruitless speculation lasted many decades. This was also the situation for the material found in 1869 by Friedrich Miescher, who named it "nuclein" to denote its association with cell nuclei.

Nuclein and Chromatin

In 1868 Miescher received his M.D. degree at Basle, and on the advice of his uncle Wilhelm His (at that time Professor of Anatomy and Physiology there) went to Tübingen, first to study organic chemistry with Adolf Strecker, then to work with Felix Hoppe-Seyler on the chemical constitution of the cells present in pus. Miescher obtained the cells by washing used bandages from the surgical clinic, and, as he wrote his uncle in February 1869:

After the pus cells had been prepared, there was the question of the aims and methods of the research . . . . [T]here presented itself the task of giving as complete an accounting as possible of the characteristic chemical constituents from whose variety and arrangement there results the structure of the cell (Miescher, 1897, Vol. I, p. 34).

After describing some of his experimental difficulties, Miescher continued:

In the experiment with weakly alkaline fluids, I obtained, by neutralization of the solutions, precipitates that were insoluble in water, acetic acid, very dilute hydrochloric acid, or sodium chloride solution; consequently, they could not belong to any of the known albuminoid substances. Where did this substance come from? By prolonged action of very dilute hydrochloric acid on the cells, a point is reached when the acid does not take up anything more. The residue consists of partly isolated, partly shrunken nuclei . . . . Very weakly alkaline fluids (even

1/100000 sodium carbonate) cause the nuclei to become pale and to swell considerably . . . . From this fact, well-known to the histologists, the substance could belong to the nuclei, and therefore gripped my interest. The most reasonable next step was then to prepare pure nuclei (*ibid.*, pp. 34–35).

The method he eventually used involved extraction of the cells with alcohol to remove fatty material, followed by treatment of the alcohol-insoluble residue with an acidified extract of pig gastric mucosa (containing pepsin). This procedure had been described by Kühne, who described the result of the action of pepsin as follows:

The protoplasm of the cells of all edible glandular structures, such as liver, etc., largely dissolves, leaving behind small crumbs and greatly shrunken nuclei (Kühne, 1866, p. 50).

The preparation Miescher obtained contained 14% nitrogen, 2% sulfur, and 3% phosphorus. Of special interest was the relatively large amount of organically bound phosphorus; at that time this element was usually associated either with proteins like casein or with alcohol-soluble constituents such as the widely distributed lecithin, known to be a fatty acid derivative of glycerophosphoric acid, and under active study in Hoppe-Seyler's laboratory. Because the solubility properties and resistance to pepsin set his material apart from these known substances, Miescher concluded that he had isolated a new cell constituent, which he named nuclein. In the article describing his results, published in 1871, he stated:

I believe that the analyses given here, no matter how incomplete they may be, permit the conclusion that we are not dealing with a random mixture but, apart from small impurities, with a chemical individual or a mixture of very closely related substances (Miescher, 1871a, p. 458).

He also raised the question whether nuclein might be a compound of an albuminoid substance with lecithin, but this idea was inconsistent with the relatively high content of both nitrogen and phosphorus; a physiological relation to lecithin seemed possible, however:

The relation of the nuclear substances to lecithin will immediately grip our attention; the first thought will be that the nuclein, which is more closely related to the albuminoid substances, is the precursor of lecithin (*ibid.*, p. 459).

Miescher's paper was published along with four others: two of them were by Hoppe-Seyler's associates Pal Plósz (who used Miescher's method to demonstrate the presence of nuclein in the erythrocytes of birds and snakes) and Nicholas Lubavin (who examined the action of pepsin on casein to produce nuclein-like products). The fourth was by Hoppe-Seyler himself, who confirmed Miescher's claim and also reported the presence of nuclein in yeast.

> Just as the pus cells, the elementary yeast organisms contain in addition to cholesterol, lecithin, and fat, a phosphorus-containing substance that resembles albuminoid bodies, is indigestible by gastric juice, is colored brown by iodine, also belongs to the nuclei, and perhaps may play a highly important role in all cell development (Hoppe-Seyler, 1871, p. 501).

The fifth paper was by Miescher, and was based on work he had done after he left Hoppe-Seyler's laboratory in 1869. In it he described the application of his method for the isolation of nuclein to the chemical examination of the microscopic particles in the yolk of the hen's egg. He chose this material because His, among others, believed these particles (yolk platelets) to be cell nuclei on account of their strong light refraction and the fact that they stained red with the dye carmine; this insect coloring matter (cochineal) had been found during the 1850s (Hartig, Gerlach) to stain the nuclei of plant and animal cells. Miescher obtained from the yolk platelets a material that contained organic phosphorus, was soluble in alkali and was precipitated by acid, gave the biuret, Millon, and xanthoproteic tests for protein, and was resistant to the action of pepsin. He concluded therefore:

> A substance having such a composition and such properties clearly can only be ranked with the nuclein of pus. The observed deviations, the lower sulfur content and the much higher phosphorus content alter nothing as to its sharp separation from all other groups. There is a group of nucleins and it will surely increase with additional members (Miescher, 1871b, p. 507).

Miescher's work on the nuclein of pus cells and of egg yolk was a prelude to his most important biochemical investigation, during 1871–1873, after his return to Basle from a stay in Ludwig's laboratory in Leipzig—the study of the chemical composition of the spermatozoa of the Rhine salmon. Basle was particularly suitable for such research, as it was a center of the salmon fishing industry. The Rhine salmon feeds in salt water, and travels up the river to spawn; in fresh water, it does not feed, but instead there is a massive conversion of skeletal muscle to gonadal tissue. As it was already known that the sperm heads are essentially equivalent to cell nuclei, the ready availability of large amounts of such nuclear material made salmon sperm an attractive starting material for Miescher's experiments. In 1874 he reported that the sperm heads are largely composed of ". . . an insoluble salt-like combination of a very nitrogen-rich organic base with a phosphorus-rich nuclein body which assumed the role of the acid" (Miescher, 1897, Vol. II, p. 66). The organic base, which was soluble in acid, and could be crystallized in the form of its hydrochloride, he named "protamine." His analytical data suggested an empirical formula of $C_9H_{21}N_5O_3$; later analyses gave $C_{16}H_{28}N_9O_2$ (*ibid.*, p. 366). The acid-insoluble component, after reprecipitation, was found to contain 13.1% nitrogen and 9.6% phosphorus, was free of sulfur, and failed to give the color tests for protein; Miescher assigned it an empirical formula of $C_{29}H_{49}N_9P_3O_{22}$. In spite of the differences between this acidic material and the nucleins he had described earlier, Miescher named it a nuclein as well; to explain the variable composition of the nucleins from various biological sources (he also examined bull spermatozoa), especially as regards their sulfur content, Miescher suggested that

> . . . there exists a sulfur-containing nuclein which is cleaved by warm alkali into sulfur-free nuclein and a compound containing sulfur in an unoxidized state. This compound is not albumin; the sulfur content is too high for that; rather, one might consider an atomic grouping like the one involved in the structure of the keratin substances. The cleavage occurs readily with bull sperm, with greater difficulty with pus nuclei. It is most probable that both nucleins occur together in the nuclei. This may be regarded as certain for the yolk platelets of the hen's egg (*ibid.*, Vol. II, pp. 101–102).

About twenty years of work was required to sort out the confusion generated by these views.

Of particular interest in relation to work done much later were Miescher's observations regarding the physical properties of the sperm nuclein. It appeared to be a colloidal material, since it did not diffuse through a parchment membrane. It also behaved as a polybasic acid, and Miescher suggested that the salt-like "nucleoprotamine" might exist in various forms, in which the organic base was replaced by inorganic components (*e.g.*, sodium). In addition, he noted that the nuclein was unstable, and had to be isolated at low temperatures. In January 1874, Miescher wrote to His:

I have obtained the salmon nuclein entirely pure, that is undecomposed (9.6 percent P) before the end of the working day [*Thorschluss*]. Only the greatest possible speed and low temperature leads to the goal. When nuclein is to be prepared, I go to the laboratory at 5 o'clock in the morning, and work in an unheated room. No solution may be kept as long as 5 minutes, no precipitate may be left as long as one hour, before everything is preserved under absolute alcohol (*ibid.*, Vol. I, pp. 76–77).

The contemporary chemical judgment as to the validity of Miescher's claims was not entirely favorable. In Germany his conclusions were questioned by Jakob Worm-Müller (1874), and the French chemist Adolphe Wurtz, after carefully summarizing the results, concluded that they ". . . appear to be somewhat vague from a chemical point of view and seem to require new research" (Wurtz, 1880, p. 143). An English worker who had analyzed nuclein stated that it "is nothing but an impure albuminous substance" (Kingzett, 1878, p. 360), and Thudichum (1881b, p. 186) cited this report in writing that "nuclein was exploded, in this country at least."

It must be added that, from the start, Miescher was conscious of the uncertainties in the field he had entered. In his letter to His of February 1869 he wrote:

There is nothing more difficult than sharp separation in the field of the albuminoid substances. I can well understand why their definition is so variable and controversial; and it is indeed the curse of the amorphous substances that one has no assurance as to the purity of one's preparation. That is why the real chemists shun them so much (Miescher, 1897, Vol. I, p. 36).

Also, he appears to have developed, early in his career, a distaste for hasty publication. In 1870 he wrote to Hoppe-Seyler:

I would be very reluctant to agree to preliminary communications in the Centralblatt, since for various reasons I am opposed to the currently fashionable system of preliminary aphoristic publications (*ibid.*, p. 44).

These attitudes may provide a partial explanation for the fact that after 1874 Miescher published little on the chemistry of the nucleins; his later papers dealt largely with the physiology of the Rhine salmon and with breathing. During the five years before his early death in 1895, however, he resumed his chemical studies; the data were published posthumously by his friend Oswald Schmiedeberg. By that time significant advances in the study of the material Miescher discovered had been made by others, notably Albrecht Kossel (Jones, 1953).

Kossel's entry into the nuclein field appears to be related to the appearance, in 1878, of a paper by Carl von Nägeli and Oscar Loew in which they concluded that the nuclein of yeast

> . . . did not differ from albumin with a slight admixture of potassium and magnesium phosphates. In view of the considerable phosphate content of yeast, a slight contamination with "phosphorus," from whose presence Hoppe-Seyler inferred the existence of yeast nuclein, cannot be surprising (Nägeli and Loew, 1878, pp. 339–340).

This attack on Hoppe-Seyler was countered by his research associate Kossel, who confirmed the presence in yeast of ". . . a characteristic substance distinguished by a high phosphorus content and solubility properties. Further work must decide whether the nuclein as prepared is a pure substance" (Kossel, 1879, p. 288). In continuing his work on the nucleins, Kossel provided a new basis for their chemical characterization. By 1880 he had demonstrated that some of the nucleins (from yeast, pus, and goose erythrocytes) contain "hypoxanthine" as a structural unit, whereas the material from egg yolk does not. The latter was designated a "paranuclein" and turned out to be a pepsin-resistant phosphorus-rich polypeptide subsequently named "phosvitin." Similar paranucleins (Hammarsten used the term "pseudo-nucleins") were considered to be formed on the peptic digestion of casein and vitellin, and continued to be objects of great interest well into the twentieth century, when it was recognized that their phosphorus was bound in the form of phosphorylated serine units. During the 1890s there was interest in a material obtained by Kossel and Albert Neumann (1896) by treatment of nuclein (from calf thymus) with boiling water; this product ("thymic acid") was considered to be similar to the paranuclein derived from egg yolk. By 1900, however, sufficient chemical information had accumulated about the "true" nucleins (which came to be known as "nucleic acids") to allow them to be distinguished from the paranucleins. Clearly, Miescher's early conclusions about the similarity of the nucleins of pus cells, salmon sperm, and yolk platelets were influenced by his conviction that all these cell constituents were derived from nuclei, and by his failure to find chemical criteria for discriminating among them.

This failure was not a consequence of his ignorance of the existence of hypoxanthine and of compounds related to it. In his 1874 paper on the constituents of the spermatozoa of the salmon and of other vertebrates, Miescher reported that upon treatment of his protamine preparation with nitric acid there appeared a yellow color which turned red upon the addition of alkali. This color reaction was known to be given by uric acid and by several closely related animal products. The isolation of uric acid from human urine by Scheele (in 1776) and from bird excrement (guano) by Fourcroy and Vauquelin (in 1805) was followed by the work of Brugnatelli and of Prout (in 1818) who showed that nitric acid converts uric acid into a product (later named alloxan) which gives a brilliantly red ammonium salt, "murexide." The elementary composition of uric acid was established in 1834 by Liebig and by Mitscherlich, and the subsequent joint studies of Liebig and Wöhler on the chemical degradation of this compound, published in 1838, rightly evoked the admiration of their contemporaries. This work laid the basis for that of Strecker around 1860; by 1870 the relation of uric acid to xanthine (found in kidney stones by Marcet in 1817), to the sarkine (later named hypoxanthine) obtained in 1850 by Scherer from pancreas, and to guanine (isolated from guano in 1845 by Unger) had been clearly established, although the definitive proof of their structure did not come until later through the synthetic work of Emil Fischer. All these substances gave the murexide reaction, and Miescher, as a former pupil of Strecker, could not have failed to recognize the possibility that some of the relatives of uric acid might be constituents of sperm. He did not follow up his observation, however; instead he asked his colleague Jules Piccard to do so.

Piccard (1874) found that, after treatment with acid, both guanine and sarkine (hypoxanthine) were formed from salmon nuclein, and that the positive murexide test Miescher had obtained with protamine was probably a consequence of its coprecipitation with these bases. There was still uncertainty, however, whether hypoxanthine was a constituent of albuminoid substances; thus in 1878 Georg Salomon stated that it was present in fibrin, and this report was confirmed in the following year by Russell H. Chittenden working in Kühne's laboratory. Shortly thereafter, however, Kossel (1881) showed that carefully prepared protein preparations did not yield hypoxanthine, which had probably come from leucocyte material present in the crude fibrin preparations.

During the 1880s Kossel provided further evidence for the separate identity of the nucleins not only by the isolation of the "xanthine bodies" guanine and hypoxanthine after acid hydrolysis, but also by the discovery of a new member of this group (adenine) as a cleavage product of the nucleins. Furthermore, in 1882 Emil Fischer began work on the chemistry of uric acid and its relatives; during the succeeding fifteen years he established their structure as being derived from a fundamental bicyclic nucleus he named "purine." Through unequivocal synthesis, Fischer demonstrated the structure of uric acid to be (as proposed by Ludwig Medicus in 1875) 2,6,8-trioxypurine; xanthine was shown to be 2,6-dioxypurine; hypoxanthine, 6-oxypurine; adenine, 6-aminopurine; and guanine, 2-amino-6-oxypurine. As in the case of his studies on the sugars, and his later research on polypeptides, Fischer's work in the purine field was elegant in its execution, and placed this aspect of nuclein chemistry on a firmer chemical foundation. With the knowledge of the chemical structure of the various purines, it was possible around 1905 to show that the hypoxanthine found by Kossel upon degradation of the nucleins was a secondary product formed by the deamination of adenine. By that time it was generally agreed, therefore, that the two purines guanine and adenine represent authentic constituents of the nucleins.

Hypoxanthine   Xanthine   Uric acid

Adenine   Guanine

Kossel's initial evidence for the separate identity of the nucleins was not completely accepted, however; for example, during the period 1888–1892 several investigators (notably Liebermann, Pohl, and Malfatti) reiterated the claim that these materials were merely phosphorylated derivatives of ordinary proteins. A large part of the continuing uncertainty arose from the difficulty in obtaining nuclein preparations with reproducible chemical properties, and much effort was expended toward the end of the nineteenth century in modifying the original isolation methods devised by Miescher and Kossel; these improved procedures were also applied to a wide variety of animal and plant tissues. In particular, Richard Altmann (1889) succeeded in developing a method that yielded nuclein preparations apparently free of protein; he termed such protein-free nucleins "nucleic acids." We noted before that in 1874 Miescher had given the name "nucleoprotamine" to the saltlike precipitate of the acidic salmon nuclein with the basic protamine. Kossel (1884) then obtained from goose erythrocytes a peptone-like substance (he named it "histone") that readily combined with nuclein to form a "nucleohistone"; similar histones were prepared from various biological sources (*e.g.*, calf thymus), and differed from the protamines by their less basic character and the fact that, on acid hydrolysis, the histones yielded a larger variety of amino acids. It seemed likely, therefore, that the protein-containing nuclein preparations, such as the one Miescher had obtained from pus, belonged to a class of conjugated proteins in which a nucleic acid was combined with a protein-like substance to form a nucleoprotein, just as in hemoglobin an iron-porphyrin is combined with globin. Although the nucleic acids were considered to be the prosthetic groups of the various nucleoprotein preparations that had been described by 1900, it was realized that none of them approached the crystalline hemoglobins in homogeneity. Furthermore, it was known that isolated nucleic acids could precipitate a variety of proteins, and the question was raised (but could not be definitely answered) whether the nucleoproteins obtained from various biological sources were preformed cell constituents or merely artifacts of the isolation procedure. Indeed, Altmann concluded that ". . . until now no fact is known to establish the authenticity of the so-called nucleins as compounds of nucleic acids and proteins" (Altmann, 1892, p. 230).

The difficulties in defining the identity and purity of the nucleoproteins prepared from various biological sources were associated with claims regarding their physiological role. For example, around 1885, when there was considerable interest in the problem of blood coagulation, Leonard Wooldridge reported that the injection of nuclein into animals caused intravascular clotting, but by the 1890s it was clear that this property was lost when the nucleoproteins were purified more extensively. Later Wilhelm Spitzer (1897) claimed that an enzyme involved in intracellular respiration is an iron-nucleoprotein, Cornelis Pekelharing (1902) suggested that pepsin might be a nucleoprotein, and Gustav Mann concluded that ". . . the nucleoproteids are the agencies by which amino acids are built up into the cell-plasm" (Mann, 1906, p. 454). In his *The Dynamics of Living Matter*, Jacques Loeb expressed a skeptical attitude:

Inasmuch as the enzymes are necessary for the chemical processes in living matter, the formation of enzymes is one of the essential functions living matter has to perform. Spitzer and Friedenthal were inclined to assume that the nucleo-proteids act as enzymes. This view, while possible, is not yet proven (Loeb, 1906, p. 28).

We shall see later in this chapter that the possible enzymic nature of nucleoproteins continued to be discussed for many years afterward in connection with theories of heredity.

The availability of nucleic acid preparations relatively free of protein material was soon followed by the identification of degradation products other than the purines guanine and adenine. By subjecting thymus nucleic acid to hot concentrated acid, Kossel and Neumann in 1893 found among the products a new substance they named thymine; in Miescher's posthumous publication there is evidence that he also isolated this product. Thymine was recognized to be a new member of a known class of organic ring compounds, the pyrimidines; by 1901 Kossel's associate Hermann Steudel had shown it to be 5-methyl-2,6-dioxypyrimidine, and this structure was established by synthesis in Emil Fischer's laboratory (the numbering of the ring is the one used at that time). Furthermore, in 1894 Kossel and Neumann found that hydrolysis of thymus nucleic acid with strong acid gives another pyrimidine, which they named cytosine; this turned out to be 6-amino-2-oxypyrimidine, and was synthesized by Henry L. Wheeler and Treat B. Johnson in 1903. Like the adenine and guanine formed on mild acid hydrolysis, cytosine was found to be a cleavage product of all the nucleic acid preparations examined (thymus, spleen, sperm, yeast, etc.). In addition, Kossel's associate Alberto Ascoli found, in 1900, that yeast nucleic acid yields uracil (2,6-dioxypyrimidine), and a few years later, Phoebus A. Levene showed that this nucleic acid yields, as pyrimidines, only uracil and cytosine, and no thymine.

Thymine   Cytosine   Uracil

The relatively rapid elucidation of the chemical structure of thymine, cytosine, and uracil after their discovery as cleavage products of nucleic acids is noteworthy, because it marks one of the early fruits of the new synthetic organic chemistry as applied to a biological problem. The synthetic work of Robert Behrend and of Adolf Pinner around 1885 had provided valuable knowledge available to Steudel in the examination of the structure of thymine; indeed, the words "pyrimidine" and "uracil" were introduced by Pinner and Behrend, respectively, on the basis of their chemical studies.

By the first decade of this century, therefore, it was recognized that, in addition to phosphorus present in the form of a derivative of phosphoric acid, nucleic acids yielded two purines (adenine and guanine) and three pyrimidines (thymine, cytosine, and uracil). Furthermore, during the 1890s Albrecht Kossel and Olof Hammarsten showed the presence in nucleic acids of a carbohydrate component. Kossel (in 1893) identified the carbohydrate in yeast nucleic acid as a 5-carbon sugar, because, as with known pentoses, heating with strong acid produced furfural. With thymus nucleic acid, similar treatment gave levulinic acid, a product of the decomposition of hexoses. Although this behavior of the two nucleic acid preparations from yeast and thymus indicated a difference in the nature of their carbohydrate components, the chemical structure of the sugars remained unclear for many years afterward. In 1903 Levene wrote regarding the nucleic acids of animal origin:

> It remains to establish the nature of the carbohydrate present in their molecule, and further to establish the proportions of the different components in the acids of different tissues. It seemed probable from our present experience that the proportion of the purine bases to the pyrimidine bases, as well as the proportion of the different bases of each group, varies considerably in acids of different tissues (Levene, 1903, p. 211).

As we shall see, ten years later he abandoned this view in favor of a simpler structure in which the four bases are present in equal proportions. For a biography of Levene, see Van Slyke and Jacobs (1944).

Nuclein and Heredity. Before considering the subsequent development of nucleic acid chemistry, we must take note of the rise and fall of nuclein, during 1880–1910, as the chemical agent for the transmission of hereditary characters. We saw that Miescher undertook the work that led to the discovery of nuclein in the hope of determining the chemical constitution of cell nuclei. It is not clear from his published writings what Miescher considered the physiological role of the nucleus to be; although there was general acceptance of the idea advocated by Robert Remak and Rudolf Virchow in the 1850s that cells are formed by the division of preexisting cells, the function of the nucleus in cell division was not fully appreciated until the dramatic development of cytology during 1870–1890 (Coleman, 1965). A widely held view around 1870 accorded to the nucleus a role in the synthetic operations of the cell (for example, see Bernard, 1878–1879, Vol. II, p. 523).

The interplay of chemical and biological thought, especially evident in the discussion of physiological problems after 1840, found expression in relation to the nature of fertilization. Thus Theodor Ludwig Wilhelm Bischoff (1847) was unable to see the penetration of an egg by spermatozoa, and concluded that the sperm acts by "contact" or "catalytic force," with the transmission to the egg of molecular vibrations of the kind Liebig considered to be the chemical cause of alcoholic fermentation. Of special interest for our story is that in 1874 Miescher's uncle His defended the idea of "transmitted motion" in fertilization: "It is neither the form, nor the form-building material that is transmitted, but the excitation to form-developing growth" (His, 1874, p. 152). Among the consequences of His's theory was the denial that the nucleus has a special role in development, and an emphasis on the spatial orientation of the constituents of protoplasm.

The close relationship between His and his nephew may help to explain Miescher's views about the possible physiological role of the sperm nuclein. In his 1874 paper he noted that many authors

> . . . incline to the idea that the spermatozoa might be carriers of specific substances that act as fertilizing agents by virtue of their chemical properties . . . . To the extent that we wish to assume at all that a single substance, acting as a ferment or in another manner as a chemical sensitizer, is the specific cause of fertilization, one must unquestionably think above all of nuclein . . . . If we consider the sperm to be only a carrier of a specific fertilization substance, how do we explain the variations in action, from species to species, from genus to genus, from individual to individual? . . . . A variety of reasons speak against a decisive role of chemical phenomena . . . . There are no specific fertilization substances. The chemical phenomena have only a secondary significance; they are subordinate to a higher explanation. If we seek an analogy to explain all the available knowledge, it seems to me that there remains nothing but a picture of an apparatus that evokes or transforms some kind of motion (Miescher, 1897, Vol. II, pp. 94–98).

After these words were written, however, a series of remarkable cytological studies established the role of the nucleus in the transmission of hereditary characters, and assigned to nuclein an important role in this process.

These studies were greatly furthered by important technical advances in microscopy, through the use of immersion objectives and the invention of apochromatic lenses (by Abbe), the improvement of the microtome for cutting sections, and the introduction of various synthetic dyes as biological stains. The last was a by-product of a major industrial development. The appearance of the aniline dyes, after William Perkin's discovery of mauve in 1856, had great economic impact in Europe, as it marked the emergence of a powerful German chemical industry with strong ties to university chemists (Beer, 1959). It also influenced greatly the microscopic observation of cell structures, especially after 1877, when Paul Ehrlich reported the first systematic study of the biological staining properties of the new synthetic dyes. By the 1870s it was known that, when certain natural coloring matters (*e.g.*, carmine, hematoxylin) or synthetic dyes (*e.g.*, methyl green) are applied to fixed biological tissues (acetic acid or osmic acid were used as fixatives), the nuclei commonly appear deeply stained whereas the cytoplasm remains relatively pale. Such "nuclear" dyes were recognized to be basic in chemical character, and Ehrlich referred to the cell elements that took them up as "basophilic." In contrast, the cytoplasm was preferentially stained by dyes such as eosin or acid fuchsin, which are acidic in nature, and the cell elements made visible in this manner were termed "oxyphilic."

In 1882 Walther Flemming's influential book *Zellsubstanz, Kern und Zelltheilung* appeared; in it he summarized the considerable cytological knowledge that he and others had provided, and he used the term "chromatin" to denote the structural framework of the nucleus:

> The framework owes its refractile character, the nature of its reactions, and particularly its affinity for dyes to a substance that I have tentatively named chromatin because of the last-named property. Possibly this substance is identical with the nuclein bodies; at any rate, it follows from the work of Zacharias that it is their carrier, and if it is not nuclein itself, it consists of substances from which nuclein can be split off. I retain the word chromatin until a decision about it is made by chemical means, and I use it entirely empirically to denote the substance in the cell nucleus which is stained by nuclear dyes (Flemming, 1882, p. 129).

The studies of Eduard Zacharias (1881) were important, therefore, in providing a link between Miescher's results and those of the new cytology. He applied to various kinds of cells the tests Miescher had used for nuclein, namely, resistance to pepsin, solubility in alkali, and swelling in salt solution; by using pepsin to digest away the cytoplasm, Zacharias also obtained isolated nuclei that were readily stained by suitable dyes.

In his book, Flemming (1882, p. 177) also stated that "All the morphological and chemical knowledge we have of the nucleus . . . leaves us almost completely in the dark as to its biological significance." A clue was to come from cytological studies on the changes in the internal structure of the nuclei of eggs and spermatozoa during the process of fertilization. In 1875 Oscar Hertwig had observed that, in the fertilization of the sea-urchin egg, the nucleus of the spermatozoon unites with the egg nucleus to form a single "cleavage nucleus" which is the precursor of all the nuclei of the embryo [for a historical critique of this view, see Baker (1955)]; this discovery was soon confirmed and extended by Hermann Fol, and was found to apply to other animals and to plants. At about the same time, Flemming described the transformation of the nuclear structure during cell division in a process he later termed "mitosis." Of the many important studies that followed, the one published in 1883 by Edouard van Beneden was especially significant. He reported that, in the case of the worm *Ascaris megalocephala*, there occur during fertilization remarkable changes in the chromatin of the egg and sperm. The inequality of the size of their nuclei disappears and, before or during their union, each of them is transformed into a definite number of rodlike structures which appear to be of the same shape, size, and number in the two sexes. Confirmation of this discovery, and its extension to other animals and to plants, was provided by numerous workers, notably Theodor Boveri and Eduard Strasburger. The rodlike chromatin segments were given a variety of names, but the one assigned by Wilhelm Waldeyer (1888), who called them "chromosomes," won preference.

It was during this period that Wilhelm Roux proposed (in 1883) that the linear structure and the apparent longitudinal division of the chromosomes argued for their role as bearers of the hereditary material. Also, August Weismann developed his theory of the "germ-plasm," and emphasized the importance of the chromosomes of the germ cells in the transmission of hereditary characters:

> In all multicellular organisms sexual reproduction forms the basis of their multiplication, nowhere is it entirely lacking, and in the majority it is the only means of multiplication. Here reproduction is linked to particular cells, germ cells, which may be contrasted with the cells which form the body, and indeed must do so, since they play a totally different role. They are without importance for the life of their bearer (*i.e.* for the maintenance of life), but they alone maintain the species, since each one can under certain circumstances again develop into a complete organism of the same species as the parental one,

equipped with all of its possible individual characteristics to a greater or lesser degree (Weismann, 1883, pp. 5–6).

The studies on the fertilization process, together with cytological findings regarding the mechanism of cell division, profoundly influenced the discussion of the numerous theories of inheritance that had been proposed during the latter half of the nineteenth century. A common feature of many of these theories was the assumption of discrete units of inheritance (see Delage, 1895). We can mention here only the key words that epitomized some of these proposals: Herbert Spencer's "physiological units" of 1864, Charles Darwin's "gemmules" of 1868, Louis Elsberg's (and Ernst Haeckel's) "plastidules," and Francis Galton's "stirps" of 1876 were among the precursors of the "germ-plasm" of Weismann. Special attention was given to the "idioplasm," proposed by Nägeli in 1884; this was thought to be a micellar network providing the physical basis of heredity, and set in a matrix of structureless "trophoplasm" that represents the nutritive components of cells. With the experimental demonstration by Boveri and others that the cell nucleus is the carrier of hereditary characters, and the recognition by 1890 that in the maturation of the egg and sperm cells there occurs a "reduction division" (meiosis) of the chromosomes, most of the theories placed the units of heredity in the nucleus. The proliferation of new terms to denote presumed fundamental units of life continued into the 1890s: the "pangens" proposed by Hugo deVries in 1889 and the "bioblasts" of Altmann in 1890 were followed by Weismann's "biophors" in 1892. According to Weismann, the germ-plasm is a nuclear material with a specific molecular constitution, and the units of heredity are "determinants" composed of biophors needed to form cells of a certain type.

The germ-cell of a species must contain as many determinants as the organism has cells or groups of cells which are independently variable from the germ onwards, and these determinants must have a definite mutual arrangement in the germ-plasm, and must therefore constitute a definitely limited aggregate, or higher vital unit, the "id" (Weismann, 1893, p. 453).

To these terms of the 1890s may be added the "plasomes" of Julius Wiesner and the "idioblasts" of Hertwig; of the various particulate units of inheritance postulated during the period before the rediscovery of Mendel's work in 1900, the idioblasts approximated most closely the "gene" as defined by Wilhelm Johanssen in 1909.

As for the relation of Miescher's nuclein to the mechanism of heredity, by 1895 many cytologists held the view expressed by Edmund B. Wilson:

> The precise equivalence of the chromosomes contributed by the two sexes is a physical correlative of the fact that the two sexes play, on the whole, equal parts in hereditary transmission, and it seems to show that the chromosomal substance, the *chromatin,* is to be regarded as the physical basis of inheritance. Now chromatin is known to be closely similar to, if not identical with a substance known as nuclein ($C_{29}H_{49}N_9P_3O_{22}$, according to Miescher), which analysis shows to be a tolerably definite chemical compound of nucleic acid (a complex organic acid rich in phosphorus) and albumin. And thus we reach the remarkable conclusion that inheritance may, perhaps, be effected by the physical transmission of a particular compound from parent to offspring (Wilson, 1895, p. 4).

In this passage, Wilson was echoing earlier statements by Albert Kölliker (1885, p. 42) and by Oscar Hertwig; the latter had written:

> I believe that I have made it at least highly probable that nuclein is the substance that is responsible not only for fertilization but also for the transmission of hereditary characteristics and thus corresponds to Nägeli's idioplasm . . . nuclein is in an organized state before, during and after fertilization . . . fertilization is not only a physico-chemical process, as was usually assumed by physiologists, but it is also a morphological process . . . (Hertwig, 1885, pp. 290–291).

Furthermore, in 1882 Julius von Sachs had not only expressed the view that nuclein represents the hereditary material, but had also suggested that the nucleins of the egg and the sperm must be different.

Miescher's reactions to the cytological developments that followed the work of Zacharias were quite different. We have already seen that in his 1874 paper on salmon sperm he inclined against assigning to nuclein, or to any single chemical compound, a specific role in the fertilization process, and preferred a kinetic theory reminiscent of Liebig's theory of fermentation. By 1892, when he had resumed his chemical work on nuclein, Miescher had come to the view that it merely forms a structural envelope for a physiologically important iron-containing protein which he termed "karyogen"; in that year he wrote to His that "It is not the nuclein, but the iron-containing phosphorus-free substance that gives the chromatin reactions (methyl violet, safranin, etc.)" (Miescher, 1897, Vol. I, p. 110), and he found support for this conclusion in reports of

Gustav Bunge (1885) and of Archibald B. Macallum (in 1892) on the presence of iron in nucleoprotein preparations. Apparently, Miescher took a dim view of the inferences drawn by cytologists from staining reactions, for he wrote to His (in 1890):

I must again defend my skin against the guild of dyers who insist that there is nothing but chromatin (nuclein), whereas the presence of various spatially separated substances in the sperm head can be demonstrated easily (*ibid.*, pp. 107–108).

As regards the contemporary discussion of the various theories of heredity, Miescher wrote to His (in 1892):

For me, the key to sexuality lies in stereochemistry. The "gemmules" of Darwinian pangenesis are nothing but the numerous asymmetric carbon atoms in organized substances. These carbon atoms change their positions under the influence of minimal causes and external conditions, whereby there gradually arise faults in the organization. Sexuality is a mechanism for the correction of these unavoidable stereometric faults in the architecture of organized substances. A left-handed coil is corrected by a right-handed one, and the equilibrium is restored. With the enormous molecules of the albuminoid substances, or the even more complicated ones in hemoglobin, etc., the many asymmetric carbon atoms permit such a colossal variety of stereoisomers that the entire wealth and multiplicity of hereditary transmissions can find as good an expression as can the words and concepts of all languages in the 24–30 letters of the alphabet. It is therefore entirely superfluous to make of the egg or sperm cell, or the cell in general, a repository of innumerable chemical substances, of which each is supposed to be a carrier of a particular hereditary trait (deVries pangenesis). I must assume from my researches that protoplasm and nucleus do not consist of innumerable chemical substances, but of very few chemical individuals, albeit of very complicated structure (*ibid.*, pp. 116–117).

And, in the following year, Miescher again wrote to His:

The speculations of Weismann and others are afflicted with half-chemical concepts which are partly unclear, and partly correspond to an obsolete state of chemistry. If, as is easily possible, a protein contains 40 asymmetric carbon atoms, then there are 2^{40}, or about a billion, isomers. And this is only the kind of isomerism wherein the isomers of nitrogen and the unsaturated valences are not taken into account. My theory is therefore more satisfactory than any other in providing the infinite multiplicity required by the doctrine of heredity (*ibid.*, p. 122).

It should be emphasized that Miescher did not publish these ideas during his lifetime, and their impact on the discussion of the mechanism of heredity seems to have been negligible after they appeared posthumously in 1897.

During the 1890s opinion was divided among biologists regarding the role of chromatin and its nuclein as the physical basis of heredity. In part, doubts arose from the finding that

> . . . the affinity of the chromosomes for colouring matter varies markedly at different periods, and this indicates that slight changes, which are beyond our control, take place in the constitution of this substance and are sufficient to cause its most striking reaction with regard to colouring matters to disappear for a time (Weismann, 1893, p. 50).

An explanation was offered by Martin Heidenhain, who concluded that the nucleic acid (which he termed "basichromatin") and the nuclear material with an affinity for acidic dyes ("oxychromatin") "are not to be regarded as permanent unchangeable substances, but may change their affinity to dyes by the uptake or release of phosphorus" (Heidenhain, 1894, p. 548). In drawing this conclusion from his cytological studies, he relied on chemical experiments such as those of Hans Malfatti (1891) who reported that the basophilic affinity of nuclein preparations corresponded to their phosphorus content. Malfatti included in his studies not only yeast nucleic acid but also products of the phosphorylation of egg albumin with metaphosphoric acid; as we noted earlier, the latter were considered to resemble the paranucleins. This ambiguity regarding the chemical nature of chromatin continued into the twentieth century, as is indicated by the following statement made by a leading British biologist:

> The outstanding feature of the nucleus is the constant presence in abundance of nuclein and nucleoproteins. Nuclein, which is probably identical with chromatin, is a complex albuminoid substance rich in phosphorus (Minchin, 1916, pp. 18–19).

Furthermore, it was found that combinations of a nucleic acid preparation with egg albumin ("nucleoalbumin") stained less intensely with basic dyes than did the nucleic acid itself. These findings were interpreted by Wilson as follows:

> We may infer that the original chromosomes contain a high percentage of nucleic acid; that their growth and loss of staining power is due to a combination with a large amount of albuminous substance to form a lower member of the nuclein series, perhaps even a nucleoalbumin; that their final diminution in size and resumption of staining power is caused by a giving up of the albumin constituent, restoring the nuclein to its original state as a preparation for division (Wilson, 1896, p. 246).

This explanation, which later turned out to be partly correct, was considered inadequate, however, when cases were reported (*e.g.*, the "lampbrush" chromosomes) where the basichromatin apparently had disappeared completely. Such observations led leading cytologists, notably Eduard Strasburger, to conclude that the chromatin itself cannot be the hereditary substance. Earlier, he had strenuously resisted suggestions (such as that made by Hans Winkler in 1901) that nuclein provided the chemical basis of heredity:

> . . . I must speak out against every conception, even partial ones, of true fertilization as a purely chemical process, therefore against every chemical theory of heredity . . . . For me, the essence of fertilization lies in the union of organized elements (Strasburger, 1901, p. 359).

The newer cytological evidence led Strasburger to state that

> The chromatin cannot itself be the hereditary substance, as it . . . leaves the chromosomes, and the amount of it is subject to considerable variation in the nucleus, according to its stage of development (Strasburger, 1909, p. 108).

By 1910, therefore, the conclusions expressed by Hertwig and Wilson in the previous century had lost favor. Indeed, in the third edition of his great treatise on the cell, published in 1925, Wilson radically revised his view of the role of nuclein in heredity; after summarizing the cytological evidence he concluded:

These facts offer conclusive proof that *the individuality and genetic continuity of chromosomes does not depend upon a persistence of "chromatin" in the older sense (i.e., basichromatin)* (Wilson, 1925, p. 351).

and he further inferred that the loss in affinity for dyes ". . . seems to indicate a progressive accumulation of protein components and a giving up, or even a complete loss, of nuclein" (*ibid.,* p. 652).

The uncertainty about the role of the nucleins was widespread during the first decade of this century; Jacques Loeb wrote:

> In order to decide whether the nucleins or the histones or the protamines are of importance for the hereditary qualities, it would be necessary to decide whether the nuclei of the eggs of one form contain always the same base as that found in the sperm of the same species. This should be expected from the fact that the hereditary influence of egg and sperm is equal in the adult offspring, at least. It seems that the base is not always identical in the egg and spermatozoön of the same species, and this seems to indicate that the nucleic acid is of more importance for heredity than protamines or histones. Aside from the nuclein we find albumin and globulin, especially in the tail, and in the latter also lecithin, cholesterin, and fat. Miescher believed that in the head of the spermatozoön an iron compound exists. It is impossible to draw any far-reaching inference concerning the nature of the substances which transmit hereditary qualities from these meager data (Loeb, 1906, p. 180).

The unsatisfactory state of the problem was also emphasized by Richard Burian (1906), who argued against a genetic role of both the nucleic acid and protein components of the sperm nucleoproteins.

The decline of the nuclein theory of heredity during 1910–1930 was also indicated by the fact that the three most significant monographs on the chemistry of nucleic acids written at that time (Jones, 1920a; Feulgen, 1923; Levene and Bass, 1931) say little about their presumed biological role. For example, Levene and Bass noted that, by virtue of their acidic character, the nucleic acids ". . . may serve as regulators of the physicochemical properties of the cell or at least of the nucleus of the cell. This role of the nucleic acids was especially emphasized by Einar Hammarsten" (Levene and Bass, 1931, p. 280). Hammarsten (1924) had performed a thorough study of the physicochemical properties of carefully prepared calf thymus nucleic acid, and the inferences drawn from his results were apparently influenced by the considerable interest, during the 1920s, in the physiological regulation of acid-base equilibria and of osmotic pressure. In particular, much importance was attached to the "Donnan equilibrium," which describes the effect of a colloid on the distribution of diffusible ions on either side of a membrane. As for the role of nucleic acids in heredity, the physiological chemist Albert P. Mathews (who had worked with Kossel) stated:

. . . [W]here there should be the most complex chromatin in the body we find the simplest; namely a salt of protamin and nucleic acid for which a definite formula may be given . . . . Now, it is very improbable that were the chromosomes constituted of widely different genes they would show so simple and definite a composition. The nucleic acid of widely different cells appears to be the same . . . . We are, therefore, forced to seek the differences in the protein moiety of the chromatin if such differences exist. Now it is remarkable that nothing of the kind which one would expect on this theory is found in those sperms which we have examined. Always we find but one sort of protamin, containing only three or four different amino acids. While a considerable difference of kind of arrangement is possible even in this one sort of protamin, there is no chemical indication that it exists. The theory of special genes and chromosomal inheritance by unit characters is not supported by such chemical evidence as we have so far obtained (Mathews, 1924, p. 89).

In a letter written shortly before his death, Miescher stated that the questions relating to the physical basis of heredity will be "fought out between the morphologists and the biochemists during the twentieth century" (Miescher, 1897, Vol. I, p. 128). The great biologist Boveri (Baltzer, 1967), in noting this remark, added that the biochemist Miescher's

. . . entire life-work expressed sufficiently clearly the conviction that the victory will go to his field. The morphologist also, in striving for knowledge, will possess sufficient self-denial to wish his opponent the final victory; he also could think of nothing better than the possibility that morphological analysis would lead to a point where the final elements are chemical compounds. At present, however, the goal seems farther than ever; it is even questionable whether such a goal exists at all in the sense that the final essential elements of living matter are chemical substances. Be that as it may, the insight into the constitution of the nucleus gained through descriptive and experimental morphology has revealed a discrimination of such delicacy that we can scarcely imagine a means to understand these differences in chemical terms (Boveri, 1904, p. 123).

The Structure of Nucleic Acids

We have seen that Miescher intended his studies to be a contribution to the chemistry of the cell nucleus, and that during 1880–1900 several leading cytologists assigned to nuclein an important role in fertilization and inheritance. Although this view lost favor after 1900, the chemical study of nucleic acids continued to be of interest to physiological chemists, especially those with ties to medicine. The finding that the purines adenine and guanine are constituents of nucleic acids indicated a relation to the clinical problem of gout, long known to be associated with the deposition of solid uric acid in the joints and in the kidney (Coley, 1971). This relation became clearer around 1905, when animal tissues were shown to contain enzymes for the breakdown of nucleic acids to yield purines, for the conversion of adenine to hypoxanthine and of guanine to xanthine, and for the oxidation of hypoxanthine and xanthine to uric acid. Furthermore, such phenomena as the massive metabolic conversion of skeletal protein to sperm nuclein, demonstrated by Miescher for the Rhine salmon, indicated that the animal organism can make nucleic acids from other body constituents. The problems posed by these observations made the question of the chemical structure of nucleic acids a matter of physiological interest, even though the role of these substances in the organism was obscure. It would appear, however, that the greater biological importance assigned to the proteins, as well as the more extensive knowledge of their chemistry, made them more popular objects of chemical study than the nucleic acids. Indeed, as we shall see in what follows, the ideas about the structure of nucleic acids during 1910–1950 were greatly influenced by the parallel developments in protein chemistry.

The first substantial step toward the elucidation of the structure of nucleic acids was taken by Phoebus A. Levene and Walter A. Jacobs (1908, 1909) in their study of a substance isolated by Liebig (1847b) from aqueous extracts of beef muscle. In part, Liebig's work on the constituents of muscle extracts was intended to clarify the uncertain status of creatine, isolated in 1832 by Chevreul; among the by-products Liebig obtained was an acidic material he named "inosinic acid" [Greek *is* (genitive *inos*), muscle, sinew]. During the succeeding decades, there was some doubt about the existence of inosinic acid, as difficulty was encountered in its isolation, and it was not until the work of Franz Haiser in 1895 that its authenticity was generally accepted. Haiser made two important discoveries about this substance: it yields hypoxanthine on acid hydrolysis, thus indicating a relation to the nucleic acids, and it also contains phosphorus (in the form of phosphate). In addition, he concluded that the compound gives a third component on hydrolysis; by 1907 Friedrich Bauer and Carl Neuberg had shown this to be a 5-carbon sugar (a pentose). This was followed by Haiser's demonstration that another material ("carnine") obtained from meat extract was a mixture of hypoxanthine and a phosphorus-free compound he termed "inosine," composed of hypoxanthine and a pentose. Neuberg had identified the pentose of the nucleic acids from yeast and pancreas as *l*-xylose; he believed this to be the sugar of inosinic acid as well, but Bauer stated it to be *dl*-arabinose, and Haiser proposed still another isomeric pentose, *d*-lyxose. Levene and Jacobs questioned the validity of all these proposals, designated the sugar of inosinic acid as "carnose," and proceeded to isolate it from carnine in crystalline form. With the work of Emil Fischer on the stereochemistry of the sugars as a background, it was possible for them to show carnose to be a pentose that had not yet been isolated from natural sources or prepared synthetically, and that it was the optical antipode of the known *l*-ribose, namely, *d*-ribose (it is now denoted D-ribose). Furthermore, Levene and Jacobs established unequivocally the arrangement of the three components of inosinic acid as hypoxanthine-ribose-phosphate; the linkage of the purine to the sugar was glycosidic in nature, since inosinic acid did not reduce Fehling's solution until after the purine had been removed by mild acid hydrolysis, and the resulting sugar phosphate was shown to be a phosphate ester involving the hydroxyl group at carbon 5 of the ribose.

This elucidation of the structure of inosinic acid, though not in all detail, led Levene and Jacobs to apply their findings to a substance that Olof Hammarsten (1894) had isolated from a pancreatic nucleoprotein ("β-nucleoprotein") upon treatment with warm alkali; this product was named "guanylic acid" because it contained phosphorus and gave guanine, but not adenine, upon hydrolysis. After some confusion whether guanylic acid contained glycerophosphate, a possibility attractive to those who still sought a relation of the nucleic acids to lecithin, Hermann Steudel in 1907 showed guanylic acid to yield only guanine, pentose, and phosphate on hydrolysis; shortly thereafter Levene and Jacobs demonstrated that the sugar is D-ribose, and that the arrangement of the components in guanylic acid is similar to that in inosinic acid, namely, guanine-ribose-phosphate. Furthermore, upon mild hydrolysis, the phosphate of guanylic acid was lost and the product ("guanosine") was found to be comparable to inosine. Levine and Jacobs named compounds such as guanylic acid and inosinic acid "nucleotides," and compounds such as guanosine and inosine, "nucleosides."

By 1912 Levene and Jacobs had applied the method used to prepare guanosine from guanylic acid (hydrolysis with aqueous ammonia at 135°) to the partial hydrolysis of yeast nucleic acid, and had isolated four nucleosides corresponding to the two purines and two pyrimidines known to be formed upon complete acid hydrolysis; in addition to guanosine, they obtained adenosine, cytidine, and uridine, in each of which D-ribose is linked to adenine, cytosine, or uracil by a glycosidic bond. These findings led them to conclude that the intact nucleic acid is composed of the corresponding four nucleotides (guanylic acid, adenylic acid, cytidylic acid, and uridylic acid) in equivalent proportions to form a tetranucleotide. This conclusion was derived in part from the supposed identity of yeast nucleic acid with the one prepared from wheat germ ("triticonucleic acid") by Thomas B. Osborne and Isaac F. Harris in 1902, and their report that it gave equivalent amounts of adenine and guanine upon hydrolysis; a similar finding was reported for thymus nucleic acid by Steudel in 1906. Furthermore, the yields of purines and pyrimidines obtained upon hydrolysis of the two kinds of nucleic acids were taken by Levene (1909) to indicate equivalent proportions of all four bases. It must be emphasized, however, that the tetranucleotide hypothesis, as first formulated, was not based on the quantitative estimation of the products of the hydrolysis. In 1912 this problem had not been solved for the nucleic acids; as in the case of the proteins, the solution came after 1945, through the use of chromatography as an analytical tool. It is difficult to escape the impression, however, that, whereas the protein chemists sought, during the period 1910–1940, to develop analytical methods for the quantitative estimation of the amino acids in protein hydrolysates, similar concerted efforts were not under way in the nucleic acid field, not only because of the limitations of the available methods, but also because of the widespread acceptance of the formulation of the structure of nucleic acids as tetranucleotides. Thus it did not matter if the actual analytical data corresponded to molar proportions such as 1:0.6:1.2:0.8, since the deviation from a 1:1:1:1 ratio could be attributed to the limitations of the precipitation procedures then employed. A further assumption implicit in the work of Levene and his contemporaries was that each of the nucleic acid preparations they handled represented a single chemical entity; the multiplicity and individuality of the proteins made evident by 1910 apparently was not considered to apply to the nucleic acids, presumably because of the greater simplicity of the tetranucleotide structure as compared to that of the polypeptide chains of proteins.

For this reason chemical studies dealt largely with nucleic acid preparations from two biological sources: calf thymus and yeast. By 1912 several differences were evident between these two nucleic acids: in addition to the occurrence of thymine in place of uracil among the products of acid hydrolysis, and the presence of a peculiar sugar different from that in yeast nucleic acid, thymus nucleic acid was completely resistant to hydrolysis by aqueous ammonia under conditions that cleaved yeast nucleic acid to nucleosides and, under milder conditions, to nucleotides. This resistance of thymus nucleic acid to alkaline hydrolysis was circumvented by the use of enzyme preparations; such "nucleases" had been studied by Araki, Ivanov, and Sachs during 1903–1905. In 1911 Levene compared the enzymic cleavage of thymus and yeast nucleic acids, concluded that they possess a similar general structure, and proposed (Levene and Jacobs, 1912) a tetranucleotide structure for thymus nucleic acid as well. In 1920 Walter Jones summarized the view held by most of the nucleic acid chemists in stating that ". . . there are but two nucleic acids in nature, one obtainable from the nuclei of animal cells, and the other from the nuclei of plant cells" (Jones, 1920a, p. 9), and that the designation of the animal tissue or plant from which either is derived "is as superfluous as would be the application of a similar nomenclature to lecithin" (*ibid.*, p. 11). Thus the term "thymus nucleic acid" was frequently applied to any nucleic acid preparation from an animal source (*e.g.*, fish sperm) and "yeast nucleic acid" often denoted a material from an organism other than yeast.

Although the sugar present in thymus nucleic acid was considered by most investigators to be a hexose related to glucose, numerous efforts to establish its nature were unsuccessful. Robert Feulgen (1914) reported that a neutralized acid hydrolysate of thymus nucleic acid restores the color of a fuchsin solution previously decolorized by sulfur dioxide; this reaction, introduced by Hugo Schiff in 1866, was given by ordinary aldehydes but not by glucose. Feulgen suggested that the carbohydrate of thymus nucleic acid is "glucal" (a glucose derivative described by Fischer in 1914), but this view had to be abandoned when Max Bergmann showed in 1920 that the original preparation of glucal was contaminated with aldehyde material, and that the pure substance does not give the Schiff reaction. The situation was not clarified until 1929–1930, when Levene succeeded in isolating the nucleosides from an enzymic digest of thymus nucleic acid, and found that the sugar component (which he named "thyminose") belongs to a group of pentoses lacking an oxygen atom, and hence named desoxypentoses (later, deoxypentoses). Such sugars had been studied by Heinrich Kiliani around 1910, and Bergmann had shown during the 1920s that 2-deoxy sugars are readily formed from members of the glucal family. This knowledge served as the background for Levene's identification of the sugar of thymus nucleic acid as 2-deoxyribose, which gives the aldehyde reaction.

$$HO\overset{5}{C}H_2 \quad O \quad H \qquad\qquad HOCH_2 \quad O \quad H$$

D-Ribose

2-Deoxy-D-ribose

In addition to the aldehyde reaction (which he named the "nucleal" reaction), Feulgen also used a test in which a pine shaving impregnated with a solution of hydrolyzed nucleic acid was exposed to moist hydrogen chloride vapor. With thymus nucleic acid, a green color appeared; in the presence of ammonia, the pine splinter turned red. These reactions were not given by the nucleic acids from yeast or wheat germ, known to contain ribose units, and provided the basis for a distinction between ribonucleic acids and desoxyribonucleic acids (Levene and Bass, 1931); the latter were subsequently renamed deoxyribonucleic acids, and both terms were abbreviated to the currently used RNA and DNA. In what follows, we shall employ these abbreviations, but it should be noted that they did not come into general usage until about 1955. We saw before that in 1920 ribonucleic acids were considered to be restricted to plant tissues, and those resembling thymus nucleic acid were termed "animal nucleic acids"; during the succeeding decade this separation according to biological sources was found to be incorrect. The introduction of the aldehyde reaction as a cytochemical tool (Feulgen and Rossenbeck, 1924) showed that both plant and animal cells contain DNA. Furthermore, it was realized that a nucleic acid obtained from beef pancreas is a ribonucleic acid, indicating that animal cells, as well as plant cells, contain RNA; some investigators (*e.g.,* Feulgen) had believed the pancreatic nucleic acid to be a compound of thymus nucleic acid and guanylic acid, but this was disproved by Erik Jorpes in 1928.

After 1910, with the widespread, but by no means universal, acceptance of the tetranucleotide structure of nucleic acids, there was much speculation concerning the mode of linkage of the nucleotides to each other. A variety of structures were proposed: in some, sugar units were joined to each other to form ether bridges, while, in others, the nucleotides were joined solely through the phosphoryl groups; various combinations of these two types of linkage were also suggested. Most of these hypotheses had to be discarded, however, when the presumed partial degradation products (dinucleotides) on which they were based proved to be mixtures of mononucleotides. By 1930 the structure advanced by Levene (1921) was generally accepted; in this formulation, the individual nucleotide units were joined to each other by phospho diester bridges:

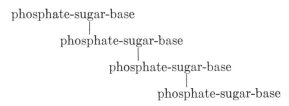

More definitive evidence regarding the nature of the internucleotide linkages came during the 1930s, in large part as a consequence of the isolation, by Siegfried Thannhauser and Willibald Klein, of the mononucleotides derived from thymus nucleic acid by enzymic digestion. By 1935 deoxyribose had been identified as the sugar component, and the techniques of carbohydrate chemistry had shown that it is present in nucleotides in the form of a five-membered "furanose" ring, leaving only the hydroxyl groups at carbons 3 and 5 for joining the nucleotide units to each other. On the basis of the tetranucleotide formula, Levene concluded that ". . . in *desoxy-ribose nucleic acid* the positions of the phosphoric acid radicles are carbon atoms (3) and (5) of the desoxyribose" (Levene and Tipson, 1935, p. 625). The problem of the mode of linkage of the mononucleotide units in yeast nucleic acid proved to be more difficult because the furanose form of the ribose unit has three free hydroxyl groups (at carbons 2, 3, and 5), and it was not readily evident which two were involved. During the 1930s preference was given to a phospho diester bridge linking carbons 2 and 3 (Gulland, 1938), and the alkali-lability of ribonucleic acids was attributed to this mode of linkage, but later observations were incompatible with this idea; the solution of the problem came after 1950 from the study of the degradation of yeast nucleic acid with the enzyme ribonuclease.

The presence, in extracts of swine pancreas, of a heat-stable enzyme that degraded yeast nucleic acid to mononucleotides was reported by Walter Jones in 1920; he observed no action on thymus nucleic acid, and noted: "This is curious. An active agent is present in animal pancreas which is specifically adapted to plant nucleic acid. It suggests evolutionary matters" (Jones, 1920b, p. 204). Although Levene stated that he was unable to confirm Jones's report, the existence of the enzyme was accepted after the work of René Dubos and Robert Thompson (in 1938), who named it ribonuclease. This designation was questioned by Levene and Schmidt; they preferred "ribonucleodepolymerase" because, in their view, the enzyme caused a depolymerization of yeast nucleic acid, without the formation of mononucleotides. It is difficult to escape the impression that this idea was influenced by the contemporary discussion about the mode of action of pepsin on proteins, to which we referred on p. 155. As in the case of pepsin, the answer came from two advances: the crystallization of ribonuclease (from beef pancreas, by Moses Kunitz in 1939) and the synthesis of well-defined substrates for this enzyme by Alexander Todd and his associates after 1950.

Before discussing the post-World War II work on the action of ribonuclease, it is necessary to summarize some of the findings during the 1930s about the nucleotides obtained from yeast nucleic acid. We noted that cleavage with alkali had given nucleotides considered to be similar in structure to the inosinic acid obtained from muscle. The validity of this conclusion came into question when Gustav Embden and Gerhard Schmidt reported (in 1929) that an adenylic acid isolated two years earlier in Embden's laboratory from rabbit muscle differed chemically from the adenylic acid obtained from yeast. This advance was made possible by Schmidt's discovery of an enzyme that converts muscle adenylic acid into inosinic acid (by deamination), but does not act on yeast adenylic acid. Since inosinic acid was already known to be a nucleoside-5'-phosphate, it was clear that the phosphate of muscle adenylic acid is linked to the hydroxyl group at carbon 5 of the ribose unit. By 1933 Levene had reported that in "yeast" adenylic acid the phosphate is linked to the hydroxyl at carbon 3 of the ribose unit, and the three other nucleotides obtained upon alkaline cleavage of yeast nucleic acid were also considered to be nucleoside-3'-phosphates (see the accompanying formulas). Embden's isolation of muscle adenylic acid was soon followed by the discovery of the closely related adenosine-5'-triphosphate (ATP), whose central importance in metabolism became evident during the 1930s. Furthermore, it was also discovered that the biologically active forms of several vitamins are nucleotide-like coenzymes. We shall return to these important findings in the next chapter, but it should be noted here that they provided a stimulus to the study of the detailed chemical structure of nucleotides. Before 1940 several investigators (notably, Levene, Gulland, and Bredereck) made significant contributions, but it was the work of Todd and his associates during 1940–1955 that yielded the most decisive results.

Through unequivocal chemical synthesis of nucleosides, Todd estab-

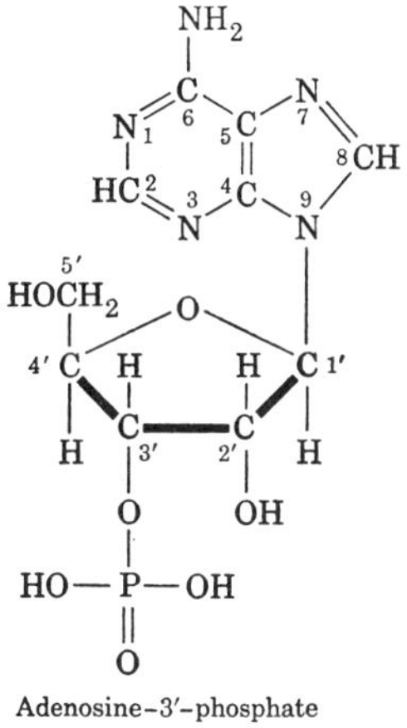

Adenosine-3′-phosphate

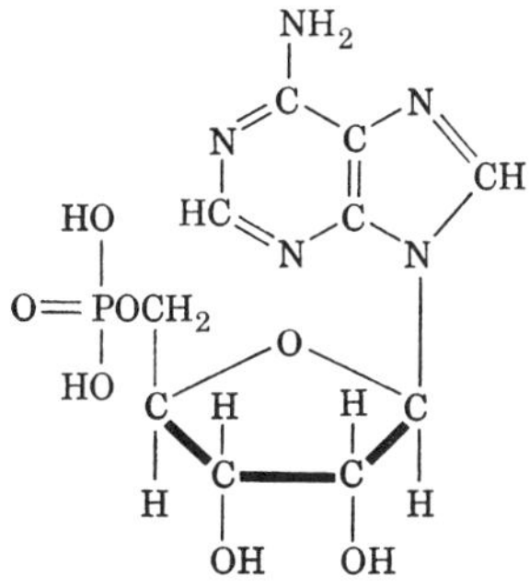

Adenosine-5′-phosphate

lished the validity of earlier conclusions that the sugar is linked to the nitrogenous base by a β-glycosidic bond, and that the site of attachment is the nitrogen in position 9 of the purine ring (not 7, as thought before Gulland's work) or in position 3 of the pyrimidine unit. This was followed by the synthesis of nucleotides in which a phosphate was attached to the hydroxyl group at position 2, 3, or 5 of the sugar unit of various nucleosides. By demonstrating the identity of the synthetic products with those obtained upon degradation of nucleic acids, Todd provided a sounder basis for the subsequent consideration of the problem of the mode of linkage between the nucleotide units in nucleic acids. In particular, the synthetic methods developed in his laboratory permitted the synthesis of artificial substrates for ribonuclease; together with the parallel work of Roy Markham, Todd's results clearly demonstrated that in ribonucleic acids (*e.g.*, yeast nucleic acid) the individual mononucleotides are joined to each other by phospho diester bonds linking the hydroxyl at carbon 3 of one ribose unit to the hydroxyl at carbon 5 of another ribose unit. By 1952 it was clear that the mode of linkage proposed by Levene and Tipson for thymus nucleic acid also applied to yeast nucleic acid, and the lability of the latter to alkali was fully explained in terms of the participation of the hydroxyl group at carbon 2 of the ribose unit.

The Demise of the Tetranucleotide Hypothesis. In 1931 Levene had written that

   . . . it must be borne in mind that the true molecular weight of nucleic acids is as yet not known. The tetranucleotide theory is the minimum molecular weight and the nucleic acid may as well be a multiple of it (Levene and Bass, 1931, p. 289).

During the 1930s, however, chemical studies on nucleic acids were based on the assumption that they are simple linear tetranucleotides, and not macromolecules. For example, when Makino reported in 1935 that, per atom of phosphorus, yeast nucleic acid behaved like a monobasic acid, he supported the idea advanced a few years earlier by Takahashi that it is not a linear, but a cyclic, tetranucleotide; Makino's result was also consistent with a long-chain polynucleotide structure. Soon afterward it was recognized that nucleic acids, if carefully prepared, are materials of very high molecular weight. This conclusion came from physicochemical studies, to which we shall return shortly. The discussion then revolved about the question whether the presumed polynucleotide chain represents a random sequence of nucleotide units or a succession of tetranucleotide units. In summarizing the situation in 1946, after a careful review of the conflicting chemical data for various nucleic acid preparations, John Masson Gulland stated:

> There is at present no indisputable chemical evidence that the nucleotides are arranged in other than a random manner in the polynucleotides. It must be realized that the existence of tetranucleotide units, repeated throughout the molecule, would limit the potential number of isomers and hence diminish the possibilities of biological specificity. Thus, the immense number of variations presented by a polynucleotide in which the nucleotides occur in random sequence is reduced, in the case of an unbranched polynucleotide, to a single possible structure, varying only in its length, if the same, uniform tetranucleotide unit occurs throughout (Gulland, 1947a, p. 12).

The decisive evidence on this question came during 1946–1950 from the first quantitative analyses of the composition of nucleic acids, performed by Erwin Chargaff and his associates.

We mentioned on p. 169 that the introduction, in 1944, of partition chromatography on paper strips had a profound influence on the development of biochemistry after World War II. Although initially applied to the study of the amino acid composition of protein hydrolysates, the method was immediately adopted for the separation and identification of a large variety of biochemical substances, including the purines, pyrimidines, nucleosides, and nucleotides derived from nucleic acids. For their quantitative analyses of the individual components separated by paper chromatography, Chargaff and his associates were greatly aided by the general availability, after 1945, of the photoelectric ultraviolet spectrophotometer. It had been known since the work of Jacques Soret (*ca.* 1880), Walter N. Hartley (in 1905), and Charles Dhéré (in 1906) that purines and pyrimidines strongly absorb ultraviolet light near 2600 Å, and Dhéré had described the absorption spectrum of yeast nucleic acid (see Holiday, 1930). Before World War II, however, analytical spectrophotometers with quartz optics were highly specialized instruments and not generally available in biochemical laboratories; thus treatises on nucleic acid chemistry, such as that of Levene and Bass (1931), did not mention ultraviolet spectrophotometry as an analytical tool. During the 1930s this technique was applied by Torbjörn Caspersson to the cytochemical study of nucleic acids, and his findings contributed greatly to a resurgence of interest in their biological role, but the general use of the method came only after reliable photoelectric ultraviolet spectrophotometers became available as standard items of laboratory equipment. The general method used by Caspersson had been introduced by August Köhler, who showed in 1904 that cell nuclei absorb ultraviolet light strongly; this was noted by Bayliss (1924, p. 9) who added that "Unfortunately, the method has not as yet been made much use of, owing to the necessarily elaborate nature of the apparatus required."

By 1950 Chargaff was able to summarize the extensive analytical data he had collected for a number of carefully prepared DNA samples obtained from various biological sources (*e.g.*, beef thymus, spleen, and liver; human sperm; yeast; turbercle bacilli). Before presenting the results, he stated:

> We started in our work from the assumption that nucleic acids were complicated and intricate high-polymers, comparable in this respect to the proteins, and that the determination of their structures and their structural differences would require the development of methods suitable for the precise analysis of all constituents of nucleic acids prepared from a large number of different cell types (Chargaff, 1950, pp. 202–203).

The analytical data showed unequivocally that the adenine, guanine, cytosine, and thymine were not present in equimolar proportions, as demanded by the tetranucleotide hypothesis. Moreover, the composition of the hydrolysates of DNA preparations from different types of organisms differed widely, and it was necessary to abandon the assumption that the deoxyribose nucleic acid of calf thymus is identical with the DNA obtained from other biological sources. From his results, Chargaff concluded:

The deoxypentose nucleic acids extracted from different species thus appear to be different substances or mixtures of closely related substances of a composition constant for different organs of the same species and characteristic of the species.

The results serve to disprove the tetranucleotide hypothesis. It is, however, noteworthy—whether this is more than accidental, cannot yet be said—that in all the deoxypentose nucleic acids examined thus far the molar ratios of total purines to total pyrimidines, and also of adenine to thymine and of guanine to cytosine, were not far from 1 (*ibid.*, p. 206).

Although Chargaff's analyses of hydrolysates of various preparations of RNA (*e.g.*, pancreas, yeast) did not show the same regularities of composition, it was evident that the ribose nucleic acids from different biological sources contain different proportions of the four nitrogenous bases, and that yeast nucleic acid could not be considered to represent the RNA present in all biological forms.

Subsequent work strengthened and extended these conclusions, but it was already clear in 1950 that the idea of the multiplicity and individuality of the nucleic acids, long accepted for the proteins, had received its first reliable experimental basis from Chargaff's analytical data. Taken together with the parallel recognition that the nucleic acids represent long-chain molecules of high molecular weight, the problem of the elucidation of their chemical structure was seen to be fundamentally similar to that for the proteins, with the nucleotide units of the nucleic acids as the counterparts of the amino acid units in long polypeptide chains. In addition to Chargaff's achievement, the development of new chromatographic techniques (especially ion-exchange chromatography) led to other significant advances in the study of nucleic acids. For example, in addition to the four major nitrogenous bases, some DNA preparations (notably that from wheat germ) were found to contain 5-methylcytosine; the natural occurrence of this pyrimidine had been postulated by Treat B. Johnson (in 1925) on the ground that a 5-methyl derivative of cytosine should be a counterpart of the known 5-methyl derivative of uracil (thymine). Subsequently, other purines and pyrimidines were found by chromatography to be minor components of some ribonucleic acids. The remarkable selectivity of ion-exchange chromatography was made evident by Waldo Cohn's demonstration, in 1949, that the "yeast adenylic acid" produced upon treatment of yeast nucleic acid with alkali is a mixture of two nucleotides, subsequently shown by Todd to be adenosine-2'-phosphate and adenosine-3'-phosphate; similar pairs of the three other

DNA   RNA

nucleotides from yeast RNA also were identified. This development, together with Cohn's isolation of nucleoside-5′-phosphates from enzymic digests of ribonucleic acids, played an important role in the elucidation of the mode of the linkage of the nucleotides in intact RNA.

We turn now to the findings that led to the recognition that nucleic acids are macromolecules. Miescher had found that his preparation of salmon nuclein did not diffuse through a membrane, but, when nucleic acids were considered to be simple tetranucleotides, this colloidal behavior was attributed to the association of the small units to form non-diffusible aggregates. Such views were reflections of the debate during the 1920s regarding the macromolecular nature of proteins; by 1950 it was clear that, like the proteins, both deoxyribonucleic acids and ribonucleic acids are long-chain molecules of high molecular weight.

The macromolecular nature of calf thymus DNA first became evident during the 1930s through studies of the physical properties (viscosity, rate of diffusion, rate of sedimentation) of solutions of the carefully prepared material described by Einar Hammarsten (1924). Thus, in 1938, Signer, Caspersson, and Hammarsten reported that this DNA behaved like thin rods, whose length was about 300 times its width, and whose apparent molecular weight was between 500,000 and 1,000,000. Subsequent work by Gulland (in 1947) gave a preparation found to have a molecular weight of about 3,000,000 and a ratio of length to width of about 700. These enormous fibers were considered to represent aggregates of polynucleotide chains, but no estimates could be made of the number or lengths of the individual chains believed to compose the highly asymmetric DNA particles. Because of this asymmetry, the determination of the molecular weight of the DNA "molecule" by measurement of the physical properties of DNA solutions presented exceptional difficulties; thus, around 1949, one report stated that calf thymus DNA is composed of units of about 35,000 (Jungner *et al.*), and another that DNA is a true macromolecule of molecular weight about 820,000 (Cecil and Ogston). Indeed, the extent to which the term "molecule" can be used in speaking of such long polynucleotide chains has continued to be a subject of discussion (Chargaff, 1968, pp. 320–323). As Miescher had recognized qualitatively during the course of his work on salmon nuclein, the physical properties of DNA preparations depended greatly on the methods used to isolate them; a succession of studies during the 1940s demonstrated that extremes of acidity or alkalinity, or the presence of nucleases, could lead to degraded nucleic acid preparations. In particular, Gulland (1947b) concluded that, in undegraded calf thymus DNA, polynucleotide chains are held together by hydrogen bonds whose rupture by acid or alkali causes a decrease in molecular size, and he likened this change in structure to the process of protein denaturation.

Although it was long known that, in contrast to calf thymus DNA, yeast RNA is cleaved by alkaline hydrolysis, most of the older methods used for the preparation of yeast RNA involved some treatment with alkali. By 1945 it was evident that such preparations represented partially degraded polynucleotides, and that, in addition, the presence of ribonucleases in the tissue extracts could cause extensive cleavage. Some indication of the uncertainty introduced by these factors is provided by the fact that during 1935–1945 the estimates of the molecular weight of yeast RNA ranged from about 1500 (close to the theoretical value for a tetranucleotide) to about 23,000 (corresponding to about 70 nucleotide units). During this period, however, there came a dramatic development, one by-product of which was the clear demonstration of the macromolecular nature of the ribonucleic acids.

In 1935 Wendell Stanley reported that he had isolated, from the juice of Turkish tobacco plants infected with tobacco-mosaic virus, a crystalline protein which caused the appearance of the disease in healthy plants. It had been known since the 1890s, through the work of Ivanovski and Beijerinck, that the characteristic leaf mottling associated with this disease is caused by a factor that can pass through filters designed to retain bacteria. During the succeeding decades, many diseases of plants, animals, and man were found to be caused by such filterable factors, and a variety of names were assigned to this class of infective agents: "contagious living fluid" (Beijerinck), "filterable viruses" (Wolbach), "ultraviruses" (Levaditi), among others. By the 1930s, however, these terms had been replaced by "viruses," defined as submicroscopic, filterable pathogenic agents which could multiply only within living cells. It was also recognized that the "bacteriophage" discovered by Frederick Twort (in 1915) and Félix d'Herelle (in 1917) represents a class of viruses that infect bacterial cells.

In his isolation of the apparently crystalline tobacco-mosaic virus, Stanley used procedures (*e.g.*, precipitation with ammonium sulfate) that had enabled Northrop and Kunitz (also working at The Rockefeller Institute in Princeton, N. J.) to crystallize several enzymes. The virus protein had the solubility properties of a globulin, and its apparent molecular weight as determined in the ultracentrifuge was found to be very high (about 40 million). In a paper describing the details of his work, Stanley wrote:

> The evidence presented in this paper indicates that the virus of tobacco mosaic . . . is not alive. It is exceedingly difficult to conceive of a living agent that may be not only crystallized, but recrystallized repeatedly, with retention of constant physical, chemical, and biological properties. Such characteristics are regarded as those of pure chemical compounds. Hence, it seems reasonable to assume that the crystalline protein is a chemical compound and, therefore, inanimate (Stanley, 1936, p. 317).

Stanley's report on the isolation of the virus protein was quickly followed by that of Bawden and Pirie (see Pirie, 1970), who made the additional important discovery that their preparation contained

. . . 0.5 per cent phosphorus and 2.5 per cent carbohydrate. The last two constituents can be isolated as nucleic acid of the ribose type from protein denatured by heating (Bawden *et al.*, 1936, p. 1051).

It soon became clear that the infective protein is a nucleoprotein, and subsequent work led to the isolation of several other plant viruses as crystalline nucleoproteins having a nucleic acid of the RNA type. The question whether the nucleic acid was essential for infectivity was in doubt for many years. Thus Stanley reported that "It was possible to re-move the nucleic acid and to obtain phosphorus-free protein possessing virus activity" (Stanley, 1937, p. 329). Later improvements in the methods for the separation of the RNA and protein components of the plant viruses led to a revision of the view suggested by this finding, and by 1957 it had been clearly established that the infectivity of the nucleoprotein is asso-ciated with its nucleic acid component. This shift in emphasis from the protein part of the virus to its nucleic acid paralleled the growing accept-ance, after 1945, of the central role of nucleic acids in heredity; this de-velopment will be considered later in this chapter.

After the discovery that the plant viruses are nucleoproteins, it was demonstrated that they provide excellent sources for the isolation of RNA preparations in a relatively undegraded state. In an important paper on the physical properties of the RNA derived from tobacco-mosaic virus, Seymour Cohen and Wendell Stanley concluded that the data for their preparation

. . . yielded a molecular weight of approximately 300,000, a value cor-responding to about one-eighth of the total nucleic acid per mole of virus. Since the isolation procedure only involved heating to 100° for 1 minute, it is possible that the nucleic acid thereby obtained closely approximated its native size, since there is no evidence that such treat-ment greatly affects the state of polymerization of any type of nucleic acid (Cohen and Stanley, 1942, p. 596).

Although subsequent studies, with improved methods, showed that the apparent molecular weight of the RNA of the plant viruses is even higher, the work of Cohen and Stanley clearly indicated the macromolecular na-ture of some ribonucleic acids. Together with the analytical data provided by Chargaff and others around 1950, it became evident that all the nu-cleic acids, whether of the DNA or RNA type, represent polynucleotides of considerable length. The remarkable subsequent development of nu-cleic acid chemistry during the period 1955–1970 falls outside the scope of this chapter; mention must be made, however, of the discovery of a class of relatively small ribonucleic acids (the so-called transfer RNAs or tRNAs) which have been shown to be single polynucleotide chains consisting of about 75 nucleotide units. In 1965 Robert Holley succeeded in establishing the sequence of the nucleotide units in one of the many tRNAs of yeast in a manner analogous to that employed by Frederick Sanger for the determination of the amino acid sequence of insulin.

The recognition during the 1930s that the crystalline plant viruses are large nucleoproteins came soon after the development of a new instrument, the electron microscope. After its invention in 1931 by Max Knoll and Ernst Ruska (Freundlich, 1963), several years passed before a resolving power of about 10 mμ (1 mμ = 1 nm = 10 Å = 10^{-7} cm) was attained; since plant viruses such as tobacco-mosaic virus had been estimated to consist of particles about 400 mμ long and about 12 mμ wide, the electron microscope offered obvious opportunities for the direct measurement of the size and shape of viruses. Such studies were described in 1941 by Wendell Stanley and Thomas Anderson, whose electron micrographs gave dimensions similar to those inferred from physicochemical measurements with solutions of the crystalline viruses. During succeeding years the electron microscope became a basic instrument for the study, not only of viruses, but of other macromolecules and of intracellular structures as well. In particular, it was applied to the examination of DNA preparations. We noted earlier that relatively undegraded DNA behaved in solution like highly elongated fibrous particles of enormous size; after the initial studies of Jesse Scott (in 1948), and the development of the "shadow-casting" technique by Robley Williams, electron microscopy became during the 1950s an important method for the examination of the size and shape of DNA preparations. Thus careful studies by Cecil Hall showed that suitably prepared samples of salmon sperm DNA are very long fibrils having a diameter of about 20 Å. Among the many other important results yielded by electron microscopy during the 1940s was the remarkable finding by Helmuth Ruska and by Salvador Luria and Thomas Anderson that some of the bacteriophages, belonging to the group of "T" phages which infect *Escherichia coli* B, are tadpole-like structures consisting of an oval body (about 65 × 80 mμ) to which is attached a straight tail measuring about 20 × 120 mμ. These observations were soon followed by the demonstration that such phages (the one denoted T2 was studied most extensively) are composed almost entirely of protein and DNA, the latter constituting about 40 percent of the total weight. Furthermore, chemical analysis of the phage DNA led to the discovery (by Gerard Wyatt and Seymour Cohen) that the cytosine usually found in DNA preparations is replaced by 5-hydroxymethylcytosine. These electron-microscope and chemical studies provided much of the background for the dramatic experiments of Alfred Hershey and Martha Chase (in 1952) demonstrating the dominant genetic role of the DNA portion of the bacteriophage.

To complete the summary of the findings that led to the recognition of the macromolecular character of DNA, we must return to the 1930s. In addition to the evidence derived from studies on the physicochemical properties of carefully prepared calf thymus DNA, the examination of the diffraction of X-rays by fibers prepared from such preparations led William Astbury and Florence Bell to infer that they were dealing with long polynucleotide chains; furthermore,

> . . . the natural conclusion from the X-ray data is that the spacing of 3.3_4A. along the fibre axis corresponds to that of a close succession of flat or flattish nucleotides standing out perpendicularly to the long axis of the molecule to form a relatively rigid structure (Astbury and Bell, 1938, p. 747).

In 1947 Astbury concluded that the X-ray data also indicated

> . . . that the pattern repeats along the axis of the molecule at a distance corresponding to the thickness of eight nucleotides or a multiple of eight nucleotides—most probably eight or sixteen nucleotides. The least possible value of the fibre period is 27 A., which would make the effective average thickness of a nucleotide about 3.4 A . . . . It hardly seems likely . . . that the fact that the intramolecular pattern is found to be based on a multiple of four nucleotides is unrelated to the conclusion that has been drawn from chemical data that the molecule is composed of four different kinds of nucleotides in equal proportions. It seems improbable, too, to judge by the degree of perfection of the X-ray fibre diagram, that these four different kinds of nucleotides are distributed simply at random (Astbury, 1947b, p. 67).

At the time these words appeared, however, Chargaff had begun his analytical studies on DNA preparations, and was providing the experimental basis for the abandonment of the tetranucleotide hypothesis.

We noted the impact on protein chemistry of the early X-ray studies on cellulose fibers (see p. 154); similarly, after 1935, ideas about the structure of nucleic acids were increasingly influenced by the results obtained with fibrous proteins, especially Astbury's X-ray data for keratin. Furthermore, the gradual acceptance of the view that nucleic acids are complex polynucleotides, rather than simple tetranucleotides, depended heavily on results obtained with methods that were being developed for the study of proteins as macromolecules. This influence was especially great in speculations about the three-dimensional structure of DNA.

The Double Helix. After World War II, with the growing appreciation of the biological importance of the nucleic acids, the attention of X-ray crystallographers was drawn to the problem of defining as precisely as possible the molecular dimensions, *i.e.*, the bond lengths and bond angles, of their fundamental constituents. In the protein field, such data had been provided by Robert Corey during 1939–1950, and served as the basis for the ideas developed by Pauling and Corey (1950) regarding the stereochemically possible modes of the folding of peptide chains. Their work demonstrated that a significant contribution could be made to the understanding of protein structure by the construction of molecular models in which careful account was taken of the interatomic distances and bond angles of hypothetical peptide chains, and by examination of the ways in which such models could be folded so as to permit maximum hydrogen bonding, either between parts of a single chain or between adjacent chains. In particular, Pauling and Corey called attention to the stability of a helical structure ("α-helix") which was subsequently found by John Kendrew to be a distinctive feature of the myoglobin molecule (p. 178). Although speculations had been offered earlier regarding the helical coiling of long peptide chains, these proposals either were vague or were incompatible with the molecular dimensions determined through studies on crystalline peptides (Corey and Pauling, 1953).

By 1951 X-ray studies had given the molecular dimensions of adenine and guanine, whose rings were found by June Broomhead to be almost identical in their dimensions, and Sven Furberg had determined the three-dimensional structure of cytidine; the latter work showed that in this nucleoside the pyrimidine ring and the ribofuranose ring are approximately perpendicular, thereby invalidating Astbury's assumption that the nucleotide units of nucleic acids are flat structures. In an important paper, Furberg (1952) described molecular models for a hypothetical polynucleotide in which most of the atoms constituting each nucleotide (including the phosphorus atom) lie in planes 3.4 Å apart, thus explaining the strong 3.3_4 Å reflection reported by Astbury and Bell. In one of Furberg's two models, the ribofuranose rings and the phosphorus atoms form a spiral that encloses a column with the purines and pyrimidines almost stacked on top of each other; in the other, the ribose phosphate chain forms a central column from which the purines and pyrimidines stand out perpendicularly. These proposals were followed by those of Pauling and Corey (1953), who suggested a structure for DNA in which each of three polynucleotide chains is coiled into a helix, and that the three helices are intertwined; in their model, the phosphoryl groups are closely packed about the axis of the column, and the nitrogenous bases project radially, as in Furberg's second model. A few months later, James Watson and Francis Crick published their famous note (see Watson, 1968; Olby, 1970) in which they pointed out the deficiencies of the model proposed by Pauling and Corey, and

> . . . put forward a radically different structure for the salt of deoxy-ribose nucleic acid. This structure has two helical chains each coiled round the same axis . . . . We have made the usual chemical assumptions, namely, that each chain consists of phosphate diester groups joining β-D-deoxyribofuranose residues with 3′,5′ linkages. The two chains (but not their bases) are related by a dyad perpendicular to the fibre axis. Both chains follow right-handed helices, but owing to the dyad the sequences of the atoms in the two chains run in opposite directions. Each chain loosely resembles Furberg's model No. 1; that is, the bases are on the inside of the helix and the phosphates on the outside. The configuration of the sugar and the atoms near it is close to Furberg's 'standard configuration', the sugar being roughly perpendicular to the attached base. There is a residue on each chain every 3.4 A. in the z-direction. We have assumed an angle of 36° between adjacent residues in the same chain, so that the structure repeats after 10 residues on each chain, that is, after 34 A. The distance of a phosphorus atom from the fiber axis is 10 A. As the phosphates are on the outside, cations have easy access to them . . . .

The novel feature of the structure is the manner in which the two chains are held together by the purine and pyrimidine bases. The planes of the bases are perpendicular to the fibre axis. They are joined together in pairs, a single base from one chain being hydrogen-bonded to a single base from the other chain, so that the two lie side by side with identical z-co-ordinates. One of the pair must be a purine and the other a pyrimidine for bonding to occur . . . .

If it is assumed that the bases only occur in the structure in the most plausible tautomeric forms (that is, with the keto rather than the enol configurations) it is found that only specific pairs of bases can bond together. These pairs are adenine (purine) with thymine (pyrimidine), and guanine (purine) with cytosine (pyrimidine).

In other words, if an adenine forms one member of a pair, on either chain, then on these assumptions the other member must be thymine; similarly for guanine and cytosine. The sequence of bases on a single chain does not appear to be restricted in any way. However, if only specific pairs of bases can be formed, it follows that if the sequence of bases on one chain is given, the sequence on the other chain is automatically determined (Watson and Crick, 1953a, p. 737).

The experimental basis for "the novel feature" of the model—the specific base-pairing—lay in the analytical data that had been provided by Chargaff and others on the base composition of various DNA preparations.

As regards the X-ray evidence for the double-helical structure of DNA, better photographs than those obtained by Astbury and Bell had been reported by Maurice Wilkins and John Randall early in 1953, but Watson and Crick noted that these data

> . . . are insufficient for a rigorous test of our structure. So far as we can tell, it is roughly compatible with the experimental data, but it must be regarded as unproved until it has been checked against more exact results. Some of these are given in the following communications. We were not aware of the details of the results presented there when we devised our structure, which rests mainly though not entirely on published experimental data and stereochemical arguments (*ibid.*, p. 737).

The first of two "following communications" was by Wilkins, Stokes, and Wilson, who discussed the available X-ray data in terms of a helical structure of DNA; they also reported that diffraction patterns similar to those given by fibers of calf thymus DNA had been obtained with nucleoprotein preparations (sperm heads, T2 bacteriophage), indicating that such structures are present in biological systems and are not artifacts of preparation. The second paper, by Rosalind Franklin and R. G. Gosling (their colleagues at King's College, London), presented striking X-ray photographs of two forms (A and B) of calf thymus DNA, and the authors concluded that the diffraction pattern of the B form was compatible with the Watson-Crick model; a few months later they showed that the data for the more regularly ordered A form also were consistent with the double-helical structure of DNA. A valuable account of Franklin's contributions has been prepared by Klug (1968). Subsequent studies (largely by Wilkins and his associates) have provided additional X-ray evidence in favor of the Watson-Crick model; although the limitations of the methods employed in these studies have not established the three-dimensional structure of DNA preparations with the same precision as in the case of myoglobin (or other crystalline proteins), the fact that thus far no one has yet offered an alternative model that fits the X-ray data equally well has served to strengthen the conviction that the deoxyribonucleic acids have the double-helical structure proposed by Watson and Crick in

1953. A more detailed account of the considerations which led to the formulation of the structure appeared a few months later in a paper (Crick and Watson, 1954) whose title stressed the idea of complementariness in the structure of DNA; the general importance of complementary structural relations had been emphasized several years before by Linus Pauling and Max Delbrück:

> It is our opinion that the processes of synthesis and folding of highly complex molecules in the living cell involve, in addition to covalent-bond formation, only the intermolecular interactions of van der Waals attraction and repulsion, electrostatic interactions, hydrogen-bond formation, etc., which are now rather well understood. These interactions are such as to give stability to a system of two molecules with *complementary* structures in juxtaposition . . . [I]n order to achieve maximum stability, the two molecules must have complementary surfaces, like die and coin, and also a complementary distribution of active groups (Pauling and Delbrück, 1940, p. 78).

This idea has its antecedents in Emil Fischer's "lock-and-key" hypothesis of enzyme-substrate interaction (p. 79), and also in the speculations of Leonard Troland (1917).

In their first joint paper, Watson and Crick stated that "It has not escaped our notice that the specific pairing we have postulated immediately suggests a possible copying mechanism for the genetic material" (Watson and Crick, 1953a, p. 737), and this idea was developed in a paper that appeared a few weeks later:

> The phosphate-sugar backbone of our model is completely regular, but any sequence of the pairs of bases can fit into the structure. It follows that in a long molecule many different permutations are possible, and it therefore seems likely that the precise sequence of bases is the code which carries the genetical information. If the actual order of the bases on one of the pair of chains were given, one could write down the exact order of the bases on the other one, because of the specific pairing. Thus one chain is, as it were, the complement of the other, and it is this feature which suggests how the deoxyribonucleic acid might duplicate itself (Watson and Crick, 1953b, p. 965).

Within a few years, striking experimental evidence in support of the Watson-Crick theory of DNA replication was provided by Matthew Meselson and Franklin Stahl (1958) and by Arthur Kornberg (1960); a leading geneticist later wrote:

I have said many times that I regard the working out of the detailed structure of DNA one of the great achievements of biology in the twentieth century, comparable in importance to the achievements of Darwin and Mendel in the nineteenth century. I say this because the Watson-Crick structure immediately suggested how it replicates or copies itself with each cell generation, how it is used in development and function, and how it undergoes the mutational changes that are the basis of organic evolution (Beadle, 1969, p. 2).

To see this judgment in historical perspective, it is necessary to sketch the development of genetics during the first half of this century, and to consider the speculations and experimental results relating to the chemical nature of the genetic material.

The Chemical Nature of the Gene

After the publication of Charles Darwin's *Origin of Species* in 1859, and of his *Variation of Animals and Plants under Domestication* in 1868, there was a succession of hypotheses in which material factors were assumed to be the bearers of the specific characters transmitted from parents to offspring. Among the contributions to the discussion was the book *Intracelluläre Pangenesis* (published in 1889) by the botanist Hugo de Vries, who had turned his attention from osmotic phenomena in plants to the problem of organic evolution. In writing about the hereditary factors, he stated:

> These factors are the units which the science of heredity has to investigate. Just as physics and chemistry go back to molecules and atoms, the biological sciences have to penetrate to these units in order to explain, by means of their combinations, the phenomena of the living world (de Vries, 1910, p. 13).

The hypothesis he put forward in 1889 included the following points:

> . . . [H]ereditary units are independent units, from the numerous and various groupings of which specific characters originate. Each of these units can vary independently from the others; each one can of itself become the object of experimental treatment in our culture experiments . . . . According to the hypothesis concerning their nature, these units have been given different names. For the one adopted by me I have chosen the name, pangen . . . . The pangens are not chemical mole-

cules, but morphological structures, each built up of numerous molecules. They are life-units, the characters of which can be explained in an historical way only . . . . At each cell-division every kind of pangen present is, as a rule, transmitted to two daughter cells . . . . An altered numerical relation of the pangens already present, and the formation of new kinds of pangens must form the two main factors of variability (*ibid.*, pp. 69–74).

During the succeeding years de Vries conducted experiments on plant hybridization to test his hypothesis, and in 1900 he reported the discovery of a *Spaltungsgesetz* or *loi de disjunction* ("law of segregation") of hybrids; he also stated that his conclusions had been "put forward a long time ago by Mendel for a special case (peas)" and, after citing Gregor Mendel's 1865 paper (published in 1866), he added the footnote:

This important report is cited so seldom that I only first learned of it after I had completed most of my experiments and had derived the conclusions communicated in the text (de Vries, 1900, p. 85).

There immediately followed a paper by Carl Correns who reported his hybridization experiments with peas and maize, with the statement that he had convinced himself that

. . . in the sixties, the abbot Gregor Mendel in Brünn, through very extensive experiments lasting many years with peas, had not only obtained the same result as de Vries and I, but had given exactly the same explanation, so far as it was at all possible to do so in 1866 (Correns, 1900, p. 158).

A few weeks later there appeared another report on the hybridization of peas, by Erich Tschermak von Seysenegg, with the postscript:

The experiments published by Correns just now, also dealing with the artificial crossing of different varieties of *Pisum sativum* and observations of hybrids allowed to undergo self-fertilization for several generations, confirm as do mine the Mendelian theory. The simultaneous "discovery" of Mendel by Correns, de Vries and myself is especially pleasing to me. Even during the second year of research, I thought that I had found something quite new (Tschermak, 1900, p. 239).

These papers, as well as others dealing with the emergence of Mendelian genetics, have been reprinted in the valuable book edited by Kříženecký (1965).

What Mendel had found is told in textbooks of biology: When the pollen of a variety (*e.g.*, tall plants) of *Pisum sativum* was artificially placed on the stigma of a variety with a "differentiating character" (*differirendes Merkmal*) such as short plants, the resulting seeds produced hybrids (F_1 generation) that were tall (the dominant character). If the hybrids were allowed to undergo self-fertilization, the seeds gave rise to plants (F_2 generation) of the two original parental varieties in the ratio of 3 tall to 1 short. Furthermore, upon self-fertilization, the short F_2 plants only produced seeds leading to the same variety, whereas the tall F_2 plants gave rise to both varieties in the ratio 3 tall to 1 short. The latent character (in this case, short plants) was termed the recessive member of an antagonistic pair. Mendel investigated other pairs of carefully selected characters, such as round vs. wrinkled peas, green vs. yellow peas, and concluded that

> . . . it is now evident that the hybrids having each of the two differentiating characters form seeds of which one half again develops the hybrid form, whereas the other [half] gives the other plants, which remain constant, and receive in equal parts the dominant and recessive character (Mendel, 1865, p. 17).

Mendel considered that the "constant law" he had discovered ". . . is founded on the material composition and arrangement of the elements in the [fertilized egg] cell in a viable union" (*ibid.*, p. 41). His work succeeded, where earlier attempts had failed, because he began by studying carefully the inheritance of one character at a time, and because he counted the numbers of each kind of progeny resulting from hybridization. The possible reasons for the neglect of Mendel's paper have been the subject of extensive discussion (see Dunn, 1966; Stern and Sherwood, 1966); special emphasis has been placed on his correspondence with Carl von Nägeli, who wrote Mendel that his results were "only empirical, because they cannot be proved rational" (quoted in Coleman, 1965, p. 144), in keeping with Nägeli's view that "a 'rational' explanation of heredity could only be founded upon mechanical or causal modes of explanation" (*ibid.*, p. 144).

In 1902 William Bateson proposed a terminology that was quickly adopted:

> By crossing two forms exhibiting antagonistic characters, cross-breds were produced. The generative cells of these cross-breeds were shown to be of two kinds, each being pure in respect of *one* of the parental characters. This purity of the germ-cells, and their inability to trans-

mit both of the antagonistic characters, is the central fact proved by Mendel's work. We thus reach the conception of unit-characters existing in antagonistic pairs. Such characters we propose to call *allelomorphs* [later shortened to "alleles"], and the zygote formed by the union of a pair of opposite allelomorphic gametes, we shall call a *heterozygote*. Similarly, the zygote formed by the union of gametes having similar allelomorphs, may be spoken of as a *homozygote* (Bateson and Saunders, 1902, p. 126).

It was also Bateson who first established the validity of Mendel's principle of segregation in animals (poultry); shortly afterward, in 1902–1903, the work of Lucien Cuénot and of William Castle showed its applicability to the inheritance of hair color in mice.

Of special importance to our story is the 1902 report of the clinician Archibald Garrod on the inheritance of the human condition known as alcaptonuria, described by Alexander Marcet in 1822 and by Carl Bödeker in 1859. Garrod noted:

> All the more recent work on alkaptonuria has tended to show that the constant feature of that condition is the excretion of homogentisic acid, to the presence of which substance the special properties of alkapton urine, the darkening with alkalies and on exposure to air, the power of staining fabrics deeply, and that of reducing metallic salts, are alike due . . . . Further observations, and especially those of Mittelbach, have also strengthened the belief that the homogentisic acid excreted is derived from tyrosin, but why alkaptonuric individuals pass the benzene ring of their tyrosin unbroken and how and where the peculiar chemical change from tyrosin to homogentisic acid is brought about, remain unsolved problems.
>
> There are good reasons for thinking that alkaptonuria is not the manifestation of a disease but rather of the nature of an alternative course of metabolism, harmless and usually congenital and lifelong (Garrod, 1902, p. 1616).

After collecting data on the incidence of the condition among members of individual families, he concluded that "a very large proportion are children of first cousins" (*ibid.*, p. 1617) and offered a Mendelian explanation by quoting Bateson:

> . . . [W]e note that the mating of first cousins gives exactly the conditions most likely to enable a rare, and usually recessive, character to show itself. If the bearer of such a gamete mate with individuals not

bearing it the character will hardly ever be seen; but first cousins will frequently be the bearers of similar gametes, which may in such unions meet each other and thus lead to the manifestation of the peculiar recessive characters in the zygote (Bateson and Saunders, 1902, p. 133).

In his paper, Garrod suggested that two other "chemical abnormalities"—albinism and cystinuria—might belong to the group of congenital conditions represented by alcaptonuria, and in lectures that he gave in 1908 he added a fourth, pentosuria (see Knox, 1958). These lectures formed the basis of a book *Inborn Errors of Metabolism* (1909; 2nd edition, 1923), in which he summarized the available evidence on the metabolic breakdown of phenylalanine and tyrosine in the mammalian organism, and concluded that in alcaptonuria the metabolic error lies

> . . . in the penultimate stage of the catabolism of the aromatic protein fractions, which is in accord with the fact that all the tyrosin and phenyl-alanine, both exogenous and endogenous, is swept into the net and goes to contribute to the excreted homogentisic acid in alkaptonuria.
>
> We may further conceive that the splitting of the benzene ring in normal metabolism is the work of a special enzyme, that in congenital alkaptonuria this enzyme is wanting . . . . (Garrod, 1909, p. 80).

Thus, not only did Garrod provide the first evidence of the applicability of Mendelian genetics to man, but he also perceived the relationship of hereditary factors to enzymes. As we shall see later, there were to be other indications of the relation of a Mendelian character to the presence or absence of an enzyme, before Garrod's concept was elaborated by George Beadle and Edward Tatum in 1941.

To return to the situation immediately after the rediscovery of Mendel's results, we must recall that his law of the segregation of differentiating characters was only one of the two generalizations he drew from his data, the second being what has come to be called the "law of independent assortment." Thus, if hybrids were obtained by the mating of two parents differing in two characters (*e.g.*, tall plants and round seeds vs. short plants and wrinkled seeds), the F_1 generation was tall and had round seeds, the latter being dominant over wrinkled seeds. Upon self-fertilization, the 3:1 ratio in the F_2 generation was maintained for each antagonistic pair of characters, and per 16 plants there were 9 tall-round, 3 tall-wrinkled, 3 short-round, and 1 short-wrinkled. In 1905, however, Bateson and his associates showed that this independent assortment of two characters did not apply in the case of the hybridization of sweet peas with respect to the characters for pollen shape and flower color; many other apparent exceptions to Mendel's "Second Law" were found subsequently. Although Bateson offered an explanation seemingly consistent with Mendelian principles, a more convincing answer came from the researches of Thomas Hunt Morgan in the succeeding decade.

By 1910 Mendel's rules had been widely accepted, and the area of science dealing with heredity and variation had been named "genetics" by Bateson (in 1906). It was recognized that Mendel's approach required the use of large numbers of carefully selected organisms if statistically significant quantitative results were to be obtained. This attribute of the new genetics was emphasized by Wilhelm Johannsen in his influential book *Elemente der exakten Erblichkeitslehre,* first published in 1909; he stated, however, that "we must pursue the study of heredity *with* mathematics, not *as* mathematics" (Johannsen, 1909, p. 2). To denote more precisely the concepts emerging from the new science, he introduced a new terminology:

> . . . [O]ne may designate the statistically appearing type as the appearance type or, briefly and clearly, as the phenotype. Such phenotypes are in themselves measurable realities: just what can be typically observed; hence for series of variations the centers about which the variants group themselves. In using the word phenotype the necessary reservation is only made that no further conclusion may be drawn from the appearance itself. A given phenotype may be the expression of a biological unit; but it does not need at all to be so. The phenotypes found in nature through statistical studies are certainly not so in most cases! (*ibid.,* p. 123).

To denote the "something" in the egg and sperm cells that determines the character of the progeny, he invented the word *Gen* (English, gene) by dropping the first syllable of the pangen of Darwin and de Vries, and he emphasized the distinction between the concept of the "genotype" (the genetic constitution of the individual) and that of the phenotype. Although Johannsen noted that "No definite idea about the nature of the 'genes' is at present sufficiently well-grounded" (*ibid.,* pp. 124–125), he suggested that they

> . . . may be tentatively considered to be chemical factors of various kinds . . . . One may expect principally from general physical chemistry the viewpoints for theories regarding the action of chemical hereditary factors (*ibid.,* p. 485).

In this statement, Johannsen reflected the popularity of the new physical chemistry among biologists, a trend markedly furthered by the writings of Wilhelm Ostwald, Svante Arrhenius, and others. Several decades were to pass before Johannsen's expectation was to be realized, and during that time the decisive ideas came from breeding experiments and cytological studies, rather than from physical chemistry.

The Theory of the Gene. We noted earlier in this chapter that the remarkable cytological discoveries made during 1870–1890 assigned to the chromosomes a central role in the transmission of hereditary characters. At the turn of the century, when Mendel's work became widely appreciated, much was known about the chromosomes: it had been demonstrated for several species that the number of chromosomes per nucleus is constant and usually even, with equal numbers coming from the sperm and the egg; at cell division, during mitosis, each chromosome appears to divide longitudinally, followed by nuclear division to give two nuclei each having a complete (diploid) set of chromosomes; in the special case of the germ cells, there occurs during meiosis a reduction in chromosome number during the last two divisions before the production of mature gametes, so that each egg or sperm has one-half (haploid) the full set of chromosomes; during meiosis, the maternal and paternal members of homologous sets of chromosomes form pairs (synapsis). Immediately after 1900 several investigators (notably Theodor Boveri and Walter Sutton) attempted to define the mechanism of Mendelian segregation in terms of chromosomal changes; the clearest and most detailed formulation was offered by Sutton, who wrote:

> We have seen reason, in the foregoing considerations, to believe that there is a definite relation between chromosomes and allelomorphs or unit characters but we have not before inquired whether an entire chromosome or only a part of one is to be regarded as the basis of a single allelomorph. The answer must unquestionably be in favor of the latter possibility, for otherwise the number of distinct characters possessed by an individual could not exceed the number of chromosomes in the germ-products; which is undoubtedly contrary to fact. We must, therefore, assume that some chromosomes at least are related to a number of different allelomorphs. If then, the chromosomes permanently retain their individuality, it follows that all the allelomorphs represented by any one chromosome must be inherited together (Sutton, 1903, p. 240).

It should be noted here that the first suggestion of the association of an individual character with a particular chromosome came from Sutton's teacher Clarence E. McClung. He proposed in 1901 that an accessory chromosome (later denoted "X chromosome") carries the factor for maleness; this chromosome had been described by Henking (in 1891) who found that it divides at only one of the two meiotic divisions. Within a few years, however, a "Y chromosome" was discovered, and by 1905 it had been shown that McClung's assignment was incorrect since, in various animals, the females were XX in nature whereas males were XY. These sex chromosomes were to play a large role in the study of the mechanism of Mendelian inheritance, and the ideas expressed by Sutton were to find their first decisive experimental basis in the study by Thomas Hunt Morgan of the transmission of sex-linked characters.

> In a pedigree culture of Drosophila which had been running for nearly a year through a considerable number of generations, a male appeared with white eyes. The normal flies have brilliant red eyes.
>
> The white-eyed male, bred to his red-eyed sisters, produced 1,238 red-eyed offspring, (F_1), and 3 white-eyed males. The occurrence of these three white-eyed males (F_1) (due evidently to further sporting) will, in the present communication be ignored.
>
> The F_1 hybrids, inbred, produced: 2,459 red-eyed females, 1,011 red-eyed males, 782 white-eyed males. *No white-eyed females appeared.* The new character showed itself therefore to be sex limited in the sense that it was transmitted only to the grandsons (Morgan, 1910, p. 120).

So began a paper that marked the start of a memorable series of investigations by Morgan, with Alfred H. Sturtevant, Calvin B. Bridges, and Hermann J. Muller, on the genetics of the vinegar fly *Drosophila melanogaster*. Morgan explained his results by means of the hypothesis that the factor for red eyes (R) is closely associated with the X chromosome, and that the white-eyed male had arisen through the fertilization, by male-producing sperm, of an egg that had undergone mutation with the loss of R. Shortly afterward, another sex-linked mutant type ("miniature wing") was found; there followed Morgan's famous experiment in which a female fly having the white-eye mutant gene in one X chromosome and the miniature wing gene in its homolog was found to produce male offspring that included a double mutant with white eyes and miniature wings. To explain this finding, he drew on the cytological evidence of F. A. Janssens (in 1909) indicating that, during meiosis, there were exchanges between homologous chromosomes, and Morgan suggested that the recombination of genes leading to the double mutant was a consequence of "crossing-over." Furthermore, as other sex-linked mutants were found, it was observed that some genes (*e.g.*, yellow body color and white eyes) exhibited linkage with respect to each other. These results led Morgan to formulate the theory of a linear order of genes in the chromosome, and to propose that their relative position might be ascertained by determining the frequency of crossing over for pairs of sex-linked factors. This theory was tested by Sturtevant; in the summary of his important paper he stated:

> It has been possible to arrange six sex-linked factors in Drosophila in a linear series, using the number of cross-overs per 100 cases as an index of the distance between any two factors. This scheme gives consistent results, in the main . . . . These results are explained on the basis of Morgan's application of Janssens' chiasmatype hypothesis to associative inheritance. They form a new argument in favor of the chromosome view of inheritance, since they strongly indicate that the factors investigated are arranged in a linear series, at least mathematically (Sturtevant, 1913, p. 59).

In this work, Sturtevant not only provided experimental evidence for the validity of Morgan's hypothesis, but he also produced the first chromosome "map" and defined the basic methodology involved in such gene mapping. By 1925 Morgan's group had identified over 100 gene loci in the four chromosomes of *Drosophila melanogaster*.

During the period 1910–1920 Morgan's associates produced additional evidence of lasting importance. For example, in his 1910 paper, Morgan mentioned the appearance of three white-eyed males in the F_1 generation, and had attributed this to further mutation ("sporting"). Other cases of the same kind continued to appear, however, and the detailed studies of Bridges (published in 1916) led to the recognition that the females giving rise to such offspring were XXY in composition, as a consequence of the failure of chromosomes paired at meiosis to separate ("nondisjunction") and of the passage of both members of the pair to the same gamete. Morgan and his associates summarized their remarkable findings in a book, *The Mechanism of Mendelian Heredity* published in 1915, and during succeeding years they made further refinements in the argument. As apparent exceptions to the linear order were observed, they were not only shown to be consistent with the theory, but also evidence was provided to strengthen it. At the close of Morgan's work in genetics, he formulated the "theory of the gene" as follows:

The theory states that the characters of the individual are referable to paired elements (genes) in the germinal material that are held together in a definite number of linkage groups; it states that the members of each pair of genes separate when the germ-cells mature in accord-

ance with Mendel's first law, and in consequence each germ-cell comes to contain one set only; it states that the members belonging to different linkage groups assort independently in accordance with Mendel's second law; it states that an orderly interchange—crossing-over—also takes place, at times, between elements in corresponding linkage groups; and it states that the frequency of crossing-over furnishes evidence of the linear order of the elements in each linkage group and of the relative position of the elements with respect to each other.

These principles, which, taken together, I have ventured to call the theory of the gene, enable us to handle problems of genetics on a strictly numerical basis, and allow us to predict, with a great deal of precision, what will occur in any given situation. In these respects the theory fulfills the requirements of a scientific theory in the fullest sense (Morgan, 1928, p. 25).

In this formulation, no statement is made about the nature of genes although, as we have seen, cytological studies on chromosomes played a decisive role in the development of Morgan's theory. During the period 1920–1940, however, important clues about the possible nature of the genetic material came from two lines of research: the experimental study of mutation and cytochemical studies on chromosomes.

Before 1900 there had been much discussion of "discontinuous variation" as a factor in the appearance of new species, but it was Hugo de Vries in his *Mutationslehre* (published in 1901–1903) who explicitly linked evolution to the new genetics by attributing mutation to changes in the number and kind of "pangens." Although his evidence on the mutation of the evening primrose *Oenothera Lamarckiana* was not supported by later work, his ideas focused the attention of geneticists on the experimental study of mutation. We have already seen how much came from the appearance of the white-eyed mutant in Morgan's Drosophila population, and from the subsequent study of the other mutants that arose spontaneously. The frequency with which most of these mutants appeared was too low, however, to permit quantitative studies on the mutation process. In 1918 Hermann J. Muller undertook the systematic study of spontaneous mutation in Drosophila, and sought ways to increase their frequency. A few years later, in a summary of the status of the problem, he emphasized the importance of studying "mutations as directly as possible" (Muller, 1922, p. 39) and of determining how different conditions affect the occurrence of mutations. He went on to say that

. . . it appears most promising to employ organisms in which the genetic composition can be controlled and analyzed, and to use genetic methods that are sensitive enough to disclose mutations occurring in the control as well as in the treated individuals. In this way relatively slight variations in mutation frequency, caused by special treatments, can be determined, and from the conditions found to alter the mutation rate slightly we might finally work up to those which affect it most markedly (*ibid.*, pp. 46–47).

In 1927 Muller reported that treatment of Drosophila sperm with X-rays

. . . induces the occurrence of true "gene mutations" in a high proportion of the treated germ cells. Several hundred mutants have been obtained in this way in a short time and considerably more than a hundred of the mutant genes have been followed through three, four or more generations. They are (nearly all of them, at any rate) stable in their inheritance, and most of them behave in the manner typical of the Mendelian chromosomal mutant genes found in organisms generally (Muller, 1927, p. 84).

Independent confirmation of Muller's report came in the following year through the work of Lewis J. Stadler, who showed that the irradiation of barley seeds with X-rays (or exposing them to radium) induced mutations; among Stadler's important results was the observation that the mutation rate was proportional to the radiation dosage. In 1929 Muller and Painter described the effects of X-rays on Drosophila chromosomes, and reported extensive changes in their appearance, as evidenced by breakage and translocation.

The discovery of the mutagenic action of X-rays was soon followed by studies with other kinds of radiation (*e.g.*, ultraviolet light), and during the 1930s there began a collaboration of geneticists and physicists in the study of the gene (Muller, 1935a; Waddington, 1969). In particular, Nikolai Timoféev-Ressovsky (a geneticist) and Karl Zimmer (a physicist) applied to the mutational process the "hit" and "target" theories introduced into radiobiology in the previous decade (see Hollaender, 1961). Together with the theoretical physicist Max Delbrück, they published in 1935 a paper in which the problem of the "gene molecule" suggested by their data was treated from a quantum-mechanical point of view. This paper was later given prominence through the book *What is Life?* by Erwin Schrödinger, one of the founders of quantum mechanics (see Olby, 1971). He stated:

Delbrück's molecular model, in its complete generality, seems to contain no hint as to how the hereditary substance works. Indeed, I do not expect that any detailed information on this question is likely to come from physics in the near future. The advance is proceeding and will, I am sure, continue to do so, from biochemistry under the guidance of physiology and genetics . . . . From Delbrück's general picture of the hereditary substance it emerges that living matter, while not eluding the 'laws of physics' as established up to date, is likely to involve 'other laws of physics' hitherto unknown, which, however, once they have been revealed, will form just as integral a part of this science as the former (Schrödinger, 1944, pp. 68–69).

Although the scientific concepts developed in Schrödinger's book do not appear to have had significant direct influence on research in genetics (see the reviews by Haldane, 1945; Muller, 1946), there can be no doubt that the book itself, and the possibility of finding "other laws of physics," appealed to some young physicists "suffering from a general professional malaise in the immediate post war period" (Stent, 1968, p. 392).

In addition to the emergence of radiation genetics during the 1930s, there was a marked resurgence of cytological work; this was initiated by John Belling's studies on the Jimson weed in 1926, but was especially fruitful in the case of the giant chromosomes present in the salivary glands of Drosophila. Although these large strands, with many discrete bands of basophilic material, had been observed in the salivary gland nuclei of dipteran larvae as early as 1881, it was only in 1933 that Emil Heitz and Hans Bauer showed them to be chromosomes. Later in the same year, Theophilus S. Painter reported that the succession of stained bands in the salivary X chromosome of Drosophila corresponded to the linear sequence of gene loci as determined in crossing-over experiments, and by 1935 Bridges had performed such an analysis with remarkable detail for all four chromosomes: he recognized 725 bands for the X chromosome, 1320 for the second, 1450 for the third, and 45 for the fourth. Aside from the increased precision with which gene loci of linkage maps could be identified with only a few chromatin bands, Bridges demonstrated the recurrence of certain sequences of bands, and interpreted such "duplication" as a means "for evolutionary increase in lengths of chromosomes with identical genes which could subsequently mutate separately and diversify their effects" (Bridges, 1935, p. 64). Bridges also made an important cytological contribution to the understanding of a phenomenon studied by Sturtevant in 1925: changes in the Bar mutation (a sex-linked dominant that reduces the size of the eye of Drosophila) such as reversion to wild type or the appearance of a more extreme type ("double-Bar") were interpreted by Sturtevant to be a consequence of crossing-over near the map locus of Bar to yield one chromosome that had no Bar gene, and a homologous chromosome with two Bar genes. The more pronounced phenotypic effect of double-Bar was attributed to a "position effect," thus introducing into genetic theory the important idea that genes do not act as independent units, and that their activity may be influenced by their location on the chromosome. Bridges showed that the Bar mutation itself was due to a repeat of a chromosomal section corresponding to about seven bands, and that in double-Bar this section was present in triplicate. Hence the Bar effect turned out to be due, not to a gene mutation, but to a duplication or triplication of the same genic material, and the intensity of the effect was related to the position of the genes on the chromosome. The subsequent study of position effects has played a large role in the development of theories of gene action, but falls outside the scope of our story. Mention should be made, however, of the conclusions drawn by two distinguished geneticists from the discovery of the position effect. Thus Theodosius Dobzhansky stated that the genetic system of the chromosome

> . . . is a continuum of a higher order, since the independence of the units is incomplete—they are changed if their position in the system is altered. A chromosome is not merely a mechanical aggregation of genes, but a unit of a higher order (Dobzhansky, 1936, p. 382).

A more drastic conclusion was offered by Richard Goldschmidt:

> . . . [G]ene mutation and position effect are one and the same thing. This means that no genes are existing but only points, loci, in a chromosome which have to be arranged in a proper order or pattern to control normal development. Any change in this order may change some detail of development, and this is what we call a mutation. We might of course call a change of arrangement at a locus, a gene. But then there are no genes in the normal chromosome, and the mutant gene has no wild type allele, as the whole wild type chromosome is the allele for all mutant genes in the chromosome. Better, then, give up the conception of the gene except for simple descriptive purposes (Goldschmidt, 1938, p. 271).

This opinion may be contrasted with the view that

> . . . a gene is a minute organic particle, probably a single large molecule, possessing the power of reproduction, which power is one of the main characteristics of living matter. Changes in genes (mutations) are visualized as changes or rearrangements within molecular groups of a gene molecule (Demerec, 1935, p. 133).

The widely divergent views of leading geneticists about the nature of the gene led to lively discussion for many years; as Stadler (1954) pointed out, part of the difficulty arose from differences in the definition of the gene as an operational concept.

No summary, however brief, of the advances in genetics during the 1930s can omit mention of the important work on maize (*Zea mays*). During the period 1915–1935 Rollins A. Emerson and his associates (among them, Beadle) had worked out the linkage maps of maize, and Barbara McClintock's cytological studies (in 1931) clearly demonstrated the correlation between chromosome structure and genetic crossing-over; for later developments, see Rhoades (1954). The work on the genetics of maize thus complemented that on Drosophila in establishing the chromosome theory.

It was against the background of the efflorescence of cytogenetics during the early 1930s that cytochemistry began to make significant contributions to the problem of the nature of the genetic material. Of special importance were the investigations of Torbjörn Caspersson (1936) who showed, by means of ingenious ultraviolet microspectrophotometry of chromosomes, that the so-called "euchromatin" bands (those which stain intensely upon the application of the Feulgen nucleal reaction) were also regions of high absorbance at 2600 Å, where nucleic acids absorb strongly (see p. 213). Caspersson found proteins to be present as well, and noted that chromosomes are completely digested by trypsin. Subsequent work indicated, however, that treatment with pepsin, while reducing the chromosomal volume, does not destroy the continuity of the chromatin strand; this was interpreted to indicate that the chromosome has a continuous fibrous structure of histonelike nature. Furthermore, treatment with crude nuclease preparations (deoxyribonuclease was crystallized by Kunitz only in 1948) destroyed the Feulgen-reactive material without disrupting the chromosome, suggesting that it consists of long polypeptide chains to which nucleic acid molecules are attached (Darlington, 1947). There was considerable uncertainty about this picture, which was further confused by the assumed existence of material called "heterochromatin" with which the chromosomal histone was thought to be associated (for a history of heterochromatin, see Brown, 1966). There was no doubt, however, that the chromosomes represent the principal site of DNA in the cell. In 1948 André Boivin and Roger Vendrely as well as Alfred Mirsky and Hans Ris found that the DNA content of haploid and diploid nuclei is roughly in the ratio 1:2. This finding was confirmed by Arthur Pollister (among others), and it was clear that the constancy of the chromosome number, known since the 1880s, could be correlated with the DNA content. Furthermore, it was in agreement with Boveri's "law of proportional cell growth," enunciated in 1903, that the chromatin doubles with each cell division.

We saw earlier that during the 1920s Feulgen had shown the presence of DNA in the nuclei of plant cells, thus invalidating the designation of deoxyribonucleic acid as "animal nucleic acid"; by 1940 the Feulgen reaction had also been used to show the presence of DNA in a large variety of bacteria and fungi. Furthermore, ribonucleic acid had been obtained from animal tissues, so that the idea of a "plant nucleic acid," with yeast nucleic acid as a typical representative, also had to be discarded. The question of the cellular localization of RNA began to be answered in 1938, when Feulgen's associate Martin Behrens devised a method for separating cell nuclei from the cytoplasmic matter, where he found most of the RNA. Subsequent work in which broken cells were subjected to differential centrifugation to yield subcellular particles or granules (mitochondria, microsomes, zymogen granules, etc.), showed that nearly all the RNA of animal cells is present in these cytoplasmic bodies, although some RNA was also found in the nucleolus. In the study of the intracellular distribution of RNA, the use of ribonuclease as a cytochemical reagent (introduced by Jean Brachet in 1940) provided further evidence for the cytoplasmic localization of RNA. By 1946 it was clear that the RNA/DNA ratio in various animal tissues depended on the relative volume of cytoplasmic to nuclear matter; thus in rat liver, where the nuclei are relatively small, the RNA/DNA ratio was found to be about 4, whereas in calf thymus (the traditional source of DNA) it was found to be about 0.2 (Davidson, 1947). These analytical data underlined the uncertainties about the homogeneity of the nucleic acid preparations used in the chemical studies that had led to the formulation of the tetranucleotide hypothesis.

The Gene as Nucleoprotein. The accumulation of cytogenetic and cytochemical evidence during the 1930s gave impetus to speculations about the chemical nature of the gene. Such speculation had begun earlier in the development of genetics: we have already seen that during 1880–1900 many leading cytologists had assigned to Miescher's nuclein a significant role in heredity, although Miescher himself rejected this idea, and found the stereoisomerism of proteins more satisfying as an explanation of hereditary characters. We have also seen that around 1890 there was considerable preoccupation with the concept of ferments as "living proteins"; this appears to have encouraged some biologists (de Vries, Driesch) to suggest that the factors of heredity resemble the ferments. After the turn of the century, those who liked the idea that genes are enzymes, and continued to believe that nuclein represents the material basis of heredity, found encouragement in the possibility that some enzymes might be nucleoproteins. Indeed, there was considerable interest in the view that

> . . . nuclein is in a chemical sense the formative centre of the cell, attracting to it the food-matters, entering into loose combination with them, and giving them off to the cytoplasm in an elaborated form (Wilson, 1896, p. 247).

In the face of the uncertain state of the chemistry of enzymes, proteins, and nucleic acids, however, the various hypotheses usually reflected current chemical fashion but offered little hint how they might be tested experimentally.

The discussion of the relation of genes to enzymes reached a peak during 1910–1920, and revolved about the question whether genes *are* enzymes, or only *make* enzymes. The latter possibility was given support by the studies of Garrod, and his colleague Bateson wrote that

> . . . we may draw from Mendelian observations the conclusion that in at least a large group of cases the heredity of characters consists in the transmission of the power to produce something with properties resembling those of ferments. It is scarcely necessary to emphasize the fact that the ferment itself must not be declared to be the factor or thing transmitted, but rather, the power to produce that ferment, or ferment-like body (Bateson, 1909, p. 268).

> We must not lose sight of the fact that though the factors operate by the production of enzymes, of bodies on which these enzymes act, and of intermediary substances necessary to complete the enzyme-action, yet these bodies themselves can scarcely be themselves genetic factors, but consequences of their existence. What then are the factors themselves? Whence do they come? How do they become integral parts of the organism? (Bateson, 1913, p. 86).

During 1910–1920 several investigators reported genetic data consistent with the view that

> . . . the hereditary factor is a determiner for a given mass of ferments; and we can demonstrate it by the fact that a quantitative difference in the potency of hereditary factors causes a parallel, quantitatively different, enzyme formation (Goldschmidt, 1916, p. 100).

Among such studies were those of Huia Onslow, who reported in 1915 that recessive whiteness in rabbits was due to the absence of tyrosinase, an enzyme shown a few years earlier to catalyze the conversion of tyrosine into colored products (melanins); another example was the work of Muriel Wheldale (later Mrs. Onslow) on the genetics of flower pigments. It is significant that both were working in the Biochemical Laboratory in Cambridge, and that during the 1920s J. B. S. Haldane initiated his researches (with Rose Scott-Moncrieff) on biochemical genetics in the same laboratory (Haldane, 1966; Pirie, 1966; Clark, 1968). Onslow's studies on hair color in mammals were extended by Sewall Wright, who wrote in 1917:

> The very fact that it has been relatively easy to isolate unit factors in work on color inheritance suggests that in this case the chain of processes between germ cell and adult may be relatively simple. Observations which indicate that melanin pigment is formed in the cytoplasm of cells by the secretion of oxidizing enzymes from the nucleus suggest that the chain may be very short indeed when it is remembered that genetic factors are probably characters of the chromosomes and that these seem to be distributed unchanged from the germ cell to all other cells (Wright, 1917, p. 224).

It was, however, one thing to say that genes control the formation of enzymes, and another to suggest that the gene for a given character is the specific enzyme involved in the chemical reaction which produces that character. During 1910–1920 several noted biologists, including Beijerinck (1917), made statements that implied their acceptance of the latter possibility. In large part, the view that genes might be enzymes grew out of the idea that the replication of hereditary factors is effected by an autocatalytic process; in 1908 Wilhelm Ostwald and T. Brailsford Robertson had applied the term "autocatalysis" to biological growth, and in 1911 Arend Hagedoorn had postulated:

> Each of the several transmittable genetic factors for the development of an organism is a definite chemical substance which has the property of being a ferment for its own formation (an autokatalyzer) (Hagedoorn, 1911, p. 33).

This concept was developed by Leonard Troland, who sought to apply the ideas of the currently fashionable colloid chemistry to the problem of gene action; during the course of his argument he stated:

If, as now seems probable, the genetic enzymes must be identified with the nucleic acids, we shall be forced to suppose that these substances, although homogeneous—in animal or plant—from the point of view of ordinary chemical analysis, are actually built up in the living chromatin, into highly differentiated colloidal, and colloidal-molar, structures. The apparent homogeneity results from the fact that ordinary chemical analysis provides us only with the *statistics of the fundamental radicles* which are involved.

To some minds, the idea that a portion of matter as small as a germ-cell can contain sufficient catalytic substance to control the destinies of a complex organism, seems hardly plausible. However, considering the slowness of such processes as growth, it is clear that the quantity of catalyzer required will usually be smaller than that used in laboratory experiments; and it is a truism in chemistry that radical alterations of reaction velocities can be caused by the presence of almost infinitesimal amounts of catalytic material. From the nature of the process, it is evident that only a *few molecules* of substance will be required to furnish the basis for an *auto*catalytic reaction which may eventually result in the production of any desired amount of this substance (Troland, 1917, pp. 342–343).

Furthermore, Troland called attention to "heterocatalysis"

. . . as an extension of the process of autocatalysis. It is obvious that exact similarity of the force patterns of the catalyzing and catalyzed systems is not essential. Indeed, the catalytic effect which is based upon direct similarity of structure between the two systems should be much weaker than that which accompanies certain types of structural *correspondence,* such as that existing between a body and its mirror-image, or between a lock and a key . . . . It is easily conceivable that the patterns of certain surfaces may be capable of distorting other special configurations which come under their influence, so that they fall into new equilibrium figures, without these figures being of necessity identical with those of the catalytic system. The general principles of the action, however, remain the same (*ibid.,* pp. 333–334).

The concept of the gene as an autocatalyst was further elaborated by Hermann J. Muller during the 1920s (see Carlson, 1971). Although he rejected the idea that the gene *is* an enzyme, Muller expressed the view that, in the self-propagation of the gene,

. . . it reacts in such a way as to convert some of the common surrounding material into an end-product identical in kind with the original gene itself. This action fulfills the chemist's definition of "auto-catalysis" . . . . But the most remarkable feature of the situation is not this oft-noted autocatalytic action in itself—it is the fact that, when the structure of the gene becomes changed, through some "chance variation," the catalytic property of the gene may become correspondingly changed, in such a way as to leave it still *autocatalytic*. In other words, the change in gene structure—accidental though it was—has somehow resulted in a change of exactly *appropriate* nature in the catalytic reactions, so that the new reactions are now accurately adapted to produce more material just like that in the new changed gene itself . . . . What sort of structure must the gene possess to permit it to mutate in this way? Since, through change after change in the gene, this same phenomenon persists, it is evident that it must depend upon some feature of gene construction—common to all genes —which gives each one a *general* autocatalytic power—a "carte blanche"—to build material of whatever specific sort it itself happens to be composed of. This general principle of gene structure might, on the one hand, mean nothing more than the possession by each gene of some very simple character, such as a particular radicle or "side-chain"—alike in them all—which enables each gene to enter into combination with certain highly organized materials in the outer proto-plasm, in such a way as to result in the formation, "by" the protoplasm, of more material like this gene which is in combination with it. In that case the gene itself would only initiate and guide the direction of the reaction. On the other hand, the extreme alternative to such a conception has been generally assumed, perhaps gratuitously, in nearly all previous theories concerning hereditary units; this postulates that the chief feature of the autocatalytic mechanism resides in the structure of the genes themselves, and that the outer protoplasm does little more than provide the building material. In either case, the question as to what the general principle of gene construction is, that permits this phenomenon of mutable autocatalysis, is the most fundamental question of genetics (Muller, 1922, pp. 33–34).

The questions raised by Muller in 1922 were extensively discussed by him and others during the succeeding two decades, and were formulated with increasing precision as new knowledge became available during the 1930s. Whereas in 1926 Muller stated that "At present any attempt to tell the chemical composition of the genes is only guesswork" (Muller, 1929, p. 907), around 1940 many investigators agreed with the view that the material of the genes "must be protein, and may very likely be nucleo-protein" (Gulick, 1939, p. 241).

The new knowledge in support of the view that genes are proteins came from various sources. Of special importance was the isolation of tobacco-mosaic virus (and of other plant viruses) in the form of apparently crystalline nucleoproteins whose infectivity was thought to be associated with the protein component. Because these nucleoproteins were considered to possess the power of enzymic autosynthesis, the analogy to the presumed autocatalytic capacity of genes was striking. Furthermore, some of the plant viruses were found to occur in the form of closely related strains, and variants of tobacco-mosaic virus could be produced by treatment with X-rays. Also, by 1940 a bacteriophage (from *Bacillus megaterium*) had been partially purified by John H. Northrop, who found it to have the composition of a nucleoprotein; a few years earlier, Max Schlesinger had reported that phage preparations gave a positive Feulgen reaction. The possible relation of the phages to genes had been suggested repeatedly; indeed, soon after Félix d'Herelle's report in 1917, Muller wrote that

> . . . if these d'Herelle bodies were really genes, they would give us an utterly new angle from which to attack the gene problem. They are filterable, to some extent isolable, can be handled in test tubes, and their properties, as shown by their effects on bacteria, can then be studied after treatment. It would be very rash to call these bodies genes, and yet at present we must confess that there is no distinction known between the genes and them (Muller, 1922, p. 48).

During the 1930s it was reported that the susceptibility of genes and viruses to X-rays was roughly similar, and crude estimates of the dimensions of individual genes gave values near those determined for some viruses. Thus from the mutagenic effect of ionizing radiations it was inferred that genes are either single macromolecules or only a few such molecules; for example, according to Muller's estimate,

> . . . the gene length would be between 6 and 30 times as great as its diameter . . . . This is in agreement with the fact that proteins and other complex organic molecules in general are chain-like, being much larger in one dimension than in the other two (Muller, 1935b, pp. 410–411).

We mentioned earlier the important cytochemical studies of Caspersson, which led to the concept of the chromosome as being composed of long polypeptide chains to which nucleic acid molecules are attached; he suggested:

If the genes are considered to be chemical substances, there could be only one known class of substances to which they could belong, namely the proteins because of their inexhaustible possibility of variation (Caspersson, 1936, p. 138).

Similar views were echoed by many investigators. For example, Beadle wrote:

The gene is made up of protein or nucleoprotein. It may correspond to a single giant molecule, or it may be a discrete unit of higher order made up of a group of protein or nucleoprotein molecules, with or without the addition of other substances (Beadle, 1945a, p. 18),

and in lectures that he gave in 1940 Haldane stated:

The size of a gene is roughly that of a protein molecule, and it is very probable that the genes are proteins (Haldane, 1942, p. 22). . . . [T]he gene is within the range of size of protein molecules, and may be a nucleo-protein molecule like a virus. If so, the chemists will say, we must conceive reproduction as follows. The gene is spread out in a flat layer, and acts as a model, another gene forming on top of it from pre-existing material such as amino acids. This is a process similar to crystallization or the growth of a cellulose wall (*ibid.*, p. 44).

Indeed, the widespread interest in the idea that genes are proteins encouraged speculation about autocatalytic mechanisms whereby such proteins are replicated. One hypothesis was that of Max Delbrück (1941) who based his model on Kaj Linderstrøm-Lang's suggestion that in biological systems proteins are made, not from free amino acids, but from amino aldehydes. Delbrück proposed that each successive peptide bond is formed by an oxidation reaction catalyzed by an enzyme whose polypeptide chain is copied in the process.

Although many investigators (see Schultz, 1941) recognized that the nucleic acid portion of the nucleoprotein complex of chromosomes "plays some important role in the process of gene reduplication" (Waddington, 1939, p. 313), the principal obstacle in considering nucleic acids as possible "autosynthetic molecules" was the apparent uniformity and simplicity of their chemical structure. Thus it was known that the efficiency of ultraviolet light in inducing mutations paralleled the absorption spectrum of nucleic acids, but it was considered likely that the energy so absorbed was transferred to a gene protein. Before World War II, the tetranucleotide hypothesis of the structure of RNA and DNA was seriously questioned only after the macromolecular character of calf thymus DNA and of tobacco-mosaic virus RNA had been demonstrated experimentally. On the other hand, by 1940 it had been well-established that proteins are long-chain polypeptides composed of a large variety of amino acids, thus offering the possibility of diversity and specificity unimaginable for simple tetranucleotides. For example, in a lecture given in 1945, Muller speculated about the mode of gene action in terms of variations in protein structure:

> . . . [I]t seems likely on general considerations that there is a limited number of possible types of building blocks in the gene, and that genes differ only in the arrangements and numbers of these. Under differences in arrangement may here be included not only changes in the linear sequence of amino-acids in the polypeptide chains but also changes in shape, involving folding of the chains, and attachments between R groups, whereby their active surfaces would acquire very different chemical properties (Muller, 1947, p. 26).

We have already seen that a few years later the analytical studies of Chargaff had removed the tetranucleotide hypothesis as an obstacle to the consideration of the possibility that genes might be deoxyribonucleic acids; as he expressed it in 1950:

> . . . as far as chemical possibilities go, they could well serve as one of the agents, or possibly the agent, concerned with the transmission of inherited properties (Chargaff, 1950, p. 209).

The Gene as DNA. Even before Chargaff's analytical data became available, however, there appeared from the laboratory of Oswald T. Avery a series of publications that obliged investigators to reexamine the view that genes are proteins or nucleoproteins. For many years, Avery and his colleagues at the Rockefeller Institute for Medical Research had been studying the chemical basis of the differences among strains of the bacteria (*Diplococcus pneumoniae*) that cause pneumonia in man (Hotchkiss, 1965). By 1920 it was known that the various known types of pneumococcus can be differentiated serologically through differences in the ability of bacteria-free filtrates of young cultures to precipitate antibodies elicited in rabbits by injection of the individual strains of this organism. Among these types were two designated II and III, and during the 1920s Michael Heidelberger and Avery showed that the serological specificity of these types resides in the slimy polysaccharide capsule that surrounds the virulent bacterium. These encapsulated forms grow in glistening smoothly rounded (S) colonies; for each type, a corresponding variant was known which is markedly attenuated in virulence and gives matt rough (R) colonies of organisms that are not encapsulated. Subsequent studies during the 1930s by Avery's associate Walther F. Goebel led to the chemical characterization of the capsular polysaccharides of the individual pneumococcal types, and it became clear that one of the genetic characters of each of the virulent types was the ability to make a structurally distinctive polysaccharide composed of particular sugar units.

In 1928 Fred Griffith reported that, when mice were injected with living Type II (R) pneumococci together with heat-inactivated Type III (S) cells, many of the animals died, and their blood contained living Type III organisms. This dramatic discovery indicated that the living avirulent R bacteria had acquired something from the dead Type III (S) cells which led to the transformation of Type II (R) into virulent organisms having the polysaccharide of the Type III (S) cultures. Griffith does not appear to have interpreted his results in terms of the transfer of hereditary material; instead, he suggested:

When the R form of either type is furnished under suitable conditions with a mass of the S form of the other type, it appears to use that antigen as a pabulum from which to build up a similar antigen and thus to develop into an S strain of that type (Griffith, 1928, p. 153).

Within a few years, Griffith's observation was confirmed by several investigators, and the transformation had been effected in the test tube (Dawson and Sia). By 1933 work by James Alloway in Avery's laboratory had shown that crude aqueous solutions of the transforming principle could be prepared, and, during the succeeding years, efforts were made to purify it. In 1944 there appeared a paper that began as follows:

Biologists have long attempted by chemical means to induce in higher organisms predictable and specific changes which thereafter could be transmitted in series as hereditary characters. Among microorganisms the most striking example of inheritable and specific alterations in cell structure and function that can be experimentally induced and are reproducible under well defined and adequately controlled conditions is the transformation of specific types of Pneumococcus (Avery *et al.*, 1944, p. 137).

There followed the description, in meticulous detail, of the isolation of the active transforming principle from Type III pneumococci in the form of "a highly polymerized and viscous form of sodium desoxy-ribonucleate" (*ibid.*, p. 152) that gave negative reactions when tested for the presence of protein. Aside from the evidence from chemical analysis, the strongest arguments for the identification of the transform-ing principle as DNA were its inactivation by crude preparations of "desoxyribonucleodepolymerase" (the nuclease studied by Araki and Sachs around 1905), and its resistance to the action of ribonuclease or of proteinases such as trypsin and chymotrypsin. Because of the impor-tance of this enzymic inactivation of the transforming principle, Maclyn McCarty purified the DNA "depolymerase" extensively (from beef pancreas); in 1948 the enzyme (now termed deoxyribonuclease) was crystallized by Moses Kunitz, and these highly purified preparations also were found to inactivate the transforming factor.

After noting that others had likened the transforming factor to genes and viruses, Avery, MacLeod, and McCarty stated:

> It is, of course, possible that the biological activity of the substance described is not an inherent property of the nucleic acid but is due to minute amounts of some other substance adsorbed to it or so inti-mately associated with it as to escape detection. If, however, the biologically active substance isolated in highly purified form as the sodium salt of desoxyribonucleic acid actually proves to be the trans-forming principle, as the available evidence strongly suggests, then nucleic acids of this type must be regarded not merely as structurally important but as functionally active in determining the biochemical activities and specific characteristics of pneumococcal cells. Assuming that the sodium desoxyribonucleate and the active principle are one and the same substance, then the transformation described represents a change that is chemically induced and specifically directed by a known compound. If the results of the present study are confirmed, then nucleic acids must be regarded as possessing biological specificity the chemical basis of which is as yet undetermined (*ibid.*, p. 155).

Not only were the results quickly confirmed, but also within a few years transformations induced by DNA preparations from other bacteria (*Escherichia coli, Hemophilus influenzae, Bacillus subtilis*, etc.) were described, and a variety of heritable properties were transferred. Further-more, work by Rollin Hotchkiss established that such bacterial trans-formations represent transfer of genetic material; subsequent studies (by Norton Zinder and Joshua Lederberg) showed that heritable bac-terial characters could also be transferred through their association with certain bacteriophages ("transduction"). In 1947 André Boivin (who worked with *E. coli*) wrote:

In bacteria—and, in all likelihood, in higher organisms as well—each gene has as its specific constituent not a protein but a particular desoxyribonucleic acid which, at least under certain conditions (directed mutations of bacteria), is capable of functioning *alone* as the carrier of hereditary character; therefore, in the last analysis, each gene can be traced back to a macromolecule of a special desoxyribonucleic acid (Boivin, 1947, p. 12).

Also, shortly after the appearance of the 1944 Avery paper a leading geneticist stated:

The great possible significance of this observation in the interpretation of the role of the nucleic acids of chromosomes and of other self-duplicating entities is obvious . . . . The results suggest the chemical isolation and transfer of a gene rather than induction of mutation (Wright, 1945, pp. 79–83.)

It should be added that, by 1947, clear evidence had been obtained that a DNA-containing structure comparable to the nuclei of the cells of higher organisms is present in bacteria, and that Lederberg and Tatum had reported the first instances of gene recombination in bacteria (*E. coli*) undergoing cell conjugation. Such cell conjugation could be considered to be analogous to the sexual union and chromosomal interaction known for higher organisms; consequently, the demonstration that bacterial transformation and transduction involves the transfer of genetic material without direct cell contact represented a major advance in genetic analysis, and had considerable impact on biological thought. The extensive studies on the genetics of bacteria that soon followed have been summarized in recent textbooks (*e.g.*, Hayes, 1968). It may also be added that the development of this field was greatly aided by the finding of chemical mutagens. Such agents had been sought for many years, but it was only in 1941 that the chemical-warfare agent mustard gas was found to produce a high mutation rate in Drosophila (Auerbach *et al.*, 1947); since then many other chemical substances have been shown to be mutagenic in a large variety of organisms (Auerbach, 1967).

For several years after the isolation of the transforming principle as a highly purified DNA preparation, the cautious reservation that the activity might be "due to minute amounts of some other substance" was emphasized by other investigators. Thus Mirsky wrote in 1950:

It is quite possible that DNA, and nothing else, is responsible for the transforming activity, but this has not been demonstrated conclusively. In purification of the active principle more and more protein attached to DNA is removed, as indeed in the preparation of DNA from any other source. It is difficult to eliminate the possibility that the minute quantities of protein that probably remain attached to DNA, though undetectable by the tests applied are necessary for activity—itself an exceedingly sensitive test . . . . There is accordingly, some doubt whether DNA is itself the transforming agent, although it can be regarded as established that DNA is at least part of the active principle (Mirsky, 1951, pp. 132–133).

Such doubts were reiterated by others, and are reminiscent of the earlier debates about the protein nature of enzymes and of the hormone insulin. Because of these doubts, much effort was expended (notably by Rollin Hotchkiss and Stephen Zamenhof) to purify transforming factors as completely as possible, and by 1952 active preparations were available whose protein content was estimated to be no more than 0.02%.

In that year there appeared a report by Alfred Hershey and Martha Chase, who concluded that

> . . . when a particle of bacteriophage T2 attaches to a bacterial cell, most of the phage DNA enters the cell, and a residue containing at least 80 percent of the sulfur-containing protein of the phage remains at the cell surface. This residue consists of the material forming the protective membrane of the resting phage particle, and it plays no further role in infection after the attachment of phage to bacterium.
>
> These facts leave in question the possible function of the 20 percent of sulfur-containing protein that may or may not enter the cell . . . . We infer that the sulfur-containing protein has no function in phage multiplication, and that DNA has some function.
>
> Our experiments show clearly that a physical separation of the phage T2 into genetic and non-genetic parts is possible . . . . The chemical identification of the genetic part must wait, however, until some of the questions asked above have been answered (Hershey and Chase, 1952, p. 54).

Although the chemical evidence for the cautiously worded conclusion that the genetic material of the phage is associated with its DNA was subject to the same kind of doubt, perhaps even in greater measure, as that expressed in regard to the transforming factor, the Hershey-Chase data were quickly accepted as proof of the idea that genes are made of DNA.

One factor in this favorable response was the emergence, after World War II, of a tightly knit group of investigators, led by Max Delbrück, Alfred Hershey, and Salvador Luria, who set themselves the task of studying bacterial viruses (in particular, the T phages of *E. coli* B) in a thorough and quantitative manner. After the discovery of the bacteriophages by Twort and d'Herelle, extensive studies were conducted during the 1920s in the hope that these agents might be useful in the treatment of infectious diseases. Although this expectation was not fulfilled, much was learned about the process whereby bacteria are dissolved upon phage infection. By 1926 it was known that the adsorption of the phage particle to the host cell, followed by the penetration of the particle into the bacterium, leads to the intracellular multiplication of the virus and lysis of the host cell with the release of phage progeny. During the 1930s significant advances were made (notably by Frank Burnet, André Gratia, and Max Schlesinger) in the development of methods for studying the infective cycle; of special importance was a technique described by Emory Ellis and Max Delbrück in 1939 permitting the determination of the effect of changes in the physical and chemical environment on the time of the cycle and on the number of virus particles produced per host cell. The introduction to the paper describing this technique reflected the views of its time:

> Certain large protein molecules (viruses) possess the property of multiplying within living organisms. This process which is at once so foreign to chemistry and so fundamental to biology, is exemplified in the multiplication of bacteriophage in the presence of susceptible bacteria.
>
> Bacteriophage offers a number of advantages for the study of the multiplication process not available with viruses which multiply at the expense of more complex hosts. It can be stored indefinitely in the absence of a host without deterioration. Its concentration can be determined with fair accuracy by several methods, and even the individual particles can be counted by d'Herelle's method. It can be concentrated, purified, and generally handled like nucleoprotein, to which class it apparently belongs . . . . (Ellis and Delbrück, 1939, p. 365).

During the 1940s important studies by Luria and Delbrück demonstrated that a single bacterial cell can be infected with more than one virus, thus providing the groundwork for detailed studies by Hershey on genetic recombination with mutant strains of T2 phage, and introducing the concepts developed by Morgan and Sturtevant into phage genetics. An important later consequence of the use of the bacteriophages for genetic analysis was the revision of the widely held belief that crossing-over occurs only "between" genes on a chromosome; in 1955 Seymour Benzer showed the genetic unit of function (which he termed "cistron") associated with a particular phenotypic character of T4 phage could be mapped in terms of smaller units of mutation and of recombination, thus providing experimental evidence for views expressed earlier by Richard Goldschmidt, Barbara McClintock, and Lewis Stadler.

We noted that Salvador Luria and Thomas Anderson had shown, by electron microscopy, that the T2 phage is a tadpole-shaped structure; by 1951 it was known from the work of Anderson and of Roger Herriott that the nucleic acid forms the core in the head of the phage, and that the overall shape of the particle is formed by a membrane largely composed of protein material which protects the DNA from degradation by deoxyribonuclease. In addition, electron microscopy had indicated that the T2 phage attaches itself to the host cell by means of its tail, and an essential technical aspect of the Hershey-Chase experiment was violent agitation in order to break off the head piece. Furthermore, by 1948 Seymour Cohen had not only shown that the nucleic acid of T2 phage is exclusively of the DNA type, but that, after a delay (7–10 minutes) following infection of *E. coli* cells, the metabolic activity of the host cells is channeled into the production of virus DNA in amounts much greater than the DNA of the normal bacteria; in this work he used the radioactive isotope of phosphorus (^{32}P) to show that the virus DNA is made by the enzymic apparatus of the host cell and that the phosphorus of this DNA is largely derived from inorganic phosphate in the culture medium. In other experiments, Cohen had also shown that, although the rate of protein formation after phage infection is essentially the same as before infection, most of the newly formed protein is of the type associated with the phage head membrane.

In their experiments, Hershey and Chase used ^{32}P to label the nucleic acid of phage DNA, and another radioisotope (^{35}S) to label the phage proteins. The operations in the removal of the labeled protein material from infected cells introduced factors of uncertainty, and, although numerous control experiments were performed to eliminate some of these factors, the results of Hershey and Chase did not constitute proof of the identity of the infective material with DNA, nor did they claim it. Nevertheless, after the appearance of their 1952 paper, the idea that DNA constitutes the genetic material was generally accepted, and there was no further dispute about Avery's claim that, in bacterial transformation, the hereditary factors of the donor strain are transferred to the recipient strain solely through DNA molecules. In this connection, it may be noted that in 1951 Northrop concluded from his studies on the formation of phage in *Bacillus megaterium* that:

The nucleic acid may be the essential, autocatalytic part of the molecule, as in the case of the transforming principle of pneumococcus (Avery, MacLeod, and McCarty, 1944), and the protein may be necessary only to allow entrance to the host cell (Northrop, 1951, p. 732).

Various explanations have been offered for the apparent neglect, by leading members of the "phage group," of the evidence provided by Avery and his colleagues eight years earlier; some of the possible reasons may be found in the retrospective collection of memoirs *Phage and the Origins of Molecular Biology* (edited by Cairns, Stent, and Watson), published in 1966 (see Kendrew, 1967), and in more recent articles by Stent (1970), Hotchkiss (1970), and Wyatt (1972).

By 1952, therefore, when James Watson and Francis Crick began to develop the double-helical model of DNA, the "phage group" (of which Watson was then a junior member, having received his doctorate in 1950 for work done with Luria) had accepted the idea that this substance constitutes the material basis of heredity. After describing their model, Watson and Crick offered a hypothesis regarding the biological replication of DNA:

Previous discussions of self-duplication have usually involved the concept of a template, or mould. Either the template was supposed to copy itself directly or it was to produce a "negative," which in its turn was to act as a template and produce the original "positive" once again. In no case has it been explained how it would do this in terms of atoms and molecules.

Now our model for deoxyribonucleic acid is, in effect, a *pair* of templates, each of which is complementary to the other. We imagine that prior to duplication the hydrogen bonds are broken, and the two chains unwind and separate. Each chain then acts as a template for the formation onto itself of a new companion chain, so that eventually we shall have *two* pairs of chains, where we had only one before. Moreover, the sequence of the pairs of bases will have been duplicated exactly (Watson and Crick, 1953b, p. 966).

As we noted before, this hypothesis received experimental support within a few years, although at the present writing there is still considerable uncertainty about the cellular mechanisms involved in the unwinding of the double helix and in the formation of the new chains complementary to the templates. The hypothesis provided a conceptual basis for the interpretation of gene mutation in terms of structural alteration of individual bases, of loss or gain of nucleotide units, or of more complex chemical changes such as the inversion of a nucleotide sequence. According to this model, genetic recombination may be considered to be a consequence of the breaking of internucleotide bonds in homologous DNA strands, and the rejoining of the pieces in a crossing-over process analogous to that postulated by Morgan for the chromosomes of Drosophila; the chemical mechanisms and enzymic apparatus involved in such recombination have been subjects of continuing investigation, largely with bacteria and phages. Furthermore, the Watson-Crick model, and especially the idea of specific base-pairing, has had far-reaching consequences in the experimental study of the manner in which the hereditary characters of an organism are expressed in its phenotypic properties, and gave rise to the concept of the "translation" of a "genetic code" represented by a linear sequence of nucleotide units into the linear sequences of the polypeptide chains of enzymes. The manifold fruitful applications of the Watson-Crick model in the interpretation of the extensive data gathered since 1953 on the genetics of bacteria and of phages may be found in present-day textbooks, and fall outside the scope of this chapter. We shall consider, however, some historical aspects of the problem whose beginnings we saw in the studies of Garrod—the role of genes in the formation of enzymes.

Genes and Enzymes. After Garrod's initial report on alcaptonuria, numerous other metabolic disorders in man were recognized to be hereditary, and shown to be transmitted as Mendelian recessive characters (see Harris, 1963). Among them was the condition known as phenylketonuria, because of the urinary excretion of phenylpyruvic acid; this disease was described by Ivar Følling (in 1934) who called attention to its association with serious mental deficiency. Subsequent work showed that, in phenylketonuric subjects, the liver lacks the enzyme system for the conversion of phenylalanine into tyrosine, the normal route for the degradation of phenylalanine via homogentisic acid; because of this metabolic block, phenylalanine is extensively deaminated to phenylpyruvic acid. Since 1950, a variety of human metabolic disorders (*e.g.*, galactosemia, glycogen storage diseases) have been attributed to deficiencies in individual enzymes that are involved in the normal metabolism of body constituents. In the case of alcaptonuria, a clinical finding helped to illumine a general metabolic pathway; after the elucidation of other pathways, whose history we will consider in a later chapter, clinical conditions of uncertain etiology have been shown to be associated with an inherited inability to make individual enzymes that function in these pathways of metabolism. The correctness of Garrod's surmise that in such conditions "a special enzyme . . . is wanting" has been borne out many times over.

We noted earlier that, before 1920, Huia Onslow and Sewall Wright had studied the genetics of hair color in experimental animals from the point of view indicated by Garrod. During the 1920s J. B. S. Haldane worked on the genetics of flower pigmentation in the hope of correlating differences in the chemical structure of the pigments with differences in the enzymic constitution of the various strains (Haldane, 1942, p. 78); it may be added that Haldane's important theoretical contributions to enzyme chemistry at that time were in part a consequence of his interest in the relation of genes to enzymes. These experimental approaches were followed by the significant contribution of George Beadle and Boris Ephrussi (1936), through their studies on the genetic control of eye color in Drosophila. They showed that a larval eye taken from a mutant ("vermilion"), and implanted in the abdomen of a wild-type larva, developed into an adult structure having the dull-red color characteristic of the wild-type eye, indicating that a hormonelike substance made by the wild-type organism is a precursor of the normal red pigment. This substance was found to be derived from the amino acid tryptophan, and later work showed it to be kynurenine, a known intermediate in the metabolic breakdown of tryptophan in mammals. Another of the eye-color mutants of Drosophila ("cinnabar") turned out to be unable to convert kynurenine into the red eye pigment, and it was clear that each of the two steps in the metabolic sequence tryptophan → kynurenine → red pigment is under the control of a separate gene.

In the continuation of these studies, Beadle was joined by Edward Tatum; as Beadle described it many years later, they

> . . . hit upon the idea of reversing the procedure we had been using to identify specific genes with particular chemical reactions. We reasoned that, if the one primary function of a gene is to control a particular chemical reaction, why not begin with known chemical reactions and then look for the genes that control them? In this way we could stick to our specialty, genetics, and build on the work chemists had already done. The obvious approach was first to find an organism whose chemical reactions were well-known and then induce mutations in it that would block specific identifiable reactions (Beadle, 1963, p. 13).

The organism they selected was the mold Neurospora, because the work of Bernard Dodge and Carl Lindegren had shown previously that its relatively short life cycle, its mode of reproduction, and its genetics made it especially suitable for biochemical genetic studies (Dodge, 1939). Furthermore, the nutritional requirements of this fungus for growth and reproduction were known to be quite simple; it grew in pure culture on a chemically defined medium consisting of sucrose, inorganic salts (including phosphate, nitrates, sulfates, etc.), and biotin (in 1941 a newly discovered member of the B-group of vitamins). From these materials the organism makes all the amino acids for its proteins, the purines and pyrimidines for its nucleic acids, as well as its many other cell constituents, including various vitamins (thiamine, pyridoxine, etc.). In their first report on Neurospora, Beadle and Tatum described studies with a mutant strain (obtained by irradiation with X-rays) that had lost the ability to make pyridoxine (a member of the vitamin B_6 group), and they concluded:

> It is clear from these rather limited data that this inability to synthesize vitamins B_6 is transmitted as it should be if it were differentiated from normal by a single gene.
>
> The preliminary results summarized above appear to us to indicate that the approach outlined may offer considerable promise as a method of learning more about how genes regulate development and function. For example, it should be possible, by finding a number of mutants unable to carry out a particular step in a given synthesis, to determine whether only one gene is ordinarily concerned with the immediate regulation of a given specific chemical reaction (Beadle and Tatum, 1941, p. 505).

There followed a series of important studies in which mutants of *Neurospora crassa* were used to determine the sequence of chemical reactions in the biosynthesis of a variety of cell constituents, especially several amino acids (arginine, tryptophan, valine, isoleucine). We shall return to this aspect of the work with Neurospora in a later chapter. For the problem of the relation of genes to enzymes, however, the most significant early finding was:

> When metabolically deficient strains are investigated by the above procedure, the great majority of them prove to differ from the original wild-type in the alteration of a single gene. Occasionally double mutants are obtained that require two substances for growth. In all but one instance, these have proved to have two mutant genes . . . . The

one exceptional case so far encountered involves a mutant strain which differs from the wild type by a single gene but which requires both valine and isoleucine for normal growth . . . . It is supposed that these related amino acids have a common step in their synthesis and that it is this step which the gene in question normally controls (Beadle, 1945b, p. 186).

In discussing the bearing of the results then available on the problem of the relation of genes to enzymes, Beadle suggested that

. . . the gene can be visualized as directing the final configuration of a protein molecule and thus determining its specificity. A given protein molecule, patterned after a particular gene, might become a component of a new gene like the one from which it was copied or it might become an antigenically active protein, an enzyme protein, or a storage protein . . . . If genes in some way direct the configurations of protein molecules during their elaboration, it is not necessary to assume that they function in any other way . . . . The protein components of enzymes would . . . have their specificities imposed fairly directly by genes and the one-to-one relation observed to exist between genes and chemical reactions should be a consequence. It should follow, indeed, that every enzymatically catalyzed reaction that goes on in an organism should depend directly on the gene responsible for the specificity of the enzyme concerned. Furthermore, for reasons of economy in the evolutionary process, one might expect that with few exceptions the final specificity of a particular enzyme would be imposed by only one gene (*ibid.*, pp. 191–192).

In the statement of his hypothesis, Beadle inclined to the then-current view of genes as proteins or nucleoproteins, but the idea itself [termed by Norman Horowitz (1948) the one-gene one-enzyme hypothesis] had considerable fruitful impact on biological thought, after some initial doubts [see, for example, the comments after a lecture by Bonner (1946)].

With the acceptance, after 1952, of the concept that genes are strands of DNA, the problem posed by the one-gene one-enzyme hypothesis became one of explaining the manner in which the nucleotide sequence of DNA is expressed in the specificity of enzymes. As we have seen, there was general agreement by that time with regard to the protein nature of enzymes, and Sanger's work on insulin removed the last lingering doubts about the polypeptide structure of proteins. Furthermore, the idea that a mutation in a single gene might lead to a single change in a polypeptide chain of a protein received strong support when Vernon Ingram reported a "difference in the amino-acid sequence in one small part of one of the polypeptide chains" (Ingram, 1956, p. 794) of normal human hemoglobin A and the hemoglobin S of patients with sickle-cell anemia (see p. 175); James Neel had already provided (in 1949) genetic evidence for the view that the formation of hemoglobin S is due to a mutation in a single gene. After these advances in protein chemistry, Beadle's suggestion became the "one-gene one-polypeptide" hypothesis.

During the 1950s there was extensive speculation about the manner in which the nucleotide sequence of a DNA strand might find expression in the amino acid sequence of a polypeptide chain. One of the early proposals was that offered by the astrophysicist George Gamow, who stated that

> . . . the hereditary properties of any given organism could be characterized by a long number written in a four-digital system. On the other hand, the enzymes (proteins), the composition of which must be completely determined by the desoxyribonucleic acid molecule, are long peptide chains formed by about twenty different kinds of amino-acids, and can be considered as long "words" based on a 20-letter alphabet. Thus the question arises about the way in which four-digital numbers can be translated into such "words" (Gamow, 1954, p. 318).

Gamow suggested that each amino acid can interact specifically with a triplet of adjacent nucleotides, because, if one has four different units and takes them three at a time, there are twenty possible combinations, their order being disregarded. Although this simple mathematical answer to a very complex problem has been superseded by ideas based on experimental data, the concept of a genetic code composed of triplets became firmly established during the 1960s, largely through the work of Marshall Nirenberg, Severo Ochoa, and Gobind Khorana; this "cracking of the code" represents one of the most exciting developments in the history of the interplay of chemistry and biology, and is described in present-day textbooks. Before these achievements, however, there had been several remarkable deductions, later supported by experiment. The work of Brachet and Caspersson during the 1930s had indicated that RNA plays a significant role in cellular protein synthesis; in 1957 Mahlon Hoagland, Paul Zamecnik, and Mary Stephenson reported that a particular fraction of the cytoplasmic RNA (now called transfer RNA) is directly involved in the incorporation of amino acids into polypeptides. In that year, Francis Crick proposed that protein synthesis occurs on

> . . . an RNA template in the cytoplasm. The obvious place to locate this is in the microsomal particles, because their uniformity of size

suggests that they have a regular structure. It also follows that the synthesis of at least some of the microsomal RNA must be under the control of the DNA of the nucleus . . . . Granted that the RNA of the microsomal particles, regularly arranged, is the template, how does it direct the amino acids into the correct order? . . . It is . . . a natural hypothesis that the amino acid is carried to the template by an "adaptor" molecule, and that the adaptor is the part which actually fits on to the RNA. In its simplest form one would require twenty adaptors, one for each amino acid.

What sort of molecules such adaptors might be is anybody's guess . . . . [T]here is one possibility which seems inherently more likely than any other—that they might contain nucleotides. This would enable them to join on to the RNA template by the same "pairing" of bases as is found in DNA (Crick, 1958, pp. 153–155).

Subsequent work with bacteria and viruses provided experimental evidence for the view that the "RNA template" is a form of RNA made by the enzyme-catalyzed "transcription," through a base-pairing mechanism, of the nucleotide sequence of DNA, and transferred to cytoplasmic particles (ribosomes) as "messenger RNA" (Jacob and Monod, 1961), while Crick's "adaptors" turned out to be transfer RNA molecules that have triplet "anticodons" which interact specifically by base-pairing with complementary triplets of the messenger RNA.

Crick based his deductions, in part, on a principle which he denoted "The Central Dogma":

This states that once "information" has passed into protein *it cannot get out again.* In more detail, the transfer of information from nucleic acid to nucleic acid, or from nucleic acid to protein may be possible, but transfer from protein to protein, or from protein to nucleic acid is impossible. Information means here the *precise* determination of sequence, either of bases in the nucleic acid or of amino acid residues in the protein (Crick, 1958, p. 153).

Our historical account stops here, on the threshold of the remarkable developments during the 1960s in the analysis of the chemical basis of gene expression in bacteria. The chemical understanding of chromosome behavior and of the control of gene action in higher organisms is less advanced, however.

We have seen in this chapter something of the interplay between chemistry and biology in the search for the physical basis of heredity. It should be recognized, however, that in stressing the place of Miescher's nuclein in this search we have perhaps taken undue advantage of hindsight. All the chemists, from Kossel to Levene, who laid the groundwork for the double-helical model of DNA either were indifferent to its biological function or were skeptical about its role in heredity. Moreover, the mechanism of inheritance was defined by cytologists and geneticists, many of whom, like Boveri and Morgan, either refrained from conjectures about the chemical nature of the genetic material or affirmed that chemistry could contribute little to the understanding of the mechanism. If, before 1940, some biologists and physicists speculated about the chemistry of the gene, or some chemists guessed at the biological role of the nucleic acids, most of their surmises could not be tested experimentally. By the mid-1930s the ambiguous choice between nucleic acids and proteins as the "gene molecules" had been made in favor of the proteins. These had long held the center of the biochemical stage, and had now emerged as well-defined linear macromolecules whose structural complexity seemed to be consistent with the manifold variety of genetic expression. The dimensions of the "gene molecule" were estimated from X-ray mutation studies to be similar to those of proteins, the infectivity of the plant virus nucleoproteins appeared to be associated with their protein component, and enzymes were now known to be proteins. On the other hand, the tetranucleotide structure of the nucleic acids was too simple, and the cytological evidence for their role in heredity was not considered to be compelling. These speculations of the 1930s about the chemical nature of the gene, though indecisive, helped to set the tone of later discussion of the problem, after new chemical knowledge had shifted attention back to the nucleic acids. It would seem, therefore, that from 1880 to 1940 the interplay between chemistry and biology in the study of the physical basis of heredity was tentative and uncertain; during that period, it was extremely difficult to evaluate the biological significance of individual chemical contributions at the time they were made.

After World War II there was a significant change in the relation of chemistry to genetics as a consequence of several essentially independent developments, some of which were the outcome of parallel work on proteins. They included the recognition that DNA and RNA are linear macromolecules, Avery's identification of the pneumococcal transforming factor as DNA, and Chargaff's chromatographic analysis of DNA preparations. The emergence of phage biochemistry led to the Hershey-Chase experiment, and Pauling's work on the helical structure of proteins was followed by the Watson-Crick model of DNA. The acceptance of this model, together with the formulation of the one-gene one-enzyme hypothesis stemming from the work of Beadle and Tatum, profoundly influenced biological thought and transformed the tentative interplay of chemistry and genetics into a conceptually unified area of research. This branch of science is sometimes called "molecular genetics"; it is noteworthy, however, that, in contrast to the ideas of the 1930s, the gene is not viewed as a discrete chemical molecule, but as a limited sequence of nucleotides in a linear polynucleotide chain. In current parlance, this inherited sequence provides the "information" (an older term was "organic memory") for the elaboration and integrated operation of the chemical assembly that constitutes the living organism. Furthermore, accidental changes in the sequence lead to alterations in the assembly and, during the course of many generations, to the emergence of new biological species.

The phenomenon of heredity thus links the physiology of individual organisms to the process of biological evolution through natural selection. The vast area of biology that seeks to describe the behavior of living organisms in relation to each other and to their physical environment, and to understand evolutionary change in terms of mutations that have affected the reproduction of individual species, has until recently remained outside the arena in which chemistry has directly influenced biological thought. It is otherwise with the problems posed by the physiological functions of individual organisms. Here, under the impetus of medical and agricultural practice, there has long been an intimate interplay of chemistry and biology. In the next two chapters we shall examine some historical aspects of this relationship.

Intracellular Respiration

In a famous memoir dated 1789, Armand Séguin and Antoine Lavoisier wrote of the respiration of animals that

> . . . in general, respiration is nothing but a slow combustion of carbon and hydrogen, which is entirely similar to that which occurs in a lighted lamp or candle, and that, from this point of view, animals that respire are true combustible bodies that burn and consume themselves.
>
> In respiration, as in combustion, it is the atmospheric air which furnishes oxygen and caloric; but since in respiration it is the substance itself of the animal, it is the blood, which furnishes the combustible matter, if animals did not regularly replace by means of food aliments that which they lose by respiration, the lamp would soon lack oil, and the animal would perish, as a lamp is extinguished when it lacks nourishment.
>
> The proofs of this identity of effects in respiration and combustion are immediately deducible from experiment. Indeed, upon leaving the lung, the air that has been used for respiration no longer contains the same amount of oxygen; it contains not only carbonic acid gas but also much more water than it contained before it had been inspired. Now since the vital air can only convert itself into carbonic acid by the addition of carbon; since it can only convert itself into water by the addition of hydrogen; since this double combination cannot occur without the loss, by the vital air, of a portion of its specific caloric, it follows that the effect of respiration is to extract from the blood a portion of carbon and hydrogen, and to deposit there a portion of its specific caloric which, during circulation, distributes itself with the blood in all parts of the animal economy, and maintains that nearly constant temperature observed in all animals that breathe.
>
> One may say that this analogy between combustion and respiration has not escaped the notice of the poets, or rather the philosophers of

antiquity, and which they had expounded and interpreted. This fire stolen from heaven, this torch of Prometheus, does not only represent an ingenious and poetic idea, it is a faithful picture of the operations of nature, at least for animals that breathe; one may therefore say, with the ancients, that the torch of life lights itself at the moment the infant breathes for the first time, and it does not extinguish itself except at death.

In considering such happy agreement, one might sometimes be tempted to believe that the ancients had indeed penetrated further than we think into the sanctuary of knowledge, and that the myth is actually nothing but an allegory, in which they hid the great truths of medicine and physics (Lavoisier, 1862, Vol. II, pp. 691–692).

The memoir appears to have been read to the French Academy of Sciences on 17 November 1791, and was published in 1793. This was a period of increasing desperation and cruelty in the French Revolution; Lavoisier was guillotined on 8 May 1794 (McKie, 1952; Fayet, 1960). The final paragraph in the memoir reads:

We conclude this memoir with a consoling reflection. To be rewarded by mankind and to pay one's tribute to the nation, it is not essential to be called to those public and brilliant offices that contribute to the organization and regeneration of empires. The physicist may also, in the silence of his laboratory and his study, perform patriotic functions; he can hope, through his labors, to diminish the mass of ills that afflict humanity, to increase its happiness and welfare; and if he has only contributed, through the new avenues he has opened, to prolong the average life-span of human beings by a few years, even by a few days, he could also aspire to the glorious title of benefactor of humanity (Lavoisier, 1862, Vol. II, p. 703).

As before, and since, the appeal to public authority on behalf of a scientist, because he is a "benefactor of humanity," was unavailing.

The conclusions regarding the chemical events in animal respiration were an outgrowth of Lavoisier's explanation of combustion as an uptake of oxygen. In 1777 he had concluded that during respiration, as in the combustion of charcoal, oxygen (he then called it *l'air éminemment respirable*) is removed from the air and is converted into carbon dioxide (*acide crayeux aériforme*); he suggested that this process occurs in the lungs, and that oxygen combines with the blood to give the latter its ar-terial red color. By 1780 he and Pierre Simon Laplace had performed their celebrated experiment on the amount of heat released when equal

amounts of carbon dioxide were produced in the combustion of charcoal
and in the respiration of a guinea pig; for these studies an ice calorimeter
was devised by Laplace, and the amount of ice melted in the two processes was compared. Although the use of the ice calorimeter depended
on Black's prior discovery of the latent heat of fusion of ice (McKie and
Heathcote, 1935), his name is not mentioned in the 1780 paper. Lavoisier
and Laplace found that, during ten hours, the animal had produced a
quantity of heat sufficient to melt 13 ounces of ice, whereas the burning
of charcoal (to yield an amount of carbon dioxide equal to that exhaled
by the animal) melted 10½ ounces of ice. No account was taken of the
fact that the two processes had occurred at different temperatures, and
the difference of 2½ ounces was attributed largely to the cooling of the
limbs of the animal. The oft-quoted conclusion was:

> Respiration is therefore a combustion, very slow it is true, but other-
> wise perfectly similar to that of charcoal; it occurs in the interior of the
> lungs, without producing perceptible light, because the liberated mat-
> ter of fire is immediately absorbed by the humidity of these organs
> (Lavoisier, 1862, Vol. II, p. 331).

A contemporary, and complementary, theory of animal heat was advanced
in 1779 by Adair Crawford, a student of Joseph Black and William Irvine;
he applied their studies on latent and specific heats and suggested that
the inspired air contains a great amount of "absolute heat" which is released in the pulmonary blood. For a valuable account of the development
of these early views on animal heat see Mendelsohn (1964).

In 1785, on the basis of further measurements, Lavoisier had found a
discrepancy between the amounts of oxygen removed from the air and of
the carbon dioxide exhaled, and stated:

> It is therefore clear that independently of the portion of vital air
> [oxygen] which is converted to carbonic acid, a portion of it that en-
> tered the lung did not emerge in that state; it follows that . . . either
> a portion of the vital air combines with the blood, or it combines with
> a portion of hydrogen to form water (Lavoisier, 1862, Vol. II, p. 680).

It is noteworthy that shortly before this was written Lavoisier had confirmed Cavendish's synthesis of water from hydrogen and oxygen. The
studies undertaken with Séguin were intended to clarify the situation
through gas analysis (eudiometry); Séguin (1791) improved the method
by introducing phosphorus for the selective absorption of oxygen present
in gas mixtures. From their studies, Séguin and Lavoisier concluded that

the operation of *la machine animale* is controlled by three processes: respiration, a variable transpiration through the skin depending on the need to remove heat, and digestion, which provides the blood with "chyle" for the replacement (by combustion) of the heat lost in the other two processes.

The memoir of Séguin and Lavoisier served as the starting point of the discussion, throughout most of the nineteenth century, of two central questions of animal physiology: in what parts of the animal body does the conversion of oxygen into carbon dioxide and water occur, and in what form does the carbon and hydrogen used for respiration exist? We have seen that Lavoisier repeatedly suggested that the lungs might be the site of physiological combustion. As for the material supposedly burned in the lungs, Séguin and he proposed (in a memoir dated 1790) that it is "a humor . . . principally composed of hydrogen and carbon" (Lavoisier, 1862, Vol. II, p. 708) brought by the blood, and derived from the food. Although these views were adopted by many of Lavoisier's followers, there were also contemporary criticisms; one especially deserves to be cited:

> M. de la Grange [Lagrange] considered that if all the heat which distributes itself in the animal economy were released in the lungs, it would necessarily follow that the temperature of the lungs would be increased to such an extent that one might continually fear their destruction, and if the temperature of the lungs were so considerably different from the other parts of animals, it was impossible that this had never been observed. He believed it to be very probable that all the heat of the animal economy is not solely released in the lungs, but also in all parts where the blood circulates.
>
> He supposed . . . that the blood in passing through the lungs dissolved the oxygen of the respiratory air, [and] that this dissolved oxygen was carried by the blood in the arteries and thence to the veins; that during the flow of the blood, the oxygen gradually left its dissolved state to combine partially with the carbon and hydrogen of the blood and form water and the carbonic acid which is released from the blood as soon as the venous blood leaves the heart to enter the lungs (Hassenfratz, 1791, p. 266).

Much later (in 1873) Berthelot showed that Lagrange's objection had been ill-founded.

In 1807 Jean Senebier published Lazzaro Spallanzani's findings (*Rapports de l'air avec les êtres organisés*) on the formation of carbon dioxide by a great variety of animal tissues; these results suggested that respiratory combustion is not restricted to the lungs. Furthermore, Théodore de Saussure (in 1804) extended the earlier work of Ingen-Housz and of Senebier to show that, in the dark, plant tissues take up oxygen and release carbon dioxide. In spite of these evidences that respiration is a general property of living tissues, the view that animal respiration occurs only in the lungs persisted for several decades.

Respiration as Intracellular Combustion

During the first four decades of the nineteenth century, repeated attention was given to the two major problems raised by the ideas of Séguin and Lavoisier. If Lagrange was correct in his criticism, there should be gases in the circulating blood, and a succession of investigators sought to determine whether such gases could be found there. For many years the various positive reports were insufficiently convincing to force adherents of the theories of Lavoisier and of Crawford to abandon the view that the conversion of oxygen into carbon dioxide occurs principally in the lungs, since several noted investigators (*e.g.*, Mitscherlich, Gmelin, Tiedemann) had reported failure in attempts to extract air from blood by means of an air pump. This uncertainty was removed, however, when Gustav Magnus (1837) clearly established the presence of dissolved oxygen, carbon dioxide, and nitrogen in the blood, and showed that it was possible to extract these gases by means of the higher vacuum of a mercury pump (for a biography of Magnus, see Hofmann, 1870). His analytical data suggested that arterial blood has a higher proportion of oxygen than venous blood; he concluded, as had Lagrange and others before him:

> It is very probable from these [results] that the gaseous oxygen is absorbed in the lungs and is carried by the blood around the body, so that it serves in the so-called capillary vessels for an oxidation, and probably to a formation of carbonic acid . . . it follows necessarily that, during the circulation of the blood, the carbonic acid is formed there, or is taken up by it (Magnus, 1837, pp. 602–604).

Although Magnus's results were criticized by Gay-Lussac and Magendie, they were soon confirmed in several laboratories, and around 1840 it was widely considered that the respiratory combustion occurs in the capillaries of the tiessues. By that time also, it had been reported that whole blood holds more oxygen than does serum, indicating that the "red globules carry the gas (Hünefeld, 1840, p. 103).

Most of the quantitative experiments during the period 1780–1840 were performed by men who identified themselves as physicists or chemists, and who studied animal organisms in the sense of what Lavoisier called *la machine animale*. An early critic of this approach was the brilliant young clinician Xavier Bichat who, while disavowing belief in metaphysical principles such as those advocated by eighteenth-century vitalists (Stahl, Hunter, Barthez), insisted on considering animal organisms in terms of the physiological functions of their tissues. Thus, in 1801, he wrote:

> Attempts have been made recently to determine precisely what amount of oxygen is absorbed, what amount is required to produce the water of respiration, what amount of carbonic acid gas is formed, what amount of heat is released, etc. . . . This precision would be advantageous if it could be attained; but not a single phenomenon of the animal economy can permit it in the explanations they give rise to. Chemists and physicists, accustomed to study the phenomena over which physical powers preside, have carried their mathematical spirit into the theories they have imagined to apply to the laws of vitality but it is no longer the same thing. The mode of theorizing about the organized bodies must be quite different from that of theories applied to the physical sciences (Bichat, 1801, Vol. I, pp. 534–535).

The claim that quantitative physiological measurements are pointless was roundly criticized by François Magendie, although he and his students accepted Bichat's view that the phenomena exhibited by animal organisms must be studied in their own right, and in relation to the specific structure and composition of individual tissues. This attitude was to find its clearest expression in the writings of Claude Bernard later in the century, and Bichat's negative reaction to Lavoisier's work on respiration found reflection in Bernard's attitude to the physiological theories of Justus Liebig and his followers.

Liebig's enormous reputation at midcentury, especially in Germany and England, placed his ideas at the center of discussion, and has colored many historical accounts of the development of biochemical thought. We have already seen something of his role in the controversy about the cause of alcoholic fermentation, and in the study of the nature of proteins. The ideas he expressed on these subjects formed part of a larger design which he developed in successive editions of his two popular books on the chemistry of agriculture and on animal chemistry, as well as his best-seller, *Familiar Letters on Chemistry* (first published in 1843). These views were usually presented with such complete assurance, and often without acknowledgment to his immediate predecessors, that their truth and originality seemed incontrovertible to the reading public. On the other hand, some of Liebig's distinguished scientific contemporaries found his writings on physiological subjects to be incorrect, if indeed not harmful; Berzelius's reaction to Liebig's *Animal Chemistry* was mentioned earlier (p. 97), and, in a letter (22 April 1848) to his friend Karl Ludwig, Emil duBois-Reymond wrote:

> I esteem Liebig's talent very highly, and the excellence of his earlier non-physiological theoretical studies are much too highly regarded for me, as a non-chemist, to allow myself to take it into my head to doubt it. With regard to his character, I know of course that he has been accused of unfairness in his polemics etc. . . . . I consider his physiological fantasies as worthless and pernicious, because he entirely lacked the necessary factual knowledge and critical training (duBois-Reymond, 1927, p. 19).

Liebig's view of animal respiration was based on that of Séguin and Lavoisier. In accepting the idea that respiratory oxygen is entirely converted into carbon dioxide and water, and that this combustion is the cause of animal heat, Liebig chose to dismiss the inconsistencies in the available data on the ratio of carbon dioxide output to oxygen intake, and on the relation of oxygen consumption to heat production. Thus Allen and Pepys had reported almost exact equivalence in oxygen intake and carbon dioxide output, and concluded that "there is no reason to conjecture that any water is formed by a union of oxygen in the lungs" (Allen and Pepys, 1808, p. 279). On the other hand, Dulong (in 1823) and Despretz (in 1824) stated that more oxygen was consumed during animal respiration than could be accounted for by the carbon dioxide production, and that the heat output was less than that calculated on the basis of the combustion theory of respiration, even if the excess oxygen were assumed to have been converted to water. Furthermore, Benjamin Brodie (1812) showed that despite artificial respiration the temperature of an animal with impaired brain function fell rapidly. Such experiments led physiologists to question the validity of the Lavoisier theory of animal respiration, and to favor the involvement of the nervous system in heat production; additional evidence for this view was provided by Legallois and by Chossat. Indeed, as long as heat was considered to be a material substance that came from the nutrients and the respiratory oxygen, its generation within the organism by a chemical process was difficult to conceive, and an undefined action of the nervous system seemed to be a plausible explanation. As late as 1840, in the third edition of his treatise on chemistry, Berzelius considered that the source of animal heat was unknown. Liebig, however, dismissed these doubts: "The interaction between the constituents of the food and the oxygen distributed in the body by the blood circulation is *the source of animal heat*" (Liebig, 1846b, p. 18). As for the food constituents subjected to oxidation, Liebig adopted the classification proposed by William Prout (saccharine, oily, and albuminous), and divided them into two groups: (1) the so-called "respiratory elements" of the diet included the non-nitrogenous materials (fat, starch, sugars, as well as wine, beer, spirits, etc.), and (2) the nitrogenous components, represented by the plant albumin, casein, and fibrin (eaten by herbivorous animals) and animal flesh and blood (eaten by carnivores), which he called the "plastic elements" of nutrition, and convertible into blood. In Liebig's view, as expressed in his *Animal Chemistry*, the fats and carbohydrates are the fuel of the animal body. Carbohydrates are not deposited at all, since no starch or sugar had been detected in the arterial blood, and fats are "unorganized" substances that are deposited only when there is insufficient oxygen for their complete combustion. The central place in the machinery of the animal body was occupied by the "vital" proteins of the blood and of the tissues, especially muscle. During muscular exercise or starvation, the respiratory oxygen attacks these vital proteins, with the release of nitrogenous breakdown products such as uric acid, urea, and ammonia, which are excreted in the urine.

This grand design rested on so many unfounded assumptions that it soon collapsed under experimental scrutiny. It was based on the faith that the experimental method which had led to the efflorescence of organic chemistry after 1830 could contribute decisively to the solution of physiological problems. We have seen that the development of reliable methods of elementary analysis by Gay-Lussac, Thenard, Berzelius, Dumas, and Liebig laid the groundwork for the fruitful study of such complex materials as the animal fats or uric acid. As early as 1834, Liebig had called for the same approach in research on animal respiration:

Our insight into the mysterious processes of nutrition etc. of the animal organism will gain an entirely different significance when, instead of being satisfied with decomposing the materials present in the greatest variety of organs into numerous other compounds, from whose properties we can learn nothing, we will follow step by step their changes and transformations through elementary analysis, without regard to their properties . . . .

We know that the oxygen of the air has a definite relation to the blood in the respiratory process, we determine the changes undergone by the air, and observe the phenomena in the lung; but if chemistry does not succeed in following and understanding all transformations in the organs and the accompanying interactions of the constituent substances, it is not worth the trouble to occupy oneself with such studies (Liebig, 1834, p. 3).

It was largely from the results of elementary analysis that Liebig drew the conclusions summarized above (Holmes, 1963); thus, from an examination of the data obtained by Mulder and by Liebig's assistant Scherer on the elementary composition of various protein preparations, he concluded that, in the transformation of blood proteins (albumin, fibrin) to tissue proteins (*e.g.*, tendons), oxygen is added. Furthermore, he attached physiological significance to analytical data such as the following:

> The principal components of the blood contain, per equivalent of nitrogen, 8 equivalents of carbon; no constituent of the body, possessing a characteristic form, contains less than one equivalent of nitrogen per 3 equivalents of carbon (Liebig, 1846b, p. 49).

The harsh criticism accorded the first edition of his *Animal Chemistry* (published in 1842) caused Liebig to omit from the 1846 third edition most of the speculations, derived from comparisons of elementary composition, about the transformations of chemical constituents of the animal body (see Holmes, 1964). Also, by 1846 Liebig had abandoned Mulder's protein theory, which had provided the basis for the idea that plant proteins are taken into the blood of herbivorous animals without change. There can be little doubt, however, that many felt as did Friedrich Mohr, in writing to his friend Liebig in 1842:

> Your discoveries on the economy of the animal organism are tremendous [*kolossal*], the physicians watch dumbstruck the approach of a period in which they will be shaken from their customary torpor (Kahlbaum, 1904, p. 67),

and some sense of Liebig's euphoria is suggested by the following sentence in his reply: "Chemistry, the noblest and most beautiful of all sciences, will and must gain the victory" (*ibid.*, p. 72).

Liebig was not alone among the leading chemists of his time in believing that important physiological conclusions could be drawn from comparisons of the elementary composition of chemical constituents of living organisms. During 1840–1844 he and Jean Baptiste Dumas conducted a lively argument over priority in the statement of many of the ideas in the first edition of Liebig's *Animal Chemistry*. Early in his scientific career Dumas had worked with the physiologist Prévost; among their joint successes was the demonstration (in 1823) of the accumulation of urea in the blood after surgical removal of the kidneys. As for Liebig, his only recorded physiological investigation with living subjects was the determination, in November 1840, of the amount of carbon in the food and excreta of a company of soldiers; such studies had been performed much more impressively on animals by Jean Baptiste Boussingault some years before. Despite Dumas's greater experience as a physiologist, he vied with Liebig in chemical speculation about physiological processes, including animal respiration. In a famous lecture given on 20 August 1841, Dumas used analytical data provided by Boussingault to develop his grand design:

> Green plants constitute the great laboratory of organic chemistry. It is they which, with carbon, hydrogen, nitrogen, and ammonium oxide, slowly build the most complex organic materials. They receive from the solar rays, in the form of heat or chemical radiation, the power needed for this work.
>
> Animals assimilate or absorb the organic materials made by plants. They change them bit by bit, they destroy them. In their tissues or ducts, new organic materials may arise; but these are always simpler substances, closer to the elementary state than those which they have received. They therefore decompose bit by bit these organic materials created by plants; they bring them back bit by bit toward the state of carbonic acid, of water, of nitrogen, of ammonia, the state that permits them to be restored to the air. In burning or in destroying these organic materials, animals always produce heat which, radiating from their bodies into space, serves to replace that absorbed by plants (Dumas, 1844, pp. 45–46).

During the course of the polemic with Liebig, there was also occasion for patriotic fervor:

> We are also convinced that if the views that we have summarized in this lecture retain in the future the importance accorded them today, it will have been shown that the principal studies upon which these views are founded were performed in France, and that their synthesis into a general formula goes back to Lavoisier, who gave to physiology an impetus that we have followed (*ibid.*, p. 139).

To return to Liebig's ideas about biological oxidation, it is difficult to escape the impression that he considered oxygen to be harmful to the tissues of the animal organism; thus, in writing of the role of the fats and carbohydrates in providing carbon and hydrogen for respiratory combustion, he stated that

> . . . there can hardly be any doubt that this excess of carbon alone, or of carbon and hydrogen, is used for the production of animal heat and for resistance against the external action of oxygen (Liebig, 1846b, p. 62).

Such action of oxygen on the tissues, when it occurred outside the body, Liebig (1839, p. 264) had denoted *Verwesung* or *eremacausis* (Greek *erema*, quietly + *causis*, burning); he differentiated such slow oxidative decomposition from putrefaction, in which atmospheric oxygen did not take part, and from mouldering, which occurred under conditions of low oxygen and moisture. As we saw before, Liebig had opposed the theory of fermentation advanced by Caignard-Latour, Schwann, and Kützing; he considered the process to be a form of putrefaction in which "no unpleasant odor was developed" (*ibid.*, p. 265). It is perhaps necessary to see Liebig's ideas during the 1840s about animal respiration against the background of his preoccupation with the problem of fermentation. Thus, in the first edition of *Animal Chemistry*, he wrote that

> . . . if we consider the transformation of the constituents of the animal body (the utilization of matter by the animal) as a chemical process, which proceeds under the influence of the vital force, then their putrefaction outside the animal body is a decomposition into simpler compounds, in which the vital force does not participate. The action in both cases is the same, only the products are different (Liebig, 1842, p. 110).

There is a reasonably clear connection between this statement and Liebig's view of alcoholic fermentation as the transmission, to the sugar molecules, of the molecular motion of proteins that are undergoing oxidative decomposition.

Although most of Liebig's physiological speculations were rejected by several contemporary critics, among them Jacob Moleschott (in his *Der Kreislauf des Lebens*, first published in 1852), and soon became untenable, there can be little doubt that they stimulated the study of physiological processes from a chemical point of view. The public position that he had won through his achievements in organic chemistry, both as an investigator and as a teacher, and through his popular writings on agriculture and nutrition, assured a respectful hearing to his bold hypotheses. In seeking either to prove or disprove Liebig's ideas, his defenders and critics were spurred to examine more closely substances (such as the proteins) present in living organisms, and to initiate the intensive study of the metabolic transformation of these substances. In particular, his views on the conversion of muscle protein into urea stimulated an extensive series of investigations during 1848–1860 on the fate of nitrogen compounds in the animal body; we shall return to this work in the next chapter.

The Rise of Thermodynamics. Of special importance in the development of knowledge about the chemical basis of the respiratory process was the impetus given by Liebig's speculations to the emergence of clearer ideas about the transformation of various forms of energy into each other. James Watt's development of the steam engine, during 1770–1785, and its decisive impact on European technology after 1790, made the theoretical problem of the relation of heat to motion a central question of nineteenth-century physics. There was a rapid development of thermochemistry, especially during 1820–1850; numerous measurements were made of the heat produced in chemical reactions, with better apparatus than that available to Lavoisier and Laplace in the 1780s. In 1840–1842 Germain Hess established the "law of heat summation" according to which ". . . the heat developed in a chemical change is constant, whether the change occurs directly, or indirectly in several steps" (Hess, 1840, p. 392). The subsequent work of Andrews and of Favre and Silbermann (during 1844–1853) contributed many data consistent with Hess's law; in 1853 Favre introduced the term "calorie" for the unit of heat. In his speculations, Liebig implicitly assumed the law of heat summation but, as was noted by Hermann Helmholtz in 1845, accurate calorimetric data for the heat of combustion of foodstuffs were needed, and some of Liebig's assumptions were doubtful; in particular, Helmholtz questioned the procedure of equating the heat of combustion of a substance with the sum of the heats of combustion of the carbon and hydrogen in the substance (Helmholtz, 1882, Vol. I, p. 9). During the second half of the nineteenth century, major advances were made in calorimetry, mainly by Julius Thomsen and by Marcelin Berthelot; the bomb calorimeter for combustions in oxygen was introduced by Berthelot in 1881, and he also proposed that chemical reactions be denoted *exothermic* or *endothermic*, depending on whether they were accompanied by the release or uptake of heat. Max Rubner (1894) was able to account for the heat production of a dog by measuring its carbon and nitrogen balance in a respiration calorimeter, and by determining the heats of combustion of dog fat, dog protein, and the organic constituents of dog urine; since he found that the heat produced by the animal equaled the heats of combustion of the fat and protein minus that of the urinary matter, Hess's law was taken to apply to the animal organism, as Liebig had assumed. The further development of animal calorimetry, although important for subsequent studies on human nutrition, falls outside the scope of this chapter (see Kleiber, 1961).

Aside from the stimulus given by Liebig's speculations to thermochemistry, the discussion about his ideas during the 1840s played a role in the emergence of thermodynamic theory. After the work of Rumford and Davy on the production of heat by friction, the seventeenth-century idea of heat as a mode of motion (Bacon, Boyle, Locke, Newton) returned to challenge the materiality of heat, favored in the era of phlogiston, and adopted by Lavoisier. The idea of caloric lingered on well into the nineteenth century (see Fox, 1971), as is evident from its retention by Sadi Carnot in his *Réflexions sur la Puissance Motrice du Feu* (published in 1824); he analyzed the relation of heat to motion in a steam engine in terms of a cycle of operations, and proved that the most efficient engine is one in which all the operations are reversible. During the 1830s, however, the caloric theory suffered severe blows from the demonstration by Macedonio Melloni and by Friedrich Mohr (a friend of Liebig) that heat has many of the properties of light; since the wave theory of light was firmly established, the idea that heat is also a form of motion was inescapable.

In 1842 there appeared in Liebig's *Annalen* a paper by the young physician Robert Mayer, who attempted to calculate the mechanical equivalent of heat. Although there were many obscurities in the paper, and the data available to Mayer were later refined, it represented the first explicit statement of the law of conservation of energy (later known as the First Law of Thermodynamics). It is clear from Mayer's writings that he was drawn to the consideration of the relation of heat to motion by the physiological problem of the source of animal heat:

> Starting from Lavoisier's theory, according to which animal heat is a combustion process, I considered the double color change undergone by the blood in the capillary vessels of the greater and lesser circulation as a perceptible indication, as a visible reflection, of an oxidation proceeding in the blood. To maintain the constant temperature of the human body, its heat production must necessarily be quantitatively related to a heat loss and to the temperature of the surrounding medium, and therefore the heat production and the oxidative process, as well as the differences in color between the two kinds of blood, must on the whole be less in the tropical zone than in colder regions (Mayer, 1893, p. 14).

From such musings, while serving in 1840 as a ship's doctor on a voyage to Java, Mayer came to the mechanical theory of heat at a time when it was being developed independently by Marc Séguin, Ludwig August Colding, and James Joule. Indeed, it was the series of elegant experimental papers by Joule during 1843–1849 that firmly established the validity of the First Law, long known in England as Joule's law.

In 1847 Helmholtz published his memorable article *Über die Erhaltung der Kraft,* in which the principle of conservation of energy was enunciated with precision, and applied to what he called the "force-equivalent" of heat and of electrical processes. We noted on p. 51 that Helmholtz had worked in Johannes Müller's laboratory on fermentation; in 1845 he found that the contraction of frog muscles is accompanied by changes in their chemical composition, and two years later he reported thermoelectric measurements showing the production of heat upon the contraction of isolated muscle. The 1845 paper began as follows:

> One of the most important physiological questions, intimately related to the nature of the vital force, namely whether the life of organic bodies is the effect of a self-generated purposeful force, or the result of forces also active in inanimate matter but specifically modified through the nature of their cooperation, has recently been given a much more concrete form in Liebig's effort to derive physiological conclusions from known chemical and physical laws, namely whether or not the mechanical force and the heat produced in the organism can be deduced completely from metabolic changes [*Stoffwechsel*] (Helmholtz, 1845, p. 72).

In his 1847 paper on the conservation of energy, the physiologist turned physicist applied the principle to animal respiration, as Mayer had done before him (in 1845). Helmholtz wrote that animals

> . . . take up oxygen and complex oxidizable compounds made by plants, discharge these compounds largely in form of carbonic acid and water as the products of combustion and partly as simpler reduced products, thus consuming a certain quantity of chemical potential energy, and generate thereby heat and mechanical energy. Since the latter represents a very small amount of work in relation to the quantity of heat, the question of the conservation of energy reduces itself roughly to whether the combustion and conversion of the nutritional components yields the same quantity of heat as that released by the animals. From the experiments of Dulong and Despretz, this question can be answered at least approximately in the affirmative (Helmholtz, 1882, p. 66).

Mayer had gone farther in his 1845 article, however; on the basis of some of the available data, he considered questions relating to the identity of the blood constituents that combine with oxygen, to the nature of the substances that serve as fuel, and to the site of the oxidative process. As to the last question, he concluded unequivocally that

> . . . not the hundredth part of the oxidative process occurs outside the walls of the blood vessels . . . . Unquestionably the solid parts of the organism . . . exert a powerful influence on the chemical metamorphosis of the blood,—an influence whereby in general the energy of the oxidative process or the affinity is increased . . . . [T]he blood corpuscles take up the atmospheric oxygen in the lungs, and the vital chemical process accordingly depends essentially on the combination of the oxygen absorbed by the blood corpuscles with the combustible constituents of the blood to form carbonic acid and water. In this respect, the blood corpuscles play the role of an oxygen carrier in the vital process, as does nitric oxide in the preparation of sulfuric acid (Mayer, 1893, pp. 101–103).

Although Mayer's opinions about animal respiration do not appear to have won a wide audience at the time of their publication, they reflected the widely held view that the site of respiratory oxidation is the blood. Before a clearer picture emerged, by about 1875, decisive chemical information had been obtained about the hemoglobin of the red cell, and physiological studies had been performed on the role of the tissues in biological oxidations.

Before turning to these topics, a brief digression may be added regarding Mayer's place in the development of thermodynamic theory. He has tended to attract special attention, principally in Germany, as one of the "romantic" figures in nineteenth-century science (Mittasch, 1940). In 1881 Helmholtz paid tribute to Mayer's originality, and apologized for his ignorance of Mayer's publications when he prepared his 1847 paper on the conservation of energy; he also wrote:

> Very recently, the adherents of metaphysical speculation have put the stamp of an *a priori* principle on the law of the conservation of energy, and therefore hail R. Mayer as a hero in the field of pure thought. What they regard as the pinnacle of Mayer's achievements, namely the metaphysically-formulated evidence for the *a priori* necessity of this law, will seem to every investigator accustomed to strict scientific method just the weakest aspect of his explanations, and has unquestionably been the reason for the fact that Mayer's work remained unknown

so long in scientific circles. Only after the road had been opened from another direction to the conviction that the law is valid, through the masterly researches of Mr. Joule, was attention given to Mayer's writings (Helmholtz, 1882, p. 73).

The further elaboration of thermodynamic theory, notably through the work of Clausius and Kelvin during 1850–1852 in formulating the Second Law and the concept of entropy, and of Gibbs and Helmholtz in laying the foundations of chemical thermodynamics, was to have considerable impact on the study of biological oxidation during the twentieth century; we will consider some aspects of these developments later in this chapter.

The Role of Hemoglobin. In expressing the view that the conversion of oxygen into carbon dioxide is effected solely in the blood, Mayer echoed the opinion of Liebig, in the first edition of *Animal Chemistry*. There, Liebig had noted that

> The blood corpuscles contain an iron compound. From the invariable presence of iron in red blood it must be concluded that it is absolutely essential for animal life, and since physiology has shown that the blood corpuscles do not participate in the nutritional processes, there can be no doubt that they assume a role in the respiratory process. The iron compound of the blood corpuscles behaves like an oxygen compound, since it is decomposed by hydrogen sulfide in entirely the same manner as are the iron oxides or similar iron compounds . . . . Perhaps the behavior of the iron compounds provides an explanation of the role played by iron in the respiratory process; no other metal can be compared with iron in relation to its remarkable properties (Liebig, 1842, p. 273).

He then discussed the tendency of some iron compounds to take up oxygen, and for the products to be reduced through the loss of oxygen; in particular, he noted the similar oxidation-reduction behavior of Prussian blue (ferrocyanide and ferricyanide), and called attention to the fact that, in this compound, iron is present in combination "with all the organic constituents of the animal body: hydrogen and oxygen (water), carbon and nitrogen (cyanogen)" (*ibid.*, p. 274). From the available physiological and chemical data, Liebig concluded that

> . . . blood corpuscles of the arterial blood contain an iron compound that is saturated with oxygen, and in which the living blood loses its oxygen during the passage through the capillary vessels; the same oc-

curs when blood is taken from the body and begins to decompose (begins to putrefy); the oxygen-rich compound is transformed by the loss of oxygen (reduction) into a less-oxygenated compound. One of the resultant products of oxidation is carbonic acid (*ibid.*, p. 274).

The nature of the iron compound was in doubt, however; although the association of the blood iron with the pigment that Berzelius had named hematin was established by the work of Lecanu (1838), it was not clear even during the 1850s whether this association reflected the situation in the red blood cell (Lehmann, 1855, Vol. I, p. 269).

Furthermore, there was disagreement about the reasons for the difference in color between arterial and venous blood (*ibid.*, pp. 553–560; Culotta, 1970). In lectures given in 1857, Claude Bernard (1859, Vol. I, p. 254) noted that "chemistry has not yet taught us much" about this, and he described an extensive series of physiological experiments from which he concluded that the conversion of bright red arterial blood into dark blood is a consequence of an increased proportion of carbon dioxide relative to the amount of oxygen (*ibid.*, p. 396). Within about a decade, however, some of the uncertainties had been resolved, and it was recognized that the difference in color arose from the oxygenation and deoxygenation of the hemoglobin of the blood. In part, the answers came from physiological experiments (largely in Karl Ludwig's laboratory) showing that oxygen is carried in the blood almost entirely in the form of an easily dissociable compound; in these advances the improved mercurial blood gas pumps developed by Meyer, Sprengel, and others played a major role. Of special significance was the independent discovery by Bernard (1857a) and by Felix Hoppe-Seyler that carbon monoxide can displace oxygen from arterial blood, with the appearance of a more vivid red color; during the 1860s this reaction was used by several investigators (Hermann, Dybkowsky) to determine quantitatively the oxygen-binding capacity of blood. By the end of the century, extensive studies had been conducted on the factors that influence the oxygenation of blood, notably by Paul Bert, Christian Bohr, and John Scott Haldane; as better apparatus was developed, more reliable quantitative data were obtained, and the problem of the gas exchanges in mammalian respiration was more clearly defined (Barcroft, 1914; Haldane, 1917; Henderson, 1928). We shall return to some of these later developments shortly.

The advances that were decisive in settling the question of the nature of the iron-containing compound in the blood that combines with oxygen in the respiratory process came during the 1860s from the chemical and spectroscopic studies of Hoppe-Seyler and of George Stokes. Their work established the identity of the crystalline material described by Otto

Funke and others with the respiratory pigment of mammalian blood. Hoppe-Seyler demonstrated that the isolated crystals release oxygen in an evacuated vessel, showing that they contain the gas in a loosely bound form; in 1864 he named the pigment oxyhemoglobin. In 1862 he reported that, in aqueous solution, oxyhemoglobin exhibits two absorption bands in the visible region of the spectrum, and Stokes (1864) showed that these bands disappear upon the addition of solutions known to react with oxygen (*e.g.,* an ammoniacal solution of ferrous sulfate and ammonium tartrate, later known as Stokes's solution). In place of these absorption bands, there appears a single more diffuse band, with a change in the color of the solution from that of "scarlet cruorine" to that of "purple cruorine," the names Stokes gave to oxyhemoglobin and "reduced" hemoglobin, respectively. Furthermore, he found that, when the solution is shaken with air, the two-banded spectrum of oxyhemoglobin is restored. Stokes concluded that

> . . . *the colouring matter of blood,* like indigo, *is capable of existing in two states of oxidation, distinguishable by a difference in colour and a fundamental difference in the action on the spectrum. It may be made to pass from the more to the less oxidized state by the action of suitable reducing agents, and recovers its oxygen by absorption from the air* (Stokes, 1864, p. 358).

This comparison of the behavior of cruorine to that of indigo is noteworthy, as he considered the loss of oxygen by scarlet cruorine to be "a simple reduction" (*ibid.,* p. 357). After the reduction of indigo (the coloring matter of woad) to "indigo white" had been described by Chevreul in 1808, there had been extensive studies on its chemistry; in 1823 Crum purified indigo-blue by sublimation, and Laurent (in 1841) laid the foundations for the subsequent elucidation of its structure by Baeyer. As we shall see later in this chapter, the reduction of natural pigments like indigo and the reoxidation of the reduced form were important reactions in the pioneer studies of Christian Friedrich Schönbein and of Moritz Traube on biological oxidations.

Stokes also found that treatment of blood with acid changes its absorption spectrum, and that upon reduction of the resulting pigment in alkaline solution there appear two very intense absorption bands whose positions differ from those of the oxyhemoglobin bands. He assigned the new two-banded spectrum to "reduced haematine"; in confirming Stokes's finding, Hoppe-Seyler changed the designation to "hemochromogen." In these studies the possible participation of the denatured globin was overlooked, and later work (after 1910) showed that hemochromogens are complexes of iron-porphyrins with nitrogenous bases. Furthermore, Hoppe-Seyler showed that treatment of hemochromogen with strong acid gave an iron-free pigment, to which he assigned the name "hematoporphyrin"; this finding had been anticipated in 1867 by Ludwig Thudichum, who called the pigment "cruentine." The elucidation of the chemical relations among hemoglobin, oxyhemoglobin, hemochromogen, and hematin, with respect to the nature of their iron-porphyrin component, required many decades of research, with many wrong turnings. During the 1870s the question of the valence of the iron in the various hemoglobin derivatives was entirely open, although Hoppe-Seyler correctly surmised that hematin contained ferric iron. On the other hand, the brown product (methemoglobin) he had obtained by the oxidation of hemoglobin, and whose presence he had detected in pathological blood, was not clearly identified to be the ferric form of hemoglobin until the 1920s.

Another important contribution made by Hoppe-Seyler during the 1870s was the explanation of the bright red color of the venous blood in cases of poisoning by cyanide. Bernard (1857a) had studied this phenomenon, and had concluded that the action of cyanide is the same as that of carbon monoxide; Hoppe-Seyler showed that, in cyanide poisoning, the oxyhemoglobin does not release its oxygen to the tissues, and passes unchanged into the venous blood. With the growing recognition at that time of the central importance of the tissues in animal respiration, this result pointed to the inhibition, by cyanide, of intracellular oxidation processes, a conclusion more firmly established by the work of Julius Geppert (in 1889) and others. This sensitivity of intracellular respiration to cyanide was to figure largely in the discussion, after 1920, about the mechanism of biological oxidation.

Despite the limitations of the chemical knowledge provided by Hoppe-Seyler and Stokes, their work not only gave a satisfactory explanation of the difference in color between arterial and venous blood, but also stimulated an interest in spectroscopy as a new approach to the study of biological oxidations. In particular, there emerged a group of English investigators (Sorby, Lankester, Gamgee, MacMunn) who performed extensive spectroscopic studies on biological pigments that could undergo the same kind of reversible oxidation and reduction that Stokes believed he had shown for hemoglobin. As we shall see later in this chapter, these efforts led to Charles MacMunn's discovery of the histohematins and myohematin in 1884.

By about 1875, therefore, chemical and spectroscopic studies had clarified some aspects of the relationship of hemoglobin to respiratory oxygen, and had provided important information about the nature of the pigment associated with the protein. Also, by that time, the idea that the blood represents the principal site of oxidative combustion in the mammalian organism was being replaced by the view that this process occurs primarily in the tissues. There had been many indications that this might be so since the beginning of the century; Spallanzani's results (published in 1807 by Senebier) on the oxygen uptake exhibited by a large variety of tissues, and on the formation of carbon dioxide by various organisms (*e.g.*, snails, frogs) kept in the absence of oxygen, had been confirmed and extended in 1824 by William Edwards, and most strikingly by Henri Victor Regnault and Jules Reiset (1849). The reluctance to generalize from these observations to the oxidative process in the intact mammalian organism stemmed from several factors, among which were the difficulty of imagining how gaseous oxygen could diffuse through the capillary walls, as well as the problem of the passivity of oxygen gas at the temperature of the animal body. As we have seen, even after Magnus's experiments had shown that the circulating blood contains loosely bound oxygen and carbon dioxide, but had left open the question of the site of respiratory oxidation, the general view (as expressed by Mayer in 1845) assigned to the tissues a subsidiary role in the process. Beginning about 1850, however, there ensued a lively debate about the validity of this view; the stimulus for the discussion came largely from the study of muscle.

We have already noted that the oxidative breakdown of muscle figured importantly in Liebig's theories of animal metabolism, and that the occurrence of chemical changes and the liberation of heat during muscular contraction had been demonstrated by Helmholtz. In 1850 Liebig's son Georg published an important study showing that isolated frog muscles contract for a longer period in oxygen than in other gases, and that after they have ceased to contract in the absence of oxygen, carbon dioxide is still released. He concluded that

> . . . in the process of respiration the blood really acts only as a means to effect the transport of gases to the capillaries and back, and that in the capillaries there occurs not the formation of carbonic acid, but only the exchange through the walls of the blood vessels of that already formed for the oxygen of the blood . . . . [T]he formation of carbonic acid from a part of the respiratory oxygen . . . proceeds in the body not within the capillary vessels, but outside them in the muscle tissue (Liebig, 1850, pp. 413–414).

G. Liebig's findings were soon confirmed by Gabriel Gustav Valentin, and Carlo Matteucci (1856) provided additional evidence by demonstrating that the rate of respiration is higher during contraction than at rest.

These observations provided the background for Moritz Traube's criticism, in 1861, of the idea that the principal site of respiratory oxidation is the blood, and of the theory that the energy for muscular contraction is derived solely from the oxidative breakdown of proteins. After citing the findings of Helmholtz, Regnault and Reiset, and G. Liebig, Traube concluded that

> The released oxygen passes in a dissolved state through the capillary walls and forms with the muscle fiber a loose combination that is able to transfer the oxygen to other substances, dissolved in the muscle fluid, and [the muscle fiber] can then take up new oxygen (Traube, 1899, p. 168).

This hypothesis was an application of a theory of enzyme action that Traube had advanced in 1858; in this theory, ferments act as catalysts in the transfer of molecular oxygen to suitable reducing agents. "The muscle fiber thus behaves toward reducing substances on the one hand, and to oxygen on the other, just as do indigo, indigo sulfonate . . ." (*ibid.*, p. 168). As we shall see shortly, Traube's theory of oxygen activation was to be a fruitful one; with regard to the site of biological oxidations, he wrote that

> . . . the fact that all organs of the animal body require arterial blood indicates that not only the blood, but all organs of the body respire . . . What we call respiration is therefore a very complex process. It represents the sum of the consumption of all those quantities of oxygen needed by each organ, either for its nutrition or for its maintenance. Thus, there can be an increase in the respiration of the brain, or liver and spleen, or indeed individual groups of muscle, without an accelerated respiration in other organs of the body . . . . The motive forces, however, which oxygen elicits in the muscles, nerves, spinal cord, and brain are a consequence of the characteristic construction and chemical nature of the apparatus in which the oxidative processes proceed, so that these forces do not appear in the form of heat, but in the form of their specific, as yet inexplicable, vital functions (*ibid.*, pp. 178–179).

These views were not generally accepted during the 1860s, and many gave preference to the opinion of Bernard, who stated that

> . . . it is infinitely probable that the carbonic acid of the venous blood arises from an oxidation which is effected in the blood corpuscle itself. When the blood passes through the capillaries, there might be an

exchange, not of gas, but perhaps one of liquids. As a consequence of the new conditions thereby created, the oxygen of the corpuscle would be partly used for the oxidation of the corpuscle itself (Bernard, 1859, Vol. I, p. 342).

In 1865 Alfred Estor and Camille Saint-Pierre reported that the oxygen content of blood decreased markedly with increasing distance from the heart, and concluded that respiratory combustion occurs only in the blood. Furthermore, two years later, Ludimar Hermann dismissed G. Liebig's results as having no relevance to muscular contraction in the intact animal, because he found as high an oxygen uptake with isolated muscle that had been heated to produce rigor as with unheated muscle. And in 1870 Bernard's disciple Paul Bert stated that

> . . . the quantitative importance of the metabolic activity of our tissues has been greatly exaggerated, and it is in the blood that there occurs the greater part of the changes that ultimately produce carbonic acid, urea, etc. (Bert, 1870, p. 118).

At that time, the most convincing evidence for this opinion appeared to be that provided by the careful physiological studies performed in the laboratory of Karl Ludwig, who believed that the activity of muscle is controlled largely by the rate of blood flow.

Among physiologists, the chief protagonist of the view that the tissues represent the principal site of oxidative activity was Eduard Pflüger. In an extensive series of papers published during 1866–1877, frequently dotted with acerbic comments about the work of his opponents, he and his associates provided important experimental evidence on the role of the tissues in respiratory oxidation. In his rebuttal of the evidence presented by Ludwig and others, Pflüger wrote, in characteristic style:

> Here lies, and I want to declare this once and for all, the real secret of the regulation of the oxygen consumption by the intact organism, a quantity determined only by the cell, not by the oxygen content of the blood, not by the tension of the aortic system, not by the rate of blood flow, not by the mode of cardiac action, not by the mode of respiration (Pflüger, 1872, p. 52).

Later, he stated that "For the body of rapidly-respiring vertebrates, hemoglobin is therefore only a convenient carrier with a large capacity" (Pflüger, 1875, p. 275). In stressing that the oxidative activity is localized in cells, Pflüger offered striking examples from comparative physiology;

thus he called attention to the occurrence of oxidative processes in insects, where there is no capillary system and oxygen is supplied directly to the tissues through tracheal tubes. Among the items of experimental evidence adduced in favor of Pflüger's view the most dramatic one was provided by his associate Ernst Oertmann, who determined the respiration of a frog whose blood had been replaced by a saline solution (*Salzfrosch*). The result of this experiment was that

> The oxidative processes of the frog are not altered by exsanguination, since the bloodless frog has the same metabolism as the one that contains blood. The site of the oxidative processes is therefore the tissues, not the blood (Oertmann, 1877, p. 395).

It should be added that Pflüger's conviction regarding the importance of intracellular oxidation was related to his advocacy of the idea of "living" protein molecules; an essential part of his speculations on this subject was that

> The life process is the intramolecular heat of highly unstable albuminoid molecules of the cell substance, which dissociate largely with the formation of carbonic acid, water, and amide-like substances, and which continually regenerate themselves and grow through polymerization (Pflüger, 1875, p. 343).

From ill-founded chemical considerations, which he developed at length in 1872, he proceeded to affirm that ". . . the living protein contains nitrogen, not in the form of ammonia, but in the form of cyanogen" (*ibid.*, p. 334), and attributed to the presence of the cyanogen radical the "explosive" quality of living matter, which allowed it to react with the chemically passive oxygen brought to the tissues by the circulation. As we noted in the chapter on proteins, similar hypotheses reappeared during succeeding decades, and were a reflection not only of the keen imagination of their proponents, but also of the uncertain state of protein chemistry and of the chemistry of intracellular oxidations.

Although by the end of the 1870s there was no longer any question of the greater importance of the intracellular respiration in the tissues, as compared to that in the blood, physiologists continued to debate about the relative role of various mammalian tissues (muscle, liver, lung) in the overall process. Furthermore, at the close of his life, Bernard came to the view that Lavoisier's and Liebig's identification of animal respiration with combustion was incorrect. In writing of the chemists who held this view, Bernard stated that

. . . they believed that organic combustion had as its counterpart the combustion effected outside living beings, in our furnaces, in our laboratories. On the contrary, in the organism there is probably not a single case of these supposed phenomena of combustion which is effected by the direct fixation of oxygen. All of them involve the participation of special agents, for example the ferments (Bernard, 1878–1879, Vol. I, p. 166).

We noted in our discussion of the fermentation problem that Bernard placed special emphasis on the distinction between the degradative and synthetic processes in living organisms. In his opinion,

. . . the production of carbonic acid, which is so general a phenomenon in the manifestations of life, is the consequence of a true organic destruction, of a decomposition analogous to those produced by fermentations. Moreover, these fermentations are the dynamic equivalent of the combustions; they fulfill the same purpose in the sense that they generate heat and therefore are a source of the energy necessary for life (*ibid.*, pp. 172–173).

As for the role of oxygen in the animal organism,

. . . it is not in direct combustion that this gas is used. The usual formula repeated by all the physiologists that the role of oxygen is to support combustion is not correct, because there is really no true combustion at all in the organism. What is true is that the precise role of oxygen, which we thought we knew, is still unknown to us; one can hardly guess at it . . . . It is quite certain that this gas is fixed in the organism and that it thereby becomes one of the elements of organic structure or creation. But it is not at all through its combination with the organic matter that it incites vital function. Upon contact with the tissues, it makes them excitable; they cannot exist without this contact. It is therefore as an agent of excitation that it would participate intimately in most of the phenomena of life (*ibid.*, pp. 167–171).

Thus, to maintain the principle that a degradative process leading to the formation of carbon dioxide could not be associated with the phenomena of life, and at the same time to acknowledge the essential requirement for oxygen, Bernard's physiological approach led him to question the idea that respiration is a combustion, and to leave open the role of oxygen.

We conclude this section with a brief postscript regarding more recent studies on gas exchange in the blood. In 1837 Gustav Magnus suggested that CO_2 might facilitate the release of O_2 from the blood, and in 1863 Frithiof Holmgren reported that the uptake of O_2 assists the release of CO_2. These tentative conclusions acquired more certain status from the work of Christian Bohr in 1904. His quantitative studies on the proportion of blood hemoglobin (Hb) present in the form of oxyhemoglobin (HbO_2) demonstrated the remarkable ability of the mammalian organism to regulate the proportions of Hb and HbO_2 in response to changes in the O_2 and CO_2 content of the blood. The curve obtained by plotting the percent of HbO_2 against oxygen pressure was found to be a sigmoid curve, not a rectangular hyperbola, as would have been expected from the simple equation $Hb + O_2 \rightleftharpoons HbO_2$. Furthermore, with increasing CO_2 pressure, the percent of HbO_2 at a given O_2 pressure is reduced. These two regulatory properties are of physiological advantage in unloading the oxygen of HbO_2 in the tissues, where the oxygen pressure is lower than in the lungs, and the CO_2 is liberated from intracellular oxidations. Later work by J. S. Haldane (in 1914) showed that conversely, at a given CO_2 pressure, oxygenated blood takes up less CO_2 than deoxygenated blood. Various theories were offered to explain the sigmoid nature of the saturation curve of oxyhemoglobin; in relation to later developments, that of A. V. Hill (1913) is of special interest. He explained the curve by assuming a reversible aggregation of hemoglobin units and a cooperative interaction between the units, such that the conversion of the first one to HbO_2 increases the affinity of the other units for oxygen. At that time the molecular weight of hemoglobin was considered to be about 16,000; the work of Adair and of Svedberg during the 1920s showed hemoglobin to be an oligomeric protein consisting of four subunits (see p. 136). During the past decade, the development of knowledge about the detailed chemical structure of hemoglobin has permitted a closer study of the dissociation of these subunits, and of the relation of their specific interaction to the regulatory properties of the protein (see Roughton, 1970).

The Activation of Oxygen

The central place occupied by respiratory oxygen in the survival and activity of living organisms posed a chemical dilemma throughout the nineteenth century. Although there was little doubt by 1875 that animal and plant tissues consume oxygen and release carbon dioxide in a process comparable to combustion, it was clear that this process is not identical with the rapid and complete burning of carbon-containing compounds in a furnace, as was implied by the famous dictum of Lavoisier and Laplace. To explain the "slow combustion" effected by biological systems at ordinary temperatures, a considerable variety of chemical analogies were invoked. For Liebig, who in his *Animal Chemistry* considered respiratory combustion to be limited to the blood fluid, the slow oxidation (*Verwesung*) was a consequence of the accelerated molecular motion of the substances undergoing change. From chemical experiments such as those of Davy and Döbereiner on the effect of porous platinum black in accelerating the oxidation of organic compounds it was then inferred (by Mayer, among others) that the tissues lining the capillaries promote the oxidative process by providing an extended porous surface. With the growing recognition of the primacy of the tissues in respiratory combustion, attention shifted to the role of intracellular "catalytic membranes" (Bence-Jones, 1867, p. 75) in the promotion of biological oxidations. As we shall see shortly, this idea of catalysis at intracellular surfaces was to reappear in the theory of biological oxidations developed by Otto Warburg during 1910–1925, and was greatly furthered by the rise of colloid chemistry at the beginning of this century.

For many chemists, or chemically minded physiologists, who concerned themselves with the problem of biological oxidations during 1850–1860, more definitive explanations were needed for such processes in terms of changes in the "chemical affinity" of either the substances undergoing oxidation, or of oxygen, or of both. By midcentury the spontaneous auto-oxidation of inorganic compounds such as sulfurous acid or hydrogen sulfide was recognized to have its counterpart in the oxidation of organic substances by atmospheric oxygen, especially in alkaline solution. For example, Chevreul (in 1824) and Mialhe (in 1844) showed that aqueous solutions of substances such as pyrogallol or glucose are altered by atmospheric oxygen only after alkali had been added. Observations of this kind led to speculations about the role of alkali in promoting the oxidation of glucose in the animal body (Lehmann, 1855, p. 356), and indeed to inferences about human diabetes as a consequence of a deficiency of alkali in the blood (Mialhe, 1856, pp. 75–77). Furthermore, the increased understanding of the transformations undergone by organic substances during laboratory operations served as a basis for speculations about comparable changes observed in biological systems, and led to experiments on the fate of various known organic compounds in the animal body. An early example came from the work of Friedrich Wöhler and Theodor Frerichs (1848), who fed benzaldehyde to a dog and observed an increased urinary excretion of benzoic acid (as hippuric acid); this was in accord with the known tendency of the aldehyde to take up oxygen spontaneously, with the formation of benzoic acid. Throughout the rest of the nineteenth century, many such experimental studies were to be conducted, and the results often provided the basis for later fruitful study of the pathways of animal metabolism.

During the 1860s, however, greater interest was evinced in the possibility that the slow combustion associated with biological systems involves the "activation" of the normally unreactive oxygen of the air. The leader in this development was Christian Friedrich Schönbein; in 1840 he had assigned the name "ozone" to the peculiar odor emitted during the electrolysis of sulfuric acid, and four years later he noted the same odor when moist phosphorus was exposed to air. He found that the "ozonized" air possessed a powerful oxidative capacity, as judged by such chemical reactions as the liberation of free iodine (detected with starch) from a solution of potassium iodide. After the work of Marignac and of de La Rive in 1845, Schönbein concluded that ozone represents an especially active form of oxygen. Among the chemical changes effected by ozonized air, and not by ordinary oxygen, was the rapid bluing of an alcoholic extract (tincture) of guaiac. This resin, from the West Indian plant *guayaco,* had been used in Europe for medicinal purposes (under the name *lignum vitae*) since the sixteenth century, and its chemical constituents had been studied extensively (Hadelich, 1862). Its importance in our story arises from the fact that in 1810 Louis Antoine Planche had reported the bluing of guaiac by various plant materials, such as horseradish. He described further experiments along this line in 1820, and attributed the bluing action to the presence in the horseradish of an unstable constituent he denoted "cyanogen." In the same year, Gioacchino Taddei confirmed these observations and noted the requirement for air. The clear recognition that the bluing of guaiac is an oxidative process only came, however, from Schönbein's experiments in 1845; three years later Schönbein (1848) used potatoes to confirm the observations of Planche (he referred to him as Blanche) and of Taddei, and concluded that some constituent of the potato is able to combine with atmospheric oxygen and make ozone or hydrogen peroxide. In other experiments, Schönbein showed that hydrogen peroxide (first prepared by Thenard in 1818) also can cause the bluing of guaiac, and related this finding to the previously known ability of finely divided platinum to promote the decomposition of hydrogen peroxide to water and oxygen. Berzelius had already named such reactions catalytic processes, and Schönbein accordingly suggested that the bluing of guaiac by the potato involves the participation of a catalyst. Furthermore, he demonstrated that the blue guaiac could be decolorized by the addition of such reducing agents as hydrogen sulfide or ferrous salts, and clearly interpreted the changes in the dye in terms of its alternate oxidation and reduction, as in the case of the known interconversion of indigo white and indigo blue. It may be added that Schönbein performed many of these experiments by impregnating a paper strip with the guaiac tincture, and by exposing it to the ozonized air and then to reducing agents; when, after 1850, ozone became for a time a popular item of medical investigation (Bernard, 1857a, p. 154), such paper strips (usually impregnated with iodide and starch) were widely used to detect its presence in air. In 1861 Schönbein extended the use of paper strips to the separation of chemical substances but, as we noted before (p. 168), such "paper chromatography" did not assume its present importance until after World War II.

A further brief digression may perhaps be inserted here, about both ozone and its discoverer. For many years there was skepticism in the chemical community about the validity of Schönbein's views about the nature of ozone; for example, in a letter (6 February 1857) to Liebig, Wöhler wrote that "so long as one cannot handle the thing and put it on the balance, it does not interest me greatly" (Hofmann, 1888, Vol. II, p. 38). After the work of Jacques Soret (in 1866–1868) in showing that ozone is O_3, it was isolated as a blue liquid by Hautefeuille and Chappuis in 1882, and ozone became a valuable oxidizing agent in organic chemistry. In large part, the skepticism of Schönbein's fellow chemists was a consequence of his lively imagination; he first thought nitrogen to be composed of ozone and hydrogen, then he formulated an intricate theory that made ozone negatively charged oxygen and hydrogen peroxide a compound of water and positively charged oxygen ("antozone"). For Schönbein (1860, p. 17), "the what? is more important than the how much?", and he eschewed quantitative experiments. From his ebullient letters to Faraday, Berzelius, and Liebig he emerges as a delightful person, in the "romantic" spirit of nineteenth-century German science, whose other adherents included his friend Robert Mayer, and whose philosophical spokesman was his teacher Friedrich Schelling. A less romantic aspect of Schönbein's career was his invention of guncotton, announced in 1846.

Whatever Schönbein's apparent shortcomings may have been in quantitative experimentation or in chemical thought, his empirical observations laid the groundwork for subsequent work on biological oxidations. In 1855 he described the bluing of the cut surface after the mushroom *Boletus luridus* had been broken, and he prepared a colorless alcoholic extract that turned blue upon the addition of a press juice from the plant. The similarity of this behavior to that of guaiac led Schönbein to the generalization that plants contain a material, like that present in the mushroom press juice, which activates oxygen to form ozone, and transfers the activated oxygen to oxidizable substances such as the chromogen of Boletus. Two years later he reported that red blood cells promote the bluing of guaiac upon the addition of hydrogen peroxide, and thus extended his theory of oxygen activation to animal systems. Schönbein (1863) then tested extracts of a large variety of plant and animal tissues, and found many of them to catalyze the bluing of guaiac in the presence of hydrogen peroxide, and to decompose hydrogen peroxide with the liberation of oxygen. Since this catalytic activity was lost when the extracts were heated, he concluded that he was dealing with ferments of an albuminoid nature, and considered these ozone-producing ferments to be responsible for biological oxidations.

Just as the chemists were divided in their judgment of Schönbein's ideas about the nature of ozone, his conclusions about biological oxidations elicited widely different reactions among the physiologists interested in the problem of animal respiration. In 1862 Alexander Schmidt published a widely discussed book (*Über Ozon im Blut*), in which he developed the ozone theory as an explanation of the biological oxidations effected by respiratory oxygen. A few years later, Willy Kühne stated that ". . . oxyhemoglobin really contains oxygen which gives the ozone reaction, and which is a more powerful oxidant than ordinary oxygen" (Kühne, 1866, p. 214). Such conclusions were based on the use of the chemical tests introduced by Schönbein, but efforts to demonstrate directly the presence of ozone or of hydrogen peroxide in blood were unsuccessful. This failure contributed to the demise of the ozone theory after 1870, and greater attention was given to other hypotheses about biological oxidations.

Among the leading physiologists who opposed the ozone theory was Eduard Pflüger; in a lengthy article on biological oxidations, he dismissed the claims of Schönbein and Schmidt, and concluded that

> . . . it is not the oxygen, but the protein that undergoes change, when it has become an integrating constituent of the organism. . . . As soon as this incorporation has occurred, it has lost its indifference to oxygen, in other words it begins to respire, to live (Pflüger, 1875, pp. 300–301).

Throughout the second half of the nineteenth century, this idea of "living protein" was repeatedly offered as a ready explanation of biological oxidations. Before Pflüger, Ludimar Hermann (1867) had attempted to account for the effect of oxygen in promoting the recovery of exhausted excised muscle by postulating the regeneration of a labile albuminoid material. Afterwards, Carl von Nägeli (1879, p. 117) extended his theory of fermentation to biological oxidations, and invoked the "specific states of motion of the living protoplasm," while Wilhelm Pfeffer wrote that

> The affinities developed in the functional plant suffice completely to cleave the molecule of neutral oxygen, and the maintenance of normal respiration does not require the participation of an active oxygen (Pfeffer, 1881, Vol. I, p. 374).

During succeeding years, the views of Pflüger and Nägeli reappeared several times in new dress, notably in the writings of Nencki and Sieber (1882), Loew (1896), and Verworn (1903). Although these speculations elicited lively interest among biologists, the experimental fruit was meager; the last word on the subject was said by Hopkins in 1913:

> There is, I know, a view which, if old, is in one modification or another still current in many quarters. This conceives of the unit of living matter as a definite, if very large and very labile molecule, and conceives of a mass of living matter as consisting of a congregation of such molecules in that definite sense in which a mass of, say, sugar is a congregation of molecules, all like to one another. In my opinion, such a view is as inhibitory to productive thought as it is lacking in basis. It matters little whether in this connection we speak of "molecule" or, in order to avoid the fairly obvious misuse of a word, we use the term "biogen", or any similar expression with the same connotation. Especially, I believe, is such a view unfortunate when, as sometimes, it is made to carry the corollary that simple molecules, such as those provided by foodstuffs, only suffer change after they have become in a vague sense a part of such a giant molecule or biogen. Such assumptions became unnecessary as soon as we learnt that a stable substance may exhibit instability after it enters the living cell, not because it loses its chemical identity, and the chemical properties inherent in its own molecular structure, by being built into an unstable complex, but because in the cell it meets with agents (the intracellular enzymes) which catalyze certain reactions of which its molecule is normally capable (Hopkins, 1913, p. 220).

The development of the new knowledge to which Hopkins referred was largely initiated by Moritz Traube (Bodländer, 1895; Sourkes, 1955) who proposed, during 1858–1886, a theory of biological oxidations based on the activation of molecular oxygen by intracellular enzymes.

Oxidative Enzymes. Whereas Schönbein approached the problem of biological oxidations from his chemical studies on ozone, Traube began with a new chemical theory of fermentation, whose formulation showed its debt to the ideas of Liebig:

> The true cause of the phenomena of fermentation lies in the following incontrovertibly established principles, partly based on direct experiment, partly on known facts:
> The ferments concerned with putrefaction and slow combustion [*Verwesung*] are definite chemical compounds formed by the transformation of proteins (perhaps with the cooperation of oxygen) . . . . Among the ferments made within and outside the organism there are a) those that can take up free oxygen easily and to bind it loosely (*Verwesungsfermente*); b) those that can accept oxygen that is already bound, that is, can easily deoxidize other compounds . . . (*Reductionsfermente*); c) the ferments which . . . can decompose water directly, with the liberation of free hydrogen . . . (we call this "*höchstes Fäulnissferment*").
> All these ferments possess the ability to transfer to other compounds the oxygen taken up in one of these ways, that is, to be reduced by them and to take up new quantities of oxygen that can again be transferred, and so on. In this manner, all ferments can transfer to other compounds free or bound oxygen in almost unlimited amount, that is, to effect fermentations and slow combustions (Traube, 1858, pp. 332–333).

As we noted before (p. 81), Traube generalized this theory to all reactions then considered to be catalyzed by ferments, and claimed to have shown that nearly all ferments can act as reducing agents; for example, he reported that a preparation of diastase can reduce indigo to its colorless form. Furthermore, the theory included not only the direct transfer of oxygen from the atmosphere to an acceptor (as in slow combustion), or the removal of oxygen from a compound being reduced, but also fermentations which result in the reduction of one compound and the oxidation of another, or even the case in which a compound AB is cleaved with the reduction of group A and the oxidation of group B. Thus Traube cited, as an example of the last case, the fermentation of urea to ammonia (the product of reduction) and carbon dioxide (the product of oxidation).

Traube's experimental evidence for these conclusions appeared in a book published in 1858, and reprinted in his collected works (Traube, 1899, pp. 68–147). To his contemporaries, the evidence was no more convincing as an explanation of biological oxidations than that presented in support of the ozone theory. Also, because Traube offered a chemical theory of all fermentations, including alcoholic fermentation, at a time when Pasteur was establishing the organismic theory through his elegant researches, it is perhaps not surprising that relatively little attention was given to Traube's views.

In 1874 Traube rose to challenge Pasteur; although he confirmed the finding that the multiplication and fermentation of yeast occurs in the absence of oxygen, he noted:

> Upon exclusion of air the fermentation stops after a very small part of the dissolved sugar has been decomposed, whereas upon the admission of air . . . the fermentation of the same protein-containing sugar solution proceeds with the disappearance of the sugar . . . . If, as Pasteur states, the yeast were able to take from the sugar the oxygen it needs for its growth, why does it cease to grow in the absence of air, when most of the sugar is still intact? (Traube, 1874, p. 882).

Although Traube's experimental answer to this question was indecisive, he took the opportunity to restate the theory he had advanced in 1858:

> I have shown by means of numerous examples that just as there are substances which, like platinum, nitric oxide (in sulfuric acid manufacture), indigo sulfuric acid, etc., can transfer free oxygen to other substances and to effect their oxidation (oxygen carriers, oxidation ferments), there are also substances that can transfer bound oxygen, that is, they can effect reduction of one part and oxidation of the other. If we imagine the sugar molecule to be composed of 2 atomic groupings, a reducible A and an oxidizable B, then the cleavage by the yeast ferment is effected in such a manner that it extracts oxygen from group A (the deoxidized product is alcohol) in order to transfer it to group B, which is thereby burned to carbonic acid (*ibid.*, p. 884).

A few years later, Traube had occasion once again to restate his views after Hoppe-Seyler had proposed a theory of fermentation similar in some respects to that of Traube, but which assumed that ". . . all reductions occurring in putrefying fluids are secondary processes elicited by nascent hydrogen" (Hoppe-Seyler, 1876, p. 15), and that, when oxygen is present,

> . . . instead of the reduction, there appears during the putrefaction oxidation, which can have its cause in nothing but the cleavage of the oxygen molecule by the nascent hydrogen . . . whereby the oxygen is converted to an activated state and can then act as a powerful oxidizing agent (*ibid.*, p. 16).

Clearly, Hoppe-Seyler was following Schönbein's lead in assuming that, for activation, the O_2 molecule had to be cleaved; a corollary of this idea was that the formation of hydrogen peroxide (H_2O_2) during the oxidation of organic substances by O_2 was a consequence of the addition of activated atomic oxygen to water.

In his paper, Hoppe-Seyler did not refer to Traube, and after the latter's reply (Traube, 1878), there ensued a lively exchange, the most important consequence of which was to strengthen the experimental basis of Traube's theory. In a valuable series of papers during 1882–1886, Traube presented data showing that the formula of hydrogen peroxide is HO—OH, and that it arises during the oxidation of organic compounds, not by the oxidation of water, but also by the addition of hydrogen atoms to O_2. Furthermore, he demonstrated that in *Autoxidation* (the term he used for oxidation by oxygen gas), there is no cleavage of O_2 to activated atomic oxygen, but rather the addition of molecular oxygen to the organic molecule in a loosely bound form, with the formation of what he called a "holoxide." During the 1890s several chemists interested in the problem of oxidation adopted Traube's theory (Bodländer, 1899), and in 1897 Carl Engler and Aleksei Bach independently developed it into a chemical mechanism that received impressive experimental support from subsequent studies (Milas, 1932). In particular, the work of Baeyer and Villiger (1900) on the auto-oxidation of benzaldehyde gave clear evidence for the intermediate formation of a benzoyl peroxide. It should be added that, during the second half of the nineteenth century, Schönbein's idea that the activation of oxygen involved the cleavage of O_2 into oppositely charged atoms was espoused by Clausius (in 1858) and by van't Hoff (in 1895), among others, thus contributing to the skepticism accorded Traube's chemical ideas.

From our present perspective, it is easy to see in Traube's *Verwesungsfermente* of 1858 and his *Oxydationsfermente* of 1874, as well as in his ideas about the role of these enzymes in the intracellular activation of molecular oxygen, starting points of later fruitful study; this was not so readily evident to his biological contemporaries, however, and most of the leading biologists preferred such approaches as that of Nägeli. Even among the physiological chemists (Nencki, Schmiedeberg) who were studying the oxidation of foreign organic compounds in the animal body, there was reluctance during the 1880s to accept the idea that such oxidations involve the participation of chemically defined enzymes. Not until the next decade, after Schmiedeberg's student Alfred Jaquet (1892) had shown that aqueous extracts of animal tissues contain catalysts for the oxidation of substances such as benzyl alcohol or salicylaldehyde, did the medical physiologists consider seriously the view that discrete enzymes are involved in biological oxidations. One of the important consequences of such studies was the recognition that the biological oxidation of the chemical constituents of the diet or of the tissues to carbon dioxide and water does not occur in a single explosive combustion, as envisaged by Pflüger and his contemporaries, but that it may be a stepwise process involving partially oxidized intermediates. This idea had been advanced tentatively by Dumas during the 1840s, and by Hoppe-Seyler later in the century; around 1900 it was stated more explicitly, as part of a new view of biochemical dynamics:

> In the protoplasm synthesis and breakdown occurs by way of a series of intermediate steps, whereby it is not always the same kind of chemical reaction that is involved, but rather a series of reactions of different kinds . . . . A regular reaction sequence of the chemical reactions in the cell presupposes, however, the separate activity of the individual chemical agents and a definite direction of movement of the products that are formed, in short, a chemical organization . . . that helps to explain the speed and certainty with which it functions (Hofmeister, 1901, p. 26).

A further indication of the indifference of physiological chemists to Traube's views is provided by Paul Ehrlich's highly regarded book on the oxygen requirements of the animal organism, published in 1885. In extending his studies on the staining of animal tissues by the then-new synthetic dyes, Ehrlich injected into living mice such dyes as methylene blue, alizarin blue, or indophenol blue, known to be oxidized forms of their colorless (leuco) derivatives. It was also known that these dyes differed in the ease with which they were reduced to the leuco forms by chemical reagents; thus indophenol blue was reduced more readily than alizarin blue. From the microscopic observation of the extent to which various tissues were stained, Ehrlich concluded that they differed in their "oxygen saturation," as judged by the avidity with which the dyes were decolorized, with heart muscle and brain at a high level of oxygen saturation (indophenol blue remained oxidized), with liver, lungs, and fatty tissues at a low level (even alizarin blue was reduced), and with skeletal muscle and other tissues at an intermediate level (indophenol blue was reduced, but not alizarin blue). Furthermore, Ehrlich showed that, if rabbits were injected with a mixture of α-naphthol and dimethyl-p-phenylenediamine (later known as the "Nadi" reagent), indophenol blue was synthesized in the tissues, and deposited either as the oxidized or the reduced form. These observations provided important experimental approaches to the later study of the intracellular catalysis of oxidation-reduction reactions, after it was recognized that individual enzymes might be the active agents; we shall return to this development shortly. It should be emphasized, however, that Ehrlich's interpretation of his results were in line with the prevailing ideas about protoplasm, which he embellished with his "side-chain theory"; in his view, there was present in

> . . . the living protoplasm a [chemical] nucleus of special structure which determines the specific, characteristic function of the cell, and to this nucleus there attach themselves as side chains molecules [*Atome*] and molecular complexes which are of subsidiary importance for the specific cellular activity, but not so for life as a whole. Everything points to the view that it is indeed the undifferentiated side chains which represent the points of initiation and attack in physiological combustion, in that some of them mediate the combustion by the transfer of oxygen, and the others consume it (Ehrlich, 1885, p. 10).

From the fact that the leuco derivatives of the synthetic dyes are oxidized more easily in alkaline solution than in an acid medium, Ehrlich concluded that biological oxidations depend on the alkalization of the tissues:

> If we assume that during its activity protoplasm is bathed alternately by acid and alkaline fluids, then we can readily suppose that the formation of oxygen is associated with the alkaline phase and its consumption to the acid phase, or alternatively to the decrease in alkalinity . . . . If we imagine that in a normal, alkaline-reacting cell there are present a large number of unsaturated oxygen affinities, the entering oxygen molecules will first saturate the sites of highest oxygen affinity and to a diminishing degree those of lower affinity . . . . There will finally come a point, therefore, where the oxygen is bound so loosely that under the particular conditions prevailing at that moment, any side chains can be oxidized and carbon dioxide can be formed. When this stage is reached, the phase of acidification sets in and, as already noted, the actual combustion occurs (*ibid.*, pp. 133–134).

These vague speculations elicited much discussion, and were adopted by Max Verworn in the formulation of his biogen hypothesis; it must be reiterated, however, that the experimental observations from which they were derived were important in the subsequent development of knowledge about biological oxidations.

It may also be noted that, around 1880, there was extensive discussion of the mechanism of bioluminescence (*e.g.*, of fireflies), a process long known to depend on the presence of oxygen (see Harvey, 1957). Although there was much speculation about the activation of oxygen in this process and its relation to the general problem of biological oxidations, the ideas of Traube did not figure importantly in these discussions. After the pioneer studies of Bronislaus Radziszewski (1880) on the chemiluminescence of various organic compounds, and the establishment of the enzyme theory of biological oxidations, there ensued significant progress in the study of the chemical basis of bioluminescence. In this development, the studies of Raphael Dubois (in 1887) on luciferin (the light emitter) and luciferase (the enzyme catalyzing the necessary oxidation) served as the starting point for the exciting researches of Newton Harvey and William McElroy during this century.

In contrast to the medical physiologists, some botanists were more receptive to the ideas of Schönbein and Traube, possibly because most of the known natural substances (guaiac, indigo) thought to be intracellular oxygen carriers were derived from plants. In particular, Johannes Reinke (1883) endorsed Traube's views enthusiastically, and proposed that biological oxidations depend on the auto-oxidation of specific organic substances with the formation of hydrogen peroxide, which in the presence of intracellular enzymes oxidizes the various compounds undergoing metabolism. Furthermore, the first oxidative enzymes to be clearly identified as such were obtained from plants. In 1883 Yoshida described the preparation of a "diastatic matter" which promoted the darkening and hardening of the latex of the Japanese lacquer tree, and showed that his material catalyzed the oxidation of a plant constituent (urushiol) by atmospheric oxygen. This work was extended by Gabriel Bertrand (1894) who studied the latex of the Indochinese lacquer tree, where the substrate of oxidation is laccol; he named the enzyme "laccase," and showed it to have a specificity different from that of other known enzymes. During the course of these studies, Bertrand found that, whereas laccase oxidized various polyphenols (*e.g.*, pyrogallol), it did not catalyze the oxidation of tyrosine to the black pigment melanin; such an enzyme (he named it "tyrosinase") was present in some mushrooms and in higher plants. By 1897 Bertrand had demonstrated that there were individual enzymes, which he termed "oxidases," able to catalyze the activation of molecular oxygen and to exhibit specificity with regard to the substrate undergoing oxidation. Furthermore, his chemical analysis of laccase led him to conclude that it contained manganese, and he expressed the view that this metal was bound to the enzyme and was essential for its catalytic activity; he referred to the metal as a "coenzyme." Although later work (by Keilin and Mann in 1939) showed that the metal associated with laccase is copper, not manganese, it was Bertrand who introduced the concept that the oxidases may be catalytic metalloproteins.

The work of Yoshida and Bertrand on plant oxidases thus appeared to give reality to the "oxidation ferments" of Traube; others sought specific oxidases in a variety of plant and animal tissues, and by 1910 a considerable number had been found. One of the factors that encouraged this search was, unquestionably, Buchner's preparation of zymase, for, if it was possible to separate from living cells the catalytic agent of fermentation, there was hope that the catalytic agent of intracellular oxidation could also be isolated from animal and plant tissues. The limitations of the available methods for the isolation and characterization of enzymes made the task difficult, however; much of the confusion (see Battelli and Stern, 1912) that characterizes the early literature on the oxidative enzymes is a consequence of their lability and of the grossly impure state of the available preparations.

One of the problems actively discussed during 1900–1910, and which illustrates the confusion, was the question of the role of hydrogen peroxide in biological oxidations. Schönbein had found that some plant extracts catalyze the bluing of guaiac in the presence of atmospheric oxygen only after hydrogen peroxide had been added. This effect was noted for leucocytes by Klebs (in 1868) and Struve (in 1871); with the rise of the enzyme theory of biological oxidations, it was attributed by Georges Linossier (in 1898) to an enzyme he named "peroxidase." The question of the relation of the peroxidases (similar activity was found in a variety of plants and in milk) to Bertrand's oxidases was warmly debated for many years. The theory that found most favor was the one advanced by Robert Chodat and Aleksei Bach (1903), who proposed that the oxidase preparations contain a component ("oxygenase") which binds O_2 to form an organic peroxide of the kind postulated for auto-oxidation reactions by Traube, Bach, and Engler; this peroxide was thought to participate in the peroxidase-catalyzed oxidation of substrates such as the polyphenols. Much effort was expended in attempts to isolate the oxygenase, but without success, and attention turned to the possibility that such an enzyme did not exist, but that the peroxide was generated by the auto-oxidation of suitable compounds of low molecular weight such as catechol (Moore and Whitley, 1909; Wheldale, 1911). It required over twenty years to clarify the status of the oxidases studied by Bertrand, and to establish their identity as distinct enzymes, different from the peroxidases, but by that time the focus of attention in the study of biological oxidations had shifted to other intracellular components.

An additional vexing problem was posed by the wide distribution, in animal and plant tissues, of an enzyme identified by Oscar Loew in 1901 as a catalyst in the decomposition of hydrogen peroxide to water and oxygen; he named this enzyme "catalase." After Loew's report, there was much discussion about the physiological role of catalase (Battelli and Stern, 1910); for those investigators (*e.g.*, Wieland, 1922a, p. 3647) who believed that hydrogen peroxide is a normal product of biological oxidations, the existence of this enzyme provided a ready explanation for the failure of hydrogen peroxide, known to be a very toxic substance, to accumulate in the tissues.

It should be added that the preoccupation with hydrogen peroxide as an active participant in biological oxidations led to chemical studies designed to mimic such oxidations in model reactions. Thus advantage was taken of the observations made by Henry Fenton (1894) and Wilhelm Manchot (1902) that ferrous salts catalyze the oxidation of organic substances (*e.g.*, tartaric acid) by hydrogen peroxide; in such catalytic reactions, the divalent ferrous iron was considered to act as an oxygen carrier through oxidation by hydrogen peroxide to the trivalent ferric iron, from which the ferrous form was regenerated through reduction by the organic substance. An example of the application of this approach was the work of Henry Dakin and of Carl Neuberg during 1903–1910, on the oxidation of fatty acids and of amino acids by hydrogen peroxide in the presence of ferrous salts. Comparison of the products formed in the chemical oxidations with those that appeared in the urine, or formed in the tissues, after the administration of such acids to experimental animals, provided useful clues to the nature of the intermediates formed in the oxidative breakdown of cell constituents.

A line of research which paralleled that initiated by Yoshida and Bertrand was derived from Ehrlich's studies on the oxidative synthesis of indophenol blue in the tissues of rabbits injected with the "Nadi" reagent. With the emergence of the enzyme theory of biological oxidations, extracts of animal tissues were found to effect this reaction, and Wilhelm Spitzer (1897) claimed to have isolated the active agent in the form of an iron-nucleoprotein compound; he reported that the oxidase was strongly inhibited by cyanide. We noted on p. 199 that during the 1890s there was considerable interest in the intracellular role of iron compounds, and Spitzer reflected this trend in stating that

> . . . the known oxidation processes effected by animal cells outside the organism are due to the presence in these cells of specifically active nucleoproteins or, more precisely, are attributable to oxygen transfer mediated by organically-bound iron (Spitzer, 1897, p. 653).

Although the nucleoprotein nature of the enzyme (it was later called "indophenol oxidase" by Joseph Kastle) was disproved, Spitzer's emphasis on the importance of iron in intracellular oxidations was to find forceful expression in the theory developed by Otto Warburg during 1910–1925. Before considering this theory, it should be mentioned that by 1910 several other oxidative enzymes had been found in animal tissues, among them xanthine oxidase, identified by Richard Burian (in 1905) as a catalyst in the oxidation of xanthine to uric acid, and of hypoxanthine to xanthine. Furthermore, systematic experiments had been initiated by Federico Battelli and Lina Stern, in an attempt to determine the extent to which the respiratory activity of animal tissues could be accounted for in terms of the catalytic activity of known oxidizing enzymes. Their results indicated that, for surviving minced liver, the "main respiration," corresponding to about 75% of the oxygen uptake, disappeared rapidly and was strongly inhibited by known respiratory poisons such as cyanide, whereas the more stable residual "accessory respiration" was less sensitive to such inhibitors. They concluded:

> The accessory respiration can also proceed in the absence of cells or cell fragments, whereas the main respiration only occurs in the presence of cells. The accessory respiration is probably enzymatic in nature and is effected by the action of one or more oxidases (Battelli and Stern, 1909, p. 509).

In subsequent work, they found that some of the known oxidases (*e.g.*, xanthine oxidase) could be extracted from the minced tissue with water, whereas the ability to oxidize other substrates such as succinic acid or citric acid appeared to be bound to labile water-insoluble material. Battelli and Stern gave the name "oxydones" to the catalytic agents apparently associated with the main respiration; they studied the succinoxydone (later termed succinoxidase) extensively, and stated that "The oxydones consist of protein substances similar to the nucleoproteins . . . . Up to now, however, there is no evidence whatever for the protein nature of the oxidases" (Battelli and Stern, 1914, p. 169).

Of special importance were the observations made by Battelli and Stern, and also by Horace Vernon (1911), indicating a correlation between the oxygen uptake of various tissues and their apparent content of indophenol oxidase or succinoxidase; thus the rapidly respiring heart muscle was found to be richest in indophenol oxidase, and one of the poorest in the other intracellular enzymes then known. The work of Battelli and Stern on succinoxidase later proved to be valuable in studies on intracellular respiration (see Monnier, 1942); at the time the results were published, however, their validity was questioned, especially with regard to the existence of an accessory respiration independent of cell structure.

This doubt was a reflection of the continuing debate about the significance of Buchner's zymase in relation to the problem of alcoholic fermentation. Many investigators interpreted the fact that the yeast press
juice was so much less active than an equivalent amount of yeast to
indicate that the true process of fermentation depends on the integrity
of the cell (Rubner, 1913). Similarly, when in 1904 reports came from
Vladimir Palladin's laboratory that weakly respiring press juices could
be prepared from some molds, serious doubts were expressed about the
relevance of these findings to the problem of intracellular respiration. As
for animal cells, Arthur Harden and Hugh Maclean (1911b) reported
their failure to observe any respiration whatever in press juices from
mammalian muscle.

The Role of Iron in Biological Oxidation. It was at this stage in the
development of the problem that the dominant twentieth-century figure
enters the story. In 1908 Otto Warburg initiated a series of researches
characterized by experimental ingenuity and precision, and whose results
he reported with an assurance that commanded respect. During 1908–
1914 he performed quantitative studies on the oxygen uptake of sea-
urchin eggs and of red blood cells. He began to use a small-scale version
[developed by Thomas Brodie (1910)] of the manometric apparatus
designed by Joseph Barcroft and J. S. Haldane (1902) to measure the
binding of gases by hemoglobin; in time, it came to be known as the
Warburg apparatus, and was one of the most extensively used instruments in biochemical laboratories between the two World Wars.

Warburg's initial findings bore directly on the problem of the dependence of intracellular respiration on the integrity of the cell structure. He
observed that the mechanical disintegration of red blood cells reduced
the respiration to a very low value; on the other hand, with unfertilized
sea-urchin eggs, such disruption reduced the oxygen uptake only slightly.
Furthermore, he studied the inhibition of the respiration of intact cells
by narcotics (*e.g.,* ethyl urethane); in line with the theory of Ernst
Overton (1901) that such agents act as poisons by virtue of their solubility in the lipid-containing membrane of the cell, Warburg concluded
that ". . . the oxidative processes stand in closest connection with the
physical state of the lipids" (Warburg, 1911, p. 416). A similar conclusion was reached by Vernon (1912) in relation to the indophenol oxidase
activity of minced animal tissues. Shortly afterward, Warburg abandoned
Overton's theory, and considered that the effect of the narcotics was on
". . . a more physical-chemical catalysis, associated with a specific arrangement of the substances of the cell, on the structure of the cell"
(Warburg, 1914a, p. 314). In addition to this "structural catalysis," War-

burg postulated a "chemical catalysis" caused by intracellular catalytic substances; by 1914 he had concluded from experiments with unfertilized sea-urchin eggs:

> The chemical aspect of the oxidative process in the disintegrated egg substance is elucidated to the extent that we can say that it involves the oxidation of lipids in the presence of iron salts (*ibid.*, p. 335).

The experimental basis of this assertion was that chemical analysis showed iron to be present in the egg material, that the addition of iron salts (ferrous or ferric) to the suspension increased the rate of oxygen uptake, and that extraction of the egg substance with ether gave a material (containing lecithin) that was oxidized by O_2 in the presence of iron salts. Torsten Thunberg (1911) had already reported the catalytic effect of iron salts on the auto-oxidation of lecithin, and Warburg found that this also applied to a highly unsaturated fatty acid (linolenic acid), known to be a constituent of lecithin. An additional item of evidence offered by Warburg was that the increase in oxygen uptake caused by the addition of iron salts to the broken egg cells was inhibited by ethyl urethane. Warburg then examined the effect of iron salts in promoting the auto-oxidation of a variety of organic compounds, and found significant catalysis for cysteine (and other thiol compounds) and for aldehydes. In large part, these results confirmed those obtained earlier by chemists (notably Manchot) who had developed Traube's theory of auto-oxidation during 1900–1910. On the basis of the data available to him in 1914, Warburg proposed the theory

> . . . that the oxygen respiration in the egg is an iron catalysis; that the oxygen consumed in the respiratory process is taken up initially by dissolved or adsorbed ferrous ions (Warburg, 1914b, pp. 253–254).

It should be noted that, although he had found, as had others before him, that the respiration of various cells is strongly inhibited by cyanide, and the cyanide inhibition of the auto-oxidation of substances like cysteine had been reported by Albert Mathews and Sydney Walker (1909), Warburg did not include these findings as evidence for the theory he advanced in 1914.

After World War I, when Warburg resumed work on biological oxidations, the action of cyanide became a central point in the argument, and emphasis was placed on model experiments in which oxidations were catalyzed by iron-containing charcoals; these were first prepared by the incineration of blood, and later of hemin and of impure aniline dyes contaminated with iron salts. In a series of papers during 1921–1924, Warburg and his assistants reported that amino acids (cystine, tyrosine, leucine) are extensively oxidized by oxygen in the presence of such charcoals, and that this catalysis is inhibited by cyanide and by narcotics such as ethyl urethane. These results led to the formulation of a theory of cellular respiration as a cyclic process in which

> . . . molecular oxygen reacts with divalent iron, whereby there results a higher oxidation state of iron. The higher oxidation state reacts with the organic substance with the regeneration of divalent iron . . . . [M]olecular oxygen never reacts directly with the organic substance (Warburg, 1924, p. 479).

In justifying the use of the data obtained with the charcoal models for a theory of physiological oxidation, Warburg stated:

> The experiments . . . are model experiments in so far as the conditions under which we work are simpler than those in the cell. The experiments are more than model experiments if one succeeds with the help of iron in transferring the oxygen to the combustible substances of the cell (*ibid.*, p. 483).

He believed that the results justified the reiteration of the view that he had expressed in 1914:

> Thus there arises that remarkable interplay of unspecific surface forces and specific chemical forces, characteristic for the hemin-charcoal as well as for the living substance. Both systems behave on the one hand like unspecific surface catalyses, on the other as specific metal catalyses. The specific anticatalyst is hydrocyanic acid, the unspecific anticatalysts are the narcotics (*ibid.*, p. 488).

The title of the remarkable paper (see also Warburg, 1925a) in which this theory was advanced was: "On iron, the oxygen-transferring constituent of the respiratory enzyme." At the end of the paper, Warburg defined the respiratory enzyme (*Atmungsferment*) in relation to the process of oxidation as ". . . the sum of all catalytically-active iron compounds present in the cell" (Warburg, 1924, p. 494). Beyond stating that "The catalytically-active substance in hemin-charcoal is therefore iron, but not iron in any form whatever, but iron bound to nitrogen" (*ibid.*, pp. 485–486), Warburg offered no suggestion as to the nature of the intracellular material to which iron is bound in biological systems. His reference to "hemin-charcoal" cannot be interpreted to imply the involvement of hemin-like compounds in physiological oxidations, since the hemin used to make the charcoal had been completely destroyed by incineration. In 1925, however, David Keilin drew attention to the earlier work of Charles MacMunn on intracellular iron-porphyrin compounds, and assigned to these compounds (which Keilin called cytochrome) a significant role in intracellular respiration.

Before considering the impact of this important development, it should be added that the physicochemical orientation of Warburg's theory, with its emphasis on "unspecific surface forces," reflected the popularity of colloid chemistry among cell physiologists, but neglected the organic-chemical problem of specificity with respect to the molecules undergoing oxidation. This aspect of the problem, raised in the work of Battelli and Stern (among others), was developed during 1910–1925 by Heinrich Wieland and by Torsten Thunberg, and focused attention on the specific "dehydrogenase"-catalyzed oxidation of organic substrates through the activation of their hydrogen atoms for reaction with molecular oxygen. As we shall see shortly, Warburg disputed this emphasis, and reiterated his view that "molecular oxygen never reacts directly with the organic substance." Thus, after Malcolm Dixon and Sylva Thurlow (1925) had reported that the action of liver xanthine oxidase is not inhibited by cyanide, Warburg stated that, according to his theory,

> . . . the life processes that are specifically inhibited by hydrocyanic acid are heavy-metal catalyses. If the oxidation of hypoxanthine is *not* inhibited by hydrocyanic acid, then it represents a system that plays no role in the respiration of liver, a system that is *not* a respiration system (Warburg, 1925b, p. 252).

Warburg's comment brings to mind that of Pasteur (1860c) in rejecting Berthelot's use of the term "ferment" in connection with invertase. Six years later, however, Warburg and Christian (1931) described an enzyme system that catalyzes the oxidation of glucose-6-phosphate by molecular oxygen, and that is not inhibited by cyanide. This change in attitude may be related to Warburg's abandonment of the view he expressed in the introduction to a selected collection of his papers on biological oxidations:

> Since experience teaches that the catalysts of the living substance— the ferments—cannot be separated from their inactive accompanying material, it is appropriate to forego the methods of preparative chemistry, and to study the ferments under the most natural conditions of their activity, in the living cell itself (Warburg, 1928, p. 1).

After the crystallization of pepsin by John Northrop in 1930, Warburg turned to the purification of enzymes, and thereby made his most decisive contributions to the understanding of the intracellular mechanisms of oxidation and fermentation. Around 1925, however, the ideas of Richard Willstätter regarding the nature of enzymes held sway. Apart from the general impact of Willstätter's views, he also entered the debate on biological oxidations through his work during 1918–1925 on the purification of horseradish peroxidase. It was thought from the work of Wolff and of Madelung (in 1911) that oxyhemoglobin can act as a peroxidase, though more weakly than the plant enzyme, thus suggesting that peroxidase might be an iron-porphyrin in association with a protein, but also raising the possibility that the peroxidase activity of animal tissues might be due to hemoglobin. We mentioned on p. 156 that Willstätter developed valuable adsorption techniques for the purification of enzymes; when he applied them to peroxidase, he found no correlation between the iron content of various preparations and their enzymic activity, and concluded that "Iron compounds are closely associated with peroxidase, but the enzyme does not contain iron as an integral constituent" (Willstätter, 1922, p. 3623). Moreover, many leading biochemists shared Hopkins' view that

. . . the production of hydrogen peroxide in the course of dehydrogenations and the simultaneous activity of peroxidases are together responsible for prominent oxidations in the living cell (Hopkins, 1926, p. 56).

Consequently, when Willstätter (1926) announced that he had prepared a nearly iron-free peroxidase, a powerful blow appeared to have been struck at Warburg's theory of the essential role of iron compounds in biological oxidations. As in the case of other enzymes (*e.g.*, invertase) that Willstätter had purified, the chemical tests he used for the detection of such constituents as protein or iron were far less sensitive than the test for enzymic activity; after the establishment of the protein nature of enzymes during the 1930s, the peroxidases were shown to be iron-porphyrin containing proteins, in which the prosthetic group is the same as in methemoglobin (see Theorell, 1947).

Cytochrome and the Respiratory Enzyme. In 1925 there appeared a paper by David Keilin (for a biography, see Mann, 1964) on a widely distributed intracellular pigment which he called cytochrome; the paper began as follows:

Under the names myohaematin and histohaematin MacMunn (1884–1886) described a respiratory pigment, which he found in muscles and other tissues of representatives of almost all the orders of the animal kingdom. He found that this pigment, in the reduced state, gives a characteristic spectrum, with four absorption bands occupying the following positions: 615–593/567.5–561/554.5–546/532–511/. When oxidized, the pigment does not show absorption bands (Keilin, 1925, p. 312).

We referred earlier in this chapter to the group of English biologists who were stimulated by Stokes's observations on blood to use the spectroscope for the study of animal and plant pigments; Henry Sorby (1873) called this pursuit "chromatology." After Sorby (1867) had invented the microspectroscope, it was used by E. Ray Lankester (1872) in a systematic search for hemoglobin-like pigments in a large variety of organisms. One such pigment had already been identified by Albert Kölliker and Willy Kühne in mammalian muscle, and had been named muscle hemoglobin, although its identity as a distinct protein was not established until much later, through the work of Karl Mörner (in 1897) and of Hans Günther (1921); the latter gave it its present name, myoglobin. During the course of his studies, Lankester found myoglobin in mollusks devoid of blood hemoglobin, and a new respiratory pigment (which he named chlorocruorin) in some marine worms. In such investigations, the recognition of a respiratory pigment was based on the observation of spectroscopic changes when reducing agents were added, and the reappearance of the original "oxidized" form when the solution was shaken with air. It was this approach that led Charles MacMunn to find the four-banded spectrum of the presumed respiratory pigment to which Keilin referred. In a detailed account of his work, after his initial report in 1884, MacMunn stated:

> In no animal have I succeeded as yet in isolating the histohaematins. In them the coloured constituent occurs united to a proteid in all probability, and hence the difficulty attending attempts at isolation . . . . On the other hand, oxidation and reduction can be brought about in the *solid* organs and tissues . . . . I have proved to my own satisfaction that the *banded condition belongs to the reduced state, and the bandless to the oxidised;* but . . . this oxidation and reduction are not as simple as the reduction and oxidation of hemoglobin, for example, the oxygen being apparently more firmly fixed than in the case of oxyhemoglobin. Thus, from echinoderms to man throughout the animal kingdom, we find in various tissues and organs a class of pigments whose spectra show a most remarkable resemblance to each other . . . . [T]heir bands are intensified by alkalies and enfeebled by acids, inten-

sified by reducing agents, and enfeebled by oxidizing agents; they accordingly appear to be capable of oxidation and reduction, and are therefore *respiratory* . . . . Hence the histohaematins are concerned in the *internal* respiration of the tissues and organs of invertebrates and vertebrates . . . . Why a *coloured* constituent should be more useful than a *colourless* one is not clear, but in haemoglobin, haemocyanin, and my echinochrome and Professor Lankester's chlorocruorin, as well as Sorby's aphidein, we have colouring matters which are respiratory pigments (MacMunn, 1886, pp. 279–280).

In MacMunn's view, the muscle pigment (myohaematin) was a histohematin and, though related to hemoglobin, different from it. He offered, in support of this conclusion, spectroscopic data on products derived from the muscle pigment, and which resembled, but were not identical with, the hemochromogen and hematoporphyrin identified by Hoppe-Seyler and Stokes as derivatives of hemoglobin (MacMunn, 1887). MacMunn's evidence regarding the separate identity of myohematin was not convincing, however, and was strongly attacked by Ludwig Levy (1889), a student of Hoppe-Seyler. There ensued a brief exchange between MacMunn and Hoppe-Seyler; in a footnote appended to MacMunn's final note on the subject, in the journal edited by Hoppe-Seyler, the latter terminated the discussion by reiterating the view that ". . . the investigations that have been reported provide no basis for the assumption that special pigments are present in fresh pigeon muscle" (MacMunn, 1890, p. 329). Subsequent printed references to MacMunn's work, until 1925, stressed the probable identity of his myohematin with myoglobin, and no account was taken of his evidence for the presence of histohematins in animal tissues where hemoglobin or myoglobin could not be present. In a book published posthumously, MacMunn wrote:

> The Histohaematins and Myohaematin have not found their way into text-books because they do not belong to the ordinary pigments . . . . A good deal of discussion has taken place over this pigment, and the name of Hoppe-Seyler has prevented the acceptance of the writer's views. The chemical position is undoubtedly weak, but doubtless in time this pigment will find its way into the text-books (MacMunn, 1914, pp. 72–73).

It was not only Hoppe-Seyler, but also MacMunn's fellow chromatologists in England, who found it difficult to accept his claims. The reasons have been discussed carefully by Keilin (1966, pp. 103–105); one of them merits special mention here, namely, that no hemoglobin derivative was known to exhibit a four-banded spectrum in the reduced state and only faint absorption in the oxidized state. It remained for Keilin to resolve this difficulty in 1925, and to provide new and convincing evidence justifying the inclusion of MacMunn's hemoglobin-like pigment in "the text-books."

It is important to recall that, during the forty years between the work of MacMunn and of Keilin, significant progress had been made, not only in the study of intracellular respiration, but also in the elucidation of the chemical structure of the porphyrins and of their iron complexes. Thus, in addition to proposing (in 1913) the correct cyclic tetrapyrrole structure of the porphyrins, William Küster had also suggested (in 1910) that both in hemoglobin and in oxyhemoglobin the iron is in the divalent state (ferro), whereas the iron of methemoglobin is in the ferri state. After the validity of this suggestion had been demonstrated experimentally, largely through the work of James Conant and Louis Fieser (1925), oxyhemoglobin was recognized to be the "oxygenated" form, rather than the "oxidized" form, of hemoglobin, the latter being represented by methemoglobin. The quotation from MacMunn's paper, given above, is a typical example of the confusion generated in the early literature on respiratory pigments by the use of the terms "oxidation" and "reduction" in the sense in which Stokes applied them to hemoglobin. Another important chemical advance was the clearer definition in 1910 (Dilling, Zeynek) of the term "hemochromogen" as referring to a compound in which a ferroporphyrin is combined with two additional nitrogenous groups; this was firmly established by the work of Robert Hill (1926). A corollary was the recognition that the hemochromogens, like all other known derivatives of deoxygenated hemoglobin such as its ferroporphyrin (termed heme or haem), combine reversibly with carbon monoxide (CO). The fact that the CO-hemoglobin described by Bernard and Hoppe-Seyler is sensitive to light, with the release of the carbon monoxide, had been discovered by J. S. Haldane and Lorrain Smith (1896); the photochemical decomposition of the simpler $Fe(CO)_5$ to release one of its CO groups was described in 1891 by Mond and Langer and in 1905 by Dewar and Jones. It should also be noted that in the background of these and related chemical studies on the iron-porphyrin compounds was the new inorganic chemistry formulated by Alfred Werner from 1891 onward; from his work there emerged a broadened theory of valency to include the "coordination" of atomic groups about a metal ion (Kauffman, 1966; Schwarzenbach, 1966). As applied to heme, the theory led to the recognition that the ferrous iron is hexacoordinate, with four valencies satisfied by the nitrogens of the pyrroles of the porphyrin ring and the other two available for other interactions (*e.g.*, with O_2, CO, imidazole, pyridine).

Thus, when Keilin established MacMunn's pigment as a participant in intracellular oxidations, a large number of chemical findings had transformed the mode of thought about the reactions that the "histohematins" might undergo. Keilin (1966, pp. 143–147) has described his discovery of cytochrome and his subsequent realization of its relation to MacMunn's work; from 1909, he had been studying the physiology of insects, and in 1924 he observed with the aid of a microspectroscope the presence of a four-banded spectrum in the muscles of the horse bot-fly (*Gasterophilus intestinalis*). Of special importance was his finding of a similar spectrum with a suspension of yeast, and the observation that the absorption bands disappeared when the cell suspension was shaken with air, and reappeared shortly afterward. Furthermore, cyanide was seen to inhibit the oxidation of reduced cytochrome, and narcotics to inhibit the reduction of oxidized cytochrome, suggesting that the pigment acted as a mediator in the oxidation of cell metabolites by molecular oxygen. From the effect of various chemical treatments, Keilin concluded that the four-banded spectrum was associated with three separate hemochromogens which he named cytochromes a, b, and c. Hemochromogens were known to be characterized by two absorption bands in the visible region of the spectrum; each of the three cytochrome bands at longer wavelengths (605 mμ, 565 mμ, and 550 mμ) was assigned to a different hemochromogen, and the second bands of all three cytochromes were considered to be fused into the single band near 520 mμ. By providing strong evidence for the separate identity of the three cytochromes, Keilin gave the first satisfactory explanation of MacMunn's four-banded spectrum; for example, of the three components, only cytochrome c was stable to heat and extractable with water. Furthermore, although cyanide inhibited the oxidation of all three components by air, ethyl urethane inhibited the reduction of cytochromes a and c, and the oxidation of cytochrome b. In 1926 Keilin considered the oxidized cytochromes to contain ferric iron, and he denoted the oxidized hemochromogens "parahematins."

Although it was clear from the observations reported by Keilin in 1925 that the oxidation-reduction state of the cytochromes, as seen spectroscopically with yeast suspensions or insect muscle, was intimately related to the intracellular respiration of these biological systems, there was a serious difficulty. The cytochromes did not appear to be auto-oxidizable, and therefore it was necessary to assume the existence of an oxidase that catalyzes the oxidation of the reduced cytochromes by molecular oxygen. Keilin (1927, 1929) attempted to solve this problem by turning to the indophenol oxidase system, and sought to determine the relation of the cytochromes to the oxidase studied by Vernon and by Battelli and Stern; on the basis of the light-sensitive CO inhibition of the yeast indophenol oxidase, he concluded:

These experiments show that Warburg's respiratory ferment is a poly-phenol or indophenol oxidase system, which can display its character-istic reactions even in dead cells in which the respiration is abolished. All this clearly indicates that the oxidase systems revealed by the in-dophenol test belong to respiratory catalysts essential for the oxygen uptake of the living yeast cells (Keilin, 1927, p. 671).

Furthermore, in studies with a washed heart-muscle preparation that ef-fected the cytochrome-dependent oxidation of succinate by molecular oxy-gen, Keilin (1930) found that both the conversion of the "Nadi" reagent to indophenol blue and the oxidation of cytochrome were inhibited by cyanide and carbon monoxide; he therefore considered indophenol oxi-dase to be the enzyme responsible or the oxidation of reduced cytochrome by oxygen. Later he modified this view, and used the term "cytochrome oxidase" (suggested by Dixon, 1929, p. 385) to denote his equivalent of Warburg's *Atmungsferment*.

We noted earlier that Warburg (1924, 1925a) had stressed the impor-tance of the cyanide inhibition of cell respiration, as evidence of the par-ticipation of iron in the action of his *Atmungsferment*. After the appear-ance of Keilin's first paper on cytochrome, he turned his attention to the inhibitory action of carbon monoxide, long known to be toxic to animals because it displaces oxygen from the hemoglobin of the blood; it was, however, believed to be nontoxic to cellular respiration, largely because of the conclusion reached by Haldane:

Apart from its action in putting the red cells out of action as oxygen-carriers carbonic oxide would thus appear to be a physiologically in-different gas like nitrogen (Haldane, 1895, p. 213).

In a series of brilliant papers during 1926–1929, Warburg demonstrated not only that the oxygen uptake of yeast is inhibited by carbon monoxide, but also that this inhibition is reversed in a manner consistent with the conclusion that ". . . the *Atmungsferment* is an iron-pyrrole compound, in which the iron is bound to nitrogen, as in hemoglobin" (Warburg, 1927, p. 370). After calling attention to the apparent differences between hemoglobin and the *Atmungsferment* in their relative affinities for O_2 and CO, he stated:

A substance that is widely-distributed in cells, and related to hemo-globin, is the cytochrome of Keilin. Cytochrome is also not identical with the *Atmungsferment*, since cytochrome does not react with carbon monoxide.

It is remarkable that carbon monoxide inhibits the oxidation of cyto-chrome by air, although it does not react with cytochrome. In spite of appearances, cytochrome is not auto-oxidizable; it is rather the acti-vated oxygen of the *Atmungsferment* which oxidizes cytochrome in the cells (*ibid.*, p. 371).

Indeed, Warburg subsequently questioned whether cytochrome (which he termed the "MacMunn hemins") is involved at all in normal cellular respiration:

Whether the MacMunn hemins lie on the normal path of respiration —whether therefore respiration is not a simple, but a four-fold iron catalysis—is a question that we cannot answer today. The available spectroscopic observations are also consistent with the view that the MacMunn hemins are only reduced in the cell when, as a consequence of oxygen deficiency, the concentration of the activated combustible substances exceeds the physiological level (Warburg, 1932, p. 2).

In making this statement, Warburg chose to dismiss Keilin's original evi-dence in favor of the conclusion that the three hemochromogens serve as links between metabolic substrates and molecular oxygen, and that

. . . cytochrome acts as a respiratory catalyst, which is functional in oxidized as well as in partially reduced form. The oxygen is con-stantly taken up by the pigment and given up to the cells. In the living organism the state of the cytochrome as seen spectroscopically denotes only the difference between the rates of its oxidation and reduction (Keilin, 1925, p. 323).

Warburg's doubts about the intracellular function of cytochrome were echoed by Mortimer Anson and Alfred Mirsky:

The question naturally arises what the function of cytochrome is. Ex-perimentally nothing is known. Some properties of cytochromes are known qualitatively; it is known for instance that it can be oxidized and reduced. But what cytochrome actually does accomplish in normal respiration still remains to be demonstrated. Several theoretical possi-bilities have been suggested (Keilin, 1929; Dixon and Elliott, 1929). It is of course conceivable that the ferment which combines with CO is only one catalyst in a complicated catalytic system of which cytochrome and glutathione are essential parts. There is, however, as yet no ade-quate experimental basis for a discussion of this question.

The heme pigments have been given credit for the peroxidase action of cells (Keilin, 1929) on the grounds that all hemes are to some extent peroxidases and that the peroxidase action of tissues is in a general way parallel to their content of visible heme pigments. There are, however, so many substances whose concentrations go with the intensity of tissue metabolism that parallelism of concentration is hardly sufficient evidence of identity of structure.

It has also been claimed (Keilin, 1925, 1929) that cytochrome is a respiratory pigment because it can be oxidized and reduced. The chromatologists have always called every pigment which can be oxidized or reduced a respiratory pigment—why, it is not clear. There are many colorless substances in the cell which can be and probably are oxidized and reduced. This property is not taken as proof that they are respiratory substances (Anson and Mirsky, 1930, pp. 541–542).

Within a few years, not only the cytochromes but also some "colorless substances" were to be established as participants "in a complicated catalytic system," while glutathione (which we shall consider later in this chapter) was to fall from the favor it had enjoyed during the 1920s.

The most impressive experimental achievement reported by Warburg in his studies on the *Atmungsferment* was the indirect determination of the absorption spectrum of the CO complex of its reduced form, by measurements of the relative efficiency of various wavelengths of light in counteracting the inhibition, by carbon monoxide, of the respiration of a yeast (Warburg and Negelein, 1929). The photochemical "action spectrum" showed a large band near 430 mμ, characteristic of porphyrin compounds (the so-called Soret band), as well as smaller bands in the visible region at 600 mμ, 550mμ, and 525 mμ. Furthermore, there was an absorption band at about 280 mμ, which was assigned to a protein component, thus supporting the conclusion that the *Atmungsferment* is a heme protein related to hemoglobin.

In considering the possible relation of the *Atmungsferment* to the various oxidases such as indophenol oxidase, found in cell extracts, Warburg suggested that they might be artifacts of preparation:

If one calls oxidases ferments that transport molecular oxygen, then the extracts contain oxidases, and if one classifies the oxidases, as is customary in ferment chemistry, according to the observed actions, then one has in the extracts different oxidases, glucose oxidase, alcohol oxidase, indophenol oxidase, and so on. Strictly speaking there are as many different oxidases as extraction experiments. If the extract-oxidases had been preformed in the cell, a single type of cell would contain innumerable oxidases. But the multiplicity of oxidases in the living cell would be in opposition to a sovereign principle in the living substance . . . . Therefore, if many different oxidases have been found in extracts of a cell type, these were not ferments that were already present in the living cell, but are rather products of the transformation and decomposition of a single homogeneous substance present in life (Warburg, 1929, pp. 1–2).

It is difficult to escape the impression that, at the time Warburg wrote these words, he still held the view that the intracellular oxidation of metabolites occurred by the direct action of oxygen activated by a single protoplasmic *Atmungsferment*, and that he still rejected Keilin's conclusion that the cytochromes serve as links between molecular oxygen and the enzymic catalysts that activate the substrates, thus reiterating his opposition to the ideas of Wieland and Thunberg on the importance of specific dehydrogenases. As we shall see in the next section of this chapter, however, the development of the story during the 1930s showed that the chemical mechanism of the oxidation of cell metabolites is far more complex than merely the activation of oxygen by an *Atmungsferment*. In this development the role of the cytochromes as mediators in the overall process was clearly established; moreover, at the end of the decade, Keilin and Hartree (1939) demonstrated the existence of a CO-sensitive, auto-oxidizable cytochrome, whose distinctive band near 600 mμ lies close to that of cytochrome a. This "cytochrome a$_3$" was identified as the cytochrome oxidase Keilin had previously thought to be the same as indophenol oxidase, and the spectroscopic properties of its CO complex agreed with those of the CO complex of Warburg's *Atmungsferment*. A historical account of this work may be found in Keilin (1966); an earlier book by Warburg, *Schwermetalle als Wirkungsgruppen von Fermenten* (1946; English translation, *Heavy Metal Prosthetic Groups and Enzyme Action*, 1949) emphasizes his view of the development of the subject.

Before concluding this section, it should be added that many of the difficulties in elucidating the role of the iron-porphyrin compounds in biological oxidations arose from the fact that most of them (the *Atmungsferment*, cytochrome oxidase, cytochrome a, cytochrome b) could not be extracted from animal tissues with aqueous solvents, and could only be studied spectroscopically. These limitations were brilliantly overcome, but only in part, by Keilin's use of the microspectroscope and the Hartridge reversion spectroscope, and by Warburg's remarkable achievement in determining the photochemical action spectrum of the CO complex of the *Atmungsferment*. Since about 1930, many new heme pigments have been identified spectroscopically in biological systems, and important technical improvements have made possible more precise spectroscopic measurements; nevertheless, where there has been difficulty in the isolation of a cell pigment (possibly because of its intimate association with lipid-containing intracellular structures), the initial conclusions drawn from spectroscopic studies have often required revision. For example, the question of the separate identity of cytochromes a and a_3 has been a subject of debate into the 1960s. On the other hand, in the case of cytochrome c which, as we noted before, Keilin found to be extractable with water, the situation has been different. During the 1930s the work of Karl Zeile (1935) and especially of Hugo Theorell (1947) led to the complete purification of this cytochrome, and its characterization as a protein having a molecular weight of about 13,000, with the porphyrin unit joined by covalent linkage to two cysteine residues of the protein. After World War II, crystalline cytochrome c preparations were obtained from a large variety of biological sources, and their amino acid sequences were determined. Not only has this new knowledge permitted the study of the oxidation-reduction behavior of cytochrome c with all the available techniques of modern protein chemistry, but also comparisons of the amino acid sequences for different biological forms have provided new insights into problems of biochemical evolution (Margoliash and Schejter, 1966). In this connection, the following statement made over a hundred years ago is of interest:

> The chemical differences among various species and genera of animals and plants are certainly as significant for the history of their origins as the differences in form. If we could define clearly the differences in molecular constitution and function of different kinds of organisms, there would be possible a more illuminating and deeper understanding of the question of the evolutionary relations of organisms than could ever be expected from morphological considerations (Lankester, 1871, pp. 318–319).

The Activation of Hydrogen

In 1912 Wieland published a series of papers on the catalysis, by finely divided palladium (palladium black), of the oxidation of various organic compounds, and offered evidence for the view that the oxygen entering the compound (as in the oxidation of an aldehyde RCHO to the corresponding carboxylic acid RCOOH) is not derived from molecular oxygen, and that there was no need to postulate the activation of oxygen in such processes. Instead, he proposed that oxidations of this type involve the catalytic labilization of the organic compound, with the activation of certain of its hydrogen atoms, which are transferred to molecular oxygen to form hydrogen peroxide ($2 H + O_2 = H_2O_2$). Thus, in the case of the oxidation of an aldehyde to an acid, Wieland assumed an initial addition of water to the aldehyde, followed by the catalytic dehydrogenation (*Dehydrierung*) of the aldehyde hydrate:

$$RCHO + H_2O \longrightarrow RCH(OH)_2 \overset{+O_2}{\longrightarrow} RCOOH + H_2O_2$$

In support of this theory of "hydrogen activation," as it came to be called in the ensuing debate, Wieland reported that, in the absence of molecular oxygen, substances known to be readily reduced in the presence of palladium or platinum (*e.g.*, quinone, methylene blue) could serve as hydrogen acceptors in the palladium-catalyzed oxidation of organic substances, and could be converted into their hydrogenated forms (*e.g.*, hydroquinone, leucomethylene blue).

The reports of these chemical studies were soon followed by an important paper in which Wieland stated that he had begun work on biological oxidations

> . . . with the aim of testing whether the dehydrogenation theory could contribute to the understanding of the mechanisms of these largely unexplained reactions. As is well-known, the almost universally accepted view of biologists and chemists who have worked on this subject is that the intracellular oxidations and combustions owe their rapid rate to the participation of oxygen-activating ferments (Wieland, 1913, p. 3328).

He proceeded to show that his model system could effect, in the apparent absence of molecular oxygen, the oxidation of glucose to CO_2, or the conversion of lactic acid into pyruvic acid [$CH_3—CH(OH)—COOH \rightarrow CH_3—CO—COOH$], or even the oxidation of polyphenols (*e.g.*, pyrogallol) considered to be substrates for oxidases. To demonstrate the relevance of his theory to oxidations in biological systems, Wieland reported that the oxidation of ethanol (CH_3CH_2OH) or of acetaldehyde (CH_3CHO) to acetic acid (CH_3COOH) by "acetic acid bacteria" (*Acetobacter*) could proceed anaerobically provided that quinone or methylene blue was present as a hydrogen acceptor; he was unable, however, to prepare a cell-free press juice that would effect this process, in agreement with Buchner's earlier finding that the aerobic oxidation of ethanol to acetic acid by this organism was associated with the integrity of the bacterial cell.

Wieland's demonstration that many oxidative processes in biological systems might involve the removal of hydrogen from a metabolite, rather than the addition of oxygen, brought greater clarity to a problem that had concerned physiological chemists for several decades, namely, the question of the existence of separate "reduction ferments" as counterparts of the "oxidation ferments." As we have seen, the nineteenth-century physiologists and chemists who were concerned with the problem of animal respiration tended to emphasize either the activation of oxygen or some special property of protoplasmic protein that made it reactive toward oxygen. Indeed, as noted by Bernard, it was not clear whether a physiological oxidation of bound hydrogen took place in animals:

> It was believed at the time of the first volumetric determinations that the oxygen taken up during respiration was completely transformed into carbonic acid. Later, more precise studies having shown that all of the oxygen was not represented in the carbonic acid, it was supposed that the surplus was used to burn hydrogen and form water. But as noted by MM. Regnault and Reiset, that is a gratuitous hypothesis. It has no reason for existence except to explain a deficit, itself hypothetical (Bernard, 1876, p. 27).

By that time, however, the work of Pasteur (1861) on anaerobic fermentation had led to the recognition that the long-known formation of hydrogen-rich compounds (*e.g.*, hydrogen sulfide, methane) during putrefaction is effected by living microorganisms, so that such reductions had to be viewed as biological processes. Moreover, by 1876, several metabolic transformations had been observed in animal tissues (*e.g.*, the formation of bile pigments from hemoglobin) that were recognized to be reductions, and Hoppe-Seyler was impelled to assume the occurrence ". . . of processes in which organic substances are changed and cleaved by the action of water in a manner similar to that found in the process of putrefaction" (Hoppe-Seyler, 1876, p. 5). To explain the reductions observed in biological systems, Hoppe-Seyler chose to emphasize the activation of hydrogen, and cited the earlier studies of Gottfried Osann and of Thomas Graham in support of his views.

Osann (1853) found that hydrogen gas released at a platinum electrode during the electrolysis of dilute sulfuric acid had a stronger reducing action than that of ordinary hydrogen; he considered this active hydrogen to be analogous to Schönbein's ozone, and named it "ozone-hydrogen." This phenomenon was studied further by Graham, who concluded in 1869 that

> Hydrogen (associated with platinum) unites with chlorine and iodine in the dark, reduces a persalt of iron to the state of protosalt [*i.e.*, ferric to

ferrous], converts red prussiate of potash to yellow prussiate [*i.e.*, potassium ferricyanide to the ferrocyanide], and has considerable deoxidizing powers. It appears to be the active form of hydrogen, as ozone is of oxygen (Graham, 1876, p. 299).

We saw earlier that Hoppe-Seyler's exchange with Traube on the relative importance of hydrogen activation and oxygen activation in biological oxidations had led, by the end of the nineteenth century, to the acceptance of Traube's theory, and as little interest was shown in the active hydrogen of Osann and Graham as in the ozone of Schönbein. Nevertheless, the problem of the mechanism of biological reductions continued to arise repeatedly during the period 1880–1910.

Of special importance were the studies of Ehrlich (1885) on the intracellular reduction of various synthetic dyes; among them was the easily reducible methylene blue, which was to play a significant role in later studies on biological oxidation-reduction reactions. Ehrlich's findings were followed by a succession of other reports indicating the presence, in animal tissues, of materials that catalyze reductions. Thus, Joseph de Rey-Pailhade observed the reduction of sulfur to hydrogen sulfide by extracts of yeast and of animal tissues; he gave the name "philothion" to the material apparently responsible for catalyzing the reaction, and concluded:

Acting like an enzyme [*diastase*], it serves to add one more proof of M. Berthelot's theory of fermentation. It is the first known example of a substance extracted from a living organism and possessing the property of hydrogenating sulfur (Rey-Pailhade, 1888, p. 44).

In the face of the uncertainty about the nature of enzymes, this conclusion was ill-founded, and was soon disproved, but similar reports came from other laboratories. By the end of the century, the question of the existence of reducing enzymes came to a head in the work of Abelous and Gérard, who reported the presence, in extracts of animal tissues, of soluble ferments that catalyze the reduction of nitrate to nitrite, of nitrobenzene to aniline, and of methylene blue to its colorless leuco form. Furthermore, they concluded that ". . . in the aqueous maceration from horse kidney there co-exist a soluble reducing ferment and a soluble oxidizing ferment" (Abelous and Gérard, 1899, p. 1025). A few years later (in 1903), Abelous revised this conclusion, and raised the possibility that a single "oxidation-reduction ferment" might be present. The view that reductions such as those studied by Abelous were enzymic in nature was disputed by Arthur Heffter (1908), who considered these reactions to involve the sulfhydryl (SH) groups of protoplasmic proteins. By that time, the structure of cysteine, the SH-containing constituent of proteins, had been elucidated, and a relatively specific color reaction (the nitroprusside test) for the SH group had been introduced by Karl Mörner (in 1901). Heffter showed that a purified protein such as egg albumin gave a positive nitroprusside reaction, and, since sulfhydryl compounds were known to be auto-oxidizable, he proposed that the SH groups of protoplasmic proteins play a significant role in intracellular oxidations. Thus oxygen was thought to oxidize the sulfhydryl groups:

$$R(SH)_2 + O_2 \longrightarrow R\begin{matrix} \diagup S \\ \big| \\ \diagdown S \end{matrix} + H_2O_2$$

the resulting disulfides being reduced in the presence of a suitable metabolite (M) and water:

$$R\begin{matrix} \diagup S \\ \big| \\ \diagdown S \end{matrix} + M + H_2O \longrightarrow R(SH)_2 + MO$$

For Heffter, the hydrogen atom of the sulfhydryl group was "active" in the sense of Hoppe-Seyler's theory; consequently, under anaerobic conditions, various oxidized compounds (*e.g.*, methylene blue) could be reduced without enzymic catalysis, and the assumption of the existence of reducing enzymes seemed unnecessary.

The Dehydrogenases. By 1910, however, it was clear from the work of Franz Schardinger (in 1902) and of Richard Trommsdorff (in 1909) that milk contains a heat-labile factor which catalyzes the rapid reduction of dyes such as methylene blue or indigo sulfonate to their leuco forms, provided an aldehyde (*e.g.*, acetaldehyde, CH_3CHO) is added. Also, in 1909, Georg Bredig had described a model for this enzymic catalysis; in the presence of colloidal palladium, methylene blue was reduced by aldehydes. Once again the ideas of Osann, Graham, and Hoppe-Seyler about the activation of hydrogen came to the fore, principally in the writings of Bach (1911). He proposed, in analogy to his earlier theory of the nature of the oxidases, that the reducing system (*Redukase*) of animal tissues consists of the Schardinger enzyme (the counterpart of peroxidase) and a tissue component replaceable by aldehydes (the counterpart of hydrogen peroxide). In Bach's scheme, the function of the enzyme was to activate the hydrogen of a hypothetical "perhydride" form of water, and he accordingly named the Schardinger enzyme a "perhydridase."

The studies on the Schardinger enzyme thus called attention to the possibility of the enzyme-catalyzed transfer of hydrogen atoms, and oxidants such as methylene blue became known as "hydrogen acceptors." According to this view of the oxidation of aldehydes, when oxygen was present, it was the hydrogen acceptor. Furthermore, Jacob Parnas (1910) reported that an "aldehyde mutase" present in animal tissues can catalyze the conversion of an aldehyde into the corresponding alcohol and acid ($2\ RCHO + H_2O \rightarrow RCH_2OH + RCOOH$), indicating that this "dismutation" involves a process in which one molecule of the aldehyde is acting as the hydrogen acceptor (to form the alcohol) and the other as the hydrogen donor (to form the acid); Cannizzaro had shown in 1853 that this reaction is promoted by alkali.

Even clearer evidence of the enzymic activation of hydrogen atoms emerged from the work of Battelli and Stern (1911) on the oxidation of succinic acid. They considered the product of the oxidation to be malic acid [$HOOC-CH_2-CH_2-COOH \rightarrow HOOC-CH_2-CH(OH)-COOH$], but in 1913 Hans Einbeck showed that this process is succinic acid $\rightarrow$ fumaric acid $\rightarrow$ malic acid, with the dehydrogenation of succinic acid to fumaric acid ($HOOC-CH_2-CH_2-COOH \rightarrow HOOC-CH=CH-COOH$) as the first step. Consequently, at the time Wieland advanced his dehydrogenation theory in 1913, there already was evidence that some enzymes can catalyze what appeared to be a labilization of hydrogen atoms in organic molecules, and he could write:

> If oxidative processes are considered to be dehydrogenations, as has been definitely shown for a few important cases, then they include at the same time a reductive process, since the hydrogen activated by the ferment must be taken up by some acceptor . . . . Because of this relationship, the so-called reduction ferments that have frequently appeared in the literature lose their separate status if it can be shown that their characteristic reductive action, for example the decolorization of a dye by some substrate, can also be effective in the hydrogenation of oxygen, [or] in the sense of the views held up to now, the "reductase" can also function as an "oxidase" (Wieland, 1913, p. 3339).

After offering evidence to show that this criterion was met by the Schardinger enzyme, he concluded:

> One can easily imagine that the course of a dehydrogenation process in the direction of the hydroproduct will depend on the nature of the available acceptors, so that the dehydrogenating ferments, which undoubtedly will have the most varied specificity with respect to both the dehydrogenating and hydrogenating agents, sometimes will find the

most suitable hydrogen acceptor in the substrate itself (mutase), at other times in molecular oxygen (oxidase), or finally in a dye, in nitrate, or similar substances (reductase) (*ibid.*, p. 3341).

In writing of the "dehydrogenating ferments," Wieland used the term *Dehydrase;* this was later replaced by "dehydrogenase," so as to avoid confusion with enzymes that catalyze the dehydration (removal of the elements of water) of their substrates.

After the appearance of Wieland's dehydrogenation theory, Battelli and Stern revived a hypothesis offered during the 1880s by Traube; they suggested:

In the action of the oxidizing ferments, the hydroxyl ions of water constitute the oxidizing group, whereas the hydrogen ions go to the hydrogen acceptor, for example molecular oxygen. They differ therefore from other hydrolytic or hydrating ferments solely in the fact that by the action of the oxidizing and reducing ferments, the hydroxyl ion is fixed to the molecules of one substance and the hydrogen ion to the molecules of another substance. In contrast, the action of other enzymes involves the fixation of both groups to the same molecule in producing hydration or scission of the molecule. Catalase should therefore be considered a hydrolyzing ferment (Battelli and Stern, 1920, p. 1545).

This idea was not accepted widely, however, and during the 1920s attention was focused principally on the theories of Warburg and Wieland.

Comparison of Wieland's theory with that of Warburg, as they were expressed before World War I, provides a striking example of the way in which the chemical mechanism of a biological process could be formulated in entirely different terms at a time when the extent of the chemical complexity of the process was unknown. We suggested earlier that Warburg's theory reflected the influence of the new colloid chemistry on the general physiologists of his time; although trained in organic chemistry (he had worked with Emil Fischer on amino acids), Warburg's interpretation of his observations on cellular respiration was a physicochemical theory that emphasized catalysis at surfaces. On the other hand, while Wieland's formulation was in the tradition of the best organic chemistry of his time, the limitations of the available knowledge about the nature of enzymes, and of the chemical structure of the essential participants in biological oxidations, made Wieland's theory no more convincing. In 1914, and during the period 1922–1925 when the Wieland-Warburg debate reached its peak, the choice was between two kinds of artificial models of the respiratory apparatus of living cells: the iron-containing charcoals of Warburg as models of his hypothetical oxygen-activating *Atmungsferment,* or the palladium system of Wieland as a model of the activation of hydrogen in substrates undergoing dehydrogenation in the presence of specific enzymes. The post-World War I debate between these two great scientists was largely about the shortcomings of the models; thus Wieland disputed the significance of Warburg's finding that amino acids are oxidized in the presence of iron-charcoal by noting that, ". . . in the absence of oxygen, amino acids can be dehydrogenated, with the release of ammonia, by palladium black" (Wieland, 1922b, p. 502), and Warburg (1923), in forcefully defending his model system, disproved Wieland's contention that the cyanide inhibition of respiration was a consequence of the inhibition of catalase. As it had turned out before, and since, in debates of this sort, both models were inadequate, although the discussion about them enlivened the scientific scene and stimulated fruitful experimentation. Indeed, later work of Gillespie and Liu (1931) demonstrated that, if impurities were removed from palladium black, the catalytic dehydrogenation of hydroquinone reported by Wieland could no longer be observed. It may be added here that Wieland's work on biological oxidations represented only one facet of a remarkably productive scientific career, which included the elucidation of the structure of the bile acids and of the pigments (pterins) of butterfly wings; the latter subsequently turned out to be the functional components of the widely distributed and physiologically important folic acids.

A significant consequence of Wieland's formulation of the dehydrogenation theory was the stimulus it provided to the work of Torsten Thunberg during 1917–1920. To test the validity of the theory, Thunberg developed a technique in which thoroughly washed minced tissue (*e.g.,* frog muscle) was suspended in a solution containing methylene blue, which was not decolorized by the washed tissue; after the system had been evacuated to remove oxygen (a special test tube, later known as the Thunberg tube, was devised for this purpose), organic substances were added, and the time required for the decolorization of the dye was noted. Thunberg (1920) found that several organic acids (among them lactic acid, succinic acid, malic acid, citric acid, α-ketoglutaric acid, glutamic acid, alanine) promoted the rapid reduction of methylene blue, and concluded that these substances were oxidized by specific dehydrogenases, in accordance with Wieland's theory. Thunberg offered some evidence, based on differences in the apparent stability of the various dehydrogenase activities to freezing or to heat, indicating that the dehydrogenation of the individual substrates was attributable to separate enzymes, and other investigators (*e.g.,* Bernheim, in 1928) added supporting data in favor of this view; it was not until many years later, however, that its validity was satisfactorily established through the crystallization of the dehydrogenases as well-defined catalytic proteins.

Despite these deficiencies in the characterization of the dehydrogenases, many investigators accepted the Wieland-Thunberg theory during the 1920s, and sought to reconcile it with the evidence offered by Warburg in favor of the activation of oxygen. Thus, from the fact that cyanide inhibits the dehydrogenation of succinic acid by washed muscle preparations when molecular oxygen is the "hydrogen acceptor," but not when methylene blue is used in the Thunberg technique, Alfred Fleisch and Albert Szent-Györgyi independently concluded that

. . . "activation" of both hydrogen and oxygen is necessary at least for the oxidation of succinic acid (Fleisch, 1924, p. 310).

In cellular oxidation the *activated hydrogen* is burned by the *activated oxygen*. In the terminology of the hydrogen activation theory this means that molecular oxygen is not a hydrogen acceptor; the biological hydrogen acceptor is the oxygen activated in Warburg's system (Szent-Györgyi, 1924, p. 196).

Their findings suggested that, in biological oxidations,

. . . hydrogen is transported, not directly from primary donators to oxygen, but by stages. It would appear that the path may in a given case be smoothened, and so the velocity of transport increased, by the intervention of a substance which can act alternately as an intermediate acceptor and donator. Such a substance acts therefore catalytically as a carrier of hydrogen (Hopkins, 1926, p. 52).

It should be noted that there had been anticipations of this idea. For example, Palladin (1909) concluded from his studies on plant respiration that certain polyphenols, acting as "respiratory pigments," remove hydrogen from substances undergoing oxidation, the reduced pigments then being reoxidized under the catalytic influence of the oxidases.

We have already discussed the rediscovery of cytochrome by Keilin in 1925 (not mentioned in the article by Hopkins, cited above), and have cited some of the evidence he presented to support his view that this heme pigment serves as an oxidation-reduction link between the dehydrogenation of metabolites and molecular oxygen. During the 1920s, however, equally serious attention was given to other cell constituents as possible intermediate agents in biological oxidations. Before considering their fate, it is necessary to insert a discussion of the important contributions being made at that time to the general theory of oxidation-reduction reactions.

Oxidation-Reduction as Electron Transfer. One of the sources of confusion among biologists who were debating such questions as the separate existence of "reductases" and "oxidases" was the absence of agreement on the chemical meaning of the terms oxidation and reduction. We have seen that by 1914 biological oxidation came to include the removal of hydrogen atoms from a molecule. Similarly, the ancient term reduction, originally used in the sense of the "revivification" of a metal from its ore (Macquer, 1777), could mean not only the removal of oxygen from a compound, but also the addition of hydrogen. Furthermore, some cytologists had been using dyes such as methylene blue to identify what were called intracellular "reduction places" (*Reduktionsorte*) and "oxygen places" (*Sauerstofforte*) in stained tissue preparations (Unna, 1911). This approach was criticized by Alan Drury (1914) in a paper to which William Hardy added the comment that

It must always be remembered that an oxidation place is also a reduction place, and it is to be called the one or the other according to the particular zero which is chosen. A convenient zero is the chemical potential of atmospheric oxygen, and a place would be an oxidation place if oxygen, whose chemical potential is $\geqq$ that of atmospheric oxygen, is condensed to the intramolecular state. Such a region would then be a reduction place for chemical compounds in which the oxygen potential is $\geqq$ that of atmospheric oxygen, and an oxidation place in which it is less than that of atmospheric oxygen. In the absence of some agreement as to the zero point the discussion is likely to be as confused in the future as it has been in the past (*ibid.*, pp. 175–176).

Such agreement came, however, only after the emergence of the electronic theory of valency and after the application of electrochemistry, as enriched by thermodynamic theory, to the study of the oxidation-reduction behavior of organic substances. During the period 1920–1940, as biologists began to appreciate the significance of these advances, the intracellular oxidation of a metabolite came to be considered as the loss of electrons (with or without the gain of oxygen or loss of hydrogen), and reduction the gain of electrons (with or without the loss of oxygen or the gain of hydrogen). As a consequence, in place of oxygen transfer or hydrogen transfer, the term "electron transfer" or "electron transport" was used to an increasing extent.

Because the development of electrochemistry during the nineteenth century became so important in the study of biological oxidation-reduction reactions, a brief summary of its highlights is essential. In his memorable reports on the "electrochemical decomposition" of chemical substances, published in 1834, Michael Faraday (Williams, 1965) introduced terms suggested to him by William Whewell, and used today (*e.g.*, anode, cathode, ion, anion, cation). More importantly, Faraday showed that the laws of definite and multiple proportions hold not only for chemical elements, but also for electricity:

> Although we know nothing of what an atom is, yet we cannot resist forming some idea of a small particle, which represents it to the mind; and though we are in equal, if not greater, ignorance of electricity, so as to be unable to say whether it is a particular matter or matters, or mere motion of ordinary matter, or some third kind of power or agent, yet there is an immensity of facts which justify us in believing that the atoms of matter are in some way endowed or associated with electrical powers, to which they owe their most striking qualities, and amongst them their mutual chemical affinity. As soon as we perceive, through the teaching of Dalton, that chemical powers are, however varied the circumstances in which they are exerted, definite for each body, we learn to estimate the relative degree of force which resides in such bodies; and when upon that knowledge comes the fact that the electricity, which we appear to be capable of loosening from its habitation for a while, and conveying from place to place, *whilst it retains its chemical force,* can be measured out, and being so measured is found to be *as definite in its action* as any of *those portions* which, remaining associated with the particles of matter, give them their *chemical relation;* we seem to have found the link which connects the proportion of what we have evolved to the proportion of that belonging to the particles in their natural state (Faraday, 1839, pp. 249–250).

We noted before that the question of the existence of atoms was hotly debated at midcentury; by 1881, however, when Hermann Helmholtz delivered his celebrated Faraday lecture, he could say:

> Now the most startling result of Faraday's law is this. If we accept the hypothesis that elementary substances are composed of atoms, we cannot avoid concluding that electricity also, positive as well as negative, is divided into definite elementary portions, which behave like atoms of electricity (Helmholtz, 1881, p. 290).

In the same year, George Stoney wrote that

. . . Nature presents us, in the phenomenon of electrolysis, with a simple definite quantity of electricity which is independent of the particular bodies acted on. To make this clear I shall express "Faraday's Law" in the following terms . . . : *For each chemical bond which is ruptured within an electrolyte a certain quantity of electricity traverses the electrolyte, which is the same in all cases* . . . . If we make this our unit quantity of electricity, we shall probably have made a very important step in our study of molecular phenomena (Stoney, 1881, p. 384).

To this "unit quantity of electricity," Stoney later gave the name "electron." It was the study of electrical conductivity in rarefied gases that led to the discovery of the "cathode rays," and their identification as particles of negative electricity by J. J. Thomson in 1895–1897; after 1900 Stoney's term was applied to these particles, and it was recognized that every ion, whether in an electrolytic solution or in a gas, carries an integral number of electrons.

Furthermore, by the time of Helmholtz's Faraday lecture, the experimental studies of Wilhelm Hittorf during the 1850s, and of Friedrich Kohlrausch during the 1870s, on the electrical conductivity of solutions, together with van't Hoff's theoretical treatment of osmotic pressure, had provided the basis for the formulation of the theory of electrolytic dissociation by Arrhenius, in 1886. Also, the first fruitful applications of the Second Law of Thermodynamics to the development of a quantitative theory of the electromotive force of a galvanic cell (such as those devised during the early 1840s by Daniell and Grove) were made by Josiah Willard Gibbs and by Helmholtz around 1880, and Wilhelm Ostwald's influence helped to make the new chemical thermodynamics the theoretical foundation of electrochemistry.

In 1889 Ostwald's associate Walther Nernst proposed a useful equation to describe the electrical potential between a metal and its ion in solution, although his theory of "solution pressure" at a metallic electrode was later abandoned; in a letter to Wilder Bancroft, written in May 1899, Gibbs noted that

. . . the consideration of the electrical potential in the electrolyte, and especially the consideration of the difference of potential in electrolyte and electrode, involves the consideration of quantities of which we have no apparent means of physical measurement, while the difference of potential in "pieces of metal of the same kind attached to electrodes" is exactly one of the things which we can and do measure (Gibbs, 1906, Vol. I, p. 429).

It was indeed such kinds of measurements, first performed by Ostwald's associate Rudolf Peters (1898) that provided the experimental basis for the comparison of the relative ability of chemical substances to act as oxidants or reductants. Peters measured the difference in potential between one cell containing variable proportions of ferric and ferrous salts and a reference cell containing mercury and mercuric chloride ("calomel electrode"), and showed that his data fitted the equation (rewritten in present-day symbols):

$$E_h = E^0 + 0.06 \log [(Fe^{3+})/(Fe^{2+})]$$

where E_h is the measured potential difference (in volts) and the terms in the brackets are the activities of the ferric and ferrous ions. When $(Fe^{3+}) = (Fe^{2+})$, $E_h = E^0$, which is defined as the "normal oxidation-reduction potential" for the ferrous-ferric system under the experimental conditions (temperature, acidity) of Peters' studies. Thermodynamic theory had provided a direct relationship between the magnitude of E^0, the equilibrium constant of a reversible oxidation-reduction reaction, and the maximum useful work ("free energy") that could be derived from such a reaction at constant temperature. Consequently, measurement of the normal oxidation-reduction potential of any system whose oxidized and reduced components equilibrate rapidly with a metallic electrode indicates the energy change to be expected when the oxidant of that system reacts with the reductant of another oxidation-reduction system of known potential. Thus, if the normal potentials for two oxidation-reduction systems (*e.g.*, $A_{ox} \rightleftharpoons A_{red}$, $B_{ox} \rightleftharpoons B_{red}$) have been determined, it is possible to calculate the free-energy change in a chemical reaction in which, say, the A system provides the oxidant and the B system provides the reductant

$$A_{ox} + B_{red} \rightleftharpoons A_{red} + B_{ox}$$

By about 1900, therefore, the thermodynamic theory underlying oxidation-reduction reactions had been largely developed, and soon afterward they began to be considered in terms of electron exchange. Fritz Haber (1901) made one of the initial contributions to this development, and three years later he published the first important study on the potential established at a metallic electrode by an organic oxidation-reduction system (quinone-hydroquinone). Until 1920, however, little else was done with organic systems, and attention was focused on inorganic compounds, on technical improvements in the potentiometric measurement of electrode potentials, and on the application of such measurements to analytical problems. A decisive change came after 1916, however, when Kossel and Lewis independently proposed the electronic theory of valency (Palmer, 1965), and brought the new atomic physics to bear on chemical problems. In particular, this theory led organic chemists to consider reactions in their domain as involving electronic mechanisms.

In 1920 the laboratories of Einar Biilmann, W. Mansfield Clark, and John M. Nelson independently reported electrometric studies on organic oxidation-reduction systems; both Biilmann and Nelson worked on the quinone-hydroquinone system, and Clark examined the dyes methylene blue and indigo sulfonate. All three groups recognized that in these processes, hydrogen ions are involved, and that the normal potential varies with pH (the importance of this factor in inorganic systems had been noted by Crotogino in 1900). The convention was adopted to refer the oxidation-reduction potentials of any system to

> . . . the normal hydrogen electrode. This is defined as a platinized platinum electrode held under one atmosphere of hydrogen and immersed in a solution normal with respect to the hydrogen ions. To the potential difference at such an electrode is assigned the arbitrary value zero (Clark, 1923, p. 451).

This oxidation-reduction system ($H_2 \rightleftharpoons 2 H^+ + 2e$) provided the basis for the electrometric determination of pH, and during the succeeding years many technical improvements were made in the instruments available for such measurements (*e.g.*, the use of the glass electrode; see Dole, 1941).

During the 1920s Clark and his associates reported careful determinations of the potentials for a series of organic dyes (indigo sulfonates, methylene blue, indophenols) that formed oxidation-reduction systems able to exchange electrons rapidly with a metallic electrode (*i.e.*, they were "electromotively active"). Additional dyes were introduced subsequently, notably by Leonor Michaelis. Such dyes, especially methylene blue, had long been used for the qualitative examination of the reducing or oxidizing capacity of biological systems, and we have already referred to Thunberg's use of this approach; the data provided by Clark permitted the use of these oxidation-reduction "indicators" for the estimation of the oxidation-reduction potentials of metabolite systems (*e.g.*, succinate-fumarate) with which the dyes could react under the catalytic influence of a suitable enzyme preparation. Apart from the availability of a series of such indicators having widely different normal potentials (according to the conventions employed, a much more positive potential for system A than for system B indicates that the reaction $A_{ox} + B_{red} \rightleftharpoons A_{red} + B_{ox}$ will have the tendency to go far to the right), the studies of Clark,

Biilmann, Conant, and Michaelis during the 1920s emphasized the limitations inherent in the formulation of Wieland's theory in terms of hydrogen transfer. Thus the reduction of the dyes

> . . . consists essentially in the transfer of an *electron pair* accompanied or not accompanied by hydrogen ions according to the state of acid-base equilibrium in the solution (Clark, 1925, p. 171).

A contemporary comment deserves quotation at this point:

> All students of the subject must indeed be grateful to Mansfield Clark and his colleagues for their successful endeavour to provide a number of reducible dye-stuffs for which the oxidation-reduction potentials have been accurately determined electrometrically. These can be arranged in series and used for determining the reducing power of a given solution or system, and for stating it in definite terms. Again however discrimination is necessary. Whether a given transference of hydrogen can occur, or cannot occur, is a matter determined by potentials; but because in the study of tissue oxidations we seldom deal with equilibria, observing, rather, relative velocities, we must remember that though thermodynamics decide that a given reaction may, or must occur, yet it may proceed with a velocity too slow to observe. Kinetics may then, as Wieland has somewhere said, appear to leave thermodynamics in the lurch! (Hopkins, 1926, p. 49).

Nevertheless, there was general agreement as to the importance of determining the potentials for biological oxidation-reduction systems that are not electromotively active, such as the succinate-fumarate system.

The first step in this direction was taken by Juda Quastel and Margaret Whetham (1924), who showed that, under anaerobic conditions, a resting suspension of *Escherichia coli* (known to contain succinic dehydrogenase) catalyzes the reaction

Succinate + Methylene blue $\rightleftharpoons$ Fumarate + leuco-Methylene blue

in both directions to the same equilibrium point. Shortly thereafter Thunberg (1926) reported a similar experiment, using the succinic dehydrogenase present in washed heart muscle as the catalyst; from Clark's value for the oxidation-reduction potential of the methylene blue system, Thunberg estimated a potential for the succinate-fumarate system that was consistent with the equilibrium data of Quastel and Whetham. In these studies, the extent of the reduction of methylene blue was determined colorimetrically; the subsequent important work of

Jørgen Lehmann (1930) showed that it is possible to determine the oxidation-reduction potential of the succinate-fumarate system at a metallic electrode, provided there is present, in addition to succinic dehydrogenase, a small amount of methylene blue acting as a "mediator" between the electromotively inactive metabolite system and the electrode. This method was used thereafter with other dehydrognase-catalyzed reactions, and provided valuable data on the oxidation-reduction potentials of numerous metabolite systems. Furthermore, Henry Borsook and Hermann Schott (1931) calculated the free-energy change in the reduction of fumarate to succinate by a totally independent method, using thermal and supplementary data, and obtained a value in excellent agreement with those calculated from the earlier results of Quastel, Thunberg, and Lehmann. It was clear, therefore, that the dehydrogenase accelerates the attainment of equilibrium in an oxidation-reduction reaction, but does not change the equilibrium constant, in agreement with the principle stated by van't Hoff for enzymic catalysis in general, over thirty years earlier. Although it was recognized that "kinetics may appear to leave thermodynamics in the lurch," the data obtained after 1930 on the oxidation-reduction potentials of systems considered to play a role in biological oxidations, and on the free-energy changes in the reactions catalyzed by specific enzymes, profoundly influenced biochemical thought about the nature of intracellular respiration. A critical survey of most of these data has been prepared by Clark (1960); see also Krebs and Kornberg (1957).

By the early 1930s, therefore, the measurement of the oxidation-reduction potentials of organic systems had begun to provide a basis for removing some of the confusion about which Hardy wrote in 1914. The oxidation of organic compounds was seen to involve the transfer of two electrons to the oxidant, although in some cases, as reported by Michaelis and Elema independently in 1931, this transfer could occur in one-electron steps, with the formation of an intermediate "semiquinone" (Michaelis, 1935). It was also clear that neither the Wieland-Thunberg dehydrogenases nor the Warburg *Atmungsferment* alone could account for the phenomena of cell respiration, and that there were cell constituents which served as links between the dehydrogenation of metabolites and the reduction of molecular oxygen.

Intracellular Electron Carriers. We have already mentioned Keilin's important contribution, through his work on cytochrome, to the identification of some of the links in intracellular electron transfer. During the 1920s serious attention was given to other substances; among them, the one discovered by Frederick Gowland Hopkins (1921a) occupied an important place in the discussion. Twenty years before, he had isolated tryptophan by following the course of the purification with a color test; he now used the same strategy to follow the purification of a sulfhydryl (SH) compound from yeast with the aid of the nitroprusside test. The substance he isolated appeared to be identical with the "philothion" of Rey-Pailhade, and to have the constitution of a dipeptide composed of glutamic acid and cysteine:

> Until the constitution is finally established it may be premature to suggest a name for the substance. But, provisionally, for easy reference, the name *Glutathione* will perhaps be admissible. It leaves a link with the historic *Philothion,* has the same termination as in *Peptone,* which has long served as a name for the simpler peptides, and is a sufficient reminder that the dipeptide contains glutamic acid linked to a sulfur compound (*ibid.*, p. 297).

Glutathione (GSH) was found to be auto-oxidizable in the presence of metal ions, and the resulting cystine derivative (oxidized glutathione, CSSG) was readily reduced to GSH by various tissues. The addition of GSSG to washed muscle tissue promoted the reduction of methylene blue, and the glutathione system thus appeared to "possess what are essentially catalytic properties" (*ibid.,* p. 303); the tissue constituent that caused the reduction of GSSG turned out to be stable to heat, however, indicating that it was not an enzyme. Nevertheless, Hopkins concluded:

> While there is much that is obscure in the phenomena involved the facts in our opinion fully justify the claim that a non-enzymic oxidation-reduction system represented by the thermostable residue *plus* the sulfur grouping of glutathione actually functions in the cell (Hopkins and Dixon, 1922, p. 559).

Despite extensive work in many laboratories, the role of glutathione in cell respiration has remained obscure. Moreover, during the 1920s, observations on the chemical properties of glutathione forced a revision of its original identification as a dipeptide. Although Hopkins's associates had reported in 1925 that they had confirmed by synthesis the structure γ-glutamyl-cysteine, George Hunter and Blythe Eagles questioned the evidence, and led Hopkins (1929) to reinvestigate the problem. He now isolated the compound in crystalline form, and concluded that it is a tripeptide composed of glutamic acid, cysteine, and glycine; the structure of glutathione as γ-glutamyl-cysteinyl-glycine was repeatedly confirmed through synthesis after 1930.

During the period 1925–1939 Hopkins's laboratory in Cambridge attracted many foreign investigators; among the products of their efforts was a paper that began as follows:

Eight years ago observations were made on the adrenalectomized animal which gave the impression that the adrenal cortex is in some way involved in the mechanism of biological oxidation. A detailed study of the biological oxidation was begun in the hope that this study might lead to the understanding of the function of the interrenal system (Szent-Györgyi, 1928, p. 1387).

During the course of these studies, Albert Szent-Györgyi found in the adrenal cortex a factor that retarded the peroxidase reaction (bluing of guaiac); this "reducing factor" was also abundant in plants, and its participation in biological oxidations was suggested by the finding that the

. . . oxidation of this factor is a reversible one, and that under the given conditions this factor plays the role of a catalytic hydrogen carrier between the peroxidase and the other oxidising, or reducing systems (*ibid.*, p. 1391).

Although the factor was readily reduced by glutathione or by the thermostable "fixed SH" of animal tissues, it was not reduced by succinate in the presence of succinic dehydrogenase. Thus, while there seemed to be a possibility that the factor plus peroxidase provided an alternative to the auto-oxidation of glutathione under the catalytic influence of iron, the factor itself did not appear to play a significant role in linking the dehydrogenases to molecular oxygen in cell respiration.

These studies, however, led Szent-Györgyi to the isolation of the reducing factor in crystalline form from adrenal cortex and from plants (orange, cabbage); in his 1928 paper he identified the compound as an acidic carbohydrate (a "hexuronic acid") and noted:

The reducing properties of plant juice have repeatedly attracted attention, specially from students of vitamin C . . . . The reducing substances of lemon juice have been made the object of a thorough study by Zilva . . . , who established interesting relations between vitamin C and the reducing properties of the plant juice. The main reagent employed by Zilva (1928) was phenol indophenol. Indophenol blue is readily reduced by the hexuronic acid, so that it is probable that it was this substance which has been studied by Zilva (*ibid.*, p. 1401).

Some years later, Szent-Györgyi wrote about the naming of his newly found substance that

> . . . I called it ignose, not knowing which carbohydrate it was. This name was turned down by my editor. "God-nose" was not more successful, so in the end "hexuronic acid" was agreed upon. Today the substances is called "ascorbic acid" (Szent-Györgyi, 1937, p. 73).

The identity of the hexuronic acid with vitamin C was not fully established, however, until 1932–1933, when Glen King reported the isolation of the antiscorbutic material in crystalline form from lemon juice, and Szent-Györgyi tested his preparation for vitamin activity. He found the Hungarian red pepper to be an exceptionally rich source of the substance. By 1933 also, its chemical structure had been completely determined, and its synthesis effected, largely through the efforts of W. N. Haworth and his associates.

The determination of the nature of vitamin C came during an exciting decade, when organic chemistry exhibited its powers in the rapid elucidation of the chemical structure of a series of vitamins. The term "vitamine" (the final "e" was later dropped) was suggested in 1912 by Casimir Funk, to denote what Hopkins (1912) had named "accessory factors"; the term "food hormones" had also been proposed. Interest in these dietary factors grew out of the work of Christiaan Eijkman during 1890–1897 on the effect of a constituent of rice husks in curing a neurological disease of birds (polyneuritis) that resembled the human disease known as beri-beri. In 1906 Hopkins stated that

> . . . no animal can live upon a mixture of pure protein, fat and carbohydrate, and even when the necessary inorganic material is carefully supplied the animal still cannot flourish . . . . In diseases such as rickets, and particularly in scurvy, we have had for long years knowledge of dietetic factor; but though we know how to benefit these conditions empirically, the real errors in the diet are to this day quite obscure (Hopkins, 1906, pp. 395–396).

Indeed, an anticipation of these views may be found in a paper by Lunin, who studied the effect of milk on the survival of mice fed artificial diets; he concluded that

> . . . since as shown by the above experiments they were not able to survive on albuminates, fat, sugar, salts and water, it follows that in

milk there must be other substances besides casein, fat, lactose and salt that are essential for nutrition. It would be of great interest to track them down and to study their significance in nutrition (Lunin, 1881, p. 37).

It was not until after the work of Eijkman, however, that systematic studies were conducted along this line (see Hopkins, 1930b). When Axel Holst and Alfred Fröhlich found in 1907 that the guinea pig is susceptible to scurvy, a test animal was provided for the study of the antiscorbutic factor in plant juices, whose curative value was known since James Lind's *A Treatise on the Scurvy*, published in 1753.

An account of the events leading to the identification of the various vitamins is beyond the scope of this chapter; see McCollum (1957). By 1925, five classes of vitamins had been identified and labeled alphabetically. The water-soluble antineuritic factor was termed vitamin B, vitamin A being a fat-soluble factor that prevented eye disease (xerophthalmia) and promoted growth in the rat. Since the water-soluble antiscorbutic factor differed from both of these, it was named vitamin C. When it was shown that the fat-soluble antirachitic factor is different from vitamin A, it was denoted vitamin D; a distinct fat-soluble antisterility factor (for the rat) appeared soon afterward, and was termed vitamin E. As work progressed on the so-called vitamin B, it was recognized to be a complex of many factors; after 1927 the antineuritic factor was known as vitamin B_1. We shall have occasion to mention this vitamin again later in this chapter, as well as some of the other members of what was once considered to be the vitamin B complex. Two general comments may be made here, however. Firstly, the availability of a suitable biological system for testing the activity of fractions obtained in the purification of a vitamin usually led to its isolation in pure form (or to its disappearance from the list), and to the elucidation of its chemical structure through controlled degradation and synthesis in the laboratory; aside from the highly developed state of organic chemistry, technical advances such as the microanalytical procedures of Fritz Pregl played a major role in the successes that were achieved, since at first only small amounts of the isolated vitamins were available for chemical analysis. Secondly, the elucidation of the chemical structure of the vitamins came at a time when some of them were recognized to be components of electron-carrier systems in enzyme-catalyzed biological oxidations. In this development, the achievements of Otto Warburg during the 1930s were decisive. By 1940 his work had brought together into a single sequence of electron transport the dehydrogenases of Wieland and Thunberg, the cytochromes of Keilin, and his own auto-oxidizable *Atmungsferment*. The links were provided by his discovery of electron-carrier systems of the kind that had been sought unsuccessfully during the previous decade.

The Flavoprotein Enzymes. The starting point of this aspect of Warburg's work was the report by Guzman Barron and George Harrop (1928) that, in the presence of glucose, the normally very low rate of oxygen uptake by mammalian erythrocytes is greatly increased by the addition of small amounts of oxidation-reduction indicators such as methylene blue, and that the glucose is oxidized in the process. As we shall see later in this chapter, it was known during the 1920s that the metabolic breakdown of glucose involves the formation of hexose phosphate; Barron and Harrop suggested that, in their oxidation system,

> . . . the principal point at which methylene blue acts is upon the oxidation of hexose phosphate . . . . As to the exact nature of the methylene blue effect little may be said. It is conceivable that it acts as a coenzyme or catalyst, rendering the substrate (hexosephosphate ?) more sensitive to the action of molecular oxygen. On the other hand one might consider that methylene blue plays in this system the role ascribed to iron in the oxidations produced by Warburg with his charcoal model (Barron and Harrop, 1928, p. 85).

Warburg proceeded to study the mechanism of the methylene blue effect. First, he examined the possibility that the dye oxidized the ferrous iron of oxyhemoglobin to the ferric iron of methemoglobin, and in line with his views of the 1920s he concluded that ". . . in the methylene blue respiration the oxidation of sugar is nothing but an oxidation by the hemin iron, namely by the iron of methemoglobin" (Warburg *et al.*, 1930a, p. 496), and that the reaction between methylene blue and glucose is

> . . . a surface reaction. Methylene blue, which is adsorbed on the surfaces of the blood cells, forms methemoglobin on the surfaces—that is, at the reaction sites—and therefore a small methemoglobin concentration suffices during methylene blue catalysis to cause a large oxidative effect (Warburg *et al.*, 1930b, p. 270).

Thus, despite the apparent absence of the participation of heavy metals in the effect discovered by Barron and Harrop, Warburg noted that closer investigation had shown that here also ". . . there is a heavy-metal catalysis that closely resembles the normal catalytic actions of the living substance" (*ibid.*, p. 271). Shortly thereafter, however, he found that, although disruption (cytolysis) of the erythrocytes abolished their ability to oxidize glucose in the presence of methylene blue, the so-called Robison ester (glucose-6-phosphate) was readily oxidized by such cell-free suspensions; upon fractionating the constituents of the fluid from the cytolyzed red cells (after removal of the cell debris by centrifuging the suspension), he was able to conclude that

> . . . the reaction in the blood cells between methemoglobin and hexose monophosphate or between methylene blue and hexose monophosphate occurs by the cooperation of at least two substances, of which we name one "ferment" and the other "coferment" (Warburg and Christian, 1931, p. 215).

Thus methemoglobin was not an obligatory participant in the oxidation catalyzed by methylene blue, heavy-metal catalysis was not essential, and "cell structure" was not required for the oxidation of glucose-6-phosphate by methylene blue, provided a heat-labile, nondialyzable "ferment" and a heat-stable, dialyzable "coferment" were present.

After examining various biological sources of these new factors, Otto Warburg and Walter Christian (1932) proceeded to isolate from yeast a yellow-red protein which they termed "oxygen-transporting ferment"; its pigment was decolorized to a leuco form in the presence of a reducing system composed of glucose-6-phosphate, the "coferment," and an additional "ferment" they found in yeast. They proposed that, in aerobic cells, the following oxidation pathway was operative:

$$O_2 \rightarrow \text{hemin-Fe}^{2+} \rightarrow \text{hemin-Fe}^{3+} \rightarrow \frac{\text{Leuco form}}{\text{of pigment}} \rightarrow \text{Pigment} \rightarrow \frac{\text{Reducing}}{\text{system}}$$

In the presence of cyanide, the iron-containing components were inhibited, and molecular oxygen was considered to oxidize the leuco form of the pigment directly, with the formation of hydrogen peroxide. Furthermore, since in the absence of oxygen the reduced pigment also was oxidized by methylene blue, they concluded that

> The yellow ferment is therefore not only an oxygen-transporting ferment but also a ferment of "oxygen-less respiration." . . . It is probable that in life, the yellow ferment does not transfer molecular, but "bound" oxygen. Probably, in life, it is not an oxygen-transporting ferment but an oxidation-reduction ferment (Warburg and Christian, 1933, p. 377).

Warburg's discovery of the yellow enzyme was soon followed by his brilliant chemical studies (during 1932–1933) which showed that the pigment is a small molecule which is released from the protein when the latter is denatured, and that the pigment belongs to the class of substances that Kuhn had named "flavins," one of which appeared to be identical with vitamin B_2.

In 1926 Joseph Goldberger had proved that the human disease pellagra is caused by a deficiency of a dietary factor belonging to the B-complex, but different from the antineuritic vitamin; the antipellagra factor was accordingly denoted vitamin B_2, and several laboratories embarked on its isolation. By 1933 Richard Kuhn, Paul György, and Theodor Wagner-Jauregg had obtained from milk an orange pigment, which they named "lactoflavin." Although the pigment was active as a vitamin in promoting the growth of immature rats, it had no antipellagra activity; the Goldberger vitamin was then designated vitamin B_6, the numbers 3, 4, and 5 already having been assigned to other presumed members of the B-complex (they later disappeared from the list). Pigments similar to lactoflavin had been described by numerous investigators since Wynter Blyth had reported in 1879 the presence in milk of "lactochrome," but the chemical characterization of these materials was uncertain. In 1932 Ilona Banga and Albert Szent-Györgyi found in animal tissues a yellow pigment ("cytoflave") that could undergo reversible oxidation-reduction, but whose intracellular function was unclear; also, Philipp Ellinger and Walter Koschara (1934) were led to the study of such pigments (which they named "lyochromes") by the green fluorescence of some animal tissues upon irradiation with ultraviolet light.

It was not until after Warburg's work on the degradation of the pigment from the yellow enzyme, however, and the spectroscopic identification of the product as an alloxazine derivative (Kühling, 1899), by Kurt Stern and Ensor Holiday (1934), that the chemical structure of the flavins was elucidated. Within two years, the laboratories of Richard Kuhn and of Paul Karrer had synthesized an extensive series, and vitamin B_2 was recognized to be the compound designated riboflavin. Also, Kuhn had determined the oxidation-reduction potentials for a series of flavin derivatives. Furthermore, Hugo Theorell (1935) succeeded in separating reversibly the pigment and the protein of the yellow enzyme without denaturation, and showed the pigment to be a flavin phosphate; riboflavin itself was inactive. After the chemical synthesis of riboflavin-5'-phosphate (see formula) by Kuhn, Rudy, and Weygand (1936), and their demonstration that the synthetic material combines with the separated protein part of the yellow enzyme to regenerate the catalytic agent, it was clear that a derivative of a vitamin is a prosthetic group of a catalytic conjugated protein ("flavoprotein").

Riboflavin phosphate (flavin mononucleotide)

Kuhn, as a former associate of Willstätter, felt it appropriate to state that

> R. Willstätter thought that an enzyme consists of a colloidal support and an active group. The explanation that O. Warburg and H. Theorell give for the structure of the yellow enzyme illustrates exactly this conception (Kuhn, 1935, p. 921).

The important difference between Willstätter's conception and the knowledge emerging during the 1930s about the nature of enzymes lay in the question of enzyme specificity; whereas Willstätter thought the colloidal carrier to be nonspecific, it was becoming clearer that the protein part of the enzyme system was responsible for both the catalysis and specificity characteristic of the individual enzymes. This had been stated by Thunberg (1920) for the dehydrogenases; after the crystallization of pepsin by Northrop (1930), and the rapid acceptance of the view that enzymes are proteins, striking evidence of the role of the catalytic protein in determining the specificity of chemical reaction was provided for several purified enzymes. Indeed, subsequent work in Warburg's laboratory gave proof of this for the catalytic flavoproteins. In 1938 Warburg and Christian obtained from a purified preparation of D-amino acid oxidase (an enzyme isolated from kidney by Hans Krebs in 1935) a flavin derivative whose structure was later shown to be that of riboflavin phosphate joined to adenosine-5'-phosphate (see formula). This "flavin-adenine-dinucleotide" (FAD) also was found to be the prosthetic group of a flavoprotein ("new yellow enzyme") isolated from yeast by Erwin Haas (1938); while it catalyzed the same reaction as the "old yellow enzyme," its protein portion as well as its prosthetic group was different. These findings showed, therefore, that, although the reaction $FAD_{red} + O_2 \rightarrow FAD_{ox} + H_2O_2$ is catalyzed by both the D-amino acid oxidase and the "new yellow enzyme," in the first case the reduction of FAD is specifically effected by

$$CH_2-(CHOH)_3-CH_2O-\overset{\overset{O}{\parallel}}{\underset{OH}{P}}-O-\overset{\overset{O}{\parallel}}{\underset{OH}{P}}-OCH_2$$

Flavin adenine dinucleotide (FAD)

D-amino acids in the presence of a particular protein, whereas in the second case a different protein promotes the specific reduction of FAD by the "reducing system" with glucose-6-phosphate as the substrate. It may be noted in passing that, at the present writing, the physiological role of D-amino acid oxidase in animal tissues is still unknown.

Furthermore, the studies of Erwin Negelein and Heinz Brömel (1939a) provided evidence for the partial validity of Wieland's formulation of the enzymic oxidation of amino acids by molecular oxygen as a dehydrogenation, with the formation of hydrogen peroxide. Crude preparations of D-amino acid oxidase, containing the ubiquitous catalase (which destroys H_2O_2), catalyze the reaction

$$\text{D-Alanine} + FAD_{ox} \rightarrow \text{Pyruvic acid} + NH_3 + FAD_{red}$$

With the purified enzyme, however, the hydrogen peroxide formed in the reaction $FAD_{red} + O_2 \rightarrow FAD_{ox} + H_2O_2$ oxidizes the pyruvic acid (CH_3—CO—COOH) to acetic acid (CH_3—COOH) and CO_2.

The achievements of Warburg and his associates in the study of the catalytic flavoproteins represented only a fraction of the successes they achieved during the 1930s in elucidating the chemical nature of the participants in intracellular respiration. We saw earlier that in the aerobic oxidation of glucose-6-phosphate by cytolyzed red cells, there were involved, in addition to what turned out to be a flavoprotein, a "coferment" and an additional "ferment"; Warburg and Christian (1933, p. 394) named them *Zwischen-Co-Ferment* and *Zwischenferment*, respectively, "because their area of action is between the oxygen-transporting ferments and the substrates." To see Warburg's work on these components in perspective, we must return to the beginning of the twentieth century, and consider some of the developments that followed Buchner's discovery of cell-free alcoholic fermentation.

Before doing so, we may note the following statement by Warburg:

The oxidation theory of Wieland, no less than the opposing oxidation theories of Engler and Bach, were premature because when they were proposed nothing was known about the chemical constitution of the ferments participating in respiration. They were theories regarding the mechanism of chemical reactions, proposed without knowledge of the participants in the reaction. Such theories cannot be other than erroneous, and they must disappear to the extent that the chemical nature of the reaction partners—in this case the chemical constitution of the ferments—is elucidated (*ibid.*, p. 405).

Aside from inviting the question whether Warburg intended this generalization to apply to the views he had advanced during the 1920s, on the basis of his studies on charcoal models, the passage underlines the difference in approach between investigators who attempt to explain biological phenomena by means of analogies to artificial chemical systems, and those who put the behavior of the living system first and use chemical concepts and techniques to study the processes actually observed in physiological experiments. Although Wieland's emphasis on enzyme-catalyzed dehydrogenation may have been "premature," its subsequent impact on the work of Thunberg (1920) on the dehydrogenases is undeniable. Indeed, in 1933 it seemed obvious to many that Warburg's *Zwischenferment* behaved like a dehydrogenase with specificity toward glucose-6-phosphate. Whatever one may think of Warburg's characterization of Wieland's theory as "erroneous," however, there can be no dispute about Warburg's view that the isolation and chemical characterization of the reaction partners in biochemical processes are indispensable for the understanding of biological phenomena in which such processes are involved.

Alcoholic Fermentation as Oxidation-Reduction

We suggested in an earlier chapter that the excitement generated by Eduard Buchner's discovery of cell-free alcoholic fermentation was in large part a consequence of the importance that nineteenth-century biologists and chemists attached to fermentative processes considered to occur in biological systems. With the emergence of the new organic chemistry after 1860, detailed speculations began to be offered about the chemical mechanism of the fermentation of glucose to lactic acid or to alcohol and CO_2; the most influential of these hypotheses was that offered by Buchner's teacher, Adolf Baeyer. On the assumption that

Liebig's chemical theory of fermentation was correct, Baeyer (1870) suggested that the cleavage of the sugar molecule involves the intermediate formation of an unstable linear six-carbon compound which splits in the middle to yield two molecules of lactic acid in lactic fermentation, or which undergoes further cleavage in each of the three-carbon units to yield alcohol and CO_2 in alcoholic fermentation. During the succeeding decades, as the structure of glucose and the related hexoses was elucidated, there were several reports (by Hoppe-Seyler, Nencki, and others) on the chemical cleavage of glucose to lactic acid in alkaline solution, and these results were cited in support of the view that lactic acid might be an intermediate in alcoholic fermentation. Furthermore, by the 1870s lactic acid had begun to assume importance in the study of muscle physiology. After the discovery by Emil duBois-Reymond (in 1859) that, upon muscular contraction or following death of the animal, acid appears in the muscles, Ludimar Hermann had concluded (in 1867) that this acid production is anaerobic in character, and ten years later Bernard wrote:

> This lactic ferment occurs in the blood, in the muscles, even in the liver, since I have found that muscle and animal tissues do not become acid after death unless they contain sugar or glycogen [*la matière glycogène*] which rapidly undergoes a lactic fermentation (Bernard, 1877a, p. 328).

This "lactic fermentation" in animal tissues later came to be called "glycolysis." Before 1900 there was uncertainty whether the production of lactic acid regularly accompanies muscular contraction, or is only a reflection of fatigue and incipient death. It was the work of Walter Fletcher, and especially his joint research with Hopkins (Fletcher and Hopkins, 1907), that clearly demonstrated the connection of lactic acid production with muscular contraction; they also showed that this anaerobic process was followed by an aerobic phase during which the lactic acid disappeared. This achievement, which marked the beginning of the twentieth-century study of the chemical energetics of muscular contraction (Hill, 1959), depended in large part on the use of ice-cold alcohol into which the frog muscles were plunged before the chemical determinations were made. As Hopkins described it some years later:

> We found that the confusion in the literature as to the quantitative relations of lactic acid in muscle was wholly due to faulty technique in dealing with the tissue itself. When the muscle is disintegrated as a preliminary to extraction for analytical purposes, the existing equilibrium is entirely upset . . . . Fletcher and I, however, found it quite

easy, by means of a simple method, not only to avoid starting the changes which led to the formation of lactic acid, but to arrest them at any point during their progress, and thus establish their time-relations (Hopkins, 1921b, p. 361).

This advance in technique gave results that forced the abandonment of the idea of the operation of a "living molecule" responsible for oxidations in muscle; as we noted before, such views had been advanced by Hermann and Pflüger, and were opposed by Hopkins (1913).

Around 1860, when Pasteur studied lactic fermentation (p. 54), there was uncertainty about the relation of the product formed in soured milk to the lactic acid that had been found in muscle by Berzelius (in 1807). On the basis of the elementary analysis of salts of the two acids, Liebig concluded that

> . . . the nitrogen-free acid which occurs in the animal organism is identical with the acid that arises in souring milk, and into which milk-sugar, amylose, grape-sugar, and cane-sugar are converted by contact with decomposing animal substances (Liebig, 1847b, pp. 330–331).

There were indications, however, from differences in the solubilities of comparable salts, that the two acids might be different, and for a time muscle lactic acid was thought to be a structural isomer [$CH_2(OH)$—CH_2—$COOH$] of the fermentation lactic acid. In 1873 Johannes Wislicenus showed that both acids have the same structure CH_3—$CH(OH)$—$COOH$, but differ in their effect in turning the plane of polarized light, the muscle acid being dextrorotatory, whereas the fermentation acid is optically inactive. It was in the following year that van't Hoff and LeBel explained the difference between the two lactic acids as a consequence of the two modes of arranging the four groups CH_3, H, OH, and COOH in defined positions about the tetravalent α-carbon atom; later work showed that fermentation lactic acid is initially levorotatory, and is then racemized to an optically inactive mixture of the two stereoisomers. Although Buchner's contemporaries knew, therefore, that the two acids differed in their stereochemistry, and Fischer's work had shown the importance of stereochemical configuration for enzymic catalysis, hypotheses were soon developed that made lactic acid an intermediate in alcoholic fermentation. Immediately after Buchner's discovery, attention was drawn to the similarity between muscle glycolysis and alcoholic fermentation, in order to emphasize the complexity of both processes; thus Richard Neumeister noted:

This formation of lactic acid evidently occurs as a consequence of the interaction of certain proteins present in the living muscle plasma, and is certainly no less complex a process than the cleavage of sugar to alcohol and carbon dioxide in the yeast press juice. What we understand by enzymes could not play a role in either case (Neumeister, 1897, p. 2965).

This was written at a time when nearly all the known "unorganized ferments" were catalysts of hydrolytic reactions, and the intracellular agents responsible for fermentation were associated with "living" proteins different from enzymes such as pepsin or invertase.

By 1900 Buchner's claim to have prepared a cell-free "zymase" was generally accepted, and there followed a spate of publications (notably by Julius Stoklasa) reporting the preparation, from plant and animal tissues, of cell-free juices that converted sugar into lactic acid and to alcohol and CO_2. Within a few years, the idea was widely held that the anaerobic breakdown of glucose in animals, plants, and microorganisms proceeds via lactic acid. Buchner had convinced himself that lactic acid is formed from glucose by the cell-free yeast juice, and concluded that ". . . lactic acid plays an important role in the cleavage of sugar and probably appears as an intermediate in alcoholic fermentation" (Buchner and Meisenheimer, 1904, pp. 420–421). He reiterated this conclusion more forcefully in 1905, and proposed that the term zymase be applied to the enzyme that cleaves glucose to lactic acid; the subsequent conversion of lactic acid into alcohol and CO_2 was attributed to another enzyme, "lactacidase." Thus the Gay-Lussac equation for alcoholic fermentation, $C_6H_{12}O_6 \rightarrow 2C_2H_5OH + 2CO_2$ was considered to be the sum of two consecutive processes: (1) $C_6H_{12}O_6 \rightarrow 2C_3H_6O_3$; (2) $2C_3H_6O_3 \rightarrow 2C_2H_5OH + 2CO_2$.

The lactic acid theory of alcoholic fermentation soon disappeared. Arthur Slator (1906) called attention to the fact that lactic acid was not fermented at a significant rate by yeast, while Stoklasa's finding of alcohol in juices from animal tissues was shown to be a consequence of bacterial contamination, and it could be concluded that ". . . the postmortem formation of lactic acid in animal tissues is not in any way connected with alcoholic fermentation" (Harden and Maclean, 1911a, p. 66). The demise of the lactic acid theory, which Buchner abandoned in 1910, invited speculation about the possible role of other three-carbon compounds (see accompanying formulas) that had been found upon the alkaline degradation of hexoses. Thus Alfred Wohl (1907) modified Baeyer's scheme to include glyceraldehyde and methyl glyoxal as inter-

$$
\begin{array}{c}
CH_2OH \\
| \\
H-C-OH \\
| \\
CH_2OH \\
\text{Glycerol}
\end{array}
\qquad
\begin{array}{c}
CH_2OH \\
| \\
C=O \\
| \\
CH_2OH \\
\text{Dihydroxy-}\\
\text{acetone}
\end{array}
\qquad
\begin{array}{c}
COOH \\
| \\
C=O \quad \xrightarrow{-CO_2} \\
| \\
CH_3 \\
\text{Pyruvic}\\
\text{acid}
\end{array}
\qquad
\begin{array}{c}
CHO \\
| \\
CH_3 \\
\text{Acetalde-}\\
\text{hyde}
\end{array}
$$

$\pm 2H$

$$
\begin{array}{c}
CHO \\
| \\
H-C-OH \\
| \\
CH_2OH \\
\text{Glyceralde-}\\
\text{hyde}
\end{array}
$$

$\pm 2H$

$$
\begin{array}{c}
COOH \\
| \\
H-C-OH \\
| \\
CH_2OH \\
\text{Glyceric}\\
\text{acid}
\end{array}
\qquad \pm O
\qquad \pm H_2O
\begin{array}{c}
CHO \\
| \\
C=O \\
| \\
CH_3 \\
\text{Methyl}\\
\text{glyoxal}
\end{array}
\qquad
\begin{array}{c}
COOH \\
| \\
H-C-OH \\
| \\
CH_3 \\
\text{Lactic}\\
\text{acid}
\end{array}
\qquad \pm 2H
\begin{array}{c}
CH_2OH \\
| \\
CH_3 \\
\text{Ethanol}
\end{array}
$$

mediates; several investigators, including Buchner, tested them along with the closely related dihydroxyacetone for their ability to be fermented by the yeast juice. Because he found dihydroxyacetone to be fermented more rapidly than the other two compounds, as had others before him, Buchner inserted it into his theory in place of lactic acid, and assumed the existence of some enzyme other than "lactacidase." This proposal was not accepted by his contemporaries, and after a brief debate on the subject in 1912 Buchner stopped writing about the chemical mechanism of alcoholic fermentation. By that time the focus of attention had shifted to the possible role of phosphorylated compounds and of pyruvic acid; also, the process was turning out to be much more complex than had been imagined ten years before.

In 1903 there appeared a paper by Arthur Harden, who reported that blood serum inhibited the breakdown of proteins present in Buchner's yeast juice:

The fact that yeast press-juice is able to effect the fermentation of a relatively small portion of the available sugar has generally been ascribed to the action of a proteolytic enzyme of the press-juice. It was therefore of great interest to study the effect of the addition of serum to the mixture of yeast press-juice and sugar (Harden, 1903, p. 716).

He found that the rate of CO_2 production during fermentation was increased, and concluded that this result

. . . must be attributed to an inhibitory effect which the serum exerts on the proteolytic enzyme of the press-juice; one may therefore infer that the agent responsible for alcoholic fermentation is active for a longer time. The research is being continued (*ibid.*, p. 716).

In continuing this research, Harden tested boiled and filtered solutions of autolyzed yeast juice, since it was known that the digestion products of proteins frequently inhibited the action of proteolytic enzymes. The stimulation of fermentation that he found with such solutions was not different, however, from that produced by boiled and filtered fresh juice, indicating the effect was not related to the inhibition of proteolysis. With William John Young, Harden showed that the stimulatory effect was attributable to the presence of phosphates, and that in addition ". . . the fermentation of glucose by yeast-juice is dependent upon the presence of a dialysable substance which is not destroyed by heat" (Harden and Young, 1906, p. 410). They termed this heat-stable dialyzable "substance" a co-ferment; later workers (Euler and Myrbäck, 1923) named it "cozymase," and eventually it was recognized that the co-ferment that Harden and Young had denoted a "substance" actually included several heat-stable substances, all of which are involved in the conversion of glucose into alcohol and CO_2. The elucidation of the chemical nature and the role of these cofactors did not come until the 1930s, however; we shall return to these important developments shortly.

It was the other major discovery made by Harden and Young, the effect of phosphate, that had a more immediate impact on the study of the chemical mechanism of alcoholic fermentation. Although the stimulation of zymase by phosphate had been reported earlier by Augustin Wróblewski (1901), he attributed the effect to a buffering action that protected the zymase against the deleterious action of acids or alkalis. A noteworthy feature of Harden and Young's work was the fact that they had measured the CO_2 production volumetrically rather than gravimetrically (as Buchner had done), thus permitting frequent determinations; this enabled them to find that, upon the addition of successive equal amounts of phosphate, "The extra amount of carbon dioxide evolved after each addition is the same, and is equivalent . . . to the phosphate added" (Harden and Young, 1906, pp. 415–416). They also concluded that a portion of the added phosphate had been linked to organic material, and suggested that the phosphate ". . . exists in combination with glucose, probably in the form of a phosphoric ester" (*ibid.*, p. 418). It should be added that, independently of Harden and Young, Leonid Ivanov had also found that phosphate becomes bound to organic material during the course of alcoholic fermentation.

Harden and Young isolated from the fermentation mixture an organic phosphate, which they identified as a hexose diphosphate (HDP), and from the changes in the concentration of inorganic phosphate (P_i) during fermentation they concluded in 1908 that the following two processes were occurring:

$$2 \text{ Hexose} + 2 \text{ P}_i \rightarrow 2 \text{ CO}_2 + 2 \text{ Ethanol} + 2 \text{ H}_2\text{O} + \text{HDP}$$
$$\text{HDP} + 2 \text{ H}_2\text{O} \rightarrow \text{Hexose} + 2 \text{ P}_i$$

The sum of these processes gave the Gay-Lussac equation of alcoholic fermentation.

The determination of the chemical structure of the "Harden-Young ester" only came twenty years later, through the work of Phoebus Levene and Albert Raymond (1928). In 1909 Young suggested that the sugar in the hexose diphosphate might be fructose; subsequently, Carl Neuberg showed (in 1918) that mild acid hydrolysis gave a hexose monophosphate ("Neuberg ester") which also was fermented by yeast juice. This product turned out to be different from a hexose monophosphate found in 1914 by Harden and Robert Robison in fermentation mixtures; after World War I, Robison (1922) concluded that his product was a mixture of isomeric hexose monophosphates, probably those of glucose and fructose, and by 1931 he and Earl King had isolated the glucose monophosphate ("Robison ester") in a pure state. A similar mixture of hexose mono-phosphates was obtained from muscle by Gustav Embden (in 1927), and for a time was termed "Embden ester." We should add that, during the 1920s, the uncertainty about the chemical structure of these sugar phosphates caused considerable confusion in the study of alcoholic fermentation. The various wrong turnings need not be recounted, but it is necessary to recognize the importance of the organic-chemical studies which showed the structure of the Harden-Young ester to be fructose-1,6-diphosphate, the Neuberg ester to be fructose-6-phosphate, and the Robison ester to be glucose-6-phosphate (see accompanying formulas). In these chemical studies, the new methods developed during the 1920s for the determination of the constitution of the sugars played a significant role.

One of the important consequences of the discovery of hexose diphos-

Fructose-1,6-diphosphate   Glucose-6-phosphate   Fructose-6-phosphate

phate as a product in alcoholic fermentation was the demonstration by Embden that its addition to a press juice from muscle caused a large increase in the production of lactic acid. This finding led him to suggest that the "lactacidogen" he had proposed in 1912 as a precursor of lactic acid in glycolysis is a substance related to hexose diphosphate (Embden and Laquer, 1914, 1921). To this evidence of the possible similarity of the chemical pathways in glycolysis and in alcoholic fermentation was added the discovery of Otto Meyerhof (1918) that the "co-ferment" of alcoholic fermentation is also present in animal tissues, and his later report (in 1921) that the glycolytic system of muscle requires a "co-ferment" similar to the material obtained from yeast. These discoveries by Embden and Meyerhof led to the clear recognition, before 1930, that the conversion of glycogen into lactic acid in muscle and the breakdown of glucose to alcohol and CO_2 by yeast had much in common, and that the hexose phosphates might be intermediates in both processes. Indeed, Kluyver and Donker (1926) called attention to the possibility that similar pathways were operative in the fermentation of glucose by organisms other than yeast; their concept of the "unity of biochemistry" was to be exceedingly fruitful during the succeeding decades (van Niel, 1949).

It should be stressed that Harden was not certain whether hexose diphosphate is a direct intermediate in the conversion of glucose into alcohol and CO_2. This is evident from the successive editions of his book *Alcoholic Fermentation* (in 1911, 1918, 1923, and 1932). Thus, in the third edition, he stated that "It is not impossible that the hexosephosphate is formed by combined synthesis and esterification from smaller groups produced by the rupture of the sugar molecule" (Harden, 1923, p. 109). A detailed mechanism of this sort had been proposed by Alexander Lebedev (1912), whose scheme involved the cleavage of glucose to glyceraldehyde and dihydroxyacetone, with the phosphorylation of the latter product and the synthesis of hexose diphosphate from two molecules of the triose phosphate. In part, the doubts about the role of hexose diphosphate as a direct intermediate in alcoholic fermentation arose from the finding that it was not fermented by living yeast. Furthermore, during the years immediately after 1910, attention was drawn away from hexose diphosphate by the discovery that pyruvic acid might be an important intermediate.

During the first decade of this century, the technique of perfusing surviving animal organs (*e.g.*, liver) with saline fluids (Ringer-Locke solution) containing substances of biological interest gave valuable data on the metabolic conversion of these substances. When this technique was applied to the study of the fate of protein amino acids, by Gustav Embden and Otto Neubauer, it became evident that α-amino acids could undergo oxidative deamination to the corresponding α-keto acids (R—CO—COOH). These findings were extended to yeast by Neubauer and Konrad Fromherz (1911), who suggested that the deamination of alanine would yield pyruvic acid, whose decarboxylation would yield acetaldehyde, which might be reduced to ethanol:

$$
\begin{array}{ccccc}
\text{CH}_3 & & \text{CH}_3 & & \text{CH}_3 & & \text{CH}_3 \\
| & +\text{O} & | & -\text{CO}_2 & | & +2\text{H} & | \\
\text{H—C—NH}_2 & \xrightarrow{-\text{NH}_3} & \text{C}=\text{O} & \longrightarrow & \text{C}=\text{O} & \longrightarrow & \text{CH}_2\text{OH} \\
| & & | & & | & & \\
\text{COOH} & & \text{COOH} & & \text{H} & &
\end{array}
$$

They proposed that, since they found pyruvic acid to be fermented by yeast, this compound

> . . . could be an intermediate in the alcoholic fermentation of sugar
> . . . . We ask colleagues to leave to us the further study of the role
> of pyruvic acid in the fermentation of sugar; also, it is intended to study
> the question whether it is an intermediate in the combustion of sugar
> in the higher animal organism (Neubauer and Fromherz, 1911, p. 350).

This permission was not granted, and within a few years several laboratories, notably that of Carl Neuberg, had published extensive studies on the role of pyruvic acid in alcoholic fermentation.

At the time of Neubauer's proposal there was no direct experimental evidence for the appearance of pyruvic acid during alcoholic fermentation; this was provided by Auguste Fernbach and Moise Schoen in 1913 (see Schoen, 1928). Within a few months of the appearance of the paper by Neubauer and Fromherz, however, Neuberg announced the discovery, in zymase preparations, of a new enzyme ("carboxylase") which catalyzes the decarboxylation of pyruvic acid to acetaldehyde and CO_2. By 1913 Neuberg had developed a theory of alcoholic fermentation that incorporated this new finding. Following in the tradition of Baeyer and Wohl, Neuberg and Kerb (1913) proposed that glucose is first cleaved to two molecules of methyl glyoxal, one molecule of which is reduced to glycerol, and the other is oxidized to pyruvic acid; decarboxylation of the pyruvic acid to acetaldehyde is followed by the reduction of the latter to ethanol, balanced by the oxidation of the second molecule of methyl glyoxal to pyruvic acid. There was no place in the theory for phosphorylated compounds of the kind studied by Harden and Young and by Ivanov; for many years, Neuberg considered such organic phosphates to be irrelevant to the problem of the normal pathway of alcoholic fermentation.

This theory held sway, with minor modifications, for over fifteen years;

the fact that methyl glyoxal was not readily fermented by yeast was explained by assuming that one of the many possible isomers of this compound was the "true" intermediate. An argument offered in favor of methyl glyoxal was the fact that Henry Dakin and Harold Dudley (1913) had found a widely distributed enzyme ("glyoxalase") which catalyzes the interconversion of methyl glyoxal and lactic acid; this was considered to outweigh the chemically more plausible arguments in favor of glyceraldehyde and dihydroxyacetone, as suggested by Lebedev. The most attractive features of Neuberg's theory were that separate enzyme-catalyzed reactions were provided for the liberation of CO_2 and the formation of ethanol, and that the overall process of alcoholic fermentation was formulated as a set of coupled oxidation-reduction reactions. Moreover, during World War I, Neuberg's theory received support from its successful application to the industrial manufacture of glycerol in Germany. Neuberg showed that the addition of sulfite to a fermentation mixture blocks the reduction of acetaldehyde by forming an aldehyde-sulfite addition compound, and that the overall fermentation process becomes: Glucose → Glycerol + Acetaldehyde + CO_2; the mechanism was explained by Neuberg in terms of a disturbed balance of oxidation-reduction reactions:

> Glycerol is the reduction equivalent of pyruvic acid, which decomposes to carbonic acid and acetaldehyde. If the reduction of the latter is blocked, the only remaining possibility is the increased correlative formation of glycerol (Neuberg and Reinfurth, 1919, p. 1681).

The sulfite process was patented during World War I by Connstein and Lüdecke; in their published report they stated that their experiments began in 1914, but

> . . . could not be published earlier because, during the war, the German army administration had an interest in keeping the experiments and results secret. Our work arose from the necessity of the time and owes its origin to the expectation that the supply of glycerol available to the European Central Powers would soon be insufficient, because of the blockade (Connstein and Lüdecke, 1919, p. 1385).

It may also be added that the inhibition of alcoholic fermentation by sulfite had been noted by Jean Baptiste Dumas (1874, p. 104).

If, in demonstrating the enzymic decarboxylation of pyruvic acid, Neuberg followed the lead of Neubauer, the conception of the coupling of oxidation-reduction reactions had already been stated by Jacob Parnas

(1910). We mentioned on p. 319 Parnas' discovery of the enzymic cataly-sis of the Cannizzaro "dismutation" of aldehydes; in his paper he wrote:

> In the Cannizzaro rearrangement of the aldehydes we have come to know a simple system of coupled reactions, in which through oxygen transfer and hydrogen uptake there occur simultaneous oxidation and reduction. Through an enzyme of the liver the reaction is catalyzed to such an extent that it leads to the complete disappearance of the alde-hydes . . . . Aldehydes may be regarded as general reductants for the reduction of carbonyl groups in the animal organism. Through spe-cific ferments the Cannizzaro reaction between two reacting substances is accelerated, and there are formed an alcohol (or a hydroxy acid) and a fatty acid (*ibid.*, pp. 286–287).

This view of the oxidation-reduction of aldehydes was accepted after 1910 as a key step in alcoholic fermentation; it could not be translated into reality, however, until the chemical nature of the intermediates had been established, and the enzymic catalysts responsible for their interconver-sion had been identified.

We noted before that during the 1920s the chemical pathways in alco-holic fermentation and in glycolysis were recognized to have much in common, and other fermentations (*e.g.*, the conversion of glucose into lactic acid by lactic acid bacteria) also were found to resemble those ef-fected by yeast and muscle. Furthermore, the experimental study of gly-colysis in animal tissues, and the discovery that tumors and embryonic tissues are active glycolyzing systems, were greatly furthered by Warburg's development of the tissue-slice technique as an adjunct to the manometric methods he had applied to the measurement of cell respiration (Warburg, 1926). Although by 1925 there was general agreement that pyruvic acid is a common intermediate in these processes, going either to lactic acid or to ethanol (via acetaldehyde), the role of the hexose phosphates was unclear, and the status of methyl glyoxal as an intermediate between glucose and pyruvic acid was hypothetical. Within about ten years, however, the situa-tion had been altered dramatically, and the efforts of several research groups had led to the formulation of a clear-cut sequence of enzyme-catalyzed reactions, in yeast and muscle, from glucose-6-phosphate to pyruvic acid. The impact of this development on biochemical thought was far-reaching, not only because it revealed the details of a complex meta-bolic pathway, but also because it provided a new basis for the under-standing of the cellular mechanisms for the utilization of the energy made available by oxidative processes in biological systems. In this develop-ment the new attitude toward enzymes as individual catalytic proteins, as well as the contributions of organic chemistry in the elucidation of the structure of the intermediates and of the cofactors, played decisive roles.

Of special importance as a starting point for these advances was the preparation by Otto Meyerhof (1926a) of a muscle extract which, like the yeast maceration juice prepared by Alexander Lebedev (1912), was relatively free of carbohydrate, and could therefore be used to test the role of suspected intermediates or cofactors in the overall conversion of glycogen into lactic acid. This achievement ended the discussion whether the anaerobic breakdown of carbohydrate in muscle is associated with the integrity of cellular structure, and indicated (as had the preparation of cell-free yeast juices by Buchner and Lebedev) that glycolysis could be studied as an enzymic process. Meyerhof showed that phosphate promoted the process, and that the extract converted hexose diphosphate into lactic acid; glucose was not utilized effectively, however, unless there was added, in addition to Harden's heat-stable coferment, a heat-labile factor present in yeast juice (Meyerhof named this activator "hexokinase"). A similar enzyme ("phosphatese") had been postulated by Hans von Euler in 1911 to be responsible for the initiation of the alcoholic fermentation of glucose. Before Meyerhof's work, Embden had shown in 1924 that "lactacidogen" accumulates in muscles poisoned by fluoride, and he then suggested that this accumulation is a consequence of the inhibition of the decomposition of hexose phosphate. The inhibition of glycolysis and of alcoholic fermentation by fluoride had long been known, and Embden cited his result in support of the view that hexose diphosphate is a direct intermediate in the breakdown of carbohydrate. By 1930 this view was widely accepted.

The establishment of the structure of the Harden-Young ester as fructose-1,6-diphosphate invited more modern formulation of the chemical hypotheses that had been advanced by Baeyer, Wohl, and Lebedev. Among the suggestions was one offered by Heinz Ohle (1931), who postulated, on chemical grounds, that the anaerobic breakdown of glucose begins with its phosphorylation to glucose-6-phosphate, which is converted into fructose-1,6-diphosphate (via fructose-6-phosphate); the hexose diphosphate was considered to undergo a series of oxidations and reductions leading to the formation of glyceraldehyde-3-phosphate and dihydroxyacetone phosphate. This proposal stimulated the synthetic efforts of Hermann Fischer and Erich Baer (1932), who made DL-glyceraldehyde-3-phosphate for test as an intermediate in alcoholic fermentation. In the following year, Carl Smythe and Waltraut Gerischer (Warburg's laboratory) showed this compound to be fermented by yeast; since the synthetic material was a racemate, only 50 percent was converted, and later work showed the D-isomer to be the reactive compound. It may be noted that, in their synthesis of this relatively unstable aldehyde, Fischer and Baer used a synthetic method introduced into sugar chemistry in 1928 by Karl Freudenberg (see Fischer, 1960). A few months after the report of the Fischer-Baer synthesis an ingenious variation of the method was described by Max Bergmann and Leonidas Zervas (1932) for the synthesis of peptides; we have already called attention (p. 159) to the significance of this contribution to protein chemistry.

Although the phosphorylated three-carbon compounds (triose phosphates) had reappeared in successive schemes of alcoholic fermentation and glycolysis since the work of Ivanov, it was only in 1933 that the accumulated knowledge permitted their inclusion in a scheme that proved to be fruitful for later research. In that year, the last of his life, Embden proposed that, in glycolysis, fructose-1,6-diphosphate is cleaved directly to glyceraldehyde-3-phosphate and dihydroxyacetone phosphate. These products were considered to undergo a Cannizzaro (mutase) reaction to yield 3-phosphoglycerol and 3-phosphoglyceric acid, and the latter was thought to be converted into pyruvic acid and phosphate. A second mutase reaction between pyruvic acid (which is reduced to lactic acid) and phosphoglycerol (which is oxidized to glyceraldehyde-3-phosphate) completed the balance of equations, and provided for the conversion of all the hexose into lactic acid. 3-Phosphoglyceric acid had been found (in 1929) by Ragnar Nilsson in fermenting yeast extracts to which fluoride had been added, but he did not consider it to be an intermediate in alcoholic fermentation. Soon afterward, however, 3-phosphoglyceric acid was identified in fluoride-poisoned muscle extracts undergoing glycolysis, and its importance as a precursor of pyruvic acid was generally accepted.

Furthermore, by 1935 Otto Meyerhof and Karl Lohmann had shown that fructose-1,6-diphosphate is cleaved by muscle extracts to two molecules of triose phosphate; they termed the enzyme apparently responsible for this action "zymohexase." It was renamed "aldolase" (after the aldol reaction described by Wurtz during the 1870s) when it was recognized that the immediate products of the enzymic cleavage of the hexose phosphate are glyceraldehyde-3-phosphate and dihydroxyacetone phosphate (see accompanying formulas); these are interconverted by a separate enzyme (triose phosphate isomerase). Thus the three-carbon compounds postulated on purely chemical grounds around 1910 as intermediates in alcoholic fermentation reappeared in the 1930s as their phosphorylated derivatives.

In their scheme of alcoholic fermentation, Neuberg and Kerb (1913) had assumed dismutation reactions, with the hypothetical methyl glyoxal as a participant. Whereas, as late as 1932, it was stated that "The trend of opinion seems to be that methyl glyoxal is an intermediate between hex-

| Fructose-1,6-diphosphate | Dihydroxyacetone phosphate | D-Glyceraldehyde-3-phosphate |

osephosphate and lactic acid" (Shaffer and Ronzoni, 1932, p. 258), methyl glyoxal disappeared from the pathway after 1933, with the recognition of glyceraldehyde-3-phosphate and its oxidation product 3-phosphoglyceric acid as well-defined intermediates. Meyerhof and Wilhelm Kiessling (1935) then showed that it is this oxidation of the aldehyde to the acid that balances the reduction of acetaldehyde to ethanol (in alcoholic fermentation) or of pyruvic acid to lactic acid (in glycolysis), and that phosphoglycerol is not a necessary participant in the dismutation, as Embden had proposed. Even before this advance, however, the general principle that such coupled oxidation-reduction is an essential feature of the anaerobic breakdown of glucose had been stated repeatedly since the work of Parnas (1910). In particular, Hans von Euler insisted during the 1920s that the Harden-Young co-ferment (which he named cozymase) is an obligatory cofactor for the enzymic dismutation of aldehydes, and he invoked the occurrence of such a reaction as the initial step in alcoholic fermentation (Euler *et al.*, 1928). Albert Jan Kluyver also adopted the view that oxidation-reduction reactions are key processes in fermentations, but he concluded in 1928 that a specific cofactor such as cozymase did not exist, and ascribed the activation by boiled yeast extract to the action of coagulated proteins in protecting the zymase from destruction by proteolytic enzymes. Such doubts, as well as the importance Euler attached to cozymase, led him to attempt its purification; by 1930 he had prepared an active product (from yeast) which resembled the adenylic acids from yeast nucleic acid (adenosine-3'-phosphate) and from muscle (adenosine-5'-phosphate), but was not identical with either. Although Euler provided evidence in support of his view that cozymase promotes the action of some of the dehydrogenases studied by Wieland and Thunberg, its role remained uncertain so long as its chemical identity had not been established and its relation to the action of well-defined enzymes remained unknown. It was Warburg's work on the "co-ferment" in the flavoprotein-catalyzed oxidation of glucose-6-phosphate that provided the knowledge needed to clarify the nature of Euler's cozymase, and its function in the anaerobic breakdown of glucose.

The Pyridine Nucleotides. In 1934 Warburg and Christian reported that the "co-ferment" preparation they had obtained from red blood cells released adenine on acid hydrolysis, as did Euler's cozymase. They also noted that, after removal of the adenine, other basic substances were still present in the hydrolysate; one of these ("Base I") appeared to be associated with the cofactor activity. A few months later they announced in a short note:

Mr. *Walter Schoeller* has called to our attention that the composition and melting point of Base I agree with those of nicotinic acid amide. A comparison of the two substances showed that they are identical (Warburg and Christian, 1935, p. 464).

Nicotinic acid (pyridine-3-carboxylic acid) had been known since the 1870s as an oxidation product of the plant alkaloid nicotine, and it had been isolated from rice in 1912 by Suzuki in his search for the antiberi-beri vitamin; in 1926 Vickery found it in yeast. For a time, consideration was given to the possibility that nicotinic acid might be related to the vitamin:

With regard to nicotinic acid, we ourselves lean towards the belief that its occurrence in rice-polishings has some relationship to the occurrence of the curative substance, and that it may possibly even be a degradation product of the active body (Drummond and Funk, 1914, p. 614).

This view was abandoned, however, and none of these investigators considered that they had isolated a part of a coenzyme. Shortly after the note by Warburg and Christian, Richard Kuhn reported the isolation of nicotinamide from heart muscle.

There then appeared the remarkable papers by Warburg, Christian, and Griese (1935a, b) on what they now called the "hydrogen-transporting co-ferment"; from clear-cut chemical experiments they concluded that

The pyridine component of the co-ferment is its active group, because the catalytic action of the co-ferment depends on the alternation of the oxidation state of its pyridine part (Warburg, Christian, and Griese, 1935a, p. 144).

Their chemical analyses indicated the co-ferment to be composed of adenine, nicotinamide, pentose, and phosphate in the ratio 1:1:2:3. Within a few months the chemical relation of this cofactor to Euler's cozymase had been clarified; Warburg and Christian (1936) isolated what they called the "fermentation co-ferment" from red blood cells, showed it to differ in composition from the other cofactor by having only two phosphate groups, and named the two cofactors triphosphopyridine nucleotide and diphosphopyridine nucleotide. By this time, Euler (1936) had also found nicotinamide in his yeast cozymase preparation, and reported chemical data in agreement with Warburg's results. In particular, Euler reiterated his view that cozymase is a general cofactor in dehydrogenase-catalyzed reactions; he proposed that the diphospho compound (his cozymase) and the triphospho compound (Warburg's co-ferment) be denoted *Codehydrase* I and II, respectively, and that the substrate-specific proteins with which they are associated be called *Apodehydrasen*. On the other hand, Warburg continued to eschew the terms introduced by Wieland and Thunberg, and wrote of "specific colloids" he termed *Gärungs-Zwischenfermente* that he considered to be necessary for the reduction of the "fermentation co-ferment." For many years afterward, the two cofactors were named codehydrogenases I and II, or simply coenzymes I and II. After 1945 preference was given to the use of abbreviations (DPN and TPN) of the terms introduced by Warburg. The structure of DPN, as proposed by Fritz Schlenk and Euler in 1936 (see Schlenk, 1942), and confirmed by synthesis (in Alexander Todd's laboratory in 1957), is shown in the accompanying formula. The location of the third phosphate of TPN (at the hydroxyl group marked by an asterisk) was established after 1950, through the work of Arthur Kornberg, as well as the characterization of

Diphosphopyridine nucleotide (DPN)

the adenosine-2′-phosphate and adenosine-3′-phosphate derived from the alkaline hydrolysis of ribonucleic acids. Since about 1960 the terms DPN and TPN have largely been replaced by NAD (nicotinamide adenine dinucleotide) and NADP (nicotinamide adenine dinucleotide phosphate), in accordance with a recommendation of an international commission on enzyme nomenclature.

A striking property of the nicotinamide portion of the pyridine nucleotides is the appearance of a new absorption band near 340 mμ upon their enzyme-catalyzed reduction. The discovery of this property permitted Warburg and his associates to develop rapid quantitative assays for the pyridine nucleotide-dependent enzymes, and to use this spectrophotometric method for their purification. After 1945, when reliable photoelectric quartz spectrophotometers became available commercially, this method largely replaced the use of the Warburg manometric apparatus and variants of the Thunberg methylene blue technique in studies on dehydrogenases. With regard to the reduction of the coenzymes, it may be added that work by Paul Karrer (in 1937) on the behavior of simpler pyridine derivatives led to the view that the reduction involves the addition of a hydrogen atom at the 2-position of the nicotinamide ring; during the 1950s, however, this was shown to be incorrect, and the site of hydrogen addition was recognized to be the 4-position. Warburg (1938) referred to the reduced form as *Dihydropyridin,* and later workers denoted it CoH_2 or $DPNH_2$, although it was evident from the chemistry of its reactions that only one hydrogen is added to the ring in a process involving the transfer of two electrons (the other hydrogen atom derived from the "hydrogen donor" appears in the solution as a hydrogen ion). In 1954 Birgit Vennesland and Frank Westheimer showed that, in a dehydrogenase-catalyzed reaction, the hydrogen atom added to the pyridine ring is transferred directly from the metabolite, without mixing with the hydrogen ions of the solution; this important discovery profoundly influenced the further study of the mechanism of dehydrogenase action.

By 1936 numerous dehydrogenases had been identified on the basis of their specificity toward individual metabolites, but the enzyme preparations that were actually employed in experimental studies usually represented only partially fractionated extracts. Such crude preparations from animal or plant tissues, or from bacteria, were designated "alcohol dehydrogenase" or "lactic dehydrogenase," depending on the nature of the substrate used as the hydrogen donor; for example, they were the materials employed in the determination of the apparent oxidation-reduction potentials of such systems as alcohol-acetaldehyde or lactate-pyruvate. Although many investigators agreed with the view reiterated by Thunberg (1937) that the specific dehydrogenases represent separate catalytic entities, the failure to emulate the success of Sumner, Northrop, and Kunitz in the crystallization of enzymes was a repeated source of uncertainty and controversy. The situation began to change, however, with the crystallization of the alcohol dehydrogenase from yeast by Warburg's associates Negelein and Wulff (1937). The purified protein catalyzed the reaction they wrote as:

Alcohol + Pyridine $\rightleftharpoons$ Acetaldehyde + Dihydropyridine

where "pyridine" referred to diphosphopyridine nucleotide. According to Warburg's terminology, Negelein and Wulff had isolated

. . . this colloid as a crystalline protein. The protein combines with diphosphopyridine nucleotide to form a dissociating pyridino-protein (*Pyridinproteid*), the reducing fermentation ferment, which reduces acetaldehyde to alcohol (Warburg and Christian, 1939, p. 40).

The success achieved by Negelein and Wulff was followed two years later by the crystallization, from Lebedev's yeast-juice, of what Warburg and Christian called the "oxidizing fermentation ferment." They showed that this enzyme catalyzes the oxidation of glyceraldehyde-3-phosphate, in the presence of DPN and of inorganic phosphate, to 1,3-diphosphoglyceric acid (see accompanying formula), and also noted:

Th. Bücher has found in Lebedev juice a specific protein which effects the following reaction between the end-product of the physiological oxidation reaction and adenosine diphosphate: 1,3-Diphosphoglyceric acid + Adenosine diphosphate $\rightleftharpoons$ 3-Phosphoglyceric acid + Adenosine triphosphate (*ibid.*, p. 47).

$$
\begin{array}{ccc}
\underset{\substack{\text{1,3-Diphosphoglyceric}\\\text{acid}}}{\overset{\displaystyle \text{OC—OPO}_3\text{H}_2}{\underset{\displaystyle \text{CH}_2\text{OPO}_3\text{H}_2}{\mid\ \ \text{HCOH}\ \mid}}} & + \text{ADP} \rightleftharpoons & \underset{\substack{\text{3-Phosphoglyceric}\\\text{acid}}}{\overset{\displaystyle \text{COOH}}{\underset{\displaystyle \text{CH}_2\text{OPO}_3\text{H}_2}{\mid\ \ \text{HCOH}\ \mid}}} + \text{ATP}
\end{array}
$$

This great discovery opened a new chapter in the history of biochemistry, in providing a well-defined chemical route for the coupling of the energy released in an oxidative reaction to the enzymic synthesis of adenosine triphosphate; we shall consider some of the background and consequences of this work in the next section of this chapter. At this point in our study it is sufficient to note that the availability of the two highly purified crystalline enzymes permitted Warburg and Christian to test the combined action of the "reducing" and "oxidizing" fermentation enzymes on a mixture of acetaldehyde and the triose phosphate, and to effect

> . . . a mixed dismutation. The aldehyde group . . . of the Fischer ester is oxidized to a carboxyl group, the aldehyde group of acetaldehyde is reduced to an alcohol group: $RCHO + R'CHO + H_2O = RCOOH + R'CH_2OH$. This dismutation is therefore not effected by *one* ferment, but by *two* ferments with a common prosthetic group (*ibid.*, p. 49).

This experiment brought to a close the discussion of the oxidation-reduction step in alcoholic fermentation, and gave reality to the general ideas advanced by Parnas and Wieland nearly thirty years before. It cannot be emphasized too strongly, however, that Warburg's achievements during the 1930s constituted the clearest possible evidence for the conviction that progress in the understanding of biochemical processes required not merely the formulation of plausible chemical hypotheses, but the isolation of enzymes in as pure a state as possible, and the determination of the detailed chemical structure of all the participants in the chemical reactions they catalyze.

Because of the elegance of Warburg's experiments, and the significance of his results, their impact on the biochemistry of the 1930s was decisive; his nomenclature and his views about the role of the protein components in the pyridine nucleotide-dependent reactions were a source of confusion, however. For example, before the work of Warburg and Christian in 1939, experiments with crude dehydrogenase preparations had given evidence of "coenzyme-linked" reactions in which two enzymes cooperate to catalyze the oxidation of a substrate of one enzyme by a substrate of the other enzyme. In connection with such experiments it was stated:

> The question of how the coenzyme functions in the catalytic system has now become a matter of great dispute. The concept of "Zwischenferment" introduced by Warburg implies that the coenzyme combines with the dehydrogenase to form the catalytically active complex. What is ordinarily referred to as a dehydrogenase is considered by Warburg to be merely a highly specific protein with no catalytic properties apart from its prosthetic group—the coenzyme. Euler and his school have accepted this view but they prefer to call the active complex the "holodehydrase."
> There is a good deal of evidence in favour of the view that the dehydrogenase is the seat of catalytic activity . . . . The classical concep-

tion of the dehydrogenase as the actual activating mechanism seems to be in fair agreement with the facts. No doubt the coenzyme combines with the dehydrogenase in the same way that the substrate does. But the function of the coenzyme seems to be that of a highly specific hydrogen acceptor which cannot be replaced by any other substance (Green *et al.*, 1937a, p. 948).

This point of view was developed further by Dixon and Zerfas (1940) and by Parnas (1943), among others, and continued to be a focus of discussion for many years (see Racker, 1955, p. 9). After the crystallization of numerous dehydrogenases, and the study of their properties as catalysts and as proteins, there was no longer any doubt about the fact that the dehydrogenase protein is the "seat of catalytic activity."

Before concluding this section, we return to the relation of enzymes to vitamins, made evident by Warburg's work on the catalytic flavoproteins. Shortly after Warburg's discovery that nicotinamide is the reactive constituent of the pyridine nucleotides, Elvehjem *et al.* (1937) showed that nicotinic acid was the active agent in liver concentrates effective in curing black tongue, a deficiency disease in dogs, and other investigators demonstrated the curative value of nicotinic acid in human pellagra. The designation of the antipellagra factor as vitamin B_6 was abandoned, and this term was applied to a group of pyridine derivatives related to a coenzyme denoted pyridoxal phosphate. Furthermore, nicotinic acid was recognized in 1937 (Lwoff, Knight, Mueller) to be an essential growth factor for a variety of bacteria, and it became clear that these organisms could not make the nicotinamide portion of their pyridine nucleotides at a rate commensurate with their metabolic requirements for multiplication. The subsequent development of research on vitamins has depended heavily on the use of such microbial species for the assay of fractions obtained during the course of a purification. For example, the problem of the isolation and characterization of the anti-anemia factor (vitamin B_{12}, cobalamin) proved to be intractable until a microbiological assay was found in 1947. One of the important consequences of the convergence of studies on the nutritional requirements of man and microbes with Warburg's work on the enzymes of respiration and fermentation has been the tacit assumption that a new vitamin or microbial growth factor will turn out to be part of a coenzyme. Since 1940 there have been numerous instances in which this expectation proved to be justified; we shall return to some of these cases later.

At this point, mention should be made of the fact that, during the 1930s, the antineuritic vitamin B_1 was shown to be a constituent of a coenzyme important in alcoholic fermentation. In 1932 Ernst Auhagen found that the Harden-Young co-ferment contained, in addition to Euler's cozymase, a heat-stable dialyzable organic substance that is a cofactor for yeast carboxylase, the enzyme that converts pyruvic acid into acetaldehyde and CO_2. The studies of Rudolph Peters and his associates around 1930 (see Peters, 1963) had shown that vitamin B_1 is concerned with the metabolism of pyruvic acid in the animal body, since lactic acid accumulated in the brain tissue of B_1-deficient pigeons, and the addition of vitamin B_1 (by then available in crystalline form) and pyrophosphate to the minced tissue caused the rapid disappearance of pyruvic acid without the accumulation of lactic acid. After the elucidation of the chemical structure of vitamin B_1 in 1935, through the competitive efforts of research groups in Germany (Adolf Windaus) and the United States (Robert R. Williams), it was recognized to be a thiazole derivative, and renamed thiamine (for a time it was also termed aneurin). There then followed the report of Lohmann and Schuster (1937) on the isolation of crystalline cocarboxylase from yeast, and the demonstration that it is thiamine pyrophosphate (see the accompanying formula). Subsequent studies on the pyruvate oxidation

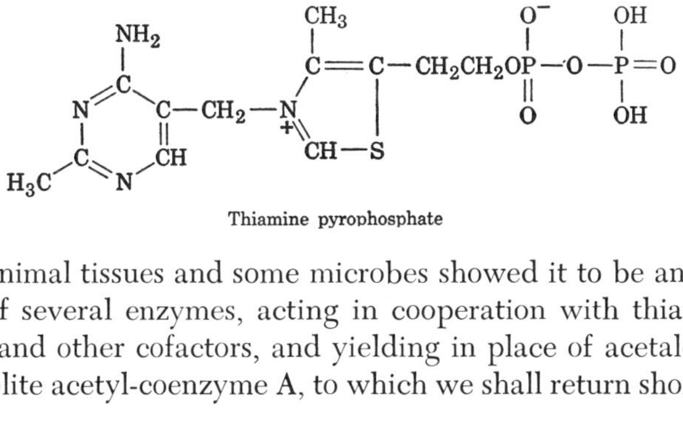

Thiamine pyrophosphate

system in animal tissues and some microbes showed it to be an integrated assembly of several enzymes, acting in cooperation with thiamine pyrophosphate and other cofactors, and yielding in place of acetaldehyde the key metabolite acetyl-coenzyme A, to which we shall return shortly.

Oxidative Generation of Biochemical Energy

In 1861 Pasteur described fermentation as a process whereby an organism, in the absence of atmospheric oxygen, can extract bound oxygen from organic substances (see p. 56); he later embellished the idea in the language of thermochemistry by stating:

The aerobic organism makes the heat that it needs by combustions resulting from the absorption of oxygen gas; the anaerobic organism makes the heat that it needs by decomposing so-called *fermentable* matter belonging to the group of explosive substances that are able to release heat by their decomposition (Pasteur, 1879, p. 141).

Thus the fact that a microbial culture used up much more glucose in anaerobic fermentation than in aerobic oxidation (the "Pasteur effect") was correlated with the finding that the heat production in the fermentation of a gram of glucose is much less than in its oxidation to CO_2 and H_2O. The view that the heat liberated during biological oxidations was somehow used for building cellular constituents was reiterated by many of Pasteur's contemporaries. It was vigorously stated by Pflüger (1875) in relation to his theory of "living albuminoid substances." To cite only one other example, when Winogradsky demonstrated that the sulfur bacteria depend on the oxidation of hydrogen sulfide to sulfuric acid for their existence, he noted that

. . . the expression "respiration" is not applicable to this process. It may be compared directly with respiration only because in both processes there is liberated heat (energy) which is essential for the survival of the organisms (Winogradsky, 1887, p. 607).

The fact that many chemically minded biologists equated the heat liberated in a chemical process with the work the process might be expected to effect in the organism was in part a reflection of the wide acceptance of Marcelin Berthelot's "principle of maximum work":

All chemical change accomplished without the intervention of external energy tends to the production of a substance or of a system of substances which liberate the most heat (Berthelot, 1875, p. 6).

After the appearance of the memorable papers by Josiah Willard Gibbs (in 1878) and by Hermann Helmholtz (in 1882) on the thermodynamics of chemical processes, it became evident to chemists that a distinction had to be made between the "total energy" derivable from an isothermal process in the form of heat (when the reaction did no work) and the "free energy" available for work. Nevertheless, although the significance of Berthelot's neglect of Clausius' concept of entropy (related to what Helmholtz called "bound energy") was increasingly appreciated by biologists and biochemists, they continued for many decades to assume the equivalence of work and heat in biological systems. The principal reason was that, whereas calorimetry readily gave values for the heat production in any chemical reaction, the determination of free-energy changes was possible only for freely reversible reactions in which the equilibrium constant could be measured experimentally. As was indicated on p. 326, the development of electrochemistry before 1900 provided such chemical processes, but most of the reactions of biological interest, such as the conversion of glucose into alcohol and CO_2, could not be studied in this manner. In 1906, however, Walther Nernst showed how it is possible to estimate equilibrium constants from thermal measurements, and his theory was applied by Julius Báron and Michael Polanyi (1913) to the calculation of the free-energy change in the combustion of glucose; they obtained a value (at 37°C) about 13 percent greater than the heat of combustion determined calorimetrically. It is also noteworthy that in 1912 A. V. Hill called the attention of his fellow physiologists to the fact that, in addition to Nernst's approach,

> In many cases it is already possible to calculate the equilibrium constant K of a reaction, and hence also . . . its free energy. The developments of ferment chemistry, especially in the case of reversible changes carried out by ferments, may make it possible to calculate directly the equilibrium constants of many breakdowns of organic material. If *e.g.* the bio-chemist can decide what chemical reactions go on in order in the process of carbohydrate breakdown, and if we can determine directly the values of K_1, K_2, K_3 . . . for these several reactions, then it will be possible not only to give the total free energy . . . but the free energy of every stage (Hill, 1912, p. 512).

Indeed, after 1920, the free-energy changes in dehydrogenase-catalyzed oxidation-reduction reactions were estimated in this manner; however, so long as the sequence of chemical reactions in carbohydrate breakdown was unknown, and the equilibria in individual enzyme-catalyzed reactions could not be determined, the method could not be applied to the study of this process.

Consequently, around 1910, when Warburg and Meyerhof began their important studies on the energetics of cellular oxidations, they made the assumption that ". . . for the oxidation of hydrogen or carbon, it is permissible to set, with some degree of approximation, the heat liberation as equal to the decrease in free energy" (Warburg, 1914a, p. 256). In large part, their studies were stimulated by the work of Jacques Loeb on the effect of oxygen lack on the division of fertilized sea-urchin eggs, and were designed to approach the problem he posed:

> The question arises, as to what connection exists between the oxidations in living tissues, and cell division and growth. We cannot answer this

question as we do not know into which form of energy chemical energy must be transformed in order to produce cell division and growth . . . . For the process of growth an increase in the quantity of living matter is required, and this requires synthetical processes (Loeb, 1906, p. 17).

It was soon found that calorimetry could give little information, since the amount of heat produced by cells undergoing division was unchanged when the cell division was artificially interrupted. Warburg therefore concluded:

A connection between visible work performance and oxidation rate is not evident or is otherwise expressed: only a small part of the work made available by respiration is used for the performance of visible work (Warburg, 1914a, p. 258).

Of special importance in the consideration of this problem was the introduction, by Wilhelm Ostwald, of the concept of coupled chemical reactions. Noting the difference between the transfer of chemical energy and the transfer of heat or electrical energy, he stated:

A direct reciprocal transformation of chemical energies is only possible to the extent that chemical energies can be set in connection with each other, that is, within such processes that are represented by a stoichiometric equation. Coupled reactions of this kind may be distinguished from those that proceed independently of each other; their characteristic lies in the fact that they can be represented by a single chemical equation with definite integral coefficients. On the other hand, chemical processes that are not coupled cannot transfer energy to each other (Ostwald, 1900, p. 250).

In applying this idea to auto-oxidation processes, Ostwald emphasized the importance of the formation of labile intermediates in coupled chemical reactions, and concluded with the statement that

. . . the appearance of labile intermediates should not be regarded as something exceptional, but rather as according to rule. It is a general principle that in chemical processes, taken as a whole, the possible products that arise first are not the stable ones, but just those that are the most unstable under the existing conditions (*ibid.*, p. 252).

This general thermodynamic statement could give no hint, however, of the nature of the coupled reactions that might be operative in the utilization, by energy-requiring processes, of the chemical energy released in cellular oxidations. The chemical nature of the reactants, labile intermediates, and products in each "stoichiometric equation" had to be identified, together with the enzymes that catalyze the coupled reactions. This challenge to biochemistry has kept many investigators busy throughout this century, and much is still unknown. During the period 1920–1940, however, great advances defined the problem more precisely, through the discovery of new cell constituents and the elucidation of their chemical structure, as well as through the purification of individual enzymes and the characterization of their mode of action. In this development, a decisive role was played by the study of the coupling to muscular contraction of the chemical energy released upon carbohydrate breakdown. For a comprehensive history of the biochemistry of muscular contraction, see Needham (1971).

The Energetics of Glycolysis. We mentioned on p. 281 some facets of the nineteenth-century research on muscle, in relation to the problem of the site of oxygen utilization in animals, and we noted Liebig's view that the energy for muscular contraction came from protein breakdown. This idea was challenged by the celebrated experiment performed in 1865 by Adolf Fick and Johannes Wislicenus. After 17 hours on a nitrogen-free diet, they walked up the Faulhorn, an Alpine peak near Grindelwald, and they determined the amount of nitrogen excreted in their urine during the trip. Their calculations showed that no more than about a third of the energy expenditure in climbing the mountain could have depended on the complete combustion of body protein. Experiments by Max Pettenkofer and Carl Voit during the 1860s also led to the conclusion that muscular work did not lead to increased nitrogen excretion, and Liebig (1870) accepted the validity of these results. The view that the energy for muscular contraction came from proteins was not abandoned, however; Pettenkofer and Voit suggested that

> . . . through the oxygen uptake of the organs and through the commensurate protein decomposition there accumulates a tension [*Spannkraft*] which is gradually used up even during rest and which we can convert at will into mechanical work (Pettenkofer and Voit, 1866, p. 572).

A similar view was expressed by Liebig (1870), and Ludimar Hermann (1867) postulated the existence of a labile albuminoid material (later termed "inogen") which undergoes anaerobic breakdown to myosin, "fixed acid" (lactic acid), and CO_2. Such energy-rich proteins also figured largely in Pflüger's speculations during the 1870s. Although there was chemical evidence (Nasse, 1869) to suggest that the appearance of lactic acid during muscular contraction paralleled a decrease in the glycogen content of the muscle, it was uncertain whether these changes reflected a vital process or merely the death of the tissue (see Fürth, 1903); this debate was not terminated until the work of Fletcher during the first decade of this century.

Furthermore, although Fick (1882) strenuously defended the view that muscular contraction is a "chemical-dynamic" process that cannot be compared to the operation of a heat engine, which depends on differences in temperature, there were repeated instances in which such temperature differences were assumed to exist in muscle. As A. V. Hill stated it,

> Unfortunately there have been, even among physiologists, many and grievous misconceptions as to the application of the laws of thermodynamics; these have been due partly to the desire to make over-hastily a complete picture of the muscle machine, partly to the completely erroneous belief that the laws of thermodynamics apply only to heat engines and not to chemical engines, and that no information can be obtained from the Second Law as to the working of a chemical machine at uniform constant temperature. The muscle fibre has been treated as a heat-engine when it is inconceivable that there are finite differences of temperature in it . . . . The muscle is undoubtedly a chemical machine working at constant temperature (Hill, 1912, p. 507).

After the experiments of Fletcher and Hopkins (1907), it was the study of muscle as an isothermal "chemical machine" that laid the groundwork for the understanding of the chemical energetics of muscular contraction, and the remarkable work of Hill on heat production in muscle had a continuing influence on biochemical studies (see Hill, 1950).

Fletcher and Hopkins found that lactic acid appeared during anaerobic contraction and disappeared upon the admission of oxygen; through the use of a sensitive microcalorimeter, Hill was able to correlate these chemical changes with two phases of heat production by surviving muscle, an "initial heat" and a "recovery heat." In 1913 Hill estimated that, under optimal conditions, the total energy measured as initial heat was approximately equivalent to the potential energy of the tension when a contracting muscle was made to do work; subsequently it was shown that such high mechanical efficiency is, in fact, not attained. Furthermore, the English investigators concluded that

> Lactic acid is therefore built up by the body into some, at present unknown, chemical combination of greater energy than glucose. It is

suggested that this body is one containing a large store of "free energy," and that it may be able to account for the mechanical work done by a muscle simply by the process of breaking down into lactic acid (Hill, 1912, p. 506).

Fletcher and Hopkins supposed that, during oxidative recovery, the lactic acid was converted into CO_2 and H_2O, but Hill's measurements indicated that the total heat liberated in a complete cycle of contractions was only about one-fifth of that expected from such complete oxidation of the lactic acid produced during the anaerobic phase.

In an important series of papers published during 1920–1922, Meyerhof established that, in surviving muscle, the lactic acid is wholly derived from glycogen, and that during oxidative recovery about four-fifths of the lactic acid is resynthesized to glycogen. In 1925 he summarized the situation as follows:

The first phase, the formation of lactic acid from carbohydrate, is anaerobic and spontaneous. This process is the immediate source of muscular force. In the second phase, with the expenditure of oxidation energy, the lactic acid is reconverted to carbohydrate. This second process corresponds to the recovery or restitution of the muscle . . . . This cyclic process permits one to interpret quantitatively the two heat phases during the performance of work, discovered by Hill. The heat liberated at the instant of contraction corresponds to the breakdown of glycogen to lactic acid and to a physical-chemical change of the muscle protein induced by the appearance of lactic acid, [the latter being] directly associated with the contractile process. The oxidative heat, on the other hand, represents the excess of oxidation energy over the endothermic reactions that occur during the reversal of the processes during the work phase (Meyerhof, 1925, p. 995).

Within a few years, however, the view that the formation of lactic acid is the cause of muscular contraction had to be abandoned. This came about through the discovery of new energy-rich compounds in muscle, and the demonstration that contraction can occur in the absence of the formation of lactic acid. The result was what Hill (1932) called "the revolution in muscle physiology"; these findings also brought new light to the problem of the anaerobic breakdown of carbohydrate, not only in muscle, but during alcoholic fermentation as well.

The "revolution" began when Fiske and SubbaRow (1927) reported that, upon applying to muscle extracts the method they had devised for the quantitative determination of phosphate, they had found an acid-labile phosphate derivative of creatine, known since 1832 (Chevreul) to be a constituent of muscle; they called this organic phosphate compound "phosphocreatine," and showed it to be hydrolyzed during contraction and resynthesized during recovery. At almost the same time there appeared a paper by Philip and Grace Eggleton reporting the presence, in muscle, of a "labile form of organic phosphate"; in a succeeding article they stated:

> There is present in the skeletal muscle of the frog an organic phosphorus compound which has hitherto been confused with inorganic phosphate, owing to its rapid hydrolysis in acid solution to phosphoric acid. There may be more than one such compound, but the hypothesis of a single compound is sufficient to explain the available facts. We have given the name "phosphagen" to this substance.
>
> The results quoted in this paper established the fact that muscular contraction is accompanied by the removal of phosphagen, and subsequent recovery in oxygen is characterized by a rapid restitution of the phosphagen—a phase of recovery apparently independent of the relatively slow oxidative removal of lactic acid (Eggleton and Eggleton, 1927, p. 159).

Although they first thought phosphagen to be a hexose monophosphate derived from Embden's "lactacidogen," its identity with phosphocreatine, later termed creatine phosphate, was soon accepted (see accompanying formula).

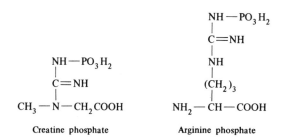

Creatine phosphate   Arginine phosphate

The discovery of creatine phosphate focused attention on acid-labile phosphate compounds; soon afterward, Meyerhof and Lohmann showed that the phosphagen in the muscles of invertebrates is arginine phosphate, and that the hydrolysis of both organic phosphates is strongly exothermic. In addition, Lohmann found the acid-labile inorganic pyrophosphate $[(HO)_2(O)P\!-\!O\!-\!P(O)(OH)_2]$ in muscle and yeast; after the report of Davenport and Sacks that fresh muscle did not contain inorganic pyrophosphate, this substance was shown by the work of Fiske and SubbaRow (1929) and Lohmann (1929) to be derived from the breakdown of a new labile organic phosphate related to the muscle adenylic acid found in Embden's laboratory two years before. The new compound was named adenylpyrophosphate; its structure was later (Lohmann, 1935) recognized to be that of adenosine-5'-triphosphate (abbreviated ATP).

Adenosine-5'-triphosphate (ATP)

The second stage in the "revolution in muscle physiology" came in 1930, when Einar Lundsgaard discovered that muscles poisoned with iodoacetic acid (ICH_2COOH) can contract without the formation of lactic acid, but with the disappearance of creatine phosphate (in crustacean muscle, arginine phosphate); he proposed that

> . . . phosphagen is the substance directly supplying the energy for contraction, while lactic acid formation in the normal muscle continually provides the energy for its resynthesis (Lundsgaard, 1930, p. 177).

After Lundsgaard's work, it was realized that Embden had been correct when he claimed (in 1924) that much of the lactic acid is actually formed after the contraction; these results were questioned at a time when the lactic acid theory of muscular contraction held sway (see Hill, 1932).

During the period 1930–1934 there was extensive debate about the nature of the chemical reaction most immediately related to muscular contraction. An important feature of the discussion was noted in a review article:

> It has been tacitly assumed, and this view is still very widespread, that the greatest significance must be assigned to that partial process, within the totality of the chemical processes associated with muscular contrac-

tion, which has undergone the greatest change after a muscle twitch (Lehnartz, 1933, p. 966).

Indeed, considerable attention was given to the extensive deamination of muscle adenylic acid (to inosinic acid), observed in muscle extracts, as a possible source of the energy for muscular contraction. As was mentioned on p. 210, Gerhard Schmidt had discovered an enzyme that catalyzes this deamination; subsequent calorimetric measurements had shown the reaction to be strongly exothermic. This line was abandoned, however, after it was recognized that the central chemical processes in muscular contraction are the cleavage and resynthesis of the pyrophosphate bonds of ATP.

In 1934 Karl Lohmann found that the apparent hydrolysis, by dialyzed muscle extracts, of creatine phosphate (CP) to creatine (C) and inorganic phosphate (P_i) was promoted by the addition of ATP, which was cleaved to adenylic acid (adenosine monophosphate, AMP) and two equivalents of P_i. He concluded that ATP acts as a "coenzyme" in the hydrolysis of CP in the following manner: (1) AMP + 2 CP → ATP + 2 C; (2) ATP + 2 H_2O → AMP + 2 P_i. The sum of the two reactions is 2 CP + 2 H_2O → 2 C + 2 P_i. Calorimetric measurements in Meyerhof's laboratory had already shown that the hydrolysis of a molecule of ATP to AMP and 2 P_i is accompanied by a large heat liberation, approximately double that found for the hydrolysis of one molecule of creatine phosphate; these data were consistent with the ready reversibility of the reaction between ATP and creatine, on the assumption that the free-energy changes parallel the changes in heat content. Also, ATP had been identified by Lohmann as the "coferment" of lactic acid formation in muscle, and he had found that magnesium ions were required. On the basis of the data available in 1931, Meyerhof and Lohmann had suggested that the role of ATP

> . . . appears to consist in the fact that the esterification of phosphate, which precedes the cleavage of carbohydrate to lactic acid, occurs with the simultaneous cleavage of adenylpyrophosphate, which is resynthesized during the further cleavage [of carbohydrate]. In this manner, the adenylpyrophosphate cycle maintains the lactic acid formation. The synthesis of phosphagen is therefore made possible . . . by the cleavage energy of the adenylpyrophosphate, while the energy of lactic acid formation (from phosphate esters) serves to resynthesize the cleaved pyrophosphate (Meyerhof and Lohmann, 1931, p. 576).

At the time this was written there was uncertainty about the relation of ATP to Euler's cozymase, also identified as a substance related to adenylic acid. This uncertainty remained until Warburg's discovery of nicotinamide as a constituent of cozymase, and the improvement of methods for the purification of ATP and DPN. Furthermore,

> As regards the question of the chemical and energetic relation of the breakdown and resynthesis of adenylpyrophosphate to the fundamental process of muscular contraction, it may be assumed that there is no direct relation (Lohmann, 1934, p. 276).

Within five years, however, came the dramatic report by Engelhardt and Lyubimova (1939) that the fibrous protein (myosin) identified with the process of muscular contraction has the properties of an enzyme which specifically catalyzes the hydrolysis of ATP, and that the contractile process is directly linked to this energy-yielding chemical reaction. Myosin had been described by Kühne in 1864; the subsequent studies of Danilevski, Halliburton, and Fürth confirmed and extended his observations, and during the 1930s the new physicochemical methods for the study of fibrous proteins were applied to the myosin problem by Muralt, Edsall, and Weber. After Vladimir Engelhardt's discovery, the further development of research on the coupling of the hydrolysis of ATP to muscular work (initially by Needham, Bailey, and Szent-Györgyi, among others) may be considered to represent a further stage in the "revolution in muscle physiology." More recent studies, using electron microscopy and X-ray diffraction observations of muscle, have shed considerable new light on the relation of ATP hydrolysis to the mechanics of contraction (Huxley, 1971).

In 1934 ATP was considered to be a "coenzyme" in glycolysis, but its function was not clear. It had also been shown to be a component of the Harden-Young coferment, in addition to Mg^{2+}, Euler's cozymase (soon to be identified as DPN), and Auhagen's cocarboxylase. Thus the increasing heterogeneity of the heat-stable dialyzable "substance" identified by Harden and Young as a cofactor in alcoholic fermentation matched the growing complexity of the zymase Buchner had considered to be a single enzyme.

At this stage, a clue to the role of ATP in glycolysis and fermentation was provided by Jacob Parnas, Pawel Ostern, and Thaddeus Mann (1934). They showed that the addition of 3-phosphoglyceric acid to iodoacetate-poisoned minced muscle caused a marked inhibition in the production of ammonia; neither fructose-1,6-diphosphate nor pyruvic acid had this effect. Since it was known that the adenylic deaminase discovered by Schmidt did not deaminate ATP, they concluded that adenylic acid had been converted into ATP, and that

. . . the resynthesis of phosphocreatine and adenosine triphosphate is not linked to glycolysis as a whole, but to definite partial processes: and this leads further to the conclusion that this resynthesis does not involve a relationship that might be termed "energetic coupling," but more probably involves a transfer of phosphate residues from molecule to molecule (Parnas *et al.*, 1934, p. 68).

In writing this passage, Parnas probably intended to refer to the conclusion drawn by Meyerhof from his measurements of the relation between oxygen uptake and glycogen resynthesis: ". . . oxidation and resynthesis do not represent a chemically-coupled process, for which one can give a stoichiometric equation, but an energetically coupled one" (Meyerhof, 1930, p. 38). This statement was criticized by Hahn (1931), who reiterated the definition of coupled reactions given by Ostwald (1900), mentioned on p. 362.

To define which molecules might be involved in the phosphate transfer postulated by Parnas, a more detailed identification of the intermediates in glycolysis and fermentation was necessary; this came quickly, largely through work in Meyerhof's laboratory. We noted earlier that, according to Embden's theory, the 3-phosphoglyceric acid formed by the oxidation of glyceraldehyde-3-phosphate is the precursor of pyruvic acid; Lohmann showed in 1935 that the formation of pyruvic acid involves the enzyme-catalyzed migration of the phosphoryl group from the 3-position to the 2-position of glyceric acid (see the accompanying formulas), followed

$$
\begin{array}{ccc}
\text{COOH} & \text{COOH} & \text{COOH} \\
| & | & | \\
\text{HCOH} \rightarrow & \text{HC—OPO}_3\text{H}_2 \rightarrow & \text{C—OPO}_3\text{H}_2 + \text{H}_2\text{O} \\
| & | & \| \\
\text{CH}_2\text{OPO}_3\text{H}_2 & \text{CH}_2\text{OH} & \text{CH}_2 \\
\text{D-3-Phosphoglyceric} & \text{D-2-Phosphoglyceric} & \text{Phosphoenol-} \\
\text{acid} & \text{acid} & \text{pyruvic acid}
\end{array}
$$

$$
\begin{array}{cc}
\text{COOH} & \text{COOH} \\
| & | \\
\text{C—OPO}_3\text{H}_2 + \text{ADP} \rightleftharpoons & \text{C=O} \quad + \text{ATP} \\
\| & | \\
\text{CH}_2 & \text{CH}_3 \\
\text{Phosphoenol-} & \text{Pyruvic} \\
\text{pyruvic acid} & \text{acid}
\end{array}
$$

by the dehydration of 2-phosphoglyceric acid to phosphoenolpyruvic acid (PEP) by the enzyme denoted enolase. This enzyme was found to be strongly inhibited by fluoride, thus explaining the effect of fluoride on alcoholic fermentation and glycolysis (p. 350). Furthermore, Meyerhof and Kiessling (1935) demonstrated that, in iodoacetate-poisoned muscle extracts, the phosphoryl group of PEP was transferred to glucose via ATP to form hexose phosphates and pyruvate; also, calorimetric measurements showed that the hydrolysis of PEP is a strongly exothermic reaction. It became clear therefore that two of the individual phosphate-transfer reactions in which ATP or its cleavage products participate are the entry of glucose into the glycolytic sequence (the hexokinase reaction identified by Meyerhof in 1927) and the dephosphorylation of PEP to form pyruvic acid. It may be added here that the stoichiometry of these reactions was not fully clarified until after Lohmann (1935) had isolated adenosine diphosphate (ADP) and after it had been shown that there is a widely distributed enzyme which catalyzes the reaction $2\ ADP \rightleftharpoons ATP + AMP$ (Kalckar, 1942). It was then recognized that, in the hexokinase reaction, ATP reacts with glucose to form glucose-6-phosphate and ADP, that in the "creatine kinase" reaction the equation is $C + ATP \rightleftharpoons CP + ADP$, and that pyruvic acid + ATP are formed in the reaction between PEP and ADP. Furthermore, with regard to the enzyme-catalyzed isomerization of the two phosphoglyceric acids, later work (in 1949) showed a requirement for 2,3-diphosphoglyceric acid, a substance that had been isolated from erythrocytes by Isidor Greenwald in 1925, but to which no physiological role could be assigned until the enzyme involved in the interconversion of the two phosphoglyceric acids had been purified.

Thus in 1935 there was clear evidence for the coupling of enzyme-catalyzed reactions in which ATP was made by the transfer of phosphate from PEP to ADP, and in which phosphate was transferred from ATP to creatine or to a carbohydrate acceptor; in neither case was there a prior hydrolysis of the phosphoryl donor followed by the condensation of inorganic phosphate with the phosphoryl acceptor. Within two years it had also been shown, both for muscle extracts and for yeast juice, that there is an anaerobic process in which ATP is synthesized by the addition of phosphate to adenylic acid, in a process coupled to dehydrogenase-catalyzed oxidation-reduction. We have seen (p. 352) that the importance of coupled dehydrogenase (mutase) reactions in carbohydrate breakdown had been emphasized by Euler during the 1920s, and that Meyerhof and Kiessling (1935) had shown that the oxidation of glyceraldehyde-3-phosphate to 3-phosphoglyceric acid is balanced by the reduction of pyruvic acid to lactic acid (in muscle) or of acetaldehyde to ethanol (in yeast). Dorothy Needham and R. K. Pillai (1937) then found that the "dismutation" between triose phosphate and pyruvic acid in muscle extracts was accompanied by the esterification of inorganic phosphate, and David Green *et al.* (1937b) showed this process to be reversible, and to depend on the presence of DPN. These observations focused attention on the relation of ATP to the DPN-dependent dehydrogenation of glyceralde-hyde-3-phosphate (G-3-P) to 3-phosphoglyceric acid (3-PG). In particular, an important study by Meyerhof *et al.* (1938) established the reversibility of the process

$$G\text{-}3\text{-}P + DPN + ADP + P_i \rightleftharpoons 3\text{-}PG + \text{reduced } DPN + ATP$$

and showed clearly that this reaction is inhibited by iodoacetate (p. 367) in the anaerobic breakdown of carbohydrate. The inhibition of the dehydrogenase was seen to be related to the known reaction of iodoacetate with the cysteine-sulfhydryl groups of proteins, and Louis Rapkine (1938) concluded that such groups are essential for the catalytic activity of the enzyme. It will be evident that the use of chemical reagents such as iodoacetate or fluoride led to significant discoveries in the study of fermentation and glycolysis; as we shall see later, the elucidation of other metabolic pathways was greatly aided by similar specific inhibition of individual enzymes.

We now return to the decisive work on what Warburg called the "oxidizing fermentation" enzyme, to which we referred on p. 356; the name glyceraldehyde-3-phosphate dehydrogenase, which he did not accept, is the one used today. In the crystallization of this enzyme from yeast extracts, Warburg and Christian (1939) used the spectrophotometric measurement of reduced DPN to follow the course of the purification, and they also took advantage of the results of Needham and Pillai (1937) on the effect of arsenate on the oxidation of triose phosphate to phosphoglyceric acid. This reagent, because of its chemical similarity to phosphate, had been tested by Harden and Young (in 1906) for its effect on the action of yeast juice; they found that the fermentation went faster, and explained the result in terms of an accelerated hydrolytic cleavage of hexose diphosphate, to regenerate glucose and phosphate (see p. 345). This explanation was questioned during the 1930s, however, when some phosphatases were found to be inhibited, not activated, by arsenate. After Needham and Pillai had shown that, in the oxidation of triose phosphate to phosphoglyceric acid, the coupled phosphorylation to form ATP is completely inhibited by arsenate, whereas the oxidation-reduction process is unaffected, it became clear that arsenate was replacing phosphate in the reaction catalyzed by glyceraldehyde-3-phosphate dehydrogenase. The nature of the chemical reaction involving arsenate was not elucidated, however, until Warburg's associates Negelein and Brömel (1939b) isolated the immediate product of the reaction catalyzed by glyceraldehyde-3-phosphate dehydrogenase in the presence of phosphate, and showed it to be 1,3-diphosphoglyceric acid (they called it "R-acid"). The isolation of this compound required exceptional chemical skill, because of its instability; in aqueous solution it undergoes rapid hydrolysis to 3-phosphoglyceric acid and inorganic phosphate. It followed, therefore, that in the presence of arsenate the corresponding 1-arseno-3-phosphoglyceric acid is formed, and is hydrolyzed in a similar manner. The difference lay in the fact that in the presence of ADP there occurs a readily reversible transfer of the 1-phosphoryl group of 1,3-diphosphoglyceric acid to form ATP (see p. 356), whereas the arseno compound only undergoes hydrolysis. As we noted before, Warburg and Christian (1939) reported that their colleague Theodor Bücher had found a separate enzyme that catalyzes the reversible reaction leading to ATP; the details of this important discovery were not published until after World War II (Bücher, 1947).

What had been achieved in these studies was the demonstration of a chemically defined enzymic mechanism whereby the energy released in the oxidation of the aldehyde group of glyceraldehyde-3-phosphate by DPN was utilized for the synthesis of a pyrophosphate bond of ATP. The key discovery was the identification of the labile intermediate in this coupling process as a compound in which the carboxyl group formed in the oxidation is combined with a phosphoryl group in an "acyl phosphate," *i.e.*, an anhydride of a carboxylic acid and phosphoric acid. This discovery was made possible by the purification of the dehydrogenase that catalyzes the oxidation, and by the fact that the crystalline enzyme preparation was relatively free of other enzymes that act on the reactants and the products of the dehydrogenase-catalyzed reaction. When seen against the background of the earlier uncertainties regarding the chemical events in glycolysis and fermentation arising from the use of yeast and muscle juices and extracts, or of crude enzyme preparations, the work of Warburg and his associates made it clear that the chemical dissection of complex biochemical processes depended on the isolation and characterization of individual enzyme proteins. Furthermore, it was evident by 1940 that the approach used by Sumner, Northrop, and Kunitz was a necessary adjunct to the elucidation of intracellular processes that had been considered to be fundamentally different from those involving hydrolytic enzymes such as urease, pepsin, and trypsin. In 1940 Bruno Straub reported the crystallization of the lactic dehydrogenase of beef heart muscle; in 1941 Warburg and Christian described the crystallization of enolase from yeast and, two years later, of aldolase (they called it zymohexase) from muscle. In the years that followed, many enzymes concerned with oxidation and fermentation, and whose activity had been thought to be associated with the integrity of cellular structure, were purified in a similar manner, and studied as chemical entities. One of the important early examples of such success was the crystallization of the glyceraldehyde-3-phosphate dehydrogenase of skeletal muscle, by Cori, Slein, and Cori (1945).

Of even greater significance in its impact on subsequent biochemical thought was the discovery that an acyl phosphate is the labile intermediate in the coupling of an enzyme-catalyzed dehydrogenation to the enzymic phosphorylation of ADP. Early in 1939 Fritz Lipmann had reported that the oxidative decarboxylation of pyruvic acid by the microorganism *Lactobacillus delbrückii* in the presence of inorganic phosphate and thiamine pyrophosphate (both of which were found to be required for the oxidation) was accompanied by the phosphorylation of adenylic acid; he suggested that the coupled phosphorylation might involve the reduction of the thiazole ring of thiamine, in analogy to the reduction of the pyridine ring of DPN. After the work of Negelein and Brömel (1939b), however, Lipmann (1939) provided evidence for the view that acetyl phosphate (the anhydride of acetic acid and phosphoric acid) is an intermediate in this process, and that its phosphate is transferred in an enzyme-catalyzed reaction to adenylic acid. There then appeared the influential review articles by Kalckar (1941) and by Lipmann (1941) in which the problem of the coupling of oxidation reactions to the generation of ATP was discussed against the background of the discoveries of the 1930s. In particular, Lipmann introduced the term "energy-rich phosphate bond" to denote linkages in compounds such as creatine phosphate, ATP, phosphoenolpyruvate, 1,3-diphosphoglyceric acid, or acetyl phosphate, whose hydrolysis is associated with a relatively large negative free-energy change (see Lipmann, 1971). As was indicated on p. 327, such large free-energy changes mean that the equilibrium constants in these reactions lie very far in the direction of the cleavage products. In enzyme-catalyzed phosphoryl-transfer reactions, such as Glucose + ATP → Glucose-6-phosphate + ADP, no hydrolysis of the "energy-rich phosphate bond" occurs; the fact that the equilibrium in this reaction lies far to the right is, in Lipmann's terminology, associated with the higher "group potential" of the phosphoryl group in ATP than the one in glucose-6-phosphate. Thus the numerical difference between the free-energy change in the hydrolysis of ATP to ADP and P_i (*ca.* −8 kilocalories or −33 kilojoules per mole) and of glucose-6-phosphate (*ca.* −3 kcal or −12 kJ per mole) is a large negative number, making the transfer reaction (the sum of the exergonic hydrolysis of ATP and the endergonic condensation of glucose with P_i) a strongly exergonic process. The terms "exergonic" and "endergonic" were used by Charles Coryell (1940) as counterparts to Berthelot's "exothermic" and "endothermic" to emphasize the distinction between free-energy changes and the changes in the heat content in chemical processes. Lipmann did not limit the "group potential" concept to phosphoryl groups, but applied it to the transfer of other groups as well; thus he proposed that, in the presence of suitable catalysts, acetyl phosphate might be an acetylating agent, and went on to state:

A broader application of this metabolic principle can be visualized if it is remembered that the formation of a mixed acyl phosphate anhydride can be effected through phosphorylation from adenyl pyrophosphate (Warburg and Christian, 1939). The major part of the constituents of protoplasma are compounds which contain the ester or the peptide linkage. In the routine procedure of organic chemistry for synthesis of compounds of this type the acyl chloride of the acid part is first prepared and then brought into reaction with the hydroxyl or amino group of the other part. In an analogous procedure the cell might first prepare the acyl phosphate with adenyl pyrophosphate as the source of energy-rich phosphate groups. The acyl phosphate of fatty acids might then condense with glycerol to form the fats. The acyl phosphate of amino acids likewise might condense with the amino groups to form the proteins (Lipmann, 1941, pp. 153–154).

These views reflected the emerging knowledge about enzyme-catalyzed group-transfer reactions, as well as the growing conviction that ATP plays a central role as a funneling agent of the chemical energy derived from the intracellular degradation of carbohydrates to "drive" endergonic processes. Although, as we shall see later, some of the specific mechanisms envisaged by Lipmann were not found to be operative in biological systems, his general approach influenced a generation of biochemists after World War II (see Kaplan and Kennedy, 1966), and the concept of the "energy-rich phosphate bond" took firm root in biochemical textbooks.

Among the enzyme-catalyzed group-transfer reactions discovered during the 1930s, special importance must be assigned to the process whereby the glucose units of glycogen enter the pathway of the anaerobic degradation of glucose to pyruvic acid. In 1936 Jacob Parnas and Thaddeus Baranowski found that, in muscle extracts, glycogen and inorganic phosphate could react to form hexose monophosphate under conditions where the oxidation-reduction process had been blocked by iodoacetate and no ATP was being generated. This "phosphorolysis" of glycogen was elucidated by Carl and Gerty Cori (1936) through the isolation of the initial product of the enzymic cleavage of the polysaccharide, and the unequivocal demonstration that it is glucose-1-phosphate; the enzyme responsible for this readily reversible cleavage was termed "phosphorylase," and was later crystallized by Arda Green and Gerty Cori (1943). Cori and Cori also demonstrated the presence of a separate enzyme that catalyzes the interconversion of glucose-1-phosphate and glucose-6-phosphate, thus linking the product of the hexokinase reaction to glycogen through two reversible enzymic reactions. We shall return to these important findings in the next chapter, but it should be noted here that the discovery of the phosphorylase reaction had a considerable impact on biochemists interested in the problem of the enzymic mechanisms in the biosynthesis of polymeric cell constituents; after the emergence of the concept of the "energy-rich phosphate bond," the reaction catalyzed by phosphorylase provided an example of the formation of a polymeric molecule in which an "energy-poor phosphate bond" was involved, indicating that a clear distinction could not be made between the two types of bonds.

By 1940 Buchner's zymase was known to include a dozen enzymic components (see scheme shown on p. 377), and it was agreed that the pathway for the anaerobic transformation of glucose-6-phosphate to pyruvic acid is the same in muscle as in yeast. Furthermore, through the work of Meyerhof, Needham, and Warburg, it was recognized that the net yield of chemical energy made available in the anaerobic breakdown of one mole of glucose was equivalent to the free energy of hydrolysis of two moles of ATP (then estimated at about -25 kcal). Since the over-all free-energy change in the conversion of glucose into lactic acid was considered to be about -66 kcal per mole of glucose (Burk, 1929), the maximum efficiency of the "chemical machine" in glycolysis appeared to be about 40 percent. As we noted on p. 360, it had been known since the work of Pasteur that the chemical energy made available to a biological organism by the aerobic oxidation of glucose to CO_2 and H_2O is much greater than that released during anaerobic fermentation; during the 1930s a value of about -700 kcal per mole was assigned to the free-energy change in the oxidation of glucose (or the glucosyl units of glycogen) to CO_2 and H_2O. By 1935 it was generally agreed that the utilization of molecular oxygen for cellular oxidations involved the action of Warburg's *Atmungsferment* on oxygen at one end of the scale of oxidation-reduction potential, and of the Wieland-Thunberg dehydrogenases acting on metabolites such as succinic acid or lactic acid at the other end, with Warburg's flavoproteins and the Keilin cytochrome system serving to link the "oxygen activation" to the "hydrogen activation." The problem was to discover how this respiratory electron-transport system is used for the aerobic oxidation of substances such as glucose to CO_2 and H_2O, and how the relatively large amount of free energy released in this process is made available to the cell for such energy-requiring processes as the resynthesis of glycogen from lactic acid during the aerobic recovery of muscle following contraction.

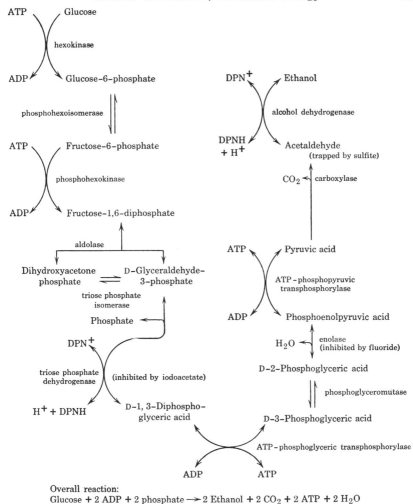

Overall reaction:
Glucose + 2 ADP + 2 phosphate → 2 Ethanol + 2 CO_2 + 2 ATP + 2 H_2O

Pathway of Anaerobic Breakdown of Glucose to Ethanol and Carbon Dioxide in Yeast

Even before the problem could be formulated in these terms, however, there had been speculation about the physiological connection between the anaerobic breakdown of glucose and the intracellular respiratory utilization of oxygen. The principal focus of discussion was Pasteur's discovery that much less glucose is metabolized by yeast in the presence of atmospheric oxygen than under anaerobic conditions. He had contented himself with a description of the phenomenon, with relatively little speculation about its mechanism.

Fermentation by yeast appears to be essentially linked to the property this little cellular plant possesses to respire, in some manner, with oxygen bound in sugar . . . . If one gives it a quantity of free oxygen sufficient for its requirements for life, nutrition, respiratory combustions, . . . it ceases to be a ferment . . . . On the other hand, if one suppresses all influence of air on the yeast, makes it develop in a sugar medium devoid of free oxygen gas, it multiplies there just as if air were present, although less actively, and it is then that its character as a ferment is most pronounced (Pasteur, 1876c, pp. 251–252).

One of the few hypotheses he offered for the effect of free oxygen was that it "gives to the yeast a great vital activity" (*ibid.*, p. 246) whereby it can assimilate nutrients and grow as rapidly as ordinary aerobic organisms.

Some of Pasteur's contemporaries went further, however. At that time, much attention was given to the views of Pflüger (1875) regarding the central role of "living proteins" as energy-rich cell constituents that react directly with molecular oxygen to cause the combustion of metabolites. This idea formed the core of hypotheses in which it was assumed that the fermentative breakdown of glucose is a normal "intramolecular respiration" occurring both in aerobic and in anaerobic cells, and which produces, in addition to CO_2, products (*e.g.*, alcohol) which then are burned by free oxygen or used for the synthesis of cellular material. For example, Wilhelm Pfeffer stated:

When we denote the intramolecular respiration as the cause of the oxygen respiration . . . . it is only intended to say that, with or without free oxygen in the living organism, there persist in the cell certain molecular rearrangements which are the initial and indispensable cause of the respiratory process (Pfeffer, 1881, Vol. I, p. 370).

Other general hypotheses of this kind were offered, among them one by Julius Wortmann (1880).

The view that there exists a close relation between the chemical processes in fermentation and respiration was later expressed by Otto Meyerhof when he discovered that the Harden-Young co-ferment present in animal tissues also acts as a respiratory catalyst; he concluded that

. . . the coenzyme of fermentation is at least partly identical with the respiratory substance of killed yeast and of muscle tissue. This result supports the already rather old hypothesis, for which there has been hitherto little evidence, that the initial phases of respiration and fermentation are closely related (Meyerhof, 1918, p. 174).

Subsequently, after he had explained the chemical changes during aerobic recovery following muscular contraction in terms of a coupling between the oxidation of about 20 percent of the lactic acid and the resynthesis of the rest to glycogen, Meyerhof stated that

> This connection between carbohydrate breakdown and oxidation is a general phenomenon of vital metabolism, and in the final analysis underlies the inhibition of the cleavages and fermentative processes by oxygen, assumed by Pasteur (Meyerhof, 1926b, p. 257).

After the discovery of ATP, Vladimir Engelhardt (1932) generalized Meyerhof's interpretation of the effect of oxygen in the lactic acid-glycogen conversion to all intracellular cyclic processes; he considered that the cleavage of the pyrophosphate is reversed by respiration, and that respiration is stimulated by ATP.

During the 1920s a new approach to the mechanism of the Pasteur effect was introduced by Warburg in connection with his studies on the metabolism of tumors and embryonic tissues, which are characterized by appreciable lactic acid formation from glucose, even in the presence of oxygen (aerobic glycolysis). From the fact that certain reagents (*e.g.*, ethyl isocyanide) which inhibited heavy-metal catalyses in the charcoal model systems also abolished the inhibition of glycolysis by free oxygen, Warburg concluded that "Respiration and fermentation are therefore linked by a chemical reaction which I term the 'Pasteur reaction' after its discoverer" (Warburg, 1926, p. 241).

For about 20 years thereafter, several investigators sought to identify the iron-containing component presumably associated with this "Pasteur reaction." For example, Lipmann (1933) reported that treatment of muscle extracts with chemical oxidizing agents inhibits glycolysis, and suggested that the Pasteur effect might be a consequence of the oxidation of the "glycolytic ferment" by an electron carrier such as one of the cytochromes. Later, Kurt Stern and Joseph Melnick (1941) reported the photochemical action spectrum of the carbon monoxide derivative of the "Pasteur enzyme" in retina and yeast. This approach, however, did not turn out to be fruitful since subsequent work showed the Pasteur effect to be a more complex physiological phenomenon than was surmised in 1940.

The Citric Acid Cycle. During the 1930s more conclusive information about the link between the anaerobic breakdown of carbohydrate and the respiratory utilization of oxygen was provided by the search for carbon compounds that might be intermediates in the intracellular oxidation of lactate or pyruvate to CO_2 and H_2O. Particular attention was initially focused on the four-carbon dicarboxylic acids succinic acid, fumaric acid, malic acid, and oxaloacetic acid. On the basis of the work of Battelli and Stern, Einbeck, Wieland, and Thunberg, to which we referred earlier (see p. 319), it had been suggested during the 1920s that the sequence:

$$\text{Succinate} \xrightarrow{-2\,H} \text{Fumarate} \xrightarrow{+H_2O} \text{Malate} \xrightarrow{-2\,H}$$
$$\text{Oxaloacetate} \xrightarrow{-CO_2} \text{Pyruvate} \xrightarrow{+O\ -CO_2} \text{Acetate}$$

might be a pathway for the oxidation of acetate (then considered to be derived from the metabolic breakdown of fatty acids and of glucose). Thunberg (1920) proposed that if there were a mechanism for linking two molecules of acetate to form succinate:

$$\text{HOOC—CH}_3 + \text{CH}_3\text{—COOH} \xrightarrow{-2\,H} \text{HOOC—CH}_2\text{—CH}_2\text{—COOH}$$

a cyclic process would be available for the oxidation of one molecule of acetate to CO_2 and H_2O. Although the assumption of the steps from succinate to acetate was supported by the presence, in animal tissues, of enzymic activities corresponding to each of the postulated conversions, the reductive condensation of acetate to succinate was entirely hypothetical, and no evidence was found to support the view that it occurred in biological systems.

Similarly, when in 1930 Toenniessen and Brinkmann reported that the perfusion of muscle with pyruvic acid gave rise to the formation of succinic acid, they invoked the operation of a sequence of enzyme-catalyzed reactions from succinate to pyruvate (as shown above), and completed the cycle by postulating the reductive condensation of two molecules of pyruvate to yield diketoadipate ($\text{HOOC—CO—CH}_2\text{—CH}_2\text{—CO—COOH}$); this hypothetical intermediate was supposed to undergo cleavage to succinate and formate, the latter giving rise to CO_2 and H_2O. Within a few years, however, Kenneth Elliott and Franz Wille had provided evidence against the role of diketoadipate and formate in the oxidation of pyruvate by animal tissues.

Of greater significance were the studies of Albert Szent-Györgyi and his associates during 1934–1937; they observed that the oxygen uptake by suspensions of minced pigeon-breast muscle (selected because of the rapid respiration of this tissue) is stabilized at a high rate by the addition of small amounts of fumarate. Since the fumarate did not disappear, it was clearly acting as a catalyst. From the ratio of the CO_2 produced to the O_2 consumed (the "respiratory quotient") in the process catalyzed by fumarate, it was inferred that carbohydrate, or a product derived from carbohydrate ("triose"), was being subjected to oxidation. These results were confirmed and extended by Frederick Stare and Carl Baumann (1936), and it was also found that the addition of the other C_4-dicarboxylic acids (succinate, malate, oxaloacetate) produced an effect similar to that shown by fumarate. Furthermore, malonate (HOOC—CH_2—COOH), which was known to inhibit succinic dehydrogenase, did not abolish the catalytic action of fumarate. The succinate-fumarate reaction, catalyzed by this dehydrogenase, was believed to react directly with the cytochrome system, whereas the malate-oxaloacetate reaction had been found to require Euler's cozymase; Szent-Györgyi (1937, p. 25) therefore proposed that the cozymase-dependent dehydrogenation of metabolites is linked to oxygen by the sequence of reactions:

"Hydrogen donor" $\rightarrow$ Oxaloacetate $\rightarrow$ Malate $\rightarrow$
 Fumarate $\rightarrow$ Succinate $\rightarrow$ Cytochrome $\rightarrow$ Oxygen

In this scheme, the role of the C_4-acids was considered to be that of "hydrogen transport," not as intermediates in the oxidation of carbohydrates. As later noted by Hans Krebs, a serious deficiency of the scheme was that it did not explain satisfactorily the inhibitory effect of malonate on cell respiration.

In the memorable paper by Krebs and Johnson (1937), there was offered a different theory which marked a new stage in the study of the link between glycolysis and respiration. Their key experimental observations were that the six-carbon tricarboxylic acid citrate exerts a catalytic effect on the respiration of minced pigeon-breast muscle, that citrate is successively converted to α-ketoglutarate and succinate, and that oxaloacetate is converted into citrate by the addition of two carbon atoms from an unidentified source, provisionally denoted as "triose." Citric acid had been known to be a major constituent of plants since its isolation by Scheele (in 1785) and Proust (in 1801); its study by Liebig during the 1830s contributed to the formulation of his important theory of polybasic acids. Later, citric acid was identified as a product of the fermentation of sugar by some molds, and also was found in small amounts in animal tissues. In considering how citrate might be formed in biological systems, Franz Knoop and Carl Martius (1936) drew attention to the chemically plausible condensation of oxaloacetate and acetate, but noted that attempts to effect this reaction by chemical means had been unsuccessful. They showed, however, that if one takes pyruvate instead of acetate and subjects the reaction mixture to oxidation with hydrogen peroxide in alkaline solution, citrate is thereby produced; we mentioned earlier (see p. 299) that, around 1905, this kind of chemical treatment was used by Dakin and Neuberg in experiments designed to mimic biological oxidations. Shortly after their chemical synthesis of citric acid, Martius and Knoop demonstrated that Thunberg's "citric dehydrogenase" (from liver) converted citrate into α-ketoglutarate anaerobically, and they offered chemical evidence in favor of the following sequence of reactions in this conversion:

Citrate → *cis*-Aconitate → Isocitrate →

Oxalosuccinate → α-Ketoglutarate

The structural formulas of these compounds are given in the scheme on p. 384. It may be added that isocitric acid had been synthesized by Rudolf Fittig in 1889, and that oxalosuccinic acid had been studied by Emile Blaise and Henry Gault in 1908; although *trans*-aconitic acid was described by Ludwig Claisen in 1891, the less stable *cis* compound was prepared only in 1928 (by Malachowski and Maslowski).

This proposed pathway of the oxidation of citrate found ready confirmation in the work of Krebs and Johnson, who showed that isocitrate and *cis*-aconitate are oxidized by muscle preparations as rapidly as citrate. With the addition of the C_4-acids of the earlier schemes, and the key step of the conversion of oxaloacetate into citrate, they accordingly proposed a "citric acid cycle":

Citrate → Isocitrate → Oxalosuccinate → α-Ketoglutarate →

Succinate → Fumarate → Malate → Oxaloacetate → Citrate

In the operation of this cycle, two carbon atoms derived from "triose" are introduced in the conversion of oxaloacetate to citrate, and one carbon atom appears as CO_2 in each of the two steps from oxalosuccinate to succinate.

The decisive experiment supporting the theory was the demonstration of the aerobic formation of succinate from fumarate under conditions where the direct interconversion of the two C_4-acids was blocked by malonate. This was brought out clearly by Krebs and Eggleston in 1940; their paper began as follows:

> We have tried to elucidate the oxidation of pyruvate in pigeon breast muscle by studying in detail the factors governing the oxidation. We find that fumarate acts as a catalyst in the oxidation of pyruvate and that malonate breaks the catalysis when one molecule of pyruvate has reacted with one molecule of fumarate. The following reaction takes place in the presence of malonate: (1) Fumarate + pyruvate + 2 O_2 = succinate + 3 CO_2 + H_2O. In the absence of malonate this action is followed by the oxidation of succinate: (2) Succinate +

$\frac{1}{2}$ O_2 = fumarate + H_2O. The net effect of 1 and 2 is the complete oxidation of pyruvate: Pyruvate + $2\frac{1}{2}$ O_2 = 3 CO_2 + 2 H_2O. In reaction 1 the succinate does not arise from fumarate by anaerobic reduction. It is formed from fumarate and pyruvate by a series of oxidative processes of the type formulated in the theory of the "citric acid cycle" (Krebs and Johnson, 1937). This theory is supported by the fact that a change in experimental conditions directs reaction 1 in such a way as to yield citrate (about 15%) or α-ketoglutarate (about 50%) instead of succinate and CO_2. Up to the present the citric acid cycle is the only theory accounting for the experimental observations in pigeon breast muscle (Krebs and Eggleston, 1940, p. 442).

Although the citric acid cycle (it was later also termed the tricarboxylic acid cycle, or the Krebs cycle) was originally proposed for pigeon-breast muscle, work in numerous laboratories soon showed it to be operative in other animal tissues, in plants, and in aerobic microorganisms. Since its enunciation in 1937, the cycle has undergone modification; for a time (Krebs, 1943, p. 235) the immediate product of the condensation of oxaloacetate and the C_2-unit derived from pyruvate was thought to be *cis*-aconitate, with citrate as a by-product of the cycle, but subsequent work restored citrate to its original place. More importantly, after World War II the cycle was elaborated considerably, as details of the individual steps were clarified through the isolation of the separate enzymes that catalyze the successive steps, through the identification of the cofactors involved in the action of these enzymic proteins, and by the use of radioactive isotopes to trace the path of labeled carbon atoms through the cycle. The scheme shown on p. 384 summarizes the status of the cycle about twenty years after it was first proposed; the present-day picture is essentially the same, with slight revisions that may be found in current biochemical textbooks. It is beyond the scope of this chapter to recount all the advances made after 1940, but the most important of them requires mention: the elucidation of the process whereby a C_2-unit is derived from pyruvate and is condensed with oxaloacetate to form citrate.

The decisive event that made this advance possible was Fritz Lipmann's discovery, after World War II, of a cofactor that promotes acetyl transfer in extracts of animal tissues. It was recognized by 1945 that the oxidation of pyruvate leads to the intermediate formation of a reactive two-carbon fragment ("active acetate"), and that thiamine pyrophosphate is a cofactor in the oxidative decarboxylation, but the chemical nature of this intermediate was unknown. Acetyl phosphate had not lived up to earlier expectations, and Lipmann turned to the study of a model system, the acetylation in liver extracts of the amino group of the drug sulfanilamide,

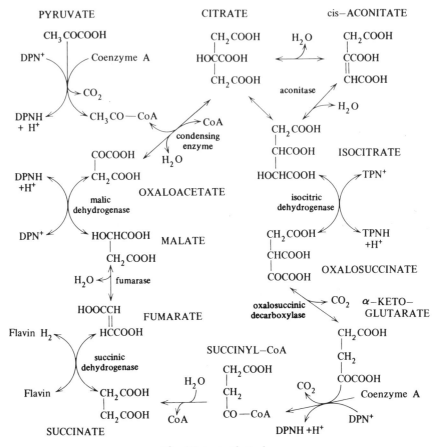

The Citric Acid Cycle

with acetate as the source of potential "active acetate." In 1938 James Klein and Jerome Harris had studied this process with liver slices, and had found it to be coupled to respiration; Lipmann (1945) showed that pigeon-liver extracts, in the presence of added ATP, also effected the acetylation. During the course of this work, the requirement for a heat-stable dialyzable cofactor was recognized, and a similar cofactor was independently found by David Nachmansohn to be required for the ATP-dependent acetylation of choline in brain tissue. Lipmann named the cofactor coenzyme A (A for acetylation); because of the numerous instances in which coenzymes had already turned out to be related to vitamins, partially purified samples of coenzyme A (usually abbreviated CoA) were tested for their vitamin activity. The hydrolysate of coenzyme A was found to contain a vitamin discovered by Roger Williams in

1933, and named pantothenic acid (Greek, *pantothen,* from every side) because of its wide distribution. By 1940 its chemical structure had been established in several laboratories and, before its identification as a constituent of coenzyme A, Williams wrote:

> The presumption, in view of the known functions of thiamin, nicotinamide and riboflavin, is that pantothenic acid fits into some enzyme system (or systems) which is essential for metabolism. What this enzyme system is or what they are is not known. There are some facts which suggest that pantothenic acid may be concerned with carbohydrate metabolism, but this is not certain (Williams, 1943, p. 267).

The structure of coenzyme A, shown in the accompanying formula, was deduced in 1952 from the chemical studies performed by Lipmann and his associates, as well as from work in the laboratories of James Baddiley and of Esmond Snell; its synthesis was effected in 1961 by John Moffatt

Coenzyme A

and Gobind Khorana. A major achievement was Feodor Lynen's isolation, in 1951, of an acetylated coenzyme A derivative from yeast, and his demonstration that the acetyl group is attached to the sulfur atom of the cofactor. This acetyl-CoA was shown by Stern, Ochoa, and Lynen (1952) to be the substance that reacts with oxaloacetate to form citrate in the presence of the "condensing enzyme" crystallized in Severo Ochoa's laboratory. As for the formation of acetyl-CoA from pyruvate, the process was found to require (in addition to thiamine pyrophosphate and coenzyme A) DPN and still another cofactor (lipoic acid) discovered in 1953. In all, the "pyruvic oxidase" of animal tissues turned out to involve four chemical reactions, the sum of which is:

$$CH_3COCOOH + CoA + DPN \rightarrow CH_3CO—CoA + CO_2 + \text{reduced DPN}$$

This process is effected by an integrated "multienzyme" system, and a similar system has been found to be operative in the oxidative decarboxylation of α-ketoglutarate to succinyl-CoA. We shall return to some of the implications of this finding at the end of this chapter.

Oxidative Phosphorylation. The formulation of the citric acid cycle, and its general acceptance after 1940 as a general mechanism for the aerobic oxidation of pyruvate to CO_2 and H_2O, provided the long-sought link between anaerobic glycolysis or fermentation and the respiratory utilization of oxygen. In the complete oxidation of a molecule of pyruvate, five atoms of oxygen are consumed, corresponding to the transfer of five pairs of hydrogen atoms (or of five electron pairs); these five oxidations are associated with about 85 percent of the free-energy change in the aerobic oxidation of glucose. In 1940, however, the available experimental knowledge did not permit the formulation of coherent schemes for the transfer of electrons, to oxygen, from any of the five metabolites undergoing dehydrogenation in the citric acid cycle. The manner in which the electron transfer was coupled to intracellular endergonic reactions was even more uncertain. The great achievements of the 1930s had revealed the existence of the pyridine nucleotides and the flavins as links in a respiratory chain from metabolites such as lactate or malate to the cytochromes and cytochrome oxidase, but the synthesis of this knowledge into a conceptual model of the chain in any defined biological system (*e.g.*, muscle) proved to be elusive. The principal difficulties appeared to be centered about the role of the flavins, since none of the flavoproteins isolated in Warburg's laboratory reacted sufficiently rapidly with both reduced DPN and cytochrome c to justify its inclusion in a physiologically significant respiratory chain. Although Hugo Theorell (1936) had concluded that the "old" yellow enzyme seemed to be the physiological link between Warburg's co-ferment (TPN) and cytochrome c, later work (see Hogness, 1942) made this doubtful. Furthermore, parallel studies in Hans von Euler's laboratory and by David Green indicated that most of the known dehydrogenases react specifically with DPN (not TPN), with the significant exception of the highly active succinic dehydrogenase, which did not require either pyridine nucleotide. For a time it seemed that a new FAD-containing flavoprotein (coenzyme factor, diaphorase) isolated from pig-heart muscle was the physiological agent for the oxidation of reduced DPN in animal tissues (Corran, Green, and Straub, 1939), but its reaction with cytochrome c was too slow. An additional source of uncertainty was introduced by Warburg and Christian (1938, p. 372), who raised the possibility that the flavin mononucleotide they had discovered in 1932 was a degradation product derived from FAD, and that all the naturally occurring flavoproteins were in fact FAD-proteins. Many of the uncertainties and frustrations that attended the efforts, around 1940, to fit catalytic flavoproteins into a consistent scheme of respiratory electron transfer were overcome after 1950 through the identification of additional electron-carrier systems (quinones, non-heme-iron proteins) that mediate oxidation-reduction reactions between DPN and cytochrome c, but active research on this problem has continued up to the present. In 1940, therefore, it could only be surmised, on thermodynamic grounds, that the respiratory chain includes an arrangement of the known pyridine nucleotides, flavins, and cytochromes in the order of their relative oxidation-reduction potentials. The estimated values (in volts, at pH 7 and 37°) were: DPN (or TPN) -0.3, flavoprotein -0.1, cytochrome b -0.04, cytochrome c $+0.26$, cytochrome a $+0.29$; the values assigned to the hydrogen and oxygen electrodes were -0.42 and $+0.81$, respectively. On the basis of such relationships, it was concluded that

> The energy liberated when substrates undergo air oxidation is not liberated in one large burst, as was once thought, but is released in stepwise fashion. At least six separate steps appear to be involved. The process is not unlike that of locks in a canal. As each lock is passed in the ascent from a lower to a higher level a certain amount of energy is expended. Similarly, the total energy resulting from the oxidation of foodstuffs is released in small units or parcels, step by step. The amount of free energy released at each step is proportional to the difference in potential of the systems comprising the several steps (Ball, 1942, p. 22).

The kind of sequence of electron transport postulated here is summarized in the accompanying scheme. During the 1950s Britton Chance developed rapid and sensitive spectrophotometric methods for the measurement of changes in the oxidation-reduction state of intracellular electron carriers, and provided direct experimental evidence in support of this sequence of reactions.

Although the limited knowledge available in 1940 only permitted thermodynamic calculations to be made of the "parcels" of energy that might be released in the individual steps of the respiratory chain, it was already clear that the coupling of respiration with phosphorylation represented an important intracellular mechanism for the utilization of these units of energy to drive endergonic reactions. After the discovery of ATP, and the work of Lundsgaard on the phosphorylation of creatine during oxidative recovery in muscle, several investigators recognized the general importance of phosphorylations in biological transformations of energy. Among the first to study the coupling of respiration with phosphorylation in animal tissues was Engelhardt (1932); his work was followed by that of Runnström and Michaelis (1935) and of Kalckar (1937). Of special importance were the experiments of Belitser and Tsibakova (1939; English translation in Kalckar, 1969), who reported that the phosphorylation of creatine in heart muscle was coupled to the oxidation of any of a variety of metabolites (citrate, α-ketoglutarate, succinate, fumarate, malate, lactate, pyruvate), and that approximately two molecules of creatine phosphate were synthesized per atom of oxygen consumed. Their findings were soon confirmed and extended in reports from several laboratories; in particular, Severo Ochoa (1941, 1943) called attention to the fact that the apparent "P/O ratio" of 2 was a minimal figure, because the molecular species synthesized in the oxidative phosphorylation is ATP, which can undergo side reactions. For this reason, he concluded that the true P/O ratio for the oxidation of a metabolite such as pyruvate is probably closer to 3. Such results were used around 1950 as a basis for thermodynamic calculations of the efficiency of oxidative phosphorylation; the free-energy change in the complete oxidation of pyruvate was taken to be about -270 kcal per mole, and the energy required for the synthesis of ATP from ADP and P_i was assumed to be about $+14$ kcal per mole (this value later decreased considerably). Since the uptake of the five atoms of oxygen required for the oxidation of a molecule of pyruvate led to the generation of about 15 molecules of ATP, corresponding to the utilization of about $+215$ kcal per mole, the efficiency of the process was estimated to be about 80 percent. It should be noted that such calculations involved assumptions of uncertain validity (see Ogston and Smithies, 1948). The principal difficulty lay in the fact that the chemical mechanism of oxidative phosphorylation was unknown. Whereas the generation of ATP from ADP and P_i in the reactions catalyzed by glyceraldehyde-3-phosphate dehydrogenase and the transphosphorylase discovered by Bücher could be explained in terms of the intermediate formation of the "energy-rich" acyl phosphate 1,3-diphosphoglyceric acid, no comparable mechanism could be invoked for the coupling of electron transfer by the respiratory chain to the phosphorylation of ADP. This chemical problem still awaits solution; significant progress has been made since 1950, however, in defining it more precisely. This advance was made possible, in large part, by contributions based on new cytological knowledge.

We recalled earlier (p. 195) some aspects of the remarkable development of cytology during the latter half of the nineteenth century, and the significance of the new microscopic observations for the understanding of the role of the cell nucleus in heredity. During this period there were recurrent reports about supposed structural elements in the cytoplasm; among them were the *microzymas* of Antoine Béchamp (1883; for a partisan account of his dispute with Pasteur, see Hume, 1923) and the *Elementarorganismen* of Richard Altmann (1890), which were considered to be fundamental units of life and homologous with bacteria. Although much of Altmann's cytological work was later recognized to be valuable, many of his contemporaries considered the cytoplasmic structures to be artifacts produced by the reagents (*e.g.*, osmic acid) used as fixatives in preparing tissue slices for staining and microscopic observation. For example, William Bate Hardy wrote:

> It is notorious that the various fixing reagents are coagulants of organic colloids, and that they produce precipitates which have a certain figure or structure. It can also readily be shown . . . that the figure varies, other things being equal, according to the reagent used. It is therefore cause for suspicion when one finds that particular structures which are indubitably present in preparations are only found in cells fixed with certain reagents, used either alone, or in particular formulae. Altmann demonstrates his granules by the aid of an intensely acid and oxidizing mixture (Hardy, 1899, pp. 160–161).

Among the cytoplasmic granules of Altmann were rodlike structures to which Carl Benda gave the name "mitochondria" in 1898. Two years later Leonor Michaelis (1900), while working in Paul Ehrlich's laboratory, described the "vital" staining of these structures with the dye Janus green; this technique, which avoided the use of fixatives, had been introduced by Ehrlich (in 1886) when he described the staining of nerves with methylene blue.

In giving reality to the mitochondria, and in providing a specific dye for their microscopic observation, Michaelis laid the groundwork for an extensive series of cytological studies during the succeeding decades. With the rise of cytochemistry after 1930, efforts were made to separate mitochondria from suspensions of mechanically disintegrated tissues ("homogenates"), and significant progress was made in this direction by Robert Bensley, in 1932–1934. Albert Claude (1946), at the Rockefeller Institute, then fractionated a suspension derived from liver cells by differential centrifugation; this method was greatly improved by George Hogeboom *et al.* (1949), also at the Rockefeller Institute, in showing that intact mitochondria could be readily isolated in this manner if the tissue was dispersed in a concentrated solution of sucrose.

Furthermore, by 1948 it was clear from the work of Hogeboom and his associates that the cytochrome oxidase and the succinic dehydrogenase of liver cells are almost exclusively localized in mitochondria; shortly afterward, Walter Schneider and Van Potter showed that mitochondria oxidize some of the intermediates of the citric acid cycle. There then followed the important report of Eugene Kennedy and Albert Lehninger (1949) that rat liver mitochondria contain the complete enzymic apparatus of the citric acid cycle, whereas the glycolytic enzymes are largely present in the cell sap (the "supernatant" fraction obtained upon high-speed centrifugation). Of special significance was their finding that isolated mitochondria are able to effect the coupling of the esterification of phosphate to the oxidation of metabolites. In discussing their results, they noted:

> The striking fact is that all the individual enzymes concerned in these complex systems should be found in one species of morphological element. These findings in some measure justify the early views of Altmann (1890) that these bodies are fundamental biological units and possess a certain degree of autonomy and certainly, together with the considerable work already done on their enzymatic and chemical composition by the Rockefeller school, Schneider, and others, provide considerable basis for the apt designation "intracellular power plants" conferred on the mitochondria by Claude.
>
> Although the mitochondria appear to be the major site of these activities, it would appear from our examination *in vitro* that these bodies are not completely autonomous with respect to their respiratory behavior, since they must be supplemented with certain cofactors such as adenosine triphosphate and Mg^{++}. It appears likely that in the cell there is a rapid interchange of these factors, substrates, and inorganic phosphate between the cytoplasm and the mitochondria. It also would appear that these bodies are dependent on the cytoplasm for certain

preparatory metabolic activities such as glycolysis, since, as our data show, they are almost completely lacking in glycolytic activity (Kennedy and Lehninger, 1949, p. 970).

It should be added that, although Altmann had suggested that his granules were responsible for intracellular oxidations, most of his contemporaries placed the site of respiratory activity in the nucleus. In retrospect, the first definite indications that cell respiration might be associated with cytoplasmic granules can be discerned in the work of Battelli and Stern (1912) and, more explicitly, of Warburg (1913). These results were interpreted, however, in terms of the prevailing interest in colloidal adsorption phenomena and, as we have seen, could not be fruitful until much more was learned about the chemical details of cell respiration. Later, as new knowledge was acquired, the association of respiratory activity with cytoplasmic granules was reiterated by several investigators, among them Keilin (1929) and Szent-Györgyi (1937). After the discovery of the citric acid cycle, Green (1951) described a particulate preparation ("cyclophorase") that effected the complete oxidation of pyruvate.

The clear demonstration that mitochondria represent the site of oxidative phosphorylation in animal cells was followed by the important experiments of Lehninger (1951), showing that the oxidation of reduced DPN by mitochondria is coupled to the synthesis of ATP with an estimated P/O ratio of 3. This indicated that the oxidative phosphorylation is associated with the respiratory chain of electron transport to oxygen. By 1951, therefore, it was evident that the pathway of oxidative glucose breakdown in animal cells, and the utilization of the resultant energy for the synthesis of ATP, begins with glycolysis in the cytoplasmic sap, and is completed by the mitochondrial citric acid cycle and respiratory chain, the latter being responsible for most of the coupled phosphorylation. Furthermore, the discovery that, within the liver cell, the glycolytic enzymes are separated from the components of the citric acid cycle and the respiratory chain showed that the Pasteur effect (p. 378) must depend on the transfer of diffusible substances from the mitochondria into the cytoplasm; although much still remains to be learned about the mechanism of the Pasteur effect, significant advances were made during the 1960s in attempts to define these substances as well as the glycolytic enzymes on which they act.

Since about 1950 there has been considerable progress in identifying the individual steps in the electron-transfer chain that are linked to the generation of ATP, but many uncertainties still remain (see Lardy and Ferguson, 1969). The experimental difficulties encountered in the chemical dissection of the mitochondrial apparatus responsible for electron transport and the coupled synthesis of ATP have been very great, in part because many of the components of the system are tightly associated with the lipid-containing constituents of the mitochondria, and the usual methods for the extraction of enzymes with aqueous solvents have proved to be unsuitable. Aside from this technical difficulty, however, the study of mitochondria and of their chemical apparatus has revived interest in the concept of the specific arrangement of enzyme proteins within subcellular structures.

We mentioned earlier (p. 219) the impact of the invention of the electron microscope on the study of DNA and of bacteriophages. Around 1950, after the development of the ultra-microtome, capable of cutting tissue sections of extreme thinness, electron microscopy became a central technique of modern cytology. One of the early fruits of this advance was the important work of George Palade (1952) on the fine structure of mitochondria, which were seen to have a highly organized anatomy characterized by a system of internal ridges (cristae) and to be bounded by a well-defined external membrane. In discussing the relation of this intramitochondrial structure to the enzymes associated with it, Palade noted:

> It may be assumed that they are arranged in the proper order in linear series or chains—a disposition comparable in design and efficiency to an industrial assembly line. Such enzymatic chains have to be built at least partially in the solid framework of the mitochondria because some of the component enzymes, namely succinic acid dehydrogenase (succinoxydase) and cytochrome c oxydase, are known to be insoluble and structure-bound (Palade, 1952, p. 439).

Thus electron microscopy offered the promise of giving reality to the long-held view of cytologists:

> It is evident that a mere random mixture of cell constituents, in the same proportions and concentrations as in protoplasm, would not give a system having the properties of a living cell. The constituents must have a definite arrangement, and must be present in a definite physical state; only under these conditions is it possible to conceive of any kind of ordered interaction as that underlying the life of the cell (Lillie, 1924, p. 171).

Since about 1950 the refinement of electron microscopy, and its manifold applications to the study of the intracellular organization of enzyme systems, have narrowed the gap between studies on the properties of individual purified enzymes and on chemical events in living cells.

Of special importance to our story is the fact that the high magnification of the electron microscope has permitted investigators to "see" the structure of aggregates of protein molecules. It was mentioned on p. 137 that many proteins were found to consist of a discrete number of subunits, each of which represents a separate coiled peptide chain in characteristic association with the other subunits of the protein. Moreover, in several enzyme systems, such as those that effect the oxidative conversion of pyruvate to acetyl-CoA or of α-ketoglutarate to succinyl-CoA, three or more different enzyme proteins have been shown to be associated by noncovalent linkages to form assemblies whose form can be discerned by electron microscopy, and whose catalytic function depends on the integrity of the specific multienzyme structure (for a recent review, see Ginsburg and Stadtman, 1970).

It is beyond the scope of our discussion to do more than to indicate the recent emphasis on the intracellular organization of enzymes into multienzyme systems. We should note, however, an important difference between this renewed interest in the "supramolecular" structure of enzyme systems and the earlier conviction of cytologists that such organization must exist. The difference lies in the availability of recently acquired chemical knowledge about individual enzymes as catalytic proteins. In concluding our survey of the development of ideas about cell respiration, we may find in this latest development further evidence of the limits placed upon biological thought by the lack of detailed chemical knowledge about the chemical structure and specific interactions of the molecular constituents of living cells. Throughout this history, proteins and enzymes have occupied a central place in the formulation of theories of biological oxidation, and an investigator's ideas about their chemical nature was closely related to his biological hypotheses. We have seen that for many of the physiologists of the period 1870–1910 (*e.g.*, Pflüger and Pfeffer), the proteins of living cells were labile energy-rich structures different from the ordinary albumins and globulins isolated from plant and animal tissues, and some biochemists (*e.g.*, Hammarsten) regarded such "simple" proteins either as nutrient materials or as artifacts of preparation, derived from true protoplasmic "proteids." At that time, only some chemists (*e.g.*, Traube) believed that biological oxidations might be caused by proteins essentially similar to the "unorganized ferments" such as pepsin, and the prevailing view among physiologists was that cell respiration involves the activity of substances fundamentally different from the hydrolytic enzymes. With the rise of colloid chemistry, after 1900, primary emphasis was placed on catalytic phenomena arising from the adsorption of chemical substances on the surfaces of colloidal particles; for many years, cell physiologists thought of proteins as colloids, rather than as chemical substances having a defined molecular structure. The uncertainty about the nature of enzymes during the 1920s was, as we have seen, closely related to the uncertain state of protein chemistry at that time. It is perhaps not surprising, therefore, that some biochemists interested in fermentation and oxidation should have questioned the existence of the separate enzymes to which distinctive names had been applied. For example, in a discussion of the role of dehydrogenation in microbial metabolism, Kluyver wrote that

> . . . it is no longer necessary to have recourse to the assumption of a large number of separate enzymes to explain the partial reactions of the dissimilation process . . . . [T]he changes which are ascribed to separate enzymes such as catalase, reductase, zymase, lactozymase, alcoholoxidase, carboxylase, carboligase, glyoxylase, aldehydomutase, Schardinger's enzyme, etc., are actually only manifestations of a definite degree of affinity of the protoplasm for hydrogen. The question as to how far one will think of the affinity of protoplasm itself for hydrogen or of the action of a single hydrogen transferring enzyme, has no significance here (Kluyver and Donker, 1925, p. 618).

A similar point of view was expressed by Kostychev (1926) in an article entitled "On the non-existence of several enzymes." Moreover, after Quastel (1926) had found that 56 substances reduce methylene blue in the presence of suspensions of *Escherichia coli,* he questioned the existence of such a large number of specific enzymes that catalyze the same type of reaction, and he suggested that this multiple dehydrogenase activity may be regarded as a function of an intracellular surface.

Nor is it surprising that during the mid-1920s the doubts expressed by Kluyver about the involvement, in respiration and fermentation, of individual enzymes of restricted specificity should have led him to question whether such enzymes could be studied as chemical entities, and to emphasize the distinction between the so-called "hydrolases" (*e.g.,* the glycosidases studied by Fischer during the 1890s) and the so-called "desmolases." In particular, he accepted the definition of Carl Oppenheimer that the desmolases

> . . . are the true metabolic ferments; they promote the decisive processes whereby the cell uses the chemical energy of its nutrients or also its own constituents to convert it into other forms of energy, above all to heat and mechanical work. For these processes, the action of the

hydrolases is of course only the indispensable preparation (Oppenheimer, 1926, p. 1213).

This distinction among the enzymes on the basis of their presumed physiological role has strongly influenced biochemical thought, even after the work of the 1930s had shown that some of the "true metabolic ferments" could be isolated and studied chemically in the manner that Sumner and Northrop had studied the hydrolases urease and pepsin.

Despite the acceptance during the 1930s of the idea that enzymes are individual proteins, and that the understanding of their biological function required their isolation in as pure a state as possible, there continued to be recurrent statements of the view that the enzymic apparatus within cells represents an organized assembly. Thus Euler (1938, p. 190) urged a search for cell constituents that had multiple enzymic functions, and Stern (1939, p. 318) proposed that intracellular electron transport might involve the action of a very large protein particle (with a weight in the millions) in which "the active groups of the various component catalysts are arranged . . . in an orderly fashion." Furthermore, after his discovery of the ATP-ase activity of myosin, Engelhardt suggested that

. . . it is highly improbable that it will be possible to obtain proteins of a completely inert nature devoid of catalytic properties from the living and active plasma of the cell. We have the right to assume that *the living protein of the plasma is primarily a catalytically active protein, a protein with the properties of an enzyme* (Engelhardt, 1942, p. 37).

It should be recalled that these views were expressed before the post-World War II studies on the chemical structure of proteins. After the work of Sanger on insulin, of Stein and Moore on ribonuclease, and of Kendrew on myoglobin, it was no longer possible to deny that proteins represent molecules whose chemical structure could be defined, by means of appropriate techniques, with the same assurance as in the case of organic substances of lower molecular weight. The fact that proteins are macromolecules introduces the possibility of cooperative chemical interactions not evident with the smaller molecules of traditional organic chemistry, and such cooperative interactions are now recognized to be of crucial biological importance. According to this view, if enzymes such as those involved in the respiratory chain and in oxidative phosphorylation form part of an integrated multienzyme assembly, it is the task of chemistry to determine the nature of the specific chemical groups in the interacting proteins and in the other substances (such as the lipids) that confer upon this assembly the characteristic structure associated with its intracellular function. This difficult problem, under study in many laboratories, today represents the principal chemical limit to biological thought about the manner in which individual enzymes interact to constitute the physiological systems of living cells. No doubt, in the decades to come, there will be further interplay between biology and chemistry in the efforts to solve this problem.

Pathways of Biochemical Change

The impact of new chemical knowledge on the study of old biological problems is perhaps most clearly evident in relation to the nature of the changes undergone by food materials when they are used by an animal to make its own tissues. This problem of biological assimilation, repeatedly posed in the Hippocratic writings, and by Aristotle and Galen, remained in the forefront of medical speculation through the centuries. With the rise of iatrochemistry, animal nutrition was explained in terms of the fermentation hypotheses of van Helmont, Sylvius, and Willis, and the "coctions" of ancient physiology, based on the analogy to cookery, were replaced by ferments analogous to the one operating in the wine vat. By 1800 the fermentation theory had declined in prestige, and assimilation was considered to be a process in which the triturated and partially dissolved food ("chyme") is transformed in the stomach and intestinal canal, under the influence of air and heat, into an undifferentiated lymphatic fluid ("chyle") that serves as the source of the blood. The blood was thought to be converted into material similar to that characteristic of each of the individual tissues; thus it was believed that the "fibrin" of muscle is formed by the solidification of the dissolved "fibrin" of the blood. The central process was considered to be that of digestion; in particular, the physiological experiments of René Antoine Réaumur (described in 1752) on the solvent power of the gastric juice, and their confirmation and extension by Lazzaro Spallanzani around 1780, focused attention on the role of the stomach in the assimilation of food materials. A contemporary report by Edward Stevens (in 1777; an English translation was appended to the English version of Spallanzani's book) stated that

> . . . it is not the effect of heat, trituration, putrefaction, or fermentation alone, but of a powerful solvent, secreted by the coats of the stomach,

which converts the aliments into a fluid, resembling the blood (Spallanzani, 1789, Vol. I, p. 389).

The emergence of the new chemistry, based on the work of Lavoisier and his contemporaries, elicited efforts to define this conversion of food into blood in terms of the elements found to be constituents of animal substances (carbon, hydrogen, oxygen, and nitrogen) and those found in plant materials (carbon, hydrogen, and oxygen).

Among the early attempts was that of Jean-Noel Hallé (1791), who sought to explain the process whereby an animal converts plant substances in its diet into animal substances. He followed Lavoisier in assigning to respiratory combustion a central role in the body economy, and combined this new approach with Claude Berthollet's identification of azote (nitrogen) as a characteristic constituent of animal substances. Hallé suggested that, under the influence of oxygen, food materials become "animalized" through the loss of carbon (as carbon dioxide) and the gain of nitrogen; he proposed that these chemical changes occur successively in the digestive organs, the lungs, and the skin, to which Séguin and Lavoisier had assigned major roles in the regulation of animal heat through nutrition, respiration, and transpiration (see p. 265). Several years later, Antoine François Fourcroy summarized the problem as follows:

. . . [T]he conversion of plant substances to animal substances, which consists simply of the fixation or addition of azote and the increase of hydrogen, must be considered the principal phenomenon of animalization, the only one which explains its mysteries; and when the mechanism of this addition of azote and of hydrogen is understood, most of the functions of animal economy that effect it or depend upon it will also be known (Fourcroy, 1806, p. 352).

By that time, Fourcroy and Vauquelin had shown that plants contain nitrogenous materials, and Fourcroy defined animalization more precisely:

It is not so much by the fixation of azote or by the addition of a new quantity of hydrogen or azote that this phenomenon occurs, but rather by the subtraction of the other elements [*principes*] which increase the proportion of the former . . . . It is believed that in releasing the excess hydrogen and carbon, respiration should necessarily increase the proportion of azote (*ibid.*, pp. 352–353).

As for the general process of assimilation in the animal body, Fourcroy considered it to represent

> . . . a complete transformation of the primitive alimentary substance into each particular organic substance; . . . this assimilation, begun in digestion, continued in respiration, almost achieved during the courses of the circulation, and entirely completed upon entry into each organ to be nourished, consists mainly in the loss of carbon and hydrogen, in an increase of azote, and a sort of transmutation hitherto termed animalization (Fourcroy, 1801, Vol. V, pp. 660–661).

The idea of the "transformation of the primitive alimentary substance" into characteristic nitrogenous constituents of animal tissues reappeared in the writings of leading medical investigators throughout the first half of the nineteenth century. Its limitations became increasingly apparent, however, after the new methods for the elemental analysis of organic substances had been applied to the "immediate principles" obtained from animal organisms (Holmes, 1963). Although in 1810 it could be stated that "A little oxygen or nitrogen more or less; therein, at the present state of science, lies the only apparent cause of these innumerable products of organized bodies" (Cuvier, 1828, p. 117), the interplay of chemistry and animal physiology showed, within a few decades, that the chemical processes in assimilation are more complex than had been envisaged in the formulations offered by Hallé and Fourcroy. This complexity was not readily acknowledged by all medical physiologists; for example, in his famous book on digestion, William Beaumont wrote:

> The ultimate principles of nutriment are probably always the same, whether obtained from animal or vegetable diet. It was said by Hippocrates, that "there are many kinds of aliments, but that there is at the same time but one aliment." This opinion has been contested by most modern physiologists; but I see no reason for scepticism on this subject . . . . The perfect chyle, or assimilated nourishment, probably contains the elements of all the secretions of the system; such as bone, muscle, mucus, saliva, gastric juice, etc. etc., which are separated by the action of the glands, the sanguiferous and other vessels of the system (Beaumont, 1833, pp. 36–37).

Foremost among these "modern physiologists" was François Magendie (Olmsted, 1944).

By 1815 many immediate principles of animals had been characterized as well-defined substances: they included crystalline nitrogenous compounds such as urea (from urine) and uric acid (from kidney stones) as well as a variety of non-nitrogenous organic compounds such as milk sugar, the sugar of diabetic urine, and several acids (*e.g.*, benzoic acid); in addition, many noncrystalline materials (albumin, fibrin, casein, olein, stearin, etc.) were included in the list. The recognition of this chemical differentiation profoundly influenced the new experimental physiology promoted by Magendie, and led to studies on the nutritional role of individual chemical components of the animal diet. Thus, in his famous textbook of physiology, Magendie called attention to the heterogeneity and variability of the chyme and chyle, and also stated:

> Since chemical analysis has made known the nature of the various tissues of the animal economy, it has been recognized that they all contain a considerable proportion of azote. Our aliments being also partly composed of this simple substance, it was probable that it was from them that the azote of the organs was derived; several eminent authors think, however, that it is derived from the respiration, and others believe that it is formed solely by the influence of life. Both parties rely particularly on the example of the herbivores, which feed exclusively on non-azotized materials; on the account of certain peoples that subsist entirely on rice and maize; on that of negroes, who can live a long time by eating nothing but sugar; finally, on what is related about caravans which, in crossing deserts, had for a long time only gum as the total nourishment. If these facts really proved that men can live a long time without azotized aliments, it would be necessary to acknowledge that the azote of the organs has a source different from that of the aliments; but the facts cited by no means lead to this conclusion. In fact, almost all the plants used by man and animals for food contain more or less azote . . . . I thought that one might acquire some accurate ideas on this subject by maintaining animals, during a necessary time period, on a diet whose chemical composition had been rigorously determined (Magendie, 1816–1817, Vol. II, pp. 388–390).

He then described his pioneer nutritional experiments with dogs fed controlled diets, showing that nitrogenous food is essential for survival, and concluded that his results

> . . . make it very probable that the azote of the organs has its original source in the aliments; they also throw light on the causes of gout and gravel [kidney stones] . . . . People suffering from these diseases are usually heavy eaters of meat, fish, milk products, and other substances

rich in azote. Most of the kidney stones, bladder calculi, and arthritic tophi are composed of uric acid, a principle that contains much azote. In reducing the proportion of azotized aliments in the diet, one succeeds in preventing these diseases (*ibid.,* p. 395).

This specific example of the application of the new chemistry to medicine was offered against the background of a general view of the assimilatory process in animal nutrition:

> The life of man and of other organisms is based on their regular assimilation of a certain quantity of matter, which we call aliment. Deprivation of this matter, even for a limited time, necessarily leads to the cessation of life. On the other hand, daily observation teaches that the human organs, as well as those of all living beings, lose at each instant a certain quantity of the matter of which they are composed; indeed, it is from the necessity of replacing these usual losses that the need for aliments arises. From these two facts . . . it has been rightly concluded that living bodies are by no means composed of the same matter at all stages of their existence; it has even been said that living bodies undergo a complete renewal. The ancients proposed that this renewal occurs within a period of seven years. Without accepting this speculative idea, we will say that it is extremely probable that all parts of the human body undergo an internal motion, which has the double effect of expelling the molecules which are no longer needed as components of the organs, and of replacing them by new molecules. This internal motion constitutes nutrition (*ibid.,* Vol. I, pp. 19–20).

As a general statement of the dynamics of animal nutrition, Magendie's summary provided an up-to-date version of the ancient physiological doctrine that living things continually undergo change. The difference lay in the conviction of the new physiologists and their chemical allies that the "internal motion" was accessible to experimental study with the aid of the new methods of chemical analysis. The success of this approach became evident during 1820–1850 through the clearer definition of many problems of digestion; we mentioned some of the results (see p. 68) in relation to the discovery of the "unorganized ferments." It should be noted, however, that even after the work of Theodor Schwann on pepsin the role of these ferments in digestion was questioned by one of the leaders of French chemistry:

> . . . [S]ince it has been proved to us that the animal does not create any organic matter at all; that it is restricted to assimilate [organic matter]

or to expend it through combustion, it is not necessary any longer to seek in digestion all those mysteries that surely could not be found there. Indeed digestion is a simple function of absorption. The soluble materials pass into the blood, for the most part unaltered (Dumas, 1841, p. 57).

The grand design that required this conclusion soon crumbled in the face of experimental studies in France during the 1840s, notably those of Apollinaire Bouchardat and Claude Sandras (1842).

The recognition of digestion as a chemical process, in which ferments and bile effect a transformation of food materials into products used by the animal organism, still left open the problem of assimilation. One chemist's view of the situation during the 1840s was that "Over the nature of assimilation the thickest darkness still hangs: there is no key to explain it, nothing to lead us to the knowledge of the instruments employed" (Thomson, 1843, p. 651), and most writers of the 1840s, whether chemists or physiologists, attributed assimilation to the action of a *"living* or *animal* principle" that "does not act according to the principles of chemistry" (*ibid.*, p. 656). Thus an influential physiological view was that assimilation

> . . . does not consist merely in the component particles of the organs attracting the fibrin, albumen, and other materials of the blood which flows through them, adding to themselves the matters similar to their own proximate principles, and changing the composition of those which are dissimilar; but the assimilating particles must infuse into those newly assimilated their own vital properties (Müller, 1843, pp. 472–473).

A notable exception was Müller's pupil Schwann (Florkin, 1960), whose formulation of the cell theory emphasized the problem of assimilation, and included the hypothesis that "organisms are nothing but the form under which substances capable of imbibition crystallize" (Schwann, 1839, p. 257). In his view, the transformation of undifferentiated extracellular matter (*blastema*), through the "metabolic power" of existing cells, is analogous to the formation of crystals in a supersaturated solution (see Mendelsohn, 1963); Schwann introduced the word "metabolic" (Greek *to metabolikon,* disposed to cause or suffer change) to denote the chemical transformations undergone by cell constituents and by the surrounding material. Although Schwann's antivitalistic hypothesis, based on the presumed specific chemical attraction of molecules by existing cellular membranes, contributed to the philosophical discussion that enlivened

nineteenth-century biology, it gave little hint of the chemical experimentation needed to test its validity.

Furthermore, during the early 1840s specific problems indicated the
complexity of the process of assimilation. One of these came from the
work of the French gelatin commission, headed by Magendie. We noted
before (p. 90) that much of the social turmoil during the decades immediately before and after 1800 was related to the price of food; the French
sought to meet the need for protein by introducing gelatin as a dietary
staple, since it was readily available from bones, and therefore cheap. By
1830, however, sufficient resistance had arisen among the poor to encourage official doubts about the nutritive value of gelatin. Magendie's
commission studied the problem experimentally (with dogs) from 1832
to 1841, and his report stated that life could not be sustained with gelatin
as the sole source of nitrogen; indeed, other "immediate principles" (albumin, fibrin, casein) considered to be good nutrients also were unsatisfactory when fed alone. He could do no more, therefore, than to report
the results and to emphasize the obscurity surrounding the problems of
animal nutrition.

If Magendie was cautious, Justus Liebig was not. During the 1840s
Liebig offered specific chemical answers to many aspects of the problem
of assimilation; as we saw on p. 269, he assigned to carbohydrates and fats
the role of "respiratory" substances whose oxidation produces animal heat,
whereas the proteins were thought to be "plastic" substances whose decomposition (as in fermentation) causes the chemical transformation
(*Stoffwechsel*, usually translated by English writers as "metamorphosis")
of body constituents. These views were presented against the background
of the careful studies of Jean Baptiste Boussingault (1839), who had used
Dumas's new method of nitrogen analysis to show that the apparent increase in the nitrogen content of animal tissues, as compared with that of
vegetable matter, could largely be accounted for by the loss of nitrogen
in the animal excreta and secretions (see Aulie, 1970). Aside from its contribution in ending the discussion about "animalization," Boussingault's
achievement marked the beginning of quantitative studies on the "nitrogen balance" in animal nutrition. This experimental line was taken up
during the 1850s, largely under the stimulus of Liebig's physiological
theories. The course that Liebig charted for the study of the chemical
transformations of body constituents was indicated in the third edition
of his *Animal Chemistry*:

> Urea and uric acid are products of changes undergone by the nitrog-
> enous constituents of the blood under the influence of water and oxy-
> gen, [and] the nitrogenous constituents of the blood are identical in

composition with the nitrogenous constituents of the diet. The relation of the latter to uric acid, urea, the oxygen of the atmosphere, and the elements of water, [as well as] the quantitative conditions of their formation, are expressed by chemistry by means of formulas and, within the limits of its domain, explains them thereby (Liebig, 1846b, p. 228).

By that time, Liebig had decided that the uncritical use of chemical formulas to explain physiological changes was a "senseless sport," but expressed the conviction that a time would come when

> . . . a numerical expression, in chemical formulas, will have been determined for all normal activities of the organism, [and] when one will measure the abnormalities in the functions of its individual parts by means of corresponding deviations in the composition of the substance constituting these parts, or of the products which it brings forth (*ibid.*, p. 231).

This apparent restriction to numerical relations based on changes in elemental composition reflected the state of organic chemistry before its transformation during 1850–1875 by the concepts of valence and structure. As we shall see shortly, this new organic chemistry found expression in the manner in which the program outlined by Liebig developed during the last quarter of the nineteenth century. Furthermore, by 1875 the site of biochemical change had been transferred from the blood to the cellular elements of the tissues (p. 283), and evidence was appearing that in addition to oxidation other important chemical processes occur during the transformation of body constituents. By that time, many of Liebig's physiological ideas had turned out to be incorrect, although they had stimulated fruitful inquiry. In view of the extremes of adulation and condemnation accorded these ideas at midcentury, the following opinion of a leading French biologist is worthy of note:

> I am far from wishing to say that the speculative views of Mr. Liebig on the transformations of organic matter in the interior of the animal economy, and the use he has made of equations to show how it would be possible to conceive the formation of the various products of chemical-physiological work, have been useless for the progress of science. On the contrary, I believe that in giving a precise form to his argument, he has rendered a real service and has accustomed physiologists to a mode of thought that is very useful for the study of the phenomena of nutrition. It is only necessary to take care in accepting these hypotheses as the expression of what is actually occurring in the organism, where

the intermediate reactions are very complex and very important to know (Milne Edwards, 1862, p. 542).

These lines were published twenty years after the appearance of the first edition of Liebig's *Animal Chemistry*; during the interval the grand design that he (as well as Dumas and Boussingault) had based on elementary analysis had disintegrated, and new discoveries indicated the existence, in the animal economy, of "intermediate reactions that are very complex." The most dramatic of these discoveries was Claude Bernard's demonstration that glucose is manufactured in the liver.

Glucose and Glycogen in the Animal Body

In 1853 Bernard published his celebrated experimental report on a "new function of the liver":

> I will establish in this work that animals, as well as plants, have the ability to produce sugar. Furthermore, I will show that this animal function, hitherto unknown, is localized in the liver (Bernard, 1853, p. 7).

There followed the account of experiments begun ten years earlier, and whose high point came in 1848, when Bernard discovered the presence of sugar in the blood leaving the liver of animals that had received no carbohydrate in their diet (see p. 60). To detect the sugar he used, in addition to a yeast fermentation test, Charles Barreswil's version of the copper reagent introduced by Carl Trommer in 1841; a better-known modification is that developed by Hermann Fehling in 1848. Bernard wrote that he had been impelled to examine the fate of sugar in the animal body, in the hope of understanding the nature of the wasting human disease diabetes mellitus, and

> . . . was led to think that there might be in the animal organism phenomena still unknown to chemists and physiologists, and able to give rise to sugar from something other than starchy substances (*ibid.*, p. 9).

For valuable accounts of the development of Bernard's ideas, as recorded in his research notes of 1848, see Grmek (1968) and Holmes (1972).

By the beginning of the 1840s it had been established that the long-known sweet taste of the urine and blood of diabetic patients was associated with the presence of glucose. Thus Chevreul (in 1815) and Peligot (in 1838) had demonstrated the chemical identity of the sugar of diabetic

urine with grape sugar, and Ambrosioni (in 1835) showed that the sugar
of diabetic blood also is glucose, as judged by its alcoholic fermentation
with yeast. Furthermore, it was considered certain that glucose could not
be produced in the animal body, and that this process is restricted to
green plants, through their power to assimilate carbon dioxide and water
in the presence of sunlight, and to convert them into starch and sugar.
That the ingestion of such carbohydrates was followed by the appearance
of fermentable sugar in the chyme and blood of normal animals had
been shown repeatedly since the work of Tiedemann and Gmelin in 1826,
and the salivary conversion of dietary starch into sugar had been established during the 1830s. It seemed during the 1840s, therefore, that the
pathological defect in diabetes might be a consequence of the failure of
the organism to utilize the glucose derived from the diet. Thus, in his
Animal Chemistry, Liebig wrote that

> In some diseases the starchy materials do not undergo the transforma-
> tions that allow them to support the respiratory process or to be changed
> into fat. In diabetes mellitus, starch is not converted beyond sugar which
> is not utilized and is removed from the body (Liebig, 1842, p. 86).

After Bernard had found that the liver can produce sugar, he concluded
that this discovery

> . . . must necessarily change the hitherto prevalent ideas about the
> nature of diabetes, based on the belief that the sugar found in the or-
> ganism is exclusively derived from the food. This opinion can no longer
> be maintained today, since we have established that the saccharine ma-
> terial is constantly renewed in the organism at the expense of the ele-
> ments of the blood and independently of the nature of the diet. Since
> this glycogenic function has been found to be localized in the liver, it
> is evident that it is in this organ we must now seek to place the seat
> of the disease (Bernard, 1855a, Vol. I, p. 85).

Furthermore, Bernard's discovery indicated that the physiological function of the liver is not limited to the production and "external secretion"
of bile, as had been thought previously, but was more complex, and included the "internal secretion" of sugar; he asked whether

> . . . the albuminoid substances of the blood, upon coming into contact
> with the hepatic cells, are cleaved into two products, a hydrocarbonated
> one which becomes the sugar, the other a nitrogenous one which be-
> comes the bile? (*ibid.,* p. 97).

It should be added that, for several years, Bernard did not accept the statements by Louis Figuier (in 1855) and by Auguste Chauveau (in 1856) that sugar is present in all parts of the circulating blood, and not only in the blood leaving the liver. By 1860, however, numerous workers had established the presence of glucose throughout the circulation.

To meet the objection that the sugar emerging from the liver might have been produced in the blood flowing through the tissue, Bernard submitted an excised dog liver to thorough lavage in order to remove the blood, and showed that by the next day sugar had again been formed; if the washed liver was boiled, the formation of sugar was abolished. The glucogenesis was therefore a property of the hepatic tissue itself, and Bernard concluded that, in addition to glucose, there is present in the liver ". . . another substance, sparingly soluble in water, . . . [which] is slowly changed to sugar by a kind of fermentation" (Bernard, 1855b, p. 467). Two years later he reported the isolation of this *matière glycogène*, and was aided by the chemist Théophile-Jules Pelouze in its characterization as a starch-like substance (*amidon animal*); this was confirmed immediately afterward by Victor Hensen. As we indicated before (p. 60), Bernard viewed the glycogenic function of the liver as involving two distinct processes:

> The first entirely vital action, so termed because it is not effected outside the influence of life, consists in the creation of the glycogenic material in the living hepatic tissue.
> The second entirely chemical action, which can be effected outside the influence of life, consists in the transformation of the glycogenic material into sugar by means of a ferment (Bernard, 1857b, p. 583),

and he called attention to the similarity of this dual process to the one occurring in plants:

> For example, in a seed which produces sugar during germination, we must also consider two distinct series of phenomena: a primary one, entirely vital, is represented by the formation of starch under the influence of the life of the plant; the other a secondary one, entirely chemical, which can occur outside the vital influence of the plant, is the transformation of the starch into dextrin and sugar by the action of diastase (*ibid.*, p. 585).

Bernard reiterated this view during the succeeding twenty years, and in 1877 he described the extraction (with glycerol) of the "diastatic ferments" of the liver and of barley, and reported that they both convert starch and glycogen into glucose. He believed that he had demonstrated

. . . the identity of the mechanism of formation of sugar in animals and plants, since we have not only seen that glycogen is identical with starch, but that the diastase of the seed is also identical with the diastase of the liver (Bernard, 1877b, p. 524).

If the mechanism of the formation of glucose seemed to be clear,

The mechanism of the formation of the amylaceous material [starch or glycogen] is on the contrary completely unknown for plants as well as for animals, and it is the problem which now presents itself for the studies of chemists and physiologists. In pursuing this study, will we find the same parallelism between animals and plants that we have established for the mechanism of the production of sugar? (*ibid.*, p. 525).

Bernard's discoveries led him to redefine the problem of assimilation as follows:

. . . [W]e might say that the aliments, dissolved and modified by the digestive juices in the intestine, only form a sort of a generative blastema in which the anatomical elements find the materials of their nutrition and of their functional activity. But the alimentary principles, once incorporated into the internal environment [*milieu intérieur*], are not assimilated at once or as such by the various tissues. Here there intervene new phenomena . . . I wish to speak of the role, as creators of reserves, played by certain internal organs which are interposed in the blood circulation coming from the intestine and which transform, produce, or store various substances, in order not to release them for consumption by the tissues except under conditions controlled by the nervous system (Bernard, 1877a, p. 436).

The term *milieu intérieur,* with which Bernard's name has been identified in twentieth-century writings, had its antecedents in the French scientific literature (*e.g., milieu de l'intérieur;* Robin and Verdeil, 1853, Vol. I, p. 14); the concept of the "constancy of the internal environment" (homeostasis) was later elaborated by numerous physiologists, notably Henderson (1928) and Cannon (1932). The control of the chemical processes in the internal environment by the nervous system was a central feature of Bernard's physiological thought; he believed that he had shown this control in his experiment of inducing glycosuria through puncture (*piqûre*) of the floor of the fourth ventricle of a dog's brain.

    In Bernard's view, therefore, the two general processes of which glycogen formation and glycogen breakdown were examples represent

. . . two kinds of phenomena, one of organization or assimilation, the other of disorganization or disassimilation . . . . Both kinds of phenomena are equally physiological, they are equally necessary for the maintenance of life. Assimilation, which is in reality nothing but a sort of reduction, cannot occur without disassimilation, which is in reality a combustion . . . . We have in the liver the two kinds of phenomena of assimilation and of disassimilation. The phenomena of assimilation correspond to the formation of glycogen, the phenomena of disassimilation correspond to its transformation into dextrin and glucose. One of these phenomena is no more post-mortem than the other; they both occur in the living organism (Bernard, 1877a, pp. 360–362).

The term "disassimilation" had appeared earlier in the writings of Bichat (1805, p. 7) and others. In particular, Louis Mialhe had stated that in the living organism

. . . there is, at each instant, a supply of new molecules and the departure of old molecules. And, following Blainville, one may term the assimilative faculty (*assimilation*) the one that . . . maintains the individual; and, in contrast, the disassimilative faculty (*désassimilation*) the one that destroys it . . . . Organized matter is therefore constantly in the presence of a chemical apparatus composed of oxygen, ferments, acids, alkalis, chlorides, [all] elements whose properties constitute unceasingly exciting forces which make the life of animals an uninterrupted series of chemical reactions (Mialhe, 1856, p. 15).

Also, the view that the liver is a site of assimilation had been expressed previously, for example by Tiedemann and Gmelin in 1820 and by Nicolas Blondlot in 1846 (see Mani, 1967).

During the period 1850–1880 the terms "progressive metamorphosis" and "regressive metamorphosis" were often used to make the distinction between assimilation as a constructive process and "disassimilation" as a degradative one. After 1880, with the recognition that both kinds of processes occur in normal cells, the term "metabolism" (derived from Schwann's usage) came into favor among English physiologists (see Bing, 1971); "anabolism" denoted the creation of cell constituents, and "catabolism" their degradation to simpler chemical compounds. In 1931 Hardy described the view held in active nineteenth-century English schools of physiology as follows:

Metabolism—that is the whole chemical cycle—consisted . . . of a phase of increased molecular complexity in which protein, fats, and carbohydrates with oxygen were built up into the living substance, and a phase of decreasing chemical complexity and liberation of energy.

The picture of two operations, anabolism and katabolism—or loading and discharge—was based in the first instance on the properties of muscle, especially on its capacity for doing work when not supplied with food or even oxygen, but it received immense support when the processes of loading and discharge of gland cells was discovered in the late eighties (Hardy, 1936, p. 898).

In German writings the terms *Stoffwechsel,* assimilation, and dissimilation continued to be widely used well into the twentieth century. For example, among the many philosophical disquisitions of German physiologists around 1890 was an essay by Ewald Hering:

> For physiological thought, the most significant feature that distinguishes living from dead matter is its metabolism [*Stoffwechsel*], *i.e.,* the internal chemical processes whereby on the one hand materials are formed that have become foreign to the living substance and are either accumulated nearby or are transferred to the circulating fluids, and on the other hand, and concurrently, food materials are taken up and appropriated by the living substance and converted into its own constituents. The latter process has been denoted assimilation [*Assimilirung*], and accordingly I have termed the first process dissimilation [*Dissimilirung*] (Hering, 1888, p. 35).

Little could be said, however, about the nature of most of the cell constituents undergoing these "internal chemical processes," or about the processes themselves, and the wide-ranging biological speculation about metabolism usually stressed the role of "living proteins." For the chemically minded physiologists, therefore, Bernard's discovery of glycogen and its role as a source of blood glucose provided a promising line for the chemical study of a metabolic process.

Indeed, Bernard himself insisted that it was the task of physiologists to study what later came to be called "intermediary metabolism." He repeatedly cited his discovery as evidence for the primacy of physiology over chemistry in the study of biological phenomena and, in the last book he wrote, Bernard stated that he had been drawn thirty years before to study the transformation of sugar in the animal body

> . . . by the conviction that the phenomena of nutrition should not be considered by the physiologist from the same point of view as that of the chemist. Whereas the latter seeks to determine a nutritional balance,

i.e., to establish the balance between the substances that enter and those that leave, the physiologist should set himself the task of following them step by step during their transit, and to study all their successive transformations in the interior of the organism. I decided to apply this method to all the substances in turn: to the albuminoids, the saccharine materials, the fats. I began with the saccharine materials, which seemed to me to present a simpler problem (Bernard, 1878–1879, Vol. II, pp. 40–41).

It may be questioned whether this account accurately describes Bernard's thought during the 1840s, and one may suspect that he was recalling old battles over the physiological ideas of Dumas and Liebig; in any case, it would appear that Bernard did not attach significance for physiology to the remarkable development of organic chemistry after 1860, or to the emergence of a group of physiological chemists who were using this new knowledge in attempts to follow metabolic changes "step by step" (see Schiller, 1967b).

By 1880 the formation of glucose by the liver was widely acknowledged to be a normal physiological function, and the objections of Frederick Pavy (among others) that it represented a post-mortem process had been largely discounted (see Young, 1937). The existence of liver glycogen as a chemical entity had been firmly established, and its starch-like character had been confirmed by a series of investigators, notably Kekulé (in 1858), Mering and Musculus (in 1878), and Külz (in 1880). Furthermore, despite Bernard's initial doubts about the occurrence of glycogen in tissues other than the liver, a starch-like material was found in other organs, especially the muscles; although muscle glycogen did not appear to be a source of blood sugar, it was considered to undergo a "lactic fermentation" with glucose as a transient intermediate (Bernard, 1877a, p. 430).

The Biosynthesis of Glycogen. Chemical studies had shown starch to be a substance formed by the union of glucose molecules with the loss of the elements of water, since the conversion of starch into glucose by means of dilute acids or by enzymes was known to involve a gain of the elements of water, and was termed a "hydratation" (later, hydrolysis). It seemed logical to suppose, therefore, that the synthesis of starch and glycogen in plants and animals involves the condensation of glucose molecules with the loss of elements of water. Syntheses of this kind had already been effected in the chemical laboratory by Marcelin Berthelot (in 1854), who produced fats by heating glycerol with fatty acids.

After the emergence of the new structural organic chemistry, the idea that the loss or gain of the elements of water might be important in metabolic processes was stated in the influential paper by Adolf Baeyer (1870), and soon afterward by Marcellus Nencki (1872). Attention was drawn to the work of Friedrich Wöhler (1842), showing that the ingestion of benzoic acid (C_6H_5COOH) by a human subject led to the urinary excretion of hippuric acid ($C_6H_5CO—NHCH_2COOH$), and the new chemical knowledge indicated that the organism had effected a condensation reaction between benzoic acid and glycine (NH_2CH_2COOH) present in the body. In 1876 Gustav Bunge and Oswald Schmiedeberg reported that, in the dog, hippuric acid is made by the kidney; five years later, Schmiedeberg found that several animal tissues (including the kidney) can hydrolyze hippuric acid, thus suggesting a metabolic balance between synthetic and degradative processes. During the 1850s the work of Adolf Strecker had shown that the glycocholic acid and taurocholic acid of the bile represent condensation products of cholic acid with glycine and taurine, respectively. In 1875 Eugen Baumann discovered that phenol (C_6H_5OH) appeared in the urine in the form of an ester of sulfuric acid extractable with ether ("ethereal sulfate"), and three years later Schmiedeberg reported that camphor was excreted in the form of a condensation product of a new sugar that was named glucuronic acid. The formation of such amides (as hippuric acid or the conjugated bile acids), phenyl sulfates, or glucuronides gave evidence of the operation in the animal organism of synthetic processes in which the elements of water are removed in condensation reactions similar to those effected in the chemical laboratory. These observations led Felix Hoppe-Seyler to suggest that

> . . . the materials of which the organs are composed, from which they build and regenerate themselves, belong to a class of substances that collectively represent *anhydrides,* and which . . . are converted or cleaved with the uptake of the elements of water by treatment with alkalies and acids, in many cases also by ferments . . . . These anhydrides usually show no striking chemical affinities, they swell in water or are sparingly water-soluble; they withstand the action of atmospheric oxygen and have, to the extent now known, very large molecules (Hoppe-Seyler, 1884, p. 17).

According to his view, the anhydrides of which "living protoplasm" is composed are of two kinds, one of which acts as a degradative ferment, but which, in the presence of "active" oxygen, can be converted into a second kind of anhydride that can effect condensation reactions. As an example, Hoppe-Seyler cited the fact that:

Upon passage of oxygenated blood through the living kidney, the combination of glycine with benzoic acid—an anhydride formation—has been demonstrated. The living organ has also been found to effect the reverse process, namely the cleavage of hippuric acid and similar compounds by the uptake of water. It may be expected that the latter process will also occur in the absence of oxygen (*ibid.*, p. 26).

This anhydride theory figures largely in the discussion during 1880–1910 about the mode of formation of glycogen. As we shall see shortly, however, Buchner's discovery of cell-free fermentation caused a shift in attitude from speculations about "living protoplasm" to the role of intracellular enzymes in synthetic processes. Furthermore, the discussions were greatly influenced by Fischer's elucidation of the chemical structure of glucose and of other monosaccharides; although the detailed structure of starch and glycogen was still a matter of conjecture in 1910, there was no doubt that they represent polyglucoses in which the monosaccharide units are joined by glycosidic linkages to form anhydrides analogous to the fats, proteins, and nucleic acids.

In the background of the debate about the mode of formation of glycogen in the liver was the uncertainty concerning the metabolic precursors of glycogen. Numerous investigators took up this question; at first, starved animals (whose supply of liver glycogen had been nearly completely depleted) were used to determine what kind of food materials would cause glycogen formation. For example, Joseph von Mering (1877) reported that the ingestion of glucose, sucrose, lactose, fructose, glycerol, and various proteins (egg albumin, casein, etc.) led to the deposition of liver glycogen. There were many contradictory reports, however, arising from deficiencies in analytical technique, the variability in the liver glycogen content of control animals, and the uncertain physiological state of the animals used for nutritional studies. In 1891 Voit summarized the situation by calling attention to two opposing theories:

According to one theory, glycogen arises by the loss of water from dietary carbohydrates, that is, glucose present or made in the intestinal canal. It is the theory of the anhydride formation of glycogen. It has been based on the fact that the greatest accumulation of glycogen in the liver occurs upon the uptake of certain carbohydrates, especially glucose.

According to the other view, glycogen arises from protein breakdown, and the only role of the readily-decomposable sugar derived from the diet is to be burned instead of glycogen, thus protecting it from decomposition; this has been called the sparing theory (Voit, 1891, pp. 245–246).

On the basis of the work of his associates, Voit concluded that, although the anhydride theory was correct, it was possible that glycogen can also arise from noncarbohydrate precursors. This possibility was denied by Eduard Pflüger; in a massive paper, he stated:

> It is generally agreed that glycogen arises in the animal organism by a synthetic pathway . . . . Many investigators now believe that the liver can make glycogen not only from sugar, but also from protein, fat, and other substances. If it were true that the liver can make the same glycogen from the most varied molecules, whose atomic groupings have no relation whatever to glycogen, one would have to assign to the liver quite mysterious synthetic capacities (Pflüger, 1903, p. 168).

After describing a valuable improvement of Bernard's method for the determination of glycogen (based on its stability in alkali), Pflüger offered data which led him to conclude that it could only arise from carbohydrates. In the following year, however, Hugo Lüthje demonstrated clearly that a diabetic dog on a carbohydrate-free diet excreted much more sugar than could have arisen from the total carbohydrate of the body. After Pflüger confirmed this finding, he accepted the view that "the liver can effect the synthesis of glycogen from protein" (Pflüger, 1910, p. 302). Moreover, by that time it was evident from other studies that some amino acids can serve as metabolic sources of carbohydrate; we shall return to this point shortly.

Lüthje induced experimental diabetes by the procedure discovered by Mering and Oscar Minkowski in 1889. They found that surgical removal of the pancreas of a dog produced a large increase in blood sugar (hyperglycemia) and a concomitant urinary excretion of glucose (glycosuria); for a valuable historical account, see Houssay (1952). This experimental production of a condition resembling human diabetes was followed by Eugene Opie's demonstration (in 1901) that, in diabetic patients, special pancreatic cells (the islets described by Paul Langerhans in 1869) were damaged, and in 1909 Jean de Meyer suggested that these cells elaborate a hormone which he termed "insuline." The search for this hormone encouraged the development of micromethods for the determination of blood sugar, so that many samples could be taken from the same animal, and with the help of such analytical procedures Frederick Banting and Charles Best (1922) succeeded in establishing the existence of insulin as a chemical entity. We have already referred to Abel's subsequent crystallization of insulin, in connection with the history of protein chemistry (see p. 171).

It should be added that three years before his collaboration with Minkowski, Mering had found that the injection of the plant glycoside phlorizin causes the urinary excretion of glucose, without an increase in the level of blood sugar. This artificial diabetes, arising from an increased permeability of blood glucose in the kidney, was used for many decades to examine the effect of various nutrients on the metabolic production of glucose. For example, Graham Lusk (1910) employed this method to show that some amino acids (glycine, alanine, glutamic acid, aspartic acid) are converted into glucose in the animal body.

By about 1910, therefore, it was widely agreed that the immediate precursor of liver glycogen is glucose, but that the metabolic capacities of the liver permitted it to convert noncarbohydrate substances such as glycerol or amino acids (*e.g.*, alanine) into glucose, and therefore into glycogen. The question of the role of proteins as metabolic precursors of glycogen had been clarified by the demise of the view that glycogen arises from carbohydrate associated with dietary proteins, not their nitrogenous components (see p. 182); highly purified proteins, free of carbohydrate, were shown to be glycogenic. Furthermore, it became clear from the work of Otto Cohnheim, Otto Loewi, and Emil Abderhalden during 1901–1905 that dietary proteins are almost completely cleaved to amino acids in the gastrointestinal tract. One had to look, therefore, to the carbon skeletons of these protein constituents for metabolic sources of body glucose. Inevitably, the chemical speculations about the possible intermediates in alcoholic fermentation and in glycolysis, to which we referred earlier (see p. 342), also embraced the question of possible routes for the conversion of the deaminated derivatives of amino acids into glucose. Leucine was first considered as a likely source, simply because it contains six carbon atoms, and fanciful reaction pathways were postulated (Leathes, 1906, p. 43); they fell by the wayside when physiological experiments showed leucine to be one of the protein amino acids that is not glycogenic in experimental animals. A more promising candidate was alanine, since it was known to be deaminated by nitrous acid to lactic acid, and in 1904 Gustav Embden showed that depancreatectomized dogs convert lactic acid into glucose. After Otto Neubauer had demonstrated, through perfusion experiments, that the liver can deaminate alanine to pyruvic acid, the metabolic relations among lactic acid, pyruvic acid, and glucose became the subject of extensive speculation, with methyl glyoxal as a preferred intermediate. We have reviewed the tortuous development, during 1910–1940, of a coherent scheme for the metabolic breakdown of glucose in yeast and in muscle; it was not until the 1930s that the synthesis of glycogen from lactic acid was considered to involve a reversal of the degradative pathway, the energy being provided by coupling to the cleavage of ATP.

Furthermore, after 1900 there was much speculation about the enzymic catalysis of the polymerization of glucose to form glycogen. For example, Franz Hofmeister wrote:

> If it should turn out that the reversibility of enzyme action has a more general validity, how simple it would then be to interpret the course of many of the most important physiological processes! Can there be a more complete self-regulation of glycogen metabolism than through the action of a diastase which converts glycogen to sugar when the supply of sugar is insufficient, but reversibly accumulates glycogen when there is an excess of sugar! (Hofmeister, 1901, pp. 21–22).

This hope was reiterated, with varying degrees of assurance, during the succeeding three decades, and was based on experiments such as those of Arthur Croft Hill (1898) on the apparent reversal of the hydrolytic action of maltase, and of Joseph Kastle and Arthur Loevenhart (1900) on the synthesis of ethyl butyrate by pancreatic lipase, as well as on the extensive work during 1900–1930 on the apparent synthesis of proteins from peptones by the action of pepsin or trypsin (Wasteneys and Borsook, 1930). It was recognized, however, that glycogen synthesis depends on the presence of oxygen, thus indicating that the process could not be simply a reversal of the action of the hydrolytic enzymes thought to be responsible for the conversion of glycogen into glucose in the liver. Nevertheless, before 1930 it could fairly be said "that the theory of enzyme synthesis of tissue compounds is the only explanation generally held and taught today to account for the building up of protein, fats, and carbohydrates (Bradley, 1913, pp. 407–408).

A complementary hypothesis about the mechanism of glycogen synthesis, much in favor during the early 1920s, was that ordinary glucose underwent a transformation to a more reactive isomeric form, which either polymerized to yield glycogen or reacted with phosphate to form hexose phosphate. A popular candidate was "γ-glucose," and a leading biochemist considered it possible that ordinary glucose

> . . . is converted under the joint influence of the tissues and of the pancreas hormone into γ-glucose, which may be polymerized to glycogen or transformed into other derivatives capable of oxidation, both of which fates are closed to the molecular configuration of ordinary glucose (Shaffer, 1923, pp. 400–401).

The origin of γ-glucose may be traced to the fact that, in his work on the constitution of glucose and other monosaccharides, Emil Fischer had left undecided the question of the size of the oxygen-containing ring formed by the interaction of the reducing group with one of the hydroxyl groups in the chain (see p. 79). This problem was attacked by Thomas Purdie and his student James Irvine, and in 1915 Irvine thought that he had shown γ-glucose to have a reactive three-membered ethylene oxide ring whereas ordinary glucose had a five-membered butylene oxide ring. Subsequent work by W. N. Haworth during the mid-1920s showed, however, that the more stable form of glucose has a six-membered amylene oxide (pyranose) ring, whereas the more reactive form is the butylene oxide (furanose) form. Although the temporary enthusiasm generated among biochemists about γ-glucose proved to be based on incorrect chemical deductions, the method introduced into carbohydrate chemistry by Purdie and Irvine, and developed by Haworth, was decisive in the elucidation of the chemical constitution of the sugars and of the polysaccharides. This method depended on the methylation of the free hydroxyl groups, and the determination of the nature of the methylated monosaccharides obtained upon acid hydrolysis.

It should also be added that, during the early 1920s, polysaccharides such as cellulose, starch (known to be composed of two different materials denoted amylose and amylopectin), and glycogen were considered to be aggregates of relatively small polyglucoses held together by noncovalent linkages. As we saw earlier (p. 154), X-ray data on cellulose fibers were taken to support this view, and Hermann Staudinger's ideas about the existence of long-chain polymers were not yet accepted. Although physicochemical measurements had indicated molecular weights for glycogen of 100,000 and more, there was reluctance to consider such values as indicating the size of a covalently bound molecule. The application of the methylation procedure to glycogen led Haworth (in 1932) to estimate that it consisted of chains of about 12 glucose units; as in starch, the monosaccharide units appeared to be linked to each other by $\alpha(1 \rightarrow 4)$ glycosidic bonds (connecting the reducing group at carbon 1 of one unit and the hydroxyl group at carbon 4 of the adjacent unit). It was not until about 1940 that the work of Haworth and of Freudenberg clarified the discrepancy between the apparent molecular weight determinations (by now in the range 300,000 to 800,000) and the initial results of the methylation studies. Glycogen was recognized to have a highly branched structure in which straight-chain arrays of 11 to 18 glucopyranose units [in $\alpha(1 \rightarrow 4)$ glycosidic linkage] are linked by means of $\alpha(1 \rightarrow 6)$ glycosidic bonds. A similar structure was found for amylopectin which, like glycogen, gives a red-brown color with iodine; the amylose portion of starch is a straight-chain $\alpha(1 \rightarrow 4)$ polysaccharide, with average chain lengths that may be about 200 glucose units, and gives the blue color reaction with iodine. These chemical advances (see Meyer, 1943) were significant for the subsequent elucidation of the details of the mechanism of the biological formation of glycogen and starch, but, before they were made, the biochemical aspects of the problem had been redefined, largely in consequence of the work of Carl and Gerty Cori.

In a series of striking experiments beginning in 1925, the Coris studied the metabolic relations among lactic acid, glucose, and glycogen in intact rats, and in 1929 they were able to conclude:

> Formation of liver glycogen from lactic acid is thus seen to establish an important connection between the metabolism of the muscle and that of the liver. Muscle glycogen becomes available as blood sugar through the intervention of the liver, and blood sugar in turn is converted into muscle glycogen. There exists therefore a complete cycle of the glucose molecule in the body which is illustrated in the following diagram:

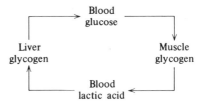

> Epinephrine was found to accelerate this cycle in the direction of muscle glycogen to liver glycogen and to inhibit it in the direction of blood glucose to muscle glycogen; the result is an accumulation of sugar in the blood. Insulin, on the other hand, was found to accelerate the cycle in the direction of blood glucose to muscle glycogen, which leads to hypoglycemia and secondarily to a depletion of the glycogen stores of the liver . . . . There is also a possibility that other hormones besides epinephrine and insulin influence this cycle (Cori and Cori, 1929, pp. 401–402).

In the same year, Bernardo Houssay provided evidence for an additional hormonal factor in the regulation of the carbohydrate metabolism of animals. He showed that, upon surgical removal of the pituitary of depancreatectomized animals, the severity of the diabetic state was markedly ameliorated, and the animals became very sensitive to the administration of insulin, indicating that the pituitary elaborates a factor (or factors) which counteract the action of insulin. One of these factors later turned out to be the adrenocorticotropic hormone (ACTH), made in the anterior pituitary; this hormone stimulates the adrenal cortex to release steroid hormones (*e.g.*, cortisol) which antagonize the action of insulin, largely by promoting the conversion of amino acids into glucose. Indeed, in 1936 C. N. H. Long showed that surgical removal of the adrenal gland of diabetic animals had an ameliorating effect similar to that demonstrated by Houssay. These observations revealed an intricate set of hormonal effects that regulate the pathways of biochemical change linking muscle and liver glycogen, blood glucose, and substances that either give rise to them in metabolism, or are formed upon their breakdown. By the mid-1930s the dissection of the enzymic apparatus in glycolysis had developed to the stage where there was hope of identifying the individual steps in carbohydrate metabolism that are under hormonal control. We will return to this question later; at this point we must consider the important contributions of Cori and Cori to the study of the enzymes involved in the metabolic interconversion of glucose and glycogen (see Cori, 1969).

During the course of their studies on the factors influencing the formation of hexose phosphates in frog muscle, Cori and Cori (1936) observed the anaerobic formation of a nonreducing sugar phosphate, which they identified as glucose-1-phosphate (see accompanying formula). A few years later they defined the metabolic role of this new compound in a paper that began as follows:

> Up to the present it has been assumed that the blood sugar is formed from liver glycogen by a diastatic enzyme. The breakdown of glycogen by animal diastases *in vitro* leads to the formation of dextrins of varying chain lengths, of nonfermentable reducing substances, and of maltose. Considerable amounts of glucose are found in the reaction mixture only after long periods of incubation. On the basis of these findings it is difficult to explain the fact that glucose is the sole product of glycogen breakdown in the liver, unless one makes the additional assumption that the diastatic activity in the intact cell is so coordinated that intermediary products between glycogen and glucose do not accumulate.
>
> Recent studies of the initial stage of glycogen degradation in various tissues (muscle, heart, liver, brain) have shown that there is present a phosphorylating enzyme which catalyzes the reaction:

$$\text{Glycogen} + \text{H}_3\text{PO}_4 \rightleftharpoons \text{glucose-1-phosphate} \qquad (1)$$

The second enzymatic step in the degradation of glycogen is the conversion of glucose-1- to glucose-6-phosphate by the phosphoglucomutase. It is at this point that the enzymatic processes in muscle and liver diverge, owing to the presence in the liver of a phosphatase which splits hexosemonophosphate to glucose and inorganic phosphate. It has been suggested that blood sugar is formed by the combined action of the liver phosphorylase and phosphatase . . . . The reversibility of reaction (1) has been demonstrated in muscle extracts after the separation of the phosphorylase from other enzymes. In the present paper glycogen synthesis from glucose-1-phosphate is shown to occur with purified liver phosphorylase (Cori, Cori, and Schmidt, 1939, pp. 629–630).

As was mentioned on p. 375, the discovery of the reversible phosphorolysis of glycogen to glucose-1-phosphate marked a turning point in the consideration of biochemical mechanisms for the synthesis of polymeric molecules. In the case of glycogen, the "anhydride" formation from glucose required initial phosphorylation with ATP to form glucose-6-phosphate, and the generation of ATP required coupling to energy-yielding oxidative reactions; the process was not simply a reversal of the action of a "diastase," as envisaged by Hofmeister in 1901.

Shortly after the discovery of the animal phosphorylases, Charles Hanes (1940) described the reversible phosphorolysis of starch by enzyme preparations from higher plants (peas, potatoes), and showed the product to

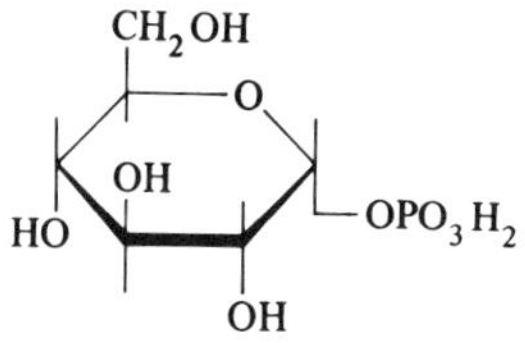

Glucose-1-phosphate

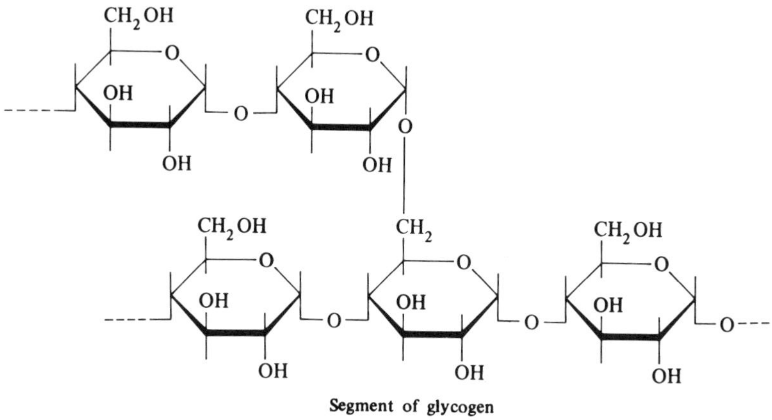

Segment of glycogen

be glucose-1-phosphate. It seemed therefore that the question Bernard had asked in 1877 about the possible parallelism in the mechanisms of the formation of starch and glycogen had been answered affirmatively.

The product of the polymerization reaction catalyzed by the plant phosphorylases behaved like the amylose (straight-chain) fraction of starch. After the elucidation of the branched-chain structure of amylo-pectin and glycogen (see formula of segment of glycogen), additional enzymes were found in plant and animal tissues that catalyze the "branch-ing" of straight-chain $\alpha(1 \rightarrow 4)$ polyglucose, with the transfer of pieces of the chain to the 6-hydroxyl of glucose units within the main chain. The crystallization of muscle phosphorylase by Arda Green and Gerty Cori (1943) was especially important, because it made possible a more pre-cise definition of the catalytic properties of the enzyme. In particular, it became evident that the enzyme uses the numerous nonreducing ends of preexisting glycogen molecules as sites of attachment of the glucosyl units derived from glucose-1-phosphate. Glycogen synthesis was thus viewed as the elongation of the branches of the "primer" glycogen, fol-lowed by further branching; conversely, the enzymic breakdown of glyco-gen was seen to involve the successive shortening (by phosphorolysis) of the branches and cleavage of the $\alpha(1 \rightarrow 6)$ branch points. By 1943, there-fore, the molecular basis for the enzyme-catalyzed interconversion of glu-cose and glycogen, as well as their relation to the pathway of anaerobic glycolysis, appeared to have been fully established.

The further development of this story is recounted in current textbooks of biochemistry. The view that glycogen phosphorylase is responsible for the synthesis of glycogen from glucose-1-phosphate has been replaced by a somewhat more complex scheme, in which the glucosyl units of glyco-gen are derived from a new compound, uridine diphosphate glucose (UDPG; see accompanying formula), discovered during the 1950s by Luis Leloir (see Leloir, 1964). This compound arises from the enzyme-catalyzed reaction of uridine-5′-triphosphate with glucose-1-phosphate,

Uridine diphosphate glucose

and the enzyme that catalyzes the polymerization of the glucosyl units of UDPG is termed "glycogen synthetase." The equilibrium in the reaction catalyzed by this enzyme is much further in the direction of polysaccharide synthesis than in the case of the phosphorylase, which is now considered to represent the physiological agent of glycogen degradation, not of synthesis. In part, the evidence for this view comes from the finding of a hereditary disease in which muscle glycogen phosphorylase is absent, but the muscle glycogen is normal. Both the phosphorylase and the synthetase have been shown to be enzymes whose catalytic activity is indirectly influenced by various hormones, which appear to act by virtue of their effect on an enzyme (adenyl cyclase) that converts ATP into 3',5'-cyclic AMP, discovered by Earl Sutherland in 1957. This compound acts on other enzymes that alter the catalytic activity of the phosphorylase and the synthetase. Further research may reveal additional complexity in this set of intricate control mechanisms. It should be added that nucleoside diphosphate sugars analogous to UDPG have been shown to be donors of the sugar units in the biosynthesis of a large variety of polysaccharides, including starch and the "peptidoglycans" of bacterial cell walls (Hassid, 1969; Strominger, 1970).

Whatever revisions there may have been in the views concerning the physiological role of glycogen phosphorylase, the impact of its discovery on biochemical thought has been lasting. The sharp distinction made by Bernard between the process of constructive assimilation and degradative dissimilation, reiterated by others for many decades, was shown to be illusory; although vestiges of this attitude continued to reappear after 1940, it was clear that metabolic synthesis, like metabolic degradation, represents an enzyme-catalyzed process, and that endergonic syntheses (*e.g.*, glucose to glycogen) are specifically coupled to exergonic reactions (*e.g.*, the cleavage of ATP). Furthermore, the work of the Coris on phosphorylase made biochemists aware of the importance of the catalysis, by specific enzymes, of the reversible transfer of a group (in this case, a glucosyl group) from one acceptor group to another (here, the phosphoryl group of glucose-1-phosphate and the 4-hydroxyl group of a glucose unit). After World War II, microbial enzymes were found that catalyze the phosphorolysis of disaccharides such as sucrose or maltose with the formation of glucose-1-phosphate, and it was recognized that these enzymes also act by a glucosyl-transfer mechanism. Moreover, this mechanism was found to apply to the long-known synthesis of polysaccharides from sucrose by certain bacteria; thus the formation of a polyglucose (dextran) was seen to represent the enzyme-catalyzed transfer of the glucosyl units of sucrose to a growing polyglucose chain, and the formation of a polyfructose (levan), by a different microorganism, was seen to involve the transfer of the fructosyl units of sucrose to a growing polyfructose chain (Hehre, 1951). Indeed, the ancient invertase, long the prototype of hydrolytic enzymes, was found to catalyze the transfer of fructosyl units not only to water, but also to various acceptors having hydroxyl groups (Edelman, 1956). It became clear that no sharp line of demarcation could be drawn between enzymes that catalyze the hydrolysis of glycosidic bonds and those that catalyze the transfer of glycosyl groups (transglycosylation), and that many highly purified enzymes could catalyze both hydrolytic and transfer reactions.

If a personal note may be added, it was the work of the Coris during the 1940s on the mechanism of the phosphorylase reaction that encouraged me to explore further the possibility that intracellular enzymes known to catalyze the hydrolysis of peptide bonds might also be effective in the catalysis of the transfer of peptidyl units (transpeptidation), and to suggest that such reactions might be important in the biosynthesis of proteins (Fruton, 1950, 1957). The remarkable development of research on the pathway of protein biosynthesis, especially after the work of Paul Zamecnik during the late 1950s (see p. 258), made my speculation superfluous, since attention was principally focused on the factors that determine the amino acid sequence of the completed peptide chain, and it was widely assumed that the intracellular condensation of amino acid units proceeds by a nonenzymic "zipper" mechanism. More recently, many efforts have been initiated to dissect the complex apparatus involved in the intracellular formation of peptide chains, and it is evident that after the ATP-dependent "activation" of free amino acids, and the attachment of the amino acyl groups to transfer RNA, transpeptidation reactions (catalyzed by "peptidyl transferase") are important steps in this process (Lipmann, 1969). At the present writing, the relation of the peptidyl transferases to known enzymes is unclear, because they have not yet been isolated in a purified state.

In this section we have seen that Bernard's discovery of glycogen in 1857 led to a more explicit formulation of the problem of assimilation in terms of the metabolic synthesis of a complex molecule from simpler ones. About 12 years later, a metabolic process long considered to be a dissimilation, namely, the formation of urea in the animal body, also was recognized to be a synthetic one. The history of research on the mechanism of this process, to which we now turn our attention, offers further evidence of the interplay of biological and chemical thought in the exploration of the pathways of biochemical change.

Urea Synthesis and Protein Metabolism

At the beginning of the nineteenth century, urea and uric acid represented the only nitrogenous "immediate principles" of the animal organism that had been isolated in crystalline form. The ready crystallization of what came to be called urea had been described by Boerhaave early in the eighteenth century, and in 1773 H. M. Rouelle noted that the material (he called it *matière savonneuse*) gave rise to much ammonia on being heated (see Kurzer and Sanderson, 1956). By 1800, through the work of several investigators, notably Fourcroy and Vauquelin (who introduced the name *urée* in 1798), urea was recognized to be the principal nitrogenous constituent of human urine, and during the years that followed it was widely believed that:

> Since urea and uric acid are the two animal substances richest in azote, one may presume that the secretion of urine is intended to remove from the blood the excess azote supplied by the food, just as the respiration removes the excess carbon (Bérard, 1817, p. 296).

Indeed, Prévost and Dumas (1823) showed the presence of urea in the blood; moreover, they demonstrated, through surgical removal of the kidneys, that these organs merely eliminate urea from the body. In considering the possible physiological site of urea formation, they stated that

> . . . all chemists know that the urine of patients with chronic hepatitis contains little or no urea; which would seem to prove that the functions of the liver are necessary for its formation (Prévost and Dumas, 1823, pp. 100–101).

This view did not appear to win much favor among physiologists (including Bernard) during 1830–1870.

In 1828 there was the dramatic meeting of "inorganic" and "organic" chemistry in the work of Friedrich Wöhler on the reaction of lead or silver cyanate (which he had studied during 1822–1824) with ammonia to produce urea. As he wrote Berzelius in his oft-quoted letter of 22 February 1828,

> . . . I cannot, so to speak, hold my chemical water and must tell you that I can make urea without the need of kidneys or even an animal, whether man or dog . . . . This artificial formation of urea, can it be

regarded as an example of the formation of an organic substance from inorganic materials? (Wallach, 1901, Vol. I, pp. 206–208).

Although this discovery justifiably aroused great enthusiasm among chemists, the attitude of physiologists was more reserved and, as we noted on p. 70, Wöhler's synthesis of urea did not sound the death knell of nineteenth-century vitalism.

The finding that urea might be made from "the already-formed living animal substance" (Marchand, 1838, p. 496), together with quantitative balance studies of Boussingault (1839), provided the basis for the physiological theories of urea formation offered by Liebig and by Dumas in 1842. In both formulations, urea was considered to arise by the oxidative breakdown of protein, and Liebig suggested that uric acid (which he and Wöhler had oxidized chemically to urea and alloxan) represented an intermediate stage of physiological oxidation. In particular, Liebig insisted that

> . . . there can be no greater contradiction than to assume that the nitrogen of the food could pass into the urine as urea without first becoming a constituent of the organized tissues, since albumin, the only blood constituent whose amount justifies consideration, cannot have undergone the slightest change upon its passage through the liver, because we find it in all parts of the body in the same condition and with the same properties (Liebig, 1843, p. 132).

This view was challenged in 1848 by Theodor Frerichs, who inferred from studies on fasting animals that the urea excreted in the urine of animals on a protein diet arises largely from the oxidation, in the blood, of the food proteins. Further opposition came from Carl Gotthelf Lehmann and especially from Friedrich Bidder and Carl Schmidt (1852), whose extensive measurements of the respiratory exchange and nitrogen excretion of fasting and fed animals led them to conclude that only a small part of the urea in the urine of fed animals could be derived from the tissue proteins; Bidder and Schmidt referred to the oxidation of the food proteins in the blood as *Luxusconsumtion.* Although they disagreed with Liebig, and drew a distinction between a small constant breakdown of tissue proteins and a variable *Luxusconsumtion,* they accepted his approach to the chemical interpretation of physiological processes:

> Almost all of the nitrogen of the groupings in albumin and collagen splits off as urea together with the equivalent C—, H—, and O—, while the remainder, representing about $\frac{5}{6}$ of the total heat-generating

material, is subjected to oxidation to carbonic acid and, after fulfilling its calorimetric functions, is eliminated through the gas exchange in the lungs (Bidder and Schmidt, 1852, pp. 386–387).

Among the chemical limitations imposed on this controversy was the lack of a reliable method for the quantitative determination of urea, and Liebig developed a procedure that was used by his adherent T. L. W. Bischoff (1853) in an attempt to refute the views of Liebig's opponents. In Bischoff's opinion, they had

> . . . through their interpretations of their observations dealt a death blow to the doctrine of metamorphosis, as it has sprung from the comprehensive labors and genius of Liebig, and had forever robbed us of the hope of finding a measure of metabolism (Bischoff, 1853, p. 74).

Bischoff's results, however, did not clarify the situation, but new support for Liebig's views came from the later experiments of Bischoff's associate Carl Voit. In 1860 Bischoff and Voit published the results, which they considered to have disproved

> . . . the assumption of a *Luxusconsumtion* of meat according to Frerichs and Schmidt, that is, of an oxidation of food protein in the blood, without entry into the nutrition and transformation of nitrogen-containing parts of the body; since the first administration and increase of a meat diet is not yet sufficient to replace the loss; upon further increase there ensues a balance of loss and replacement (Bischoff and Voit, 1860, p. 249).

Furthermore, they attributed to the more ready oxidizability of fat and carbohydrate the effect of these foodstuffs in reducing the amount of dietary protein needed to maintain body weight, by decreasing the rate of oxidation of tissue proteins. Clearly, these conclusions were based on Liebig's doctrine:

> It will and must remain forever true that only the nitrogenous substances are producers of energy, that is, only they determine the energy effects, the phenomena of motion, in the animal body; and equally it will remain incontrovertible that, upon their transformation, fat and the so-called carbohydrates produce only heat and no effects of motion (*ibid.*, p. 258).

This doctrine crumbled soon afterward; Traube (in 1861) was among the first to raise cogent arguments against it, and in 1865 the climb of

Fick and Wislicenus up the Faulhorn provided data inconsistent with Liebig's idea that the oxidative decomposition of muscle protein to urea is the cause of muscular work (see p. 363).

Shortly afterward, Voit turned against the ideas he and Bischoff had defended; from a series of experiments on dogs brought into nitrogen balance at different levels of protein intake he concluded (in 1867–1869) that there are two kinds of protein in the animal body: a variable pool of "circulating protein" (*circulirendes Eiweiss*) related the level of dietary protein, and the tissue protein (*Organeiweiss*). Of these, only the circulating protein is subjected to oxidation, whereas the tissue protein is stable, and only a small part of the tissue protein is continually regenerated from the circulating protein. There ensued a bitter exchange between Voit and Liebig, during the course of which Voit wrote that Liebig

. . . still stands on the ground he created 25 years ago; from chemical experiments he attempts to draw analogies and conclusions about the processes within the animal body; through his ideas he has had a great influence, and from his creative effort has emerged the entire movement to study the decompositions within the animal body. But he has forgotten, to the sorrow of those who know and value his high services to science better than do his flatterers, that these are only ideas and possibilities, whose validity must first be tested through experimentation on animals, and this is the ground on which I have placed myself (Voit, 1870, p. 399).

Voit's theory was widely accepted among students of human nutrition for several decades, and in 1908 it was further elaborated by his student Max Rubner. It came under immediate attack during the 1870s, however, because of its incompatibility with the growing body of evidence showing that the principal sites of physiological oxidation in the animal organism are not the circulating fluids, but the cells of the tissues.

For example, after noting that Voit's concepts had become commonplace among agricultural chemists, physiologists, and pathologists. Hoppe-Seyler asked "How in the world . . . can Voit have reached any conclusion whatever about the stability or instability of protein in the lymph or organs by weighing dogs, their diet, etc.?" (Hoppe-Seyler, 1873, p. 404), and drew attention to his own work showing that the oxygen of oxyhemoglobin cannot oxidize proteins, sugars, or fats (Hoppe-Seyler, 1866, p. 138). After reviewing the recently acquired knowledge about various metabolic transformations in the tissues, notably that of glycogen in the liver, he concluded that ". . . these organs are not stable, but remarkably variable, with the most active metamorphosis, and consequently wholly dependent on the supply of foodstuffs" (Hoppe-Seyler, 1873, p. 412). One of the chief proponents of this view was Eduard Pflüger, whose theory of physiological oxidations assigned to the tissue proteins great reactivity owing to their content of cyanogen groups (see p. 284). In a characteristically combative paper, Pflüger (1893) subjected Voit's theory to severe criticism, offered evidence in favor of the metabolic stability of circulating protein, and insisted that food protein must first be assimilated by the tissues before it can be metabolized.

The apogee (or nadir) of the metabolic balance studies came in 1905, when Otto Folin reported an elaborate series of analyses of the urine of human subjects maintained on diets of varying nitrogen content. After criticizing adversely the techniques used in the experiments offered in support of Pflüger's hypothesis, Folin used the "laws governing the composition of urine" that he had deduced from his analytical data to conclude that

To explain such changes in the composition of the urine on the basis of protein katabolism, we are forced, it seems to me, to assume that that katabolism is not all of one kind. There must be at least two kinds. Moreover, from the nature of the changes in the distribution of the urinary constituents, it can be affirmed, I think, that the two forms of protein katabolism are essentially independent and quite different. One kind is extremely variable in quantity, the other tends to remain constant. The one kind yields chiefly urea and inorganic sulphates, no kreatinin, and probably no neutral sulphur. The other, the constant katabolism, is largely represented by kreatinin and neutral sulphur, and to a less extent by uric acid and ethereal sulphates . . . . If there are two distinct forms of protein metabolism represented by two different sets of waste products, it becomes an exceedingly interesting and important problem to determine, if possible, the nature and significance of each. The fact that the kreatinin elimination is not diminished when practically no protein is furnished with the food, and that the elimination of some of the other constituents is only a little reduced under such conditions, shows why a certain amount of protein must be furnished with the food if nitrogen equilibrium is to be maintained. It is clear that the metabolic processes resulting in the end products which tend to be constant in quantity appear to be indispensable for the continuation of life; or, to be more definite, those metabolic processes probably constitute an essential part of the activity which distinguishes living cells from dead ones. I would therefore call the protein metabolism which tends to be constant, *tissue* metabolism or *endogenous* metabolism, and the other, the variable protein metabolism, I would call the *exogenous* or intermediate metabolism (Folin, 1905, pp. 122–123).

This formulation of the problem dominated the field for about thirty years, especially among American nutritionists and clinical investigators; we shall mention later the developments that brought its downfall. At this point, we may recall Claude Bernard's comments, made near the beginning of the lengthy series of nutritional studies on protein metabolism, from those of Bidder and Schmidt to those of Folin:

> One can undoubtedly establish the balance between the food consumed by a living organism and what it excretes, but these would be nothing but purely statistical results unable to throw light on the intimate phenomena of the nutrition of living things. It would be, according to the phrase of a Dutch chemist [Mulder], like trying to tell what happens in a house by watching what goes in the door and what leaves by the chimney. One can determine exactly the two extreme limits of nutrition, but if one wishes to construe the intermediate between them, one finds himself in an unknown region largely created by the imagination, and all the more easily because numbers often lend themselves admirably to the proof of the most diverse hypotheses (Bernard, 1865, p. 228).

Many years later, an analogy reflecting a higher state of technology was offered, and the nutritional balance studies were compared

> . . . with the working of a slot machine. A penny brings forth one package of chewing gum; two pennies bring forth two. Interpreted according to the reasoning of balance physiology, the first observation is an indication of the conversion of copper into gum; the second constitutes proof (Schoenheimer and Rittenberg, 1938, p. 222).

Among the "proofs of the most diverse hypotheses" that Bernard may have had in mind was the espousal, by Dumas, of Antoine Béchamp's claim (in 1856) to have produced urea by the oxidation of albumin with permanganate. Dumas welcomed this report because he had attempted unsuccessfully to oxidize albumin to urea in various ways, in order to provide support for his theory that urea is formed in the blood by the combustion of protein. For Dumas, the results of Béchamp confirmed the view that:

> The combustible constituents of the blood therefore yield, as the main products, carbonic acid, water, and urea, unless the last is replaced by products of less extensive combustion (Dumas, 1856, p. 549).

Unfortunately for Béchamp's claim, however, there followed a succession of reports denying the formation of urea upon the oxidation of albumin by permanganate.

The Biosynthesis of Urea. It will be recalled that throughout most of the latter half of the nineteenth century the chemical nature of the albuminoid substances was unclear, and shrouded by physiological speculations about their relation to "living protoplasm." Furthermore, the metabolic fate of the products of the action of pepsin and trypsin on dietary proteins also was in doubt. In particular, the chemical and metabolic relations among the products formed on the hydrolysis of proteins was obscure, and the physiologists studying protein metabolism paid little attention to the amino acids. Thus Otto Schultzen and Marcellus Nencki noted:

> It is striking that hitherto no detailed investigations have been conducted on the behavior of these cleavage products in the animal body, indeed that nobody has suggested that these substances might be the natural intermediates between protein and urea (Schultzen and Nencki, 1869, pp. 566–567).

When they fed glycine or leucine to a dog, they observed an increase in urea excretion. Since these amino acids contain only one nitrogen atom, whereas urea contains two (NH_2—CO—NH_2), Schultzen and Nencki concluded that the metabolic formation of urea must involve a synthetic process. They considered the possibility that urea production might involve the intermediate combination of two molecules of the amino acid, with the oxidation of the product in some unknown manner. This idea became superfluous, however, when Woldemar von Knieriem (1874) showed that the administration of ammonium salts also gives rise to an increased excretion of urea. In addition, he tested aspartic acid, which Ritthausen had found to be a product of protein hydrolysis (p. 109), as well as the closely related asparagine; both amino acids behaved as potential sources of urea in the animal body.

Among the numerous experimental efforts, during 1870–1900, to elucidate the metabolic synthesis of urea the work of Waldemar von Schroeder (1882) was of special significance. He clearly demonstrated that the perfusion of a dog liver with ammonium carbonate (or ammonium formate) led to the extensive formation of urea, thus adding a third function (in addition to the production of bile and of glycogen) to the liver, and providing decisive evidence for the surmise offered by Prévost and Dumas almost sixty years earlier. It may be noted that the technique of the perfusion of isolated organs was devised in Karl Ludwig's laboratory during the 1860s, and subsequently played an increasingly important role in the study of metabolic processes in animal tissues. Schroeder's results were confirmed by numerous workers, but left open the question of the nature of the intermediates between dietary proteins and the ammonia used by the liver for urea synthesis. A major difficulty was the widespread conviction during the last quarter of the nineteenth century that the peptones formed by the gastrointestinal cleavage of dietary proteins were reconstituted into proteins at or near the intestinal wall before entering the circulation. Kühne's observation that free amino acids (leucine, tyrosine) are among the products of the action of pancreatic juice on the cleavage products formed by pepsin was not considered to be significant, and Hoppe-Seyler expressed the view that:

Since acid albumin and peptone are formed in the stomach and by the pancreatic secretion, it appears to be a function of the epithelial cells of the intestine to transform these substances into serum albumin and the fibrin-generating materials (Hoppe-Seyler, 1873, p. 415).

Apparent support for this view was offered by Hofmeister (1882), on the ground that peptones disappeared when in contact with the intestinal wall; although there was a choice between resynthesis to protein or cleavage to amino acids, the latter possibility was distasteful. For example, in successive editions of the treatise by Bunge (first published in 1887) it was stated:

It may be *a priori* doubted on teleological grounds, whether under normal conditions the amount of amido-acids formed in the intestine is a large one. It would be a waste of chemical potential energy, which would serve no purpose when converted into kinetic energy by their decomposition, and a reunion of the products of such a profound decomposition is highly improbable (Bunge, 1902, p. 168).

By the time this edition appeared, however, Otto Cohnheim had shown that

. . . the disappearance of peptones in contact with the intestinal wall does not arise from their assimilation or their reconstitution to protein, but to their further cleavage to simpler products . . . . This change is effected by a special ferment elaborated by the intestinal mucosa, erepsin, which acts on peptones and some albumoses, but not on true proteins (Cohnheim, 1901, p. 465).

Shortly afterward, Cohnheim clearly demonstrated the formation of several amino acids upon the action of erepsin on peptone preparations, and this was confirmed by Emil Abderhalden, in Fischer's laboratory.

A powerful blow to the teleologically attractive view upheld by Hoppe-Seyler, Hofmeister, and Bunge (among others) was delivered by Otto Loewi (1902), who fed an extensively digested "biuret-free" autolysate of pancreatic protein to dogs, and concluded that ". . . the sum of the biuret-free end products replaces food protein, that is, can substitute for all parts of the body protein that is degraded in protein turnover" (Loewi, 1902, p. 316). About twenty years later, Loewi reported his other great experimental achievement, the identification of acetylcholine as the neurohumoral transmitter substance released from nerve endings (see Dale, 1963). Loewi's demonstration that a protein hydrolysate could maintain an animal in nitrogen balance was confirmed by others (*e.g.*, Abderhalden) who used enzymic digests, but the same results could not be obtained with acid hydrolysates of food proteins. The explanation came from the work of Gowland Hopkins (in 1907), showing that the acid-labile amino acid (tryptophan) he had discovered is indispensable for the maintenance of the nitrogen balance of animals; as we noted in the chapter on proteins, later research, notably that of William Rose (1938), established the nature of the other "essential" amino acids required in animal diets. By 1910, therefore, it was clear that proteins are nearly completely digested in the intestine to amino acids, that these amino acids can replace intact dietary proteins, and that attention had to be paid to the "quality" of a dietary protein, and not merely to its nitrogen content, in considering its nutritional value to the organism (Cathcart, 1912, p. 71).

In 1912 Abderhalden offered the idea that proteins are resynthesized in the intestinal wall from free amino acids, but this version of Hoppe-Seyler's theory was short-lived. Its demise came from the work of Donald Van Slyke during 1912–1913, through the use of an apparatus he devised for the gasometric measurement of the nitrogen produced by the action of nitrous acid on amino acids and peptides. With the aid of this apparatus, he obtained data showing that:

> Ingested proteins are hydrolyzed in the digestive tract setting free most, if not all, of their amino-acids. These are absorbed into the blood stream, from which they rapidly disappear as the blood circulates through the tissues (Van Slyke and Meyer, 1913, p. 206).

By 1913 there could be little doubt that the idea of protein resynthesis in the intestinal wall was superfluous, and that free amino acids in the blood are indeed intermediates between food proteins and urea, as had been surmised by Schultzen and Nencki.

Before returning to the problem of urea synthesis, it should be added that Van Slyke's apparatus for the determination of amino-nitrogen, and its use in his study of the amino acids in blood, were only the first in a series of remarkable achievements over a period of about forty years. Through his introduction of new quantitative microtechniques of chemical analysis, especially the versatile manometric apparatus he developed during the 1920s, Van Slyke and his associates elucidated several other important biochemical problems, and their work had a lasting effect on the development of clinical chemistry.

In the face of the uncertainty, around 1880, about the nature of proteins and their fate in metabolism, it is not surprising that the discussions about the chemical mechanism of urea synthesis in the mammalian liver were numerous and inconclusive. These discussions also included the question of the metabolic formation of uric acid, known since early in the century to be the principal nitrogen-containing product excreted by birds and many reptiles, and the counterpart of the urea in mammalian urine. It was also known that the urine of most mammals (except man and higher apes) contain allantoin, shown by Liebig and Wöhler to be a degradation product of uric acid. The work of Knieriem and Schroeder during the 1870s showed that the administration of ammonia or of amino acids (glycine, asparagine, etc.) to hens gave rise to an increased excretion of uric acid, and for many years it was thought that urea might be an intermediate in this synthesis. The fundamental problem thus appeared to be one of explaining how the ammonia (somehow derived from proteins) is used in the liver for the formation of urea. In 1877 Ernst Salkowski suggested a physiological variant of the Wöhler synthesis of urea, on the assumption that "nascent" cyanic acid (HNCO) is an intermediate; for a time it seemed plausible to suppose that cyanate reacts with amino groups to yield ureido compounds (*e.g.*, NH_2—CO—NH—CH—; then called uramido groups) which were cleaved at the NH—CH bond to yield urea. Although the possible role of cyanate was endorsed by Hoppe-Seyler, the failure of later workers to find it in the tissues led to the eclipse of the cyanate theory by alternative hypotheses. Among them was the suggestion of Schmiedeberg (1877), who invoked the new "anhydride" theory of metabolic synthesis and proposed that urea formation involves the dehydration of ammonium carbonate [$(NH_4)_2CO_3 \rightarrow NH_2$—CO—$NH_2$ + 2 H_2O]. From his experiments, he concluded that

. . . the nitrogen-containing compounds in which the N is present in the group NH_2—CH_2— break down in the body with the formation of ammonia, and the carbonate of the latter is promptly converted to urea by synthesis (Schmiedeberg, 1877, p. 14).

On the other hand, Edmund Drechsel (in 1880) preferred a process in which the ammonium salt of carbamic acid (NH_2—COOH) combines with ammonia to form urea. Furthermore, in 1896 Hofmeister revived Béchamp's idea that urea arises by the oxidation of proteins, and showed that treatment of proteins (egg albumin, gelatin) with alkaline permanganate did indeed produce urea. By that time, however, a new clue to the source of urea had emerged, through the identification of arginine as a product of the hydrolysis of proteins.

As we noted earlier (p. 110), Drechsel had isolated from an acid hydrolysate of casein a basic material ("lysatine") which later proved to be a mixture of lysine and arginine. After finding that lysatine yielded urea upon treatment with alkali, Drechsel concluded that his results

> . . . incontrovertibly show that urea arises from protein without any oxidation, simply by hydrolysis, and we can conclude from this that urea is made in the animal organism in this manner . . . . It cannot be claimed, however, that the total urea of the urine is formed in this way; on the contrary, it is only one pathway along with others that lead to the same end (Drechsel, 1890, p. 3101).

Shortly afterward, Ernst Schulze showed that the arginine he had isolated in 1886 from plants was the urea-producing component of lysatine, and by 1898 he had demonstrated that arginine yields, on treatment with alkali, the amino acid ornithine (known since 1878 from the work of Jaffé) in addition to urea. By that time arginine was recognized to be a constituent of many proteins, and it seemed possible therefore that this amino acid might be a metabolic source of urea.

Albrecht Kossel, who had found protamines to be rich in arginine, then suggested that, in addition to the amide linkage (—CO—NH—) postulated by Fischer and Hofmeister as a general structural feature of proteins, and the guanidino group (NH_2—C(=NH)—NH—) identified by Schulze in arginine, there might be combinations of the two *i.e.*, —CO—NH—C(=NH)—NH—. This idea was related to Kossel's hypothesis that the protamines represent fundamental structural units of all proteins. Accordingly, Kossel and Dakin (1904) subjected a protamine to the action of the peptone-cleaving enzyme (erepsin) described by Cohnheim, and found that, depending on the conditions of the test, either arginine or ornithine was formed; they concluded, therefore, that the enzyme preparation contained, in addition to the enzyme cleaving

CO—NH bonds, another enzyme that hydrolyzes arginine in the manner described by Schulze. This enzyme, which they named "arginase" (see accompanying scheme) was found to be especially abundant in liver. The discovery of arginase by Kossel and Dakin explained the earlier observations by Charles Richet (in 1894–1897) on the production of urea in macerated liver; together with the parallel demonstration by Ernst

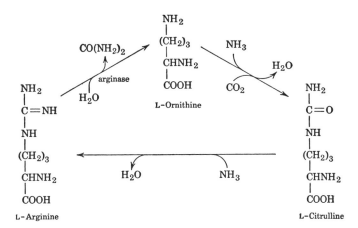

The Ornithine Cycle

Salkowski and by Martin Jacoby that animal tissues such as liver contain "autolytic" enzymes that can degrade proteins to peptones and amino acids, it seemed clear that the combined action of these enzymes and arginase could give rise to urea. It will be evident that these conclusions were consistent with the new outlook on the role of intracellular enzymes in metabolic processes, after Buchner's preparation of a cell-free zymase.

Subsequent work on arginase, especially by Antonino Clementi, strengthened the view that this enzyme plays a role in the metabolic production of urea. Of special importance was its apparent absence from the livers of birds and reptiles which excrete uric acid, in contrast to its presence in the livers of mammals, amphibians, and fishes that excrete urea. Clementi termed the first group of animals "uricotelic" and the second "ureotelic," and his generalization was largely confirmed in the subsequent work of Siegfried Edlbacher and Andrew Hunter during the 1920s. By that time the quantitative estimation of urea (and hence of arginase activity) had been made more reliable, through the use of urease as a specific reagent to convert urea into CO_2 and NH_3, either of which could readily be determined. Indeed, it was the introduction of this enzyme for the quantitative determination of urea in biological fluids that led

James Sumner, who had worked in Folin's laboratory on the problem of urea formation, to undertake work in 1917 on the purification of urease (see p. 157).

Although it seemed by 1930 that arginase is a participant in the metabolic formation of urea, it was evident that the massive conversion of ammonia into urea in the liver could not be accounted for solely in terms of the hydrolytic cleavage of the arginine derived from proteins. During the 1920s the long-discarded cyanate theory reappeared in new guise (Werner, 1923), and received respectful attention from some biochemists. More importantly, Wilhelm Löffler had shown (in 1917–1920) that the biosynthesis of urea requires the presence of oxygen, and had concluded that "Urea formation in the surviving liver is bound to the integrity of the cell structure; oxidative processes play an integrating role" (Löffler, 1920, p. 177). This view was supported by the failure of efforts (such as that of Kase in 1931) to demonstrate the formation of urea from ammonia in minced liver preparations.

In 1932 entirely new light was thrown on the problem of urea synthesis by Hans Krebs and Kurt Henseleit. Their work showed that:

> The primary reaction of urea synthesis in the liver is the addition of 1 molecule of ammonia and 1 molecule of carbonic acid to the δ-amino group of ornithine, with the elimination of 1 molecule of water and the formation of a δ-ureido acid, citrulline . . . . The second reaction of urea synthesis is the combination of 1 molecule of citrulline with an additional molecule of ammonia, with the loss of a second molecule of water and the formation of a guanidino acid, arginine . . . . The third reaction is the hydrolytic cleavage of arginine to ornithine and urea (Krebs and Henseleit, 1932, pp. 33–34).

The reactions are depicted in the scheme on p. 435.

The achievement summarized in the sentences just quoted marked a new stage in the development of biochemical thought. Not only was an explanation of a biochemical synthesis offered for the first time in terms of chemical reactions identified in the appropriate biological system, and not merely inferred by analogy to the known chemical behavior of the presumed reactants, but also the paper of Krebs and Henseleit provided a clue to the organization of metabolic pathways in living cells. This became evident to many biochemists in 1937, with the appearance of the "Krebs citric acid cycle," whose conceptual relation to the earlier "ornithine cycle" was obvious. Krebs (1947) later generalized the concepts that had emerged from these advances in terms of the physiologically unidirectional character of metabolic cycles, because of the "irreversibility" of some of the intermediate steps. Although Krebs's work on urea synthesis had far-reaching consequences, its initial objectives were more circumscribed. Many years later Krebs wrote:

> When I set out to study the mechanism of urea formation, I had no preconceived hypothesis. My idea was that the in vitro system of liver slices, incubated in a saline medium, provided a novel and easy method for measuring accurately the rate of urea synthesis under a variety of conditions. As one small rat liver supplies numerous samples, it is possible to carry out many parallel tests. No comparable method had been available before. Many biochemists and physiologists—including Otto Warburg—were quite surprised to learn that slices are capable of performing synthetic energy-requiring processes. Up to that time slices had been used solely for the study of degradative reactions—respiration and lactic acid formation. The use of tissue slices which I had learned in Otto Warburg's laboratory, so it seemed to me, opened up an entirely new kind of approach to many problems of metabolism. I chose the study of the synthesis of urea in the liver because it appeared to be a relatively simple problem (Krebs, 1970, pp. 161–162).

When such liver slices were incubated with an ammonium salt, urea was formed in amounts concordant with the equation $2 NH_3 + CO_2 = NH_2—CO—NH_2 + H_2O$. A variety of substances (including many amino acids) were tested for their effect on the rate of this process; among those that were ineffective were ammonium cyanate and ammonium carbamate. The action of most of the amino acids could be explained in terms of their deamination to yield ammonia, but unexpectedly ornithine was found to promote urea synthesis in a catalytic manner. Since the arginase reaction was already known, the effect of ornithine appeared to indicate that it was being used, together with NH_3 and CO_2, in the energy-requiring synthesis of arginine, and it was reasonable to ask whether the δ-ureido derivative of ornithine is an intermediate in this synthesis. Only two years earlier, Mitsunori Wada (1930) had proved, through chemical synthesis, that a substance isolated in 1914 from watermelon juice (and hence named citrulline) is δ-ureido-ornithine; when liver slices were found to produce urea rapidly in the presence of citrulline and ammonia (but not citrulline alone), the experimental evidence for the operation of the ornithine cycle was in hand.

We noted before that attempts to demonstrate the synthesis of urea from ammonia in disintegrated tissues were unsuccessful, and that the small urea formation observed in cell-free extracts (as in the experiments of Richet) was attributable to the action of the soluble enzyme arginase on the arginine produced by the autolysis of liver proteins. These findings were consistent with views inherited from the nineteenth century; with the recognition that animal cells can perform chemical syntheses, and that the nucleus plays a special role in cell division, many biologists considered such syntheses to be effected in the nucleus and to be characteristic of the living state. For example, when Hofmeister studied the formation of methyl groups in the animal body, and failed in attempts to obtain an active tissue extract, he concluded that "It is only a special case of the general rule that cleavages and oxidative processes can also occur in lifeless Nature, but syntheses are reserved to life" (Hofmeister, 1894, p. 214), and Loewi criticized the experiments of Richet on the ground that "The synthetic formation of urea from its precursors has not been imitated outside the living cell, any more than it has been successful for other vital syntheses" (Loewi, 1898, p. 521). Similarly, when Krebs found that, after maceration, liver tissue had lost its ability to convert ammonia into urea even in the presence of ornithine and of lactate (which greatly promoted urea synthesis by slices), he concluded that the steps leading from ornithine to arginine are "linked to life" (Krebs and Henseleit, 1932, p. 53), and dependent on the integrity of the cellular structure.

By 1946, however, there was available a cell-free liver "homogenate" that made urea from NH_3 and CO_2. During the intervening years, the metabolic importance of cofactors such as DPN and ATP had become evident, largely through the work of Warburg who, during 1910–1930, had firmly insisted on the dependence of intracellular respiration on organized cell structure. The earlier failure to obtain a cell-free system for urea synthesis turned out to be, in part, a consequence of the dilution and destruction of essential cofactors; by adding DPN, ATP, and cytochrome c, Philip Cohen and Mika Hayano (1946) succeeded where earlier workers had failed.

The subsequent developments in the study of the details of the ornithine cycle are recounted in current textbooks of biochemistry. The high points were (1) the demonstration by Sarah Ratner (1954) that in the conversion of citrulline into arginine the added nitrogen does not come directly from ammonia, but by enzyme-catalyzed ATP-dependent transfer from aspartic acid, with argininosuccinic acid as an intermediate (see the accompanying scheme), and (2) the work of Mary Ellen Jones, Leonard Spector, and Fritz Lipmann (1955) showing that the formation of citrulline involves the enzyme-catalyzed reaction of ornithine with an "energy-rich" carbamylating agent (carbamoyl phosphate), which is made in a separate enzymic reaction: $NH_3 + CO_2 + ATP \rightarrow NH_2CO—OPO_3H_2 + ADP$.

For many years, the synthesis of uric acid in uricotelic organisms was considered to proceed by way of urea; indeed, in 1904 Hugo Wiener suggested that, in uric acid formation, two molecules of urea and one of tartronic acid [HOOC—CH(OH)—COOH] were somehow combined, thus accounting neatly for the source of all the carbon and nitrogen atoms of the purine. This idea was soon disproved by Richard Burian and others; subsequent work by Clementi (in 1930) showed that urea could be used for uric acid synthesis only after prior cleavage to ammonia by the urease of the intestinal bacteria, and three years later Krebs demonstrated that hen liver slices able to use NH_3 for uric acid synthesis did not use urea for this process. The metabolic pathway leading to uric acid was not elucidated until after World War II; aside from the involvement of the well-known xanthine oxidase for the oxidation of hypoxanthine and xanthine to uric acid, the pathway turned out to be a complex sequence of enzyme-catalyzed reactions for many of which there was no precedent. After Kossel (1881) showed that nuclein yields purines on hydrolysis (see p. 189), the possibility was considered that uric acid might arise by the metabolic degradation of nucleic acids; it was soon recognized, however, that this source could not account for the massive production of uric acid in birds.

It is of considerable interest that many of the same questions encountered in the study of urea formation in animals were faced by plant physiologists concerned with the problem of asparagine formation in seedlings. As in the case of urea, asparagine was discovered (by Vauquelin and Robiquet, in 1806) because of its ready crystallization from concentrated aqueous solutions (in this case, asparagus juice). In 1844 Raffaele Piria confirmed the isolation for vetch seedlings, but reported that no asparagine could be obtained from ungerminated seeds, or from mature plants exposed to light; he suggested that "It is probable that there exists, in the seeds, an azotized material (perhaps casein) which is transformed into asparagine and other products during germination" (Piria, 1844, p. 576). Subsequently, from changes in the carbon and nitrogen content of germinating seedlings, Boussingault concluded (in 1868) that there is a respiratory combustion of seed protein to form asparagine, and compared this process to what was then believed to be the oxidative conversion of protein into urea in animals. This comparison also was made by Wilhelm Pfeffer (1872), who developed a theory that the asparagine formed during germination is derived from the seed proteins and cannot be excreted (as is the urea in animals) but diffuses in the growing plant to sites where it is resynthesized into protein. In the formulation of this theory, Pfeffer was influenced by the new findings of the agricultural chemist Heinrich Ritthausen; understandably, Ritthausen's results were more widely appreciated by the contemporary plant physiologists than by the medically inclined physiological chemists.

Pfeffer's theory became untenable after the work of Ernst Schulze, described in a series of papers (beginning in 1876) that represent a high point of nineteenth-century biochemistry (see Chibnall, 1939). Although the theory continued to be widely accepted among botanists, it was evident from Schulze's results that, during the first stages of plant growth, a variety of amino acids (leucine, tyrosine, etc.) as well as peptones are formed by the enzymic breakdown of seed proteins, in a process comparable to the gastrointestinal digestion of proteins in animals. Since asparagine was found to appear later, it clearly represented a secondary product of protein breakdown, possibly derived from ammonia formed by the decomposition of the amino acids. Schulze's approach was taken up by his student Dmitri Prianishnikov from 1897 onward; in an early paper the latter wrote in a manner reminiscent of Schultzen and Nencki (1869):

Since asparagine contains two NH_2-groups in the molecule, and the primary amino acids (leucine, aminovaleric acid, tyrosine, etc.) contain only one, it must be assumed that in the formation of an asparagine molecule the residues of two molecules of some primary amino acid (or two different amino acids) are involved (Prianishnikov, 1904, p. 42).

Twenty years later, his results on the nitrogen metabolism of plants, and the parallel studies on the metabolic fate of amino acids in animal tissues, led Prianishnikov to underline anew the similarity between the roles of asparagine and urea:

> Both amides are not primarily products of the hydrolytic cleavage of proteins, but are formed in a secondary manner, involving above all ammonia formed by the oxidation of amino acids; both amides are formed equally easily at the expense of externally supplied ammonia, with the obvious difference that for the formation of asparagine a part of the unoxidized carbon chain is necessary, whereas ammonia and carbonic acid suffice for the formation of urea. Not only the mechanism, but also the physiological purpose of the two amides is entirely similar—it consists in the detoxification of ammonia . . . . [T]he further fate of asparagine and urea is quite different: urea is excreted by the organism, since the animal does not need to economize on nitrogen, because it is abundantly supplied in the food . . . ; asparagine, on the other hand, remains in the cells as a reserve form of nitrogen which, with a more abundant supply of carbohydrate, can later contribute to various components of the protein molecule (Prianishnikov, 1924, pp. 421–423).

The work of Krebs on urea synthesis in mammalian liver, and of Prianishnikov on asparagine synthesis in plants, made it clear by the 1930s that the key to the understanding of the factors controlling the metabolism of these two amides lay in the elucidation of the metabolism of the amino acids that constitute protein molecules.

The Metabolism of Amino Acids. In 1909 Otto Neubauer had shown that amino acids are oxidatively deaminated in animal tissues to the corresponding keto acids: $R—CH(NH_2)—COOH + O \rightarrow R—CO—COOH + NH_3$. This discovery was followed by Gustav Embden's demonstration (in 1910–1912) that the perfusion of dog liver with α-keto acids and ammonia gives rise to the formation of the corresponding amino acid, *i.e.*, the reverse of the oxidative process, thus providing a metabolic link between the amino acid constituents of proteins and the nitrogen-free intermediates derived from carbohydrates and fats. Furthermore, Franz Knoop (1910) reported that, if the keto acid $C_6H_5—CH_2—CH_2—CO—COOH$ is fed to a dog, the corresponding α-amino acid (in the form of its N-acetyl derivative) appears in the urine. The use of such unphysiological phenyl derivatives for metabolic studies was introduced by Knoop

(1904) in connection with his important work on the oxidation of fatty acids; we will return to this later. The results provided by Embden and Knoop thus showed that animals, like plants, can effect the synthesis of amino acids. In particular, the conversion of lactic acid (then thought to be the normal product of carbohydrate breakdown) into alanine in the liver was considered to involve the intermediate formation of ammonium pyruvate, and gave support to the view, based on the effect of dietary carbohydrate on nitrogen balance, that ". . . carbohydrates are absolutely essential for endo-cellular synthetic processes in connection with protein metabolism" (Cathcart, 1909, p. 329). It should be added that Knoop explained the formation of the N-acetyl derivative as arising from the generation of an acetyl group by the decarboxylation of pyruvic acid, in analogy to the findings around 1900 (deJong, Erlenmeyer) that the chemical reaction of keto acids with ammonia yields acetylamino acids. The metabolic acetylation reaction continued to attract attention for many years; its mechanism was not clarified, however, until after the discovery of acetyl-coenzyme A.

With the rise of the dehydrogenation theory of Wieland, and the demonstration of amino acid dehydrogenases by the Thunberg technique (see p. 321), the intracellular deamination of amino acids began to be considered as an enzyme-catalyzed process. Furthermore, Knoop and Oesterlin (1925) offered a chemical model of the reverse reaction, by effecting the synthesis of α-amino acids through the hydrogenation (in the presence of platinum or palladium as catalyst) of a mixture of α-keto acids and ammonia. Decisive evidence on the intracellular deamination of amino acids in animal tissues only came during the 1930s, however, through the work of Hans Krebs. By use of the tissue-slice technique, he found that, in contrast to urea synthesis, the liver is much less effective in the production of ammonia from amino acids than the kidney. This finding helped to explain the observation of Thomas Nash and Stanley Benedict (in 1921) that the ammonia excreted in human urine is made in the kidney; in the urine of normal subjects, ammonia represents a small fraction of the total nitrogen, but its concentration rises in pathological states such as diabetic acidosis. In addition, Krebs (1935a) showed that the kidney contains separate oxidative deaminases for L- and D-amino acids; as was mentioned on p. 337, the D-amino acid oxidase turned out to be an FAD-containing flavoprotein. Of special significance was Krebs's observation that L-glutamic acid, which markedly enhanced the rate of oxygen uptake by kidney slices, diminished the rate of ammonia formation. Krebs (1935b) explained this phenomenon by showing that the kidney converts ammonium glutamate into glutamine, and that this energy-requiring process is also effected by brain tissue.

Although glutamine was a new component of the metabolism of animal tissues, it had long been known to be a constituent of plants, since its isolation in 1883 from beet juice by Ernst Schulze; he considered glutamine to be equivalent to the less-soluble asparagine in plant metabolism. By the 1930s, with the development of analytical methods that permitted the separate determination of the two amides, it became clear that they both occur in all plants, but in widely different proportions. After World War II, the formation of glutamine from glutamic acid and ammonia (in plants, microorganisms, and animals) was shown to be effected by an ATP-dependent enzyme (glutamine synthetase), and a similar asparagine synthetase was identified in plant tissues and in microorganisms. Subsequent work showed that glutamine represents a key participant in numerous metabolic processes (Meister, 1956).

It was already known before 1940 that the reversible dehydrogenation of glutamic acid and of aspartic acid is effected in plants and animals by pyridine nucleotide-dependent enzymes, with α-ketoglutaric acid and oxaloacetic acid as the respective products of oxidative deamination. As we saw earlier (p 382), these two keto acids are components in the citric acid cycle formulated by Krebs and Johnson (1937), thus linking glutamic acid and aspartic acid to the metabolism of pyruvate and of the carbohydrates from which it is derived. Moreover, in the same year, Alexander Braunstein and M. G. Kritzmann (1937) reported that, in minced preparations of pigeon-breast muscle, the α-amino group of glutamic acid is transferred reversibly to pyruvic acid (to form alanine) or to oxaloacetic acid (to form aspartic acid). This discovery gave reality to the suggestions arising from the biochemical studies of Dorothy Needham (1930) and the chemical work of Robert Herbst and Lewis Engel (1934). The existence of the enzyme-catalyzed "transamination" identified by Braunstein and Kritzmann not only gave further evidence of the metabolic importance of group-transfer reactions, but also emphasized the central position of glutamic acid and aspartic acid in nitrogen metabolism; as we shall see shortly, this conclusion was powerfully reinforced by the parallel work of Rudolf Schoenheimer, with ^{15}N as an isotopic tracer. It should be added that, after World War II, Fritz Schlenk and Esmond Snell found that the tissues of vitamin B_6-deficient rats have low transaminase activity; this discovery led to the identification of the cofactor of the transaminases as pyridoxal phosphate, whose parent substance (pyridoxine, vitamin B_6) had been isolated in 1938 and synthesized a year later. Subsequent work showed pyridoxal phosphate to be a component in a variety of enzyme-catalyzed reactions in which amino acids are involved (see Snell, 1958).

We mentioned earlier that the work of Sarah Ratner during the 1950s showed aspartic acid to be an obligatory reactant in the formation of arginine from citrulline. The transamination reaction discovered by Braunstein thus provides a link between the two cycles that Krebs had formulated during the 1930s (see the accompanying scheme).

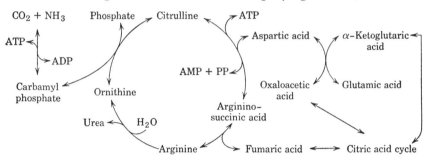

The Ornithine Cycle, and Its Relation to the Citric Acid Cycle.

To conclude this section, we may note that, in discussing criticisms made of the ornithine cycle, Krebs stated:

> The ornithine cycle is a theory; it is a supposition designed to account for a number of observations and to serve as a starting point for further investigations . . . . These observations 'prove', i.e. establish as fact, the occurrence of each step of the cycle as postulated by the theory. This of course is true only for the conditions of the experiments and it can be said with some justification that the reactions have not been proved to occur under 'physiological' conditions, i.e. in the intact liver in situ. In all experiments some facts were 'unphysiological'. The liver was taken out of the body and sliced or artificially perfused. The concentrations of the reactants were artificially raised. Hence the statement that the ornithine cycle is operative in the liver in situ implies the assumption (which some do not wish to make) that the cycle occurs under conditions different from those of the experiments. In this sense it may be formally correct to say that the cycle has not been proved to occur in 'normal' liver, but such a statement is of little scientific value for it applies to every intermediary reaction mechanism. Experiments can only prove that a tissue has the ability to perform certain chemical reactions, but if these reactions are intermediary stages, their 'normal' occurrence cannot be proved (Krebs, 1942, pp. 764–765).

As we have already seen, additional evidence supporting the validity of the ornithine cycle was offered in the years after these lines were written.

Moreover, powerful new experimental techniques became available to test hypotheses about metabolic pathways—the use of isotopic tracers (especially ^{14}C) and of mutant strains of microorganisms. It is to these developments that we now turn our attention.

Tracers and Mutants in Metabolic Research

In a famous lecture, Gowland Hopkins stated that

> . . . in the study of the intermediate processes of metabolism we have to deal, not with complex substances which elude ordinary chemical methods, but with simple substances undergoing comprehensible reactions. By simple substances I mean such as are of easily ascertainable structure and of a molecular weight within a range to which the organic chemist is well accustomed . . . . [I]t is not alone with the separation and identification of products from the animal that our present studies deal; but with their reactions in the body; with the dynamic side of biochemical phenomena (Hopkins, 1913, p. 214).

This view reflected those expressed earlier by Emil Fischer and by Henry Dakin:

> The ultimate aim of biochemistry is to gain complete insight into the unending series of changes which attend plant and animal metabolism. To accomplish a task of such magnitude complete knowledge is required of each individual chemical substance occurring in the cycle of changes and of analytical methods that permit of its recognition under conditions such as exist in the living organism. As a matter of course, it is the office of organic chemistry, especially of synthetic chemistry, to accumulate this absolutely essential material (Fischer, 1907b, p. 1753).

> A true knowledge of metabolic processes can only be obtained by the tedious unraveling of the complex system of biochemical changes into individual chemical reactions. At the present time only a few of these simple reactions have been recognized and studied, but even now it requires little imagination to realize that in the future it will be possible to construct an accurately itemized account of the animal body's chemical transactions both anabolic and catabolic. The value of such knowledge for the advancement of biology and medicine is sufficiently obvious (Dakin, 1912, p. 2).

Indeed, at the time Dakin wrote these lines, there was only fragmentary evidence for the conviction that metabolic processes (*e.g.*, the oxidative conversion of the carbon atoms of stearic acid into CO_2) involves "a complicated series of chemical reactions following upon each other in definite sequence" (*ibid.*, p. 2). A quarter-century later, there had emerged the Embden-Meyerhof pathway and the Krebs ornithine and citric acid cycles; within another quarter-century, the advance of enzyme chemistry and the use of isotopic tracers and of microbial mutants had produced so elaborate a proliferation of metabolic pathways as to boggle the minds of students of biochemistry. As new metabolic intermediates were found, and the variety of the anatomical and physiological characters of living creatures was matched by the diversity of their chemical apparatus, the goal set by Fischer of identifying completely "each individual chemical substance," or by Dakin of constructing "an accurately itemized account," increasingly seemed unattainable.

Toward the close of the nineteenth century, however, the term "intermediary metabolism" had relatively meager experimental foundation. When Franz Hofmeister wrote in his eulogy of Willy Kühne that the latter had foreseen that

> . . . the answers to the most immediate and fundamental physiological questions depend above all on three fields only accessible to the chemist, those of protein chemistry, of the action of ferments, and of intermediary metabolism (Hofmeister, 1900, p. 3877),

Hofmeister referred primarily to Kühne's work on the role of peptones as intermediates in the digestive breakdown of proteins. Similarly, the dextrins were considered to be metabolic intermediates between glycogen and glucose, and uric acid was long thought to be an intermediate in the metabolic oxidation of proteins to urea. In addition to these doubtful items of evidence, there was a series of findings, many of them haphazard, that later led to fruitful enquiry; among them were the isolation of several curious organic substances from mammalian urine, and the application of the new structural organic chemistry to the determination of their constitution.

Two of these compounds were components of dog urine; one was found in 1853 by Liebig, who named it kynurenic acid (Greek *kynos*, dog + *ouron*, urine), the other in 1874 by Max Jaffé, who called it urocanic acid (Greek *ouro* + Latin *canis*, dog). The latter compound was more elusive, as it did not appear in the urine of all dogs, and the one that produced it for Jaffé ran away; subsequent work, however, confirmed the validity of his report. Another unusual chemical compound, found in the urine of human subjects with alcaptonuria (to which we referred on p. 228), provided information that greatly influenced later research on intermediary metabolism. Extensive chemical work during three decades after Bödeker's description of the darkening of the urine of such patients showed that a strongly reducing catechol-like compound (homogentisic acid) was responsible for this phenomenon. A decisive advance was made when Michael Wolkow and Carl Baumann (1891) showed that the compound is 2,5-dihydroxyphenylacetic acid (see accompanying formula), and that its excretion by an alcaptonuric subject is

promoted by the ingestion of tyrosine. Although they inferred from the clinical report that the disease represented a "metabolic abnormality that lasted throughout the patient's life" (*ibid.*, p. 237), they also concluded that

. . . the formation of homogentistic acid from tyrosine is not effected by an inexplicable abnormal metabolic function in the tissues, but is to be regarded as the action of a particular kind of microorganism (*ibid.*, p. 277).

This statement reflected the current fashion of attributing disease to the action of microbial agents, and indeed evidence had accumulated that intestinal bacteria can degrade some amino acids.

This evidence came after the chemical work of Kekulé, Baeyer, and Fittig on benzene derivatives, and from the studies of several physiological chemists (notably Nencki, Jaffé, and Baumann) on the metabolic fate of such "aromatic" compounds in animals. For example, when Jaffé fed benzoic acid to birds, he found (in 1877–1878) that, in place of the hippuric acid excreted by mammals, the urinary product is a dibenzoyl derivative (ornithuric acid) of a new basic substance which he named ornithine (Greek *ornithos*, bird). Also, Baumann found that, upon the administration of bromobenzene to dogs, there appears a urinary product that contains both bromine and sulfur; in 1881 he showed this substance (bromophenylmercapturic acid) to be a derivative of cysteine in which the bromophenyl group is attached to the sulfur of the amino acid, and that the amino group is acetylated. Although Baumann deduced the incorrect structure for cysteine, his work provided the basis for the later correction, made by Ernst Friedmann. The metabolic site of the formation of these curious compounds was not known, but extensive studies by Nencki, Baumann, and others during 1870–1890 on the fate of aromatic compounds in the animal body had indicated that urinary products such as phenol or indole were formed by the action of putrefactive bacteria acting on substances derived from food proteins. It seemed reasonable for Wolkow and Baumann (1891) to conclude, therefore, that homogentisic acid had been made from tyrosine by bacterial action.

In 1903 Wilhelm Falta and Leo Langstein showed that phenylalanine also gives rise to the excretion of homogentisic acid in alcaptonuria, and chemical considerations inclined them to the view that phenylalanine is an intermediate in the conversion into tyrosine. A year later Neubauer and Falta tested an extensive series of aromatic acids, among them the α-keto acids corresponding to phenylalanine and tyrosine, which also turned out to be precursors of homogentisic acid. During the course of this work, they reached the conclusion that

> . . . in the normal organism the combustion of the aromatic amino acids proceeds by way of the alcapton acids and that the disorder in alcaptonuria consists only in the fact that, as a consequence of the inhibition of metabolism, the degradation stops at this point (Neubauer and Falta, 1904, p. 91).

We mentioned on p. 229 that Archibald Garrod (1902) had already noted the hereditary character of the disease, and that Bateson had interpreted the available knowledge in terms of the newly rediscoverd Mendelian genetics. After the work of Neubauer and Falta, Garrod showed that the homogentisic acid excreted by alcaptonuric subjects corresponded to the "whole of the tyrosine and phenylalanine broken down" (Garrod and

Hele, 1905, p. 205). It was clear that the formation of homogentisic acid is not a consequence of bacterial action in the digestive tract, but a step in the normal pathway of the degradation of these two aromatic amino acids in the animal body.

Subsequent studies forced a revision of the view that homogentisic acid arises from tyrosine by way of phenylalanine, and it was recognized that phenylalanine is first oxidized to tyrosine; in particular, Neubauer found that 2,5-dihydroxyphenylpyruvic acid is an excellent precursor. By 1910 the metabolic evidence, together with analogies drawn from chemical studies, led to the formulation of the sequence shown on p. 447. An essential step in the scheme is the oxidative deamination of tyrosine to the corresponding α-keto acid; it was indeed the recognition of this metabolic reaction that led Neubauer to investigate the general problem of the oxidative deamination of amino acids and to obtain, as we have already seen, results of far-reaching significance in the study of the intermediary metabolism of amino acids and of carbohydrates.

After the discovery of tryptophan by Hopkins and Cole in 1901, this new aromatic amino acid also was tested by Garrod and Neubauer as a source of homogentisic acid, with a negative result; shortly thereafter, Alexander Ellinger (1904) found, however, that, when tryptophan is administered to dogs or rabbits, kynurenic acid is excreted. By that time, this urinary product was known to be a derivative of quinoline (studied by Skraup and others during the 1880s), and a structure closely approximating that later shown to be the correct one (see accompanying formula)

Tryptophan   Formylkynurenine

Kynurenic acid   Kynurenine

had been deduced by Rudolf Camps in 1901. The excretion of kynurenic acid was considered to be associated with protein breakdown, and some biochemists took it to suggest the presence of "quinoline-like radicals in proteid molecules" (Mendel and Jackson, 1898, p. 28). This view was abandoned after Ellinger's finding; furthermore, it was evident that the metabolic transformation could occur only in a multistep process, as in the conversion of phenylalanine and tyrosine into homogentisic acid. In succeeding years, speculations were offered about the nature of the intermediates, but the pathway shown was not elucidated until after the isolation and proof of the structure of kynurenine by Adolf Butenandt in 1943, and the application of genetic and isotope techniques. Urocanic acid, after it had been reisolated, was shown by Andrew Hunter (1912) to be an imidazole derivative, and a product of the metabolic deamination of the known amino acid histidine (see accompanying formula).

$$
\begin{array}{ccc}
\text{L-Histidine} & \xrightarrow{\quad NH_3 \quad} & \text{Urocanic acid}
\end{array}
$$

L-Histidine   Urocanic acid

The Metabolism of Fatty Acids. The finding of these curious urinary products, and the recognition that they were derived from the metabolism of protein amino acids, were typical of the haphazard development of the study of intermediate metabolism before about 1900. They reflected the rise of "pathological chemistry" and the entrance of the new structural organic chemistry into clinical practice and research. In this application of chemistry to medicine, special attention was given to diabetes mellitus, considered by some to be a "purely chemical derangement of health" (Bence-Jones, 1867, p. 40); indeed, chemical observations on diabetic patients opened one of the most fruitful avenues of research on intermediary metabolism. During the period 1857–1876 the odorous material exhaled by diabetic patients was identified as acetone (CH_3—CO—CH_3), known since the seventeenth century to be formed when lead acetate was heated; acetone was found in the blood and urine of such patients, and the coma in the advanced stages of the disease was associated with the appearance of this ketone (Petters, Kaulich, Kussmaul, Markovnikov). In 1883 Rudolf Jaksch found in diabetic urine the closely related acetoacetic acid (CH_3—CO—CH_2—$COOH$), whose structure had recently been elucidated, and which was known to give rise to acetone by loss of CO_2. One year later an optically active (levorotatory) β-hydroxybutyric acid [CH_3—$CH(OH)$—CH_2—$COOH$] also was found there, by Eduard Külz and by Oscar Minkowski; the chemical relation of this compound to acetoacetic acid had already been established by Wislicenus. The production of these three compounds in diabetes was difficult to reconcile with the long-held view that the disease is associated with the inability of the body to oxidize glucose to CO_2 and H_2O (see p. 406). Acetone was first considered to be the product of the incomplete oxidation of sugar, but, with the discovery of the two acids known to be chemically related to the fatty acid butyric acid (CH_3—CH_2—CH_2—$COOH$), attention shifted to the fats as metabolic sources of β-hydroxybutyric acid (considered to be the precursor of the two other compounds). Also, by 1885 it was known that these "ketone bodies" are formed in normal human subjects kept on a diet free of carbohydrate. A brief flurry of interest during the 1890s in "tissue protein" as the source of the ketone bodies (Jaksch, Noorden) subsided after the systematic studies of Adolf Magnus-Levy and others clearly established the fatty acids as the major metabolic sources. The problem thus became one of determining the pathway of the normal oxidation of fatty acids, as this appeared to be blocked at an intermediate stage in diabetes. It will be recalled that before 1890 many physiologists shared Pflüger's view of biological oxidation as "explosive" combustions effected by "living proteins"; for them the idea of a stepwise chemical process in the complete metabolic degradation of a molecule like stearic acid [$CH_3(CH_2)_{16}COOH$] was superfluous. With the gradual acceptance among physiological chemists of the view that the living cell effects such processes through the catalytic action of specific enzymes, and with the formation of intermediates, metabolic pathways became subjects not only of chemical speculation but also of experimental study. This attitude was reflected in many writings at the turn of the century, especially in the influential lecture by Hofmeister (1901). Indeed, it was Hofmeister's associate Franz Knoop (1904) who led the way.

Knoop fed dogs the sodium salts of various straight-chain fatty acids in which the carbon atom farthest from the carboxyl group was linked to a phenyl group (*e.g.*, phenylbutyric acid, $C_6H_5CH_2CH_2CH_2COOH$). The rationale of this approach was that, whereas naturally occurring fatty acids such as butyric acid and stearic acid were known to be completely oxidized in normal animals, earlier work had shown that aromatic acids such as benzoic acid (C_6H_5COOH) and phenylacetic acid ($C_6H_5CH_2COOH$) are resistant to metabolic oxidation, and are excreted in the urine in the form of conjugates with glycine (hippuric acid and phenaceturic acid, respectively). Knoop found that, after phenylbutyric acid had been fed, the urinary product was phenaceturic acid; with phenylpropionic acid ($C_6H_5CH_2CH_2COOH$) or phenylvaleric acid ($C_6H_5CH_2CH_2CH_2CH_2$-$COOH$) the product was hippuric acid. From the finding that the phenyl derivative of an even-numbered (C_4) fatty acid chain was degraded to phenylacetic acid, whereas the phenyl derivatives of the two odd-numbered (C_3 and C_5) fatty acids gave benzoic acid, Knoop concluded that the oxidative degradation of fatty acids occurs by oxidation at the β-carbon (see accompanying scheme). This "β-oxidation theory" was powerfully buttressed by the results reported by Gustav Embden (in 1906–1910) on the perfusion of surviving liver with ordinary fatty acids;

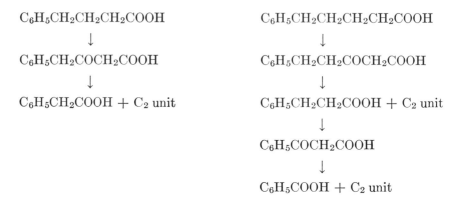

$$C_6H_5CH_2CH_2CH_2COOH$$
$$\downarrow$$
$$C_6H_5CH_2COCH_2COOH$$
$$\downarrow$$
$$C_6H_5CH_2COOH + C_2 \text{ unit}$$

$$C_6H_5CH_2CH_2CH_2CH_2COOH$$
$$\downarrow$$
$$C_6H_5CH_2CH_2COCH_2COOH$$
$$\downarrow$$
$$C_6H_5CH_2CH_2COOH + C_2 \text{ unit}$$
$$\downarrow$$
$$C_6H_5COCH_2COOH$$
$$\downarrow$$
$$C_6H_5COOH + C_2 \text{ unit}$$

those with even-numbered chains (C_4, C_6, C_8, C_{10}) gave acetone, which must have been derived from acetoacetic acid, whereas the fatty acids with odd-numbered chains (C_3, C_5, C_7, C_9) did not; furthermore, the odd-numbered fatty acids gave rise to the formation of glucose. Additional support came from the studies of Henry Dakin, who also showed that the liver contains an enzyme that interconverts acetoacetic acid and l-β-hydroxybutyric acid. Furthermore, Dakin found that phenylpropionic acid is partly converted in the animal body into the α,β-unsaturated acid $C_6H_5CH{=}CHCOOH$, but considered the direct dehydrogenation of the saturated acid "almost incredible, mainly because no chemical analogy for such a change is known" (Dakin, 1912, p. 27); such unsaturated acids were considered to be derived from the corresponding hydroxy acids by the loss of the elements of water. The hydroxy acids thus occupied a central position in the sequence of reactions in the degradation of fatty acids; as in the relation between lactic acid and pyruvic acid in carbohydrate breakdown, the hydroxy acids were thought to be the metabolic precursors of the keto acids, which were then cleaved. By 1910, therefore, the formation of the ketone bodies appeared to be a process in which long-chain fatty acids are successively shortened by the removal of —CH_2—COOH units (released as acetic acid, CH_3COOH) from the β-keto acids formed by "successive β-oxidation" (Dakin, 1909, p. 225); in diabetes or starvation, this process was thought to be blocked when acetoacetic acid was reached. In addition, from the studies of Bernard Naunyn and others around 1900 on the "antiketogenic" action of glucogenic substances it seemed that, in normal animals, there is a link between the oxidation of glucose and the complete oxidation of ketone bodies; much was made of the idea that "fats burn in the flame of the carbohydrates," and the diabetic animal was thought to be deficient in its capacity to effect oxidations. Around 1910, what seemed to be an obligatory linking of the oxidation of fats and carbohydrates thus fitted neatly with the idea of successive β-oxidation, which was widely accepted for about thirty years, despite repeated unsuccessful efforts to find the hypothetical acetic acid.

The apparent relation between the metabolism of fats and carbohydrates led to a lively but inconclusive controversy, during the early decades of this century, whether animals (like plants) can manufacture glucose from fatty acids. There was no debate, however, about the reverse process, whose occurrence in the animal body had been well established for over fifty years. By 1845 Boussingault had acknowledged that he and Dumas were wrong in denying that animals can convert sugar into fat (as Liebig had stated), and decisive data on this question were later provided by the important balance experiments of Lawes and Gilbert (1866); they also showed that fat can arise from dietary protein. During the succeeding years, chemical speculations about the intermediates in the conversion of glucose into fatty acids focused largely on two-carbon units. Thus Nencki (1878) suggested that the lactic acid derived from glucose is degraded to acetaldehyde (CH_3—CHO), two molecules of which condense to form butyric acid. During the latter half of the nineteenth century, aldehydes were popular candidates as intermediates in biosynthetic processes, because of their considerable chemical reactivity; for example, in 1861 Butlerov had found that, in alkaline solution, formaldehyde (HCHO) condenses to give a sugar-like product, and Baeyer (1870) suggested that formaldehyde might be an intermediate in the photosynthesis of glucose from CO_2 and H_2O in green plants. By the end of the century, a large number of naturally occurring substances had been found to possess a relatively complex structure, and attempts were made to guess at the intermediates involved in their biosynthesis. One of the pioneers in this effort was John Norman Collie, who wrote:

> The attempt to artificially produce naturally occurring substances, and to imitate in the laboratory some of the many processes which are perpetually being carried on around us in nature, has always been one of the chief aims of the organic chemist; and although at the present time we are almost entirely ignorant of the methods by which most of these processes are effected, still we can, with a considerable degree of certainty, guess at a solution of part of the problem.

Polymerisation and *condensation* are probably the two chief types of change which are instrumental in forming many of the multitudinous natural compounds, condensation being usually the outcome of the union of carbon atoms in consequence of the elimination of water . . . . The compounds which condense or polymerise most easily are, no doubt, the aldehydes and ketones, and they show this linking of carbon atom to carbon atom in the most pronounced manner (Collie, 1893, p. 329).

In a later paper he called ". . . attention to the manner in which the group $\cdot CH_2 \cdot CO \cdot$ . . . can be made to yield by means of the simplest reactions a very large number of interesting compounds" (Collie, 1907, p. 1806). This approach was actively pursued in the years that followed and, as we shall see shortly, the interplay of such chemical considerations with the results of metabolic studies was to be exceedingly fruitful after World War II.

Nencki's idea that fatty acids might be made from acetaldehyde was later elaborated by others, and H. S. Raper stated:

> The formation of fatty acids in animals, from carbohydrates, and the occurrence in natural fats, such as butter, of all the fatty acids containing an even number of carbon atoms, from two to twenty, suggest that their fatty acids are produced by the condensation of some highly reactive substance containing two carbon atoms and formed in the decomposition of sugar (Raper, 1907, p. 1831).

Raper's efforts to show that acetaldehyde is the "highly reactive substance" were inconclusive, and the status of Nencki's idea remained uncertain. Furthermore, some chemists (among them Emil Fischer) preferred the hypothesis that long-chain fatty acids such as the C_{18} stearic acid is produced by direct condensation of three molecules of hexose and reduction of the resulting intermediate.

The problem of the biosynthesis of fatty acids from glucose was to challenge biochemists for several decades, but an important clue was provided by the perfusion experiments of Embden and Oppenheimer (1912) showing that, in the liver, pyruvic acid is a metabolic precursor of acetoacetate; they suggested the following sequence of reactions: glucose → "active glyceraldehyde" → lactic acid → pyruvic acid → acetaldehyde → acetoacetic acid. Furthermore, perfusion experiments by Ernst Friedmann (1913) demonstrated the formation of acetoacetic acid from acetic acid; he proposed that acetaldehyde and acetic acid condense to give a C_4 acid that is successively converted into crotonic acid (CH_3—

CH=CH—COOH), β-hydroxybutyric acid, and acetoacetic acid. During the succeeding years the possible role of acetic acid as a metabolic intermediate occupied a central place in biochemical thought and, as we saw on p. 385, the search for the "active acetate" culminated in the discovery of acetyl-coenzyme A during the 1950s.

Friedmann's finding of acetoacetate production from acetate led him to call attention to a deficiency in the theory of successive β-oxidation:

> Since acetic acid increases markedly the extent of the formation of acetoacetic acid in the surviving liver, it would be expected that various normal fatty acids also form acetoacetic acid. In fact, acetoacetic acid arises only from the normal fatty acids with an even number of carbon atoms, not from the normal fatty acids with an odd number of carbon atoms. From this it follows that the pairwise degradation of normal saturated fatty acids does not proceed with the liberation of acetic acid. It also follows that it is incorrect to assume that normal saturated fatty acids are degraded by acid cleavage of β-keto acids to fatty acids having two carbons less (Friedmann, 1913, p. 442).

The uncertainty regarding the Knoop-Dakin theory continued during the period when metabolic experiments were largely conducted with normal or diabetic animals, or by the perfusion of isolated organs. The situation began to change after Krebs introduced the tissue-slice technique (see p. 437). Thus, in 1935, Juda Quastel showed that the oxidation of fatty acids with 6, 8, or 10 carbon atoms gave rise to more acetoacetic acid than was expected on the basis of the successive β-oxidation theory, and that fatty acids with 5, 7, or 9 carbon atoms also were oxidized with the production of ketone bodies. By 1940 the Knoop-Dakin theory had been challenged by Eaton MacKay's proposal that, although fatty acids are indeed cleaved into two-carbon units by β-oxidation, these fragments then join to form acetoacetic acid (see Stadie, 1945). Furthermore, the use of the tissue-slice technique indicated a relation between the formation of ketone bodies and the oxidation of pyruvate (Edson and Leloir, 1936); later work demonstrated the entry of the two-carbon units formed by the β-oxidation of fatty acids into the citric acid cycle. By 1939 the knowledge of the cofactors needed for biological oxidations had developed sufficiently to permit Luis Leloir to effect fatty acid oxidation in a fortified cell-free system containing washed liver particles. The subsequent development of this problem after World War II, and the dissection of the enzymic apparatus for the degradation and synthesis of fatty acids, is recounted in current textbooks of biochemistry. As in the study of the oxidation of pyruvate, the discovery of coenzyme A was a decisive turning point in the elucidation of the component chemical reactions in these metabolic pathways.

To return to 1904, it will be evident that Knoop's contribution was fruitful in illuminating a metabolic process of great interest to biochemists and clinicians. His achievement was significant in its impact on biochemical thought because it represented one of the first consciously planned set of experiments in which a metabolite was labeled with a group that could be recognized in the products of the physiological conversion of that metabolite. Although the phenyl derivatives that Knoop had employed were "unphysiological," the results he obtained proved to be valid for their unlabeled "physiological" counterparts. Thus, by the application of the available methods of synthetic chemistry to the preparation of these artificial metabolites, Knoop had introduced the labeling technique into the study of metabolic processes. The limitations of Knoop's approach were obvious, since the chemical modification he made was not generally applicable to a large variety of metabolites. Thirty years later, the introduction of the hydrogen isotope of mass 2 (deuterium) into metabolic studies marked the beginning of a development in which Knoop's approach became one of labeling metabolites chemically with isotopic elements. Together with the advance of enzyme chemistry and the use of microbial mutants, this development totally transformed the face of what Hopkins called the "dynamic side of biochemistry."

Isotopes in Metabolic Research. In August 1935, there appeared a short paper that began as follows:

Many attempts have been made to label physiological substances by the introduction of easily detectable groups such as halogens or benzene nuclei. However, the physical and chemical properties of the resulting compounds differ so markedly from those of their natural analogues that they are treated differently by the organism. The interpretation of metabolic experiments involving such substances is therefore strictly limited.

We have found the hydrogen isotope deuterium to be a valuable indicator for this purpose. The fact that it occurs in the same proportion (1 atom of deuterium to 5,000 atoms of protium) in the hydrogen of ordinary water and of organic matter is in itself evidence that the living body is unable to distinguish the few organic molecules which contain deuterium from those that do not. Were the reverse the case, organic matter of biological origin would display differences in isotopic ratio.

We have prepared several physiological compounds (fatty acids and sterol derivatives) containing one or more deuterium atoms linked to carbon, as in methyl or methylene groups. Their physical properties are indistinguishable from those of their naturally occurring analogues by the methods commonly employed. As, however, the deuterium content of these substances or of their physiological derivatives can readily be determined from the properties of the water formed on combustion, their fate in the body can be followed even after considerable dilution (Schoenheimer and Rittenberg, 1935, p. 156).

The paper concluded with the statement: "The number of possible applications of this method appears to be almost unlimited" (*ibid.*, p. 157). During the remaining six years of Schoenheimer's life, a succession of remarkable papers appeared from his laboratory illustrating the power of the isotope method in the study of intermediary metabolism, and laying the groundwork for even more exciting later research.

Schoenheimer's use of deuterium as a metabolic tracer was not the first application of isotopes for biological studies. This approach had been initiated by George Hevesy (1923), who used the radioactive thorium B to study the transport of lead in the bean plant. Ten years before, Hevesy and Friedrich Paneth had performed the first chemical tracer experiments, in which they demonstrated the electrochemical interchangeability of lead and the radioactive radium D; these results provided additional evidence for the views developed earlier in 1913 by Kasimir Fajans and by Frederick Soddy regarding the relation of the newly discovered radioactive elements to Mendeleev's periodic table. In a paper on this subject, Soddy wrote:

> The same algebraic sum of positive and negative charges in the nucleus, when the arithmetical sum is different, gives what I call "isotopes" or "isotopic elements," because they occupy the same place in the periodic table. They are chemically identical, and save only as regards the relatively few physical properties which depend upon atomic mass directly, physically identical also (Soddy, 1913, p. 400).

It is impossible here to summarize adequately the remarkable transformation in scientific thought that gave birth to the concept of isotopes. What began in 1896, with Henri Becquerel's discovery of the radioactivity in uranium salts, culminated in the planetary model of the atom proposed by Ernest Rutherford in 1911, and in Niels Bohr's application (in 1913) of quantum principles to this model to explain the hydrogen spectrum. Within a few years after Becquerel's discovery, a number of radioactive elements had been identified, the most famous of them being the polonium and radium isolated in 1898 by Marie and Pierre Curie. In 1902 Ruther-ford and Soddy proposed that the radioactive elements are transformed into each other by a process of radioactive decay with the emission of material particles; the mid-nineteenth century idea of unchangeable and indestructible atoms was abandoned. By 1912 much was known about the radioactive decay series, and the "displacement laws" of Fajans and Soddy clarified the relation of the products of radioactive transformation to the elements in the periodic table. The existence of isotopes was defi-nitely established in 1914, through the careful determination of the atomic weight of lead samples obtained from different minerals, by Theodore Richards and Max Ernest Lembert; it was clear that different samples can have different atomic weights despite their apparent chemical identity in other respects (see Conant, 1970). Subsequently, F. W. Aston showed (in 1920) that the occupancy of a place in the periodic table by more than one element was not limited to the heaviest ones. He identified neon iso-topes of mass 20 and 22, neither of which is radioactive (they are termed "stable" isotopes); in connection with the studies, he invented a "mass spectrograph" for the identification of the isotopic elements. Later work led to the development of the "mass spectrometer" for the quantitative estimation of the relative abundance of stable isotopes (Nier, 1955).

When Hevesy introduced the use of isotopes into biological studies, only a few elements (*e.g.*, lead, bismuth) could be labeled. During the 1920s he studied the distribution of bismuth in animals (with radium E as tracer) because of the current clinical interest in the replacement of arsenic by bismuth in the treatment of syphilis, and he also examined the partition of labeled lead between normal and tumor tissues in mice. Many years later, Hevesy wrote:

> The investigation of the partition of lead was carried out in the Institute of Physical Chemistry of the University of Freiburg. Feeling the need of an assistant trained in biology, I approached Dr. Aschoff, the direc-tor of the Institute of Pathology. He selected a Japanese scientist for this work, who however proved to have difficulties in coping with the task. It was then that Dr. Schoenheimer, associate of the Institute of Pathology, was asked by Dr. Aschoff to step in. It was in the course of these investigations that Schoenheimer became familiar with the method of isotopic indicators (Hevesy, 1948, p. 130).

With the rise of Hitler, Rudolf Schoenheimer left Freiburg in 1933 and came to Columbia University, where Harold Urey had discovered deu-terium the year before. To add a personal note, I was working for my doctorate in the Department of Biological Chemistry at Columbia when

Schoenheimer began research in the laboratory occupied by the graduate students, and I can attest to the stimulus we all derived from this imaginative and dynamic scientist, and from the warmth of his personality.

Of the elements present in biological molecules, the first for which isotopes were identified was oxygen (by Giauque in 1929), but the first to become available for biological studies was the stable isotope deuterium. After Urey and his associates had developed a practical method for the preparation of D_2O ("heavy water"), many investigators at once performed a variety of experiments to determine its effect on biological processes in animals, plants, and microorganisms (see Urey, 1934). It should be emphasized, however, that in none of these initial studies was deuterium introduced at specific loci in organic molecules by means of known chemical reactions, so as to permit one to determine the metabolic fate of the label through the isolation of well-defined biochemical substances from the organism under study. Schoenheimer, as a biochemist trained in organic chemistry (he had described a valuable method of peptide synthesis in 1926, and had worked on the chemistry of sterols), was the first to do this and he thereby transformed the labeling technique introduced by Knoop into a powerful method for the study of intermediary metabolism.

Within a few years after the discovery of deuterium, Urey developed methods for concentrating the naturally occurring stable isotopes of nitrogen (^{15}N), oxygen (^{18}O), and carbon (^{13}C). Furthermore, in 1934 Irène Curie and Frédéric Joliot discovered that radioactivity could be induced artificially by bombardment with α-rays (helium nuclei) from radium. This made available a radioactive isotope of phosphorus (^{32}P), prepared from sulfur (^{32}S); phosphate labeled in this manner was used by Hevesy during the mid-1930s in studies on the turnover of phosphorus in animal tissues. After the development of the cyclotron by Ernest Lawrence, the preparation of radioactive isotopes by bombardment with nuclear particles was simplified, and by 1938 a variety of additional tracers (*e.g.,* ^{35}S) for biological studies became available. Some of the most valuable of these, such as ^{11}C, decayed so rapidly (^{11}C has a "half-life" of 21 minutes) that biological experiments had to be of short duration, and could only be performed by investigators near a cyclotron. For this reason, most of the pioneer biological experiments with ^{11}C-labeled compounds were conducted either at Berkeley (near Lawrence's cyclotron) or in Boston (near the Harvard cyclotron). Although the more valuable long-lived isotope of carbon ^{14}C (half-life, 5570 years) had been prepared by Samuel Ruben and Martin Kamen in 1940 (see Kamen, 1963), it did not become generally available for metabolic studies until after World War II, when the cyclotron was replaced by the nuclear reactor ("atomic pile") as the source of artificial radioisotopes. Within about ten years (1945–1955), the use of ^{14}C as a metabolic tracer resolved many of the questions that had been hotly debated for a half-century, and elucidated many unknown pathways of intermediary metabolism. Also, the availability of the radioactive isotope of hydrogen (^{3}H, tritium) further broadened the scope of metabolic studies. Although the application of these radioisotopes as indicators of intermediary metabolism does not differ in principle from the method introduced by Schoenheimer, the radioisotopes have replaced the stable ones, largely because they can be determined much more easily. For many years, improved versions of the instrument devised in 1908 by Rutherford's associate Hans Geiger were used for the measurement of radioactivity; more recently, such "Geiger counters" have been replaced by scintillation spectrometers. There can be little doubt that the introduction of ^{14}C into biochemistry, along with the development of chromatographic techniques, were the decisive factors in the explosive growth of biochemical knowledge after World War II.

When Schoenheimer began his isotope experiments in the mid-1930s it was known that deuterium atoms linked to oxygen or nitrogen (as in carboxyl, hydroxyl, or amino groups) can exchange readily with protium atoms in water, and that the label will be lost from the molecule during the course of metabolic reactions. Compounds were required, therefore, that had been labeled with the formation of stable linkages, such as the carbon-hydrogen bond. For example, Schoenheimer prepared deuterated stearic acid chemically by the hydrogenation of the straight-chain C_{18} linoleic acid (which contains two double bonds). Such labeled stearic acid was included in the diet of mice for twelve days, the animals were killed, and their fatty acids were isolated and analyzed for their deuterium content. Among the labeled fatty acids found was the saturated C_{16} palmitic acid, indicating that stearic acid had been shortened by two carbons, in accordance with the Knoop-Dakin theory of β-oxidation; in later work it was found that labeled palmitic acid can be lengthened to stearic acid. Another fatty acid isolated from the tissues in the experiment with labeled stearic acid was the C_{18} oleic acid (which contains one double bond), showing that the animal organism can effect the dehydrogenation of stearic acid to oleic acid. Furthermore, when the labeled oleic acid so obtained was fed to mice, their fatty acids included labeled stearic acid. These pioneer studies clearly demonstrated the metabolic interconvertibility of the three principal fatty acids of animal fats, and showed that the body fats are in a "dynamic state" even when adequate fat is supplied in the diet.

The use of deuterium, however, could not answer the questions about the formation of ketone bodies raised in the debate about the β-oxidation theory; for this a carbon isotope was needed. After ^{13}C became available around 1940, and the value of the tissue-slice technique for metabolic studies had been recognized, Sidney Weinhouse *et al.* (1944) showed that the labeling of the acetoacetic acid produced from labeled octanoic acid [CH$_3$(CH$_2$)$_6$^{13}COOH] was inconsistent with the successive β-oxidation theory and was in accord with the view that the ketone bodies are formed by the random condensation of pairs of the four two-carbon units derived from the cleavage of the C$_8$ fatty acid. The subsequent availability of the long-lived radioactive isotope of carbon (^{14}C), and the development of the cell-free systems for the study of fatty acid oxidation (Kennedy and Lehninger, 1949), permitted Dana Crandall and Samuel Gurin (1949) to define the process more precisely, by calling attention to a metabolic difference between the terminal CH$_3$—CH$_2$— unit of a fatty acid and its other two-carbon units. These studies served as a background for the elucidation of the enzymic mechanisms in the oxidation of fatty acids, after the discovery of acetyl-coenzyme A.

In 1937 the stable isotope ^{15}N became available for metabolic studies, and Schoenheimer soon afterward initiated tracer experiments on protein metabolism. In a series of remarkable papers published in 1939–1941, he and his associates described research in which various amino acids, enriched with ^{15}N in their amino group, were administered to animals, and amino acids then were isolated from hydrolysates of the tissue proteins for determination of their ^{15}N content. The labeled amino acids were prepared by the catalytic hydrogenation of the corresponding α-keto acids in the presence of ^{15}N-enriched ammonia, according to the procedure of Knoop and Oesterlin (1925); in some cases, stably bound deuterium was also introduced into the ^{15}N-labeled amino acids to serve as an indicator of the fate of the carbon chain. A major consequence of the metabolic experiments was the demise of the Folin theory of separate endogenous ("wear-and-tear") and exogenous nitrogen metabolism, which had also been espoused by Rubner around 1910 (p. 428). Schoenheimer and his colleagues concluded:

It is scarcely possible to reconcile our findings with any theory which requires a distinction between these two types of nitrogen. It has been shown that nitrogenous groupings of tissue proteins are constantly involved in chemical reactions; peptide linkages open, the amino acids liberated mix with others of the same species of whatever source, diet, or tissue. This mixture of amino acid molecules, while in the free state, takes part in a variety of chemical reactions: some reenter directly into vacant positions left open by the rupture of peptide linkages; others transfer their nitrogen to deaminated molecules to form new amino acids. These in turn continuously enter the same chemical cycles which render the source of the nitrogen indistinguishable. Some body constituents like glutamic acid and aspartic acid and some proteins like those of liver, serum, and other organs are more actively involved than others in this general metabolic pool originating from interaction of dietary nitrogen with the relatively larger quantities of reactive tissue nitrogen (Schoenheimer, Ratner, and Rittenberg, 1939, pp. 729–730).

The Folin-Rubner theory had already been challenged during the 1930s. Henry Borsook and Geoffrey Keighley (1935) concluded from nutritional experiments that there is a "continuing" metabolism of protein, and that tissue proteins are constantly being synthesized from amino acids. George Whipple's studies on the replacement of plasma proteins after animals had been bled also led him to postulate a dynamic state of body protein (Madden and Whipple, 1940). After the work of Schoenheimer, the issue no longer seemed in doubt, and the existence of a "metabolic pool" of nitrogen in the animal body was generally accepted (see the accompanying diagram). During the 1950's, however, it was questioned whether pro-

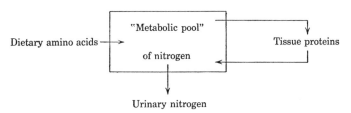

tein turnover occurs at the intracellular level, and it was suggested that in mammalian tissues protein is released only by secretion or by cellular disintegration (Hogness, Cohn, and Monod, 1955). This conclusion was based on the apparent lack of protein turnover in microbial cells, but subsequent work by Joel Mandelstam (1958) showed that the breakdown as well as the synthesis of proteins occurs in intact bacteria.

If the demolition of the Folin-Rubner theory was important in ending the era of nutritional balance studies, Schoenheimer's results were even more significant in initiating the study of the detailed pathways in the breakdown and synthesis of the individual amino acids, and their metabolic relation to other body constituents. The isotope data confirmed the special metabolic role of glutamic acid and aspartic acid, and underlined the general importance of transamination reactions, thus providing striking support for the conclusions drawn by Krebs and Braunstein from their enzymic studies. Furthermore, the labeling of arginine accorded with the Krebs ornithine cycle for urea synthesis. Among the numerous other findings, a special place must be accorded to the definition of the pathway of creatine synthesis.

The metabolic origin of creatine, and of the urinary creatinine derived from it by the loss of the elements of water, had been the object of extensive but indecisive study for decades (Hunter, 1928). After the identification of arginine as a protein constituent near the turn of the century, its metabolic relation to creatine was suspected on the ground that both compounds have a guanidino group [—NH—C(=NH)—NH$_2$], and various chemically plausible hypotheses were advanced in which arginine was thought to be degraded to creatine. The isotope experiments of Konrad Bloch and Rudolf Schoenheimer (1941) showed clearly that the metabolic synthesis of creatine involves the transfer of an amidine group [—C(=NH)—NH$_2$] from arginine to glycine, with the formation of guanidinoacetic acid (see accompanying formula) which is then methylated

$$
\begin{array}{ccc}
\overset{\displaystyle NH}{\underset{\displaystyle \|}{}} & & \\
NH\!-\!C\!-\!NH_2 & NH_2 & NH_2 \\
| & | & | \\
(CH_2)_3 & (CH_2)_3 & C\!=\!NH \\
| & | & | \\
NH_2CHCOOH \;+\; NH_2CH_2COOH \;\rightleftharpoons\; NH_2CHCOOH \;+\; NHCH_2COOH
\end{array}
$$

Arginine   Glycine   Ornithine   Guanidinoacetic acid

to give creatine. Such biological methylation reactions were extensively studied by Vincent duVigneaud and his associates after 1939, and their isotope experiments demonstrated that the methyl group of creatine can be derived from methionine in a "transmethylation" reaction (see duVigneaud, 1952). It should be added that in 1927 Max Bergmann and Leonidas Zervas effected the chemical transfer of the amidine group of arginine to glycine, thus providing a chemical model for the metabolic reaction later identified by Bloch and Schoenheimer. Furthermore, guanidinoacetic acid had been synthesized from glycine and cyanamide during the 1860s by Adolf Strecker (who called it glycocyamine) and by Jacob Volhard, and the possibility that it might be methylated to creatine was considered after the work of Franz Hofmeister (1894) on biological methylation reactions. Indeed, in 1935 Erwin Brand and Howard Lewis independently suggested that methionine (discovered during the 1920s) might serve as the metabolic source of the methyl group of creatine. These anticipations did not win acceptance, however, until the isotope experiments had clearly delineated the metabolic pathway. More importantly, the isotope data guided subsequent work on the dissection of the enzymic apparatus responsible for the overall process; in particular, later studies (largely by Giulio Cantoni and by James Baddiley during the 1950s) established the role of ATP in the enzymic synthesis of a new metabolic intermediate (S-adenosyl methionine, see accompanying formula) which proved to be the methyl donor in numerous biological transmethylation reactions.

This partial summary of Schoenheimer's achievements during 1935–1941 may convey the extent to which the initial isotope experiments on intermediary metabolism were designed to test hypotheses that had emerged from earlier biological and chemical studies. The theories influenced the choice of the labeled organic compounds that were synthesized and introduced into a biological system, as well as the products that were isolated for the determination of their isotope content. In Schoenheimer's work the theories included the Knoop-Dakin β-oxidation hypothesis, the Krebs ornithine cycle, and the Folin-Rubner theory of endogenous and exogenous nitrogen metabolism. During the early years of the development of the isotope technique, the validity of other biochemical theories was tested in a similar manner; we have already mentioned that, after the isotopes of carbon became available, they were used to test ideas about the formation of ketone bodies from fatty acids. The first use of a carbon isotope, however, was to examine hypotheses about a process whose importance had been appreciated since the eighteenth century— the conversion of CO_2 and H_2O into starch and sugar by illuminated green plants, with the liberation of O_2.

After Jan Ingen-Housz (Reed, 1949) and Jean Senebier had confirmed and extended Joseph Priestley's finding that green plants convert "fixed air" into "dephlogisticated air," it was believed that the function of light is to cleave CO_2 with the retention of carbon by the plant and the release of O_2, thus "revivifying" the atmosphere. When Théodore de Saussure showed in 1804 that, under the influence of light, carbohydrates are made in a process that involves the participation of water, he also concluded that H_2O is not the source of oxygen in photosynthesis. There was uncertainty for decades about the role of the green pigment (named chlorophyll by Pelletier and Caventou in 1817), but by the 1880s the work of Theodor Wilhelm Engelmann had shown unequivocally that chloroplasts (the intracellular structures containing chlorophyll) represent the site of photosynthesis. After Eduard Buchner's preparation of zymase, attempts were made to prepare from green leaves cell-free juices that could effect photosynthesis; the doubtful results reinforced the widely held view that the process depends on the integrity of the cell structure (see Arnon, 1955). Important clues regarding the nature of photosynthesis were provided during the 1880s by Engelmann's discovery of purple bacteria which effected light-dependent synthesis of cell material without the release of O_2, and by Sergei Winogradsky's discovery of "chemosynthetic" bacteria which contain no chlorophyll and can assimilate CO_2 in the dark. It was not until the studies of Cornelis van Niel (1931), and his application of Kluyver's idea of the unity of biochemistry, that the general significance of these clues became evident. After van Niel's work, it seemed likely that the function of light in photosynthesis, both in green plants and in photosynthetic microorganisms, is not to decompose CO_2 with the liberation of O_2, but to effect a process in which water (or some other hydrogen donor) is decomposed to produce a reducing species and O_2 (or the oxidized form of a hydrogen donor other than water). According to van Niel's theory, the overall chemical change in the photosynthetic conversion of CO_2 and H_2O into hexoses can be written as

$$6\ CO_2 + 12\ H_2A \rightarrow (CH_2O)_6 + 6\ H_2O + 12\ A$$

where A is the oxidized form of the hydrogen donor H_2A. This formulation, based on the dehydrogenation theory developed by Wieland and Thunberg, thus separated the photosynthetic process into a photochemical reaction in which light energy is used to effect the endergonic reduction of water (or H_2A), and a nonphotochemical reaction in which the endergonic synthesis of carbohydrate from CO_2 and H_2O is driven by the energy transferred by the photochemical reaction. Support for this formulation was provided in 1937 by Robert Hill's important discovery that illuminated chloroplasts can effect chemical reductions (see Hill, 1965).

Among the first metabolic experiments conducted with an isotope of carbon were those undertaken by Samuel Ruben, Martin Kamen, and Zev Hassid in 1939–1940 on the fate of $^{11}CO_2$ in leaves or in the green alga

Chlorella. The incorporation of labeled carbon into organic compounds was found to be the same in the dark as in the light, showing that the uptake of CO_2 is indeed a nonphotochemical process. Because of the long-held view that formaldehyde is an intermediate in the photosynthesis of carbohydrate, Ruben, Kamen, and Hassid attempted to isolate labeled HCHO in various ways, without success. In 1941 they showed, by the use of $H_2^{18}O$ and of $C^{18}O_2$, that in photosynthesis the released molecular oxygen comes from the water while the oxygen of CO_2 enters into organic compounds, as predicted by van Niel's theory. By the early 1940s it had also become evident from the studies of Harland Wood and Chester Werkman (1936) that the incorporation of CO_2 into organic compounds is not limited to photosynthetic or chemosynthetic organisms. They found that the so-called propionic acid bacteria (which ferment compounds such as glycerol with the formation of propionic acid) use CO_2; after ^{13}C became available, it was shown that the carbon of $^{13}CO_2$ appears in the carboxyl group of succinic acid, already known to be a component of the citric acid cycle (Werkman and Wood, 1942). Soon afterward, similar CO_2 fixation was demonstrated for animal tissues, and it was evident that the process once considered to be restricted to green plants is performed by a wide variety of organisms. It was also found around 1940 that some bacteria and animal tissues can effect the reversible reaction between pyruvic acid and CO_2 to form oxaloacetic acid. By that time, pyruvic acid was known to link the Embden-Meyerhof pathway of glucose breakdown and the citric acid cycle (p. 384); it seemed probable therefore that the metabolic pathway in the utilization of CO_2 by photosynthetic organisms might include the conversion of pyruvate into carbohydrate, in a manner analogous to the resynthesis of glycogen from lactic acid during oxidative recovery in muscle. The coupling of oxidation-reduction processes to the cleavage and formation of ATP had been demonstrated by Warburg and Christian (1939), and the general significance of their discovery had been discussed by Kalckar (1941) and by Lipmann (1941). It was against this background that, in the last year of his short life, Ruben (1943) offered the perceptive hypothesis that in photosynthesis the reduced pyridine nucleotide and ATP needed to reverse the Embden-Meyerhof pathway are generated by coupling to the light-dependent reduction of water.

After ^{14}C became available, several investigators continued along the line initiated by Ruben, Kamen, and Hassid. In particular, Melvin Calvin applied the new chromatographic techniques to the separation of the radioactive substances formed from $^{14}CO_2$ by photosynthetic algae such as Chlorella or Scendesmus either in the dark or after different periods of illumination. During the course of these studies, Calvin and his associates found that after brief illumination (5 seconds) most of the ^{14}C that is incorporated into organic material appears in the carboxyl group of 3-phosphoglyceric acid, long known as a key intermediate in the Embden-Meyerhof pathway (p. 379). Furthermore, since the radioactive hexoses that appeared after longer illumination were largely labeled in the two central carbon atoms (carbons 3 and 4), it was reasonable to conclude that they arose from 3-phosphoglyceric acid by a reversal of the glycolytic pathway. The solution of the problem of the rapid formation of the labeled 3-phosphoglyceric acid came from parallel enzymic studies in other laboratories on the metabolic fate of 6-phosphogluconic acid. We noted on p. 338 that Warburg's *Zwischenferment*, which played so large a role in the discovery of the catalytic flavoproteins and of TPN, was recognized to be a glucose-6-phosphate dehydrogenase; no place was provided, however, for the product of its oxidative action (6-phosphogluconic acid) in the schemes of glucose breakdown developed during the 1930s. After World War II, it was shown that this compound is oxidatively decarboxylated to a five-carbon sugar phosphate (ribulose-5-phosphate) which turned out to be a key intermediate in an "alternative" pathway of glucose breakdown operative in various plant and animal tissues, as well as in some microorganisms. By 1955 the main features of this "pentose phosphate pathway" had been worked out through the efforts of Frank Dickens, Seymour Cohen, Bernard Horecker, and Efraim Racker (see Horecker, 1962). These advances led Calvin to conclude that ribulose-1,5-diphosphate (see accompanying formula), made from

$$
\begin{array}{ccc}
CH_2OPO_3H_2 & & \\
| & & \\
C{=}O & & COOH \\
| & & | \\
HCOH & + CO_2 \xrightarrow{+ H_2O} & 2\ HCOH \\
| & & | \\
HCOH & & CH_2OPO_3H_2 \\
| & & \\
CH_2OPO_3H_2 & & \\
\end{array}
$$

Ribulose-1,5-diphosphate   3-Phosphoglyceric acid

ribulose-5-phosphate by enzymic phosphorylation with ATP, is the immediate CO_2 acceptor in photosynthesis (Calvin, 1956). The further development of this story is recounted in current textbooks of biochemistry; during the 1960s attention was largely focused on the mechanisms of the generation of the reduced pyridine nucleotide and of the ATP needed for the synthesis of hexoses. This more recent work led to the recognition of the role of new electron carriers (quinones, cytochromes, non-heme-iron proteins) in the coupling of electron transfer to the phosphorylation of ADP, and has revealed striking similarities between the mechanisms of ATP generation in the chloroplasts of plant cells and the mitochondria of animal cells (Arnon, 1967). In both cases, the apparatus is an integrated assembly that has proved difficult to dissect by means of the available methods for the separation of enzymes.

The discovery that the metabolic utilization of CO_2 is not limited to photosynthetic plants and certain "autotrophic" bacteria, which can derive their nourishment from inorganic materials, merits further comment. Not only did it represent one of the first indications of the power of the isotope method to reveal the operation of unsuspected biochemical processes, but it also drew the attention of biochemists concerned with problems of animal metabolism to the general importance of the extensive studies that had been conducted before 1935 on the metabolism of a large variety of microorganisms (van Niel, 1949; Woods, 1953). In particular, Marjory Stephenson's valuable book *Bacterial Metabolism* (first edition, 1930; third edition, 1949) clearly indicated the contributions microbiology had made to the study of biochemical dynamics. In 1935, CO_2 was considered to be a completely inert end product of cellular respiration in heterotrophic organisms, *i.e.*, those which require some organic compounds as nutrients. When Wood and Werkman (1936) found that the formation of succinic acid by propionic acid bacteria is related to the uptake of CO_2, little attention was given to their discovery, but, when the use of the carbon isotopes showed CO_2 fixation to be effected under conditions where no *net* uptake of CO_2 was evident, its significance came to be widely appreciated; an early example was the finding in 1941 that the carbon of $^{11}CO_2$ is incorporated by animals into their liver glycogen (see Hastings, 1970, p. 12). Moreover, studies on the metabolic utilization of labeled CO_2 by heterotrophic microorganisms revealed a variety of new biochemical reactions, whose later examination led to the discovery of new enzymes, new cofactors, and new intermediates of importance in the metabolism of animals as well.

What proved to be true for the metabolic utilization of CO_2 also became evident for other small carbon compounds. For example, the use of the isotope technique transformed the question of "active formaldehyde," so prominent in chemical speculations about biosynthetic mechanisms from 1870 onward, into a large chapter of contemporary biochemistry dealing with the metabolic interconversion and utilization of one-carbon units such as methyl ($—CH_3$), hydroxymethyl ($—CH_2OH$), and formyl ($—CHO$) groups in defined enzyme-catalyzed reactions; here the cofactor known as tetrahydrofolic acid has been shown to play a central role. Among the other small carbon compounds whose metabolic fate was illuminated by means of isotopic labeling is acetic acid

(CH_3—COOH), long considered to play a major role in the intermediary metabolism of fats and carbohydrates. Of the many early experimental studies on the metabolic fate of labeled acetic acid, we will mention here only the work on the biosynthesis of cholesterol and of the porphyrin of blood hemoglobin. Aside from its importance in providing a foundation for subsequent studies of the enzymic mechanisms in the formation of these two important cell constituents, this work had a direct relation to the earlier efforts of organic chemists to determine the structure of complex natural products, and to deduce the mode of their biosynthesis.

The Biosynthesis of Complex Natural Products. During the early decades of this century, the ingenuity of organic chemists was challenged by the structural complexity of the plant alkaloids (*e.g.*, morphine, strychnine, quinine) that had been isolated during the early 1800s, and that had acquired importance in medical practice. By 1890 the new organic chemistry was being intensively applied to the elucidation of their structure, and methods were sought for their synthesis in the laboratory. For example, Richard Willstätter's first experimental success, during the 1890s, was in the study of alkaloids related to cocaine. The structural complexity of many of the natural products encountered at that time invited attempts to discern from the architectural framework of the known members of a group of compounds the presence of a repeating structural feature that could serve as a guide in the study of other compounds whose structure was still unknown. Moreover, speculations were offered about the mode of biosynthesis of natural products from presumed intermediates thought to give rise, in the plant, to such structural units. A high point in this development was the imaginative paper by Robert Robinson (1917), who had effected the synthesis of an alkaloid (tropinone) from succinaldehyde, methylamine, and acetone in aqueous solution under relatively mild conditions. He considered that this synthesis ". . . on account of its simplicity, is probably the method employed by the plant" (*ibid.*, p. 877), and developed a theory that assigned to certain amino acids (ornithine, lysine) and to presumed products of carbohydrate breakdown (*e.g.*, acetone dicarboxylic acid, HOOC—CH_2—CO—CH_2—COOH) the role of intermediates in the biosynthesis of a large number of known alkaloids. Not only were such theories intellectually satisfying, but also they proved to be of considerable practical value in the determination of the structure and in the laboratory synthesis of many alkaloids. An outstanding example is Robert Woodward's proposal (in 1948) that the complex structure of strychnine can be derived from tryptamine (the decarboxylation product of tryptophan), phenylalanine, acetic acid, and a one-carbon unit; this idea was valuable for the elucidation of the structure of related alkaloids (Robinson, 1955). During the period 1920–1940, some organic chemists considered such "biogenetic" hypotheses to reflect the chemical processes occurring in the plant, and sought to effect the artificial synthesis of alkaloids under "physiological" conditions, but after World War II it was generally agreed that

> . . . the comparison of structures *per se* gives no information about the details of the mechanisms of biosynthetic reactions. It is the task of the biochemist to determine these by appropriate experiment (*ibid.,* p. 1).

With the availability of isotopes, "appropriate experiment" became possible; indeed, some aspects of the hypotheses Robinson had offered in 1917 regarding the biosynthesis of alkaloids were found to be correct, but new and unexpected pathways were also encountered (Leete, 1965). Moreover, the achievements of the biochemists during the 1950s in the study of the biosynthesis of other complex natural products subsequently encouraged some organic chemists to adopt the isotope technique as an aid in the solution of difficult structural problems. Among these biochemical successes of the 1950s, the elucidation of the problem of cholesterol synthesis occupies a special place, not only in providing an example of the power of the isotope technique, but also in illustrating clearly the fruitful interplay of chemistry and biology in the study of a metabolic process.

Cholesterol was the object of continued interest after 1770, when it was found in gallstones by Poulletier de la Salle. After some chemical studies by Fourcroy, among others, it was described more fully in 1815 by Chevreul, who called it *cholestérine* (Greek *chole,* bile + *stereos,* solid) and identified it as an "unsaponifiable fat" to distinguish it from the fats that could be saponified to glycerol and fatty acids. During the remainder of the nineteenth century, cholesterol turned up as a constituent of nearly all animal tissues, and was found to be specially abundant in eggs and in the brain; similar material ("phytosterine") was also found in plants and in yeast, and all these substances were later grouped under the collective term "sterols." The chemical purification and analysis of cholesterol presented some difficulty, but by 1894 Julius Mauthner had established the empirical formula to be $C_{27}H_{46}O$. The determination of the structure of this molecule was not completed, however, until the early 1930s, after three decades of work by numerous chemists, notably Adolf Windaus, Otto Diels, and Heinrich Wieland. This work also demonstrated the structural relation of cholesterol to the so-called bile acids (*e.g.,* cholic acid and deoxycholic acid) that had been described by Leopold Gmelin (in 1824) and by Adolf Strecker (in 1848). An important contribution was made by J. D. Bernal in 1932, when he reported that the X-ray diffraction pattern of ergosterol (a plant sterol known to be related to the antirachitic vitamin D) was not consistent with the structure proposed by Windaus and Wieland during the 1920s for the carbon skeleton of cholesterol and deoxycholic acid. This led Otto Rosenheim and Harold King to suggest an alternative formulation, which proved to be the correct one (see accompanying formula). Before the structure of

Cholesterol

cholesterol had been established, however, there had been important studies on its formation in the animal body. In particular, Harold Channon provided strong evidence (in 1925–1926) favoring the view that animals synthesize cholesterol, thus settling a lively debate that had proceeded inconclusively for several decades, largely because of inadequacies in the analytical methods for the determination of cholesterol. During the course of his experiments, Channon also showed that, when squalene (an unsaponifiable lipid isolated in 1916 from shark-liver oil) is fed to rats, there is an increase in the cholesterol content of the tissues. This test of squalene as a possible precursor of cholesterol was a direct consequence of the parallel chemical work of Ian Heilbron, who studied the structure of squalene because of his interest in the lipid components of fish-liver oils as sources of vitamin A. Squalene turned out to be a long-chain hydrocarbon ($C_{30}H_{50}$). In the determination of the structure of squalene, and of its metabolic relation to cholesterol, a chemical hypothesis—the "isoprene rule"—played a significant role.

The origins of the isoprene rule lie in the early decades of the nineteenth century, when there was considerable interest in the volatile products produced upon the dry distillation of rubber. Thus, in 1826 Michael Faraday isolated a low-boiling material identified in later work as an unsaturated hydrocarbon (C_5H_8) related to isopentane [CH_3—$CH(CH_3)$—CH_2—CH_3] and named isoprene; its structure was uncertain for many years, but was shown to be CH_2=$C(CH_3)$—CH=CH_2 by Vladimir Ipatiev in 1897. Furthermore, Hlasiwetz observed (in 1868) that the decomposition of turpentine oil with heat also gives rise to isoprene, and this discovery was subsequently extended to other "terpenes" present in plant oils. During the course of these studies it was noted that

> . . . from the easy transformation of 2 C_5H_8 into $C_{10}H_{16}$ and *vice versa*, it appears probable that the molecule of terpilene [dipentene] is composed symmetrically of two halves $(C_5H_8)=(C_5H_8)$ (Tilden, 1884, p. 419),

and in 1887–1888 Otto Wallach applied this idea to other $C_{10}H_{16}$ compounds ("monoterpenes") and to camphor. The hypothesis that the isoprene unit constitutes a fundamental structural unit of the terpenes only became the "isoprene rule" after 1920, when Leopold Ruzicka demonstrated its general validity for various terpenes having 15 carbon atoms ("sesquiterpenes" such as farnesol), and subsequently for more complex "diterpenes" and "triterpenes" having 20 and 30 carbon atoms, respectively; for an account of this development, see Ruzicka (1959). The synthesis of squalene in 1931 (by Paul Karrer) established the details of its structure as a triterpene composed of six isoprene units. After the acceptance of the Rosenheim-King formula for cholesterol, Robinson called attention (in 1934) to the fact that squalene contains the intact skeleton of cholesterol. Moreover, he proposed a chemical mechanism for the conversion of the linear squalene molecule to the tetracyclic sterol nucleus.

By 1934, therefore, when Schoenheimer began his isotope experiments on cholesterol metabolism, extensive chemical studies had provided the groundwork for speculations about the biosynthesis of cholesterol. Although it was clear from nutritional studies that cholesterol is made and destroyed in the animal body, the nature of the metabolic intermediates was unknown, and the role of squalene in the biosynthetic process was uncertain. Thirty years later, principally through the use of the isotope technique, but also through the discovery of a new intermediate (mevalonic acid), the pathway of the biological formation of cholesterol had largely been elucidated (see Bloch, 1965). The initial steps were tentative; when mice were kept on a cholesterol-free diet and given heavy water, about one-half of the hydrogen atoms in their tissue cholesterol was labeled with deuterium, suggesting that the sterol had been synthesized by the condensation of small molecular units, "possibly those which have been postulated to be intermediates in the fat and carbohydrate metabolism" (Rittenberg and Schoenheimer, 1937, p. 252). Furthermore, studies on the utilization of deuterium-labeled acetate by yeast led to the conclusion that "the sterols of yeast arise from acetic acid by a rather direct route" (Sonderhoff and Thomas, 1937, p. 203). By 1942, Konrad Bloch and David Rittenberg had shown that D_3C—COOH is an effective metabolic precursor of the carbon skeleton of cholesterol in the animal organism.

After the carbon isotopes became available, Bloch demonstrated that, in rat-liver slices, acetic acid can supply all 27 carbon atoms of the sterol; he then embarked on the determination of the pattern of the incorporation of the carbon atoms of acetic acid labeled either in the methyl group or the carboxyl group ([14]CH_3—COOH or CH_3—[14]COOH) or both ([14]CH_3—[13]COOH). A similar effort was also undertaken by John Cornforth and George Popják, and by 1957 the carbon-by-carbon dissection of the labeled cholesterol had been completed. By that time, moreover, there had accumulated important evidence on the biosynthesis of squalene and its metabolic relation to cholesterol, as well as on the role of isoprene units in their biosynthesis.

The story of isoprene began with nineteenth-century studies on the decomposition of rubber; the study of the metabolic utilization of isoprene units may be said to begin with the work of James Bonner and Barbarin Arreguin (in 1949) on the utilization of acetate by the guayule plant for the biosynthesis of rubber, known to be a long-chain polymer composed of isoprene units. This finding influenced the early interpretation of the labeling pattern of the cholesterol formed from ^{14}C-acetate, as well as the study of the metabolic pathway between acetate and a presumed isoprenoid precursor of squalene; by 1953 Robert Langdon and Konrad Bloch had established that labeled squalene is converted into labeled cholesterol in animals. The further development of this story is told in current textbooks of biochemistry, and need not be recounted in detail here. A few of the highlights require mention, however. The metabolic precursor of the isoprene units of squalene appeared from an unexpected quarter: in 1956, during the course of a search for a new vitamin, a research group led by Karl Folkers isolated from "distillers' solubles" a new compound which replaced acetate in the growth medium of a lactobacillus. They established the chemical structure of the compound, which they named mevalonic acid (see formula) and, because of its structural similarity to substances previously found to be precursors of cholesterol, they tested labeled mevalonic acid; it was converted into

Mevalonic acid   Squalene

labeled cholesterol by minced rat liver with great efficiency. This discovery opened the way for the study of the enzyme-catalyzed reactions in the formation of mevalonic acid and in its conversion into squalene. It also led to the identification of the pyrophosphate of isopentenol [$CH^2=C(CH_3)—CH_2—CH_2OH$] as the metabolic isoprenoid unit made from mevalonic acid in a series of ATP-dependent reactions, and from which squalene is derived. Furthermore, before the discovery of mevalonic acid, there had been important developments in regard to the problem of the conversion of squalene into cholesterol. In 1952 Ruzicka and his associates established the structure of lanosterol (a C_{30}-sterol from wool fat) and showed that it is structurally related both to cholesterol and to triterpenes such as squalene; this led to the imaginative hypothesis of Woodward and Bloch (1953) suggesting how the squalene chain might be folded to yield lanosterol as an intermediate (see the accompanying scheme). Their hypothesis predicted a different labeling pattern from

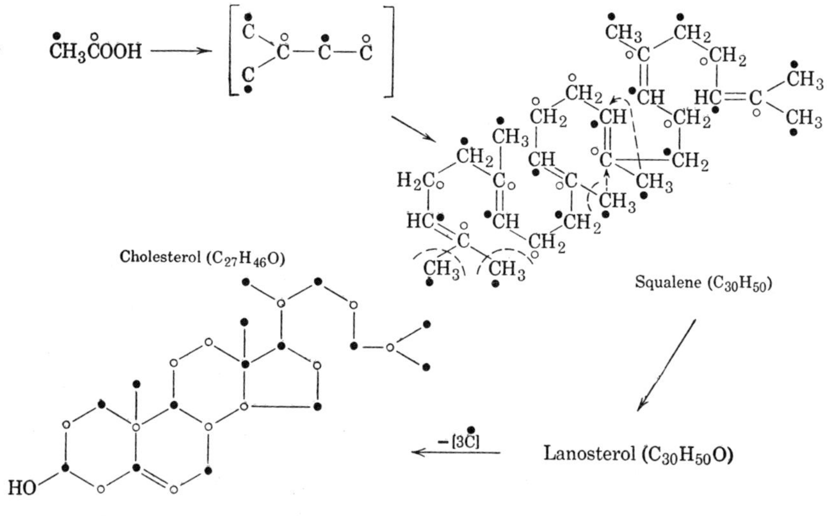

Utilization of Carbon Atoms of Acetic Acid in the Biosynthesis of Squalene and Cholesterol.

that expected for the arrangement Robinson had suggested in 1934, and the subsequent experimental data fitted the Woodward-Bloch idea. Although lanosterol was soon shown to be a metabolic precursor of cholesterol, the complex enzymic apparatus involved in this conversion has proved to be difficult to dissect, because of its tight association with water-insoluble components of the cell.

We turn now to another early demonstration of the power of the iso-

tope technique in the study of the biosynthesis of a complex cell constituent, the formation of the porphyrin (protoporphyrin IX) of blood
hemoglobin. By the 1930s the work of Hans Fischer (see p. 127) had
established the detailed chemical structure of this porphyrin (see formula), as well as many other naturally occurring members of the group

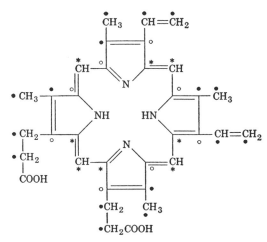

Metabolic Sources of Atoms of Protoporphyrin IX, as
Shown by Isotope Experiments

All 4 N atoms are derived from glycine; atoms
marked with asterisks are derived from the methylene
carbon of glycine; atoms marked with solid circles
are derived from the methyl carbon of acetate; atoms
marked with open circles are derived mainly from
the methyl carbon of acetate and in small part from
the carboxyl carbon of acetate; the unmarked carbon
atoms of the COOH groups are derived solely from
the carboxyl carbon of acetate.

of tetrapyrrole compounds which differ from protoporphyrin IX in the
nature and arrangement of the eight side chains attached to the four
pyrrole rings. It also was known that animals can maintain a normal level
of hemoglobin when kept on a diet free of porphyrins, and there were
speculations about the metabolic precursors. As was customary, the guid-
ing principle was one of structural similarity; thus proline and tryptophan
(which contain rings similar to the pyrrole nucleus) and glutamic acid
(which can be converted chemically into a derivative of proline) were
repeatedly considered as possible biological sources of the porphyrins.
With the application of the isotope technique to the problem after World
War II, these speculations were discarded, and it became evident that

protoporphyrin is made in the animal body from small molecules. In 1945 Bloch and Rittenberg showed that some of the carbon atoms in the side chains could be derived from deuterioacetate, and in the same year Shemin and Rittenberg (1945) made the decisive observation that the administration of ^{15}N-labeled glycine to human subjects led to the labeling of the nitrogen atoms of the hemin (ferriprotoporphyrin) isolated from the blood; ^{15}N-labeled glutamic acid was ineffective in this respect. In subsequent work, advantage was taken of the ability of the nucleated erythrocytes of birds to synthesize porphyrin, in contrast to the inability of mammalian (nonnucleated) erythrocytes to do so. After ^{14}C became available, experiments by David Shemin and by Albert Neuberger with labeled glycine (^{15}NH$_2$CH$_2$COOH, NH$_2$^{14}CH$_2$COOH, NH$_2$CH$_2$^{14}COOH) showed that all 4 nitrogen atoms of the tetrapyrrole ring are derived from glycine, whose CH$_2$ group provides 8 of the 34 carbon atoms of protoporphyrin, and that the COOH group of glycine is lost during porphyrin synthesis. Furthermore, Shemin and Wittenberg (1951) subjected the hemin obtained after incubation of duck erythrocytes with labeled acetate (^{14}CH$_3$COOH, CH$_3$^{14}COOH) to systematic chemical degradation; they succeeded not only in showing that the other 26 carbon atoms of protoporphyrin are derived from acetate, but also in assigning to each of these carbon atoms the extent to which it is derived from the methyl or carboxyl group of acetate. This achievement represented the first complete dissection of the labeling pattern of a complex compound made in a biological system from isotopic precursors. From the pattern, Shemin concluded that "... *in the biosynthesis of protoporphyrin a pyrrole is formed which is the common precursor of both types of pyrrole structure found in protoporphyrin*" (*ibid.*, p. 326). As will be seen from the formula, protoporphyrin has two kinds of pyrrole rings: one with methyl (—CH$_3$) and vinyl (—CH=CH$_2$) side chains, the other with methyl and propionic acid (—CH$_2$CH$_2$COOH) side chains. The labeling pattern also indicated that "... *in each pyrrole ring the same compound is utilized for the methyl side of the structure and for the vinyl and propionic acid sides of the structure*" (*ibid.*, p. 327), and that "the common precursor pyrrole originally formed contained acetic acid and propionic acid side chains" (*ibid.*, p. 328). Shemin proposed that the hypothetical precursor might arise by the condensation of glycine with two molecules of a succinyl compound (HOOC—CH$_2$—CH$_2$—CO—X) derived from acetate through the operation of the Krebs citric acid cycle. This "succinyl-X" later turned out to be succinyl-coenzyme A. It should be noted that the condensation reaction postulated by Shemin had precedents in synthetic organic chemistry, and was analogous to the general reaction developed by Ludwig Knorr in 1902 for the synthesis of pyrroles. Furthermore, in 1944 Hans Fischer drew attention to the ready condensation of glycine with formylacetone (CH_3—CO—CH_2—CHO) under "physiological" conditions as a possible route of pyrrole synthesis in biological systems. Among the other speculations whose validity could not be tested experimentally before the introduction of the isotope technique was Turner's proposal (in 1940) that all four pyrrole rings of protoporphyrin are derived from a common precursor pyrrole having acetic acid and propionic acid side chains (see Maitland, 1950).

The question of the existence of a common precursor pyrrole was solved in 1953 through an unexpected development in the study of the hereditary human disease known as acute porphyria (or porphyrinuria). As its name implies, this condition is associated with the appearance of the red porphyrins in the urine; the clinical manifestations, described by Günther (in 1912) and by Waldenström (in 1937), include periodic attacks of abdominal pain and mental symptoms that have occasionally led to wrong diagnoses of schizophrenia or paranoia. Indeed, a strong case has been made for the view that the "insanity" of King George III of England was a consequence of porphyria, whose recurrence has been traced in several royal houses of Europe (Macalpine, Hunter, and Rimington, 1966). During the 1930s Waldenström observed that the urine of porphyric patients contains a colorless material ("porphobilinogen") which is converted into porphyrins when the urine is acidified and heated. The porphyrins so produced were identified as uroporphyrins I and III, already shown by Fischer to have only acetic acid and propionic acid side chains; the two porphyrins differ in the way that the two kinds of side chains are arranged in the tetrapyrrole, uroporphyrin I having a symmetrical arrangement. The nature of porphobilinogen was unknown, however, before it had been isolated in crystalline form in 1952, and its structure had been established in the following year by Gerald Cookson and Claude Rimington (1954). This structure (see accompanying scheme) immediately suggested its biosynthesis from δ-aminolevulinic acid, two molecules of which could be expected to condense by the Knorr reaction to produce porphobilinogen. When Neuberger and Shemin showed independently that labeled δ-aminolevulinic acid is an excellent precursor of protoporphyrin, and that the position of the label (denoted with an asterisk in the scheme) corresponded to that previously found with methylene-labeled glycine ($NH_2{}^{14}CH_2COOH$), the main outlines of the metabolic pathway were clear. The scheme shown summarizes the status of the pathway of heme biosynthesis in about 1956; since then, extensive studies on the enzymic catalysis of the individual steps have provided additional knowledge which may be found in current biochemical treatises. Porphobilinogen has been shown to be a metabolic precursor, not

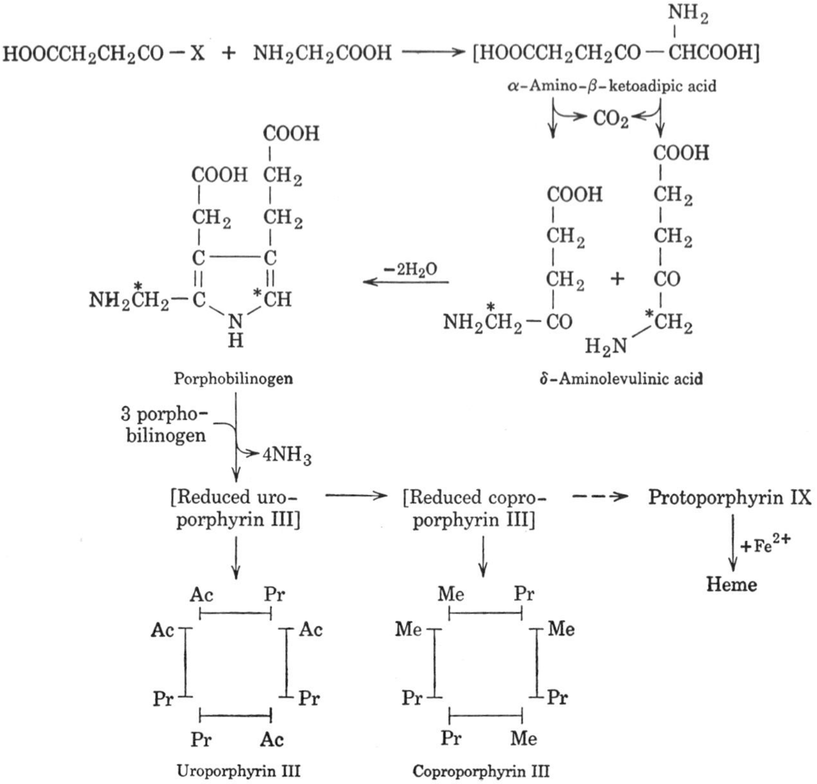

$$\text{HOOCCH}_2\text{CH}_2\text{CO} - \text{X} \;+\; \text{NH}_2\text{CH}_2\text{COOH} \longrightarrow [\text{HOOCCH}_2\text{CH}_2\text{CO} - \overset{\overset{\displaystyle \text{NH}_2}{|}}{\text{CHCOOH}}]$$

α-Amino-β-ketoadipic acid

$$\longmapsto CO_2 \longleftarrow$$

Porphobilinogen

$\xleftarrow{\;-2H_2O\;}$

δ-Aminolevulinic acid

3 porpho-
bilinogen $\longmapsto$ 4NH$_3$

[Reduced uro- $\longrightarrow$ [Reduced copro- $--\rightarrow$ Protoporphyrin IX
porphyrin III]   porphyrin III]

$\downarrow +Fe^{2+}$

Heme

Uroporphyrin III   Coproporphyrin III

Role of Aminolevulinic Acid and of Porphobilinogen in the Biosynthesis of Heme

only of the protoporphyrin in hemoglobin, but also of the chlorophylls in green plants and in photosynthetic algae, so that the first stages of the pathway may be considered to apply to all biological systems that make porphyrins. In view of the historical importance of acute porphyria in the study of porphyrin biosynthesis, it should also be added that in this disease there is a markedly increased activity of the liver enzymes that make porphobilinogen.

The work of Bloch and Shemin on the biosynthesis of cholesterol and protoporphyrin are striking early examples of the impact of the isotope technique on the study of intermediary metabolism; during the 1950s there were similar achievements, as in the determination (mainly by John Buchanan) of the metabolic sources of the carbon and nitrogen atoms of the uric acid derived from the purines of the nucleic acids (see accompanying scheme). These results revealed the capacity of the animal organism to effect the assembly of complex chemical structures from

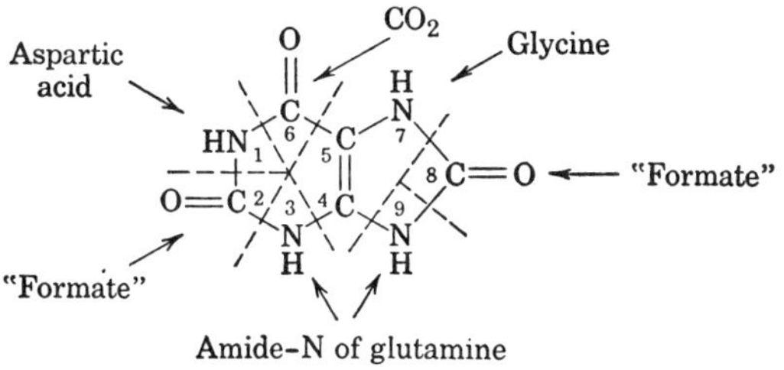

Metabolic Precursors of the Carbon and Nitrogen
Atoms of Uric Acid

small molecules such as CO_2, one-carbon units, acetic acid, and glycine, and showed that biosyntheses of this type are not limited to microorganisms. The labeling patterns that were found usually disproved hypotheses based solely on structural analogy and the chemical reactivity of the metabolites under consideration. Although, as we have seen, this chemical knowledge was essential for the metabolic studies, and the interplay of biological and chemical thought was mutually beneficial, the metabolic pathways suggested by the isotope data almost invariably turned out to be different and more complex than those guessed at from the most sophisticated chemical speculation. The new hypotheses arising from the isotope data were tested, modified, and elaborated in experiments that led to the isolation and characterization of new enzymes whose properties as catalytic proteins, and as components in multienzyme assemblies, were intensively studied. One aspect of this development requires emphasis: the important role played by the discovery of new compounds such as mevalonic acid or porphobilinogen, and the ability of organic chemists to effect the purification of these compounds (often by chromatography) and to determine their chemical structure within a short time. In many cases, the discovery of key intermediates around 1950 was no less haphazard than the finding of homogentisic acid in the nineteenth century, but their metabolic significance was appreciated much more quickly. The example of homogentisic acid illustrates the importance of knowing the structure of a key intermediate in a metabolic pathway. By 1914 the main outlines of the degradation of phenylalanine and tyrosine to ketone bodies in the animal body had been formulated on the basis of metabolic experiments; the use of the isotope method during 1945–1955 confirmed and elaborated the pathway (see the accompanying scheme). Similarly, we saw in the previous chapter that the elucidation of the steps in anaerobic glycolysis depended in large part on the discovery and characterization of the individual hexose and triose phosphates. Throughout the course of the "tedious unraveling" of metabolic processes

Oxidative Breakdown of Phenylalanine and Tyrosine

The numbering of the carbon atoms of the benzene ring is intended solely to show their metabolic fate; the numbers do not denote the position of substituents.

(to use Dakin's words of 1912), the appearance of a well-defined new compound, and the recognition of its place in a pathway under study, decisively influenced subsequent work and thought. For this reason, among others, the use of microbial mutants for metabolic studies, introduced by George Beadle and Edward Tatum (1941), constituted a powerful counterpart to the isotope technique.

Mutants in Metabolic Research. In the chapter on nucleic acids we traced some aspects of the interplay of biology and chemistry in the study of heredity, and noted that, shortly after the rediscovery of Mendel's work, Archibald Garrod described alcaptonuria as an "inborn error of metabolism" in which an enzyme concerned with the normal catabolism of tyrosine was lacking (p. 229). This idea reflected the upsurge both of Mendelian genetics, as advocated by Bateson, and of intermediary metabolism, represented by Knoop, Neubauer, and Embden. Before Garrod, it had been stated that ". . . inheritance is the recurrence, in successive generations, of like forms of metabolism" (Wilson, 1896, p. 326); Garrod gave this general view a more explicit form that invited experimental inquiry. The importance of his contribution was recognized by leading biochemists; in discussing the newly emerging methods of studying the chemical pathways of metabolism, Gowland Hopkins said:

> Extraordinarily profitable have been the observations made upon individuals suffering from those errors of metabolism which Dr. Garrod calls "metabolic sports, the chemical analogues of structural malformations." In these individuals Nature has taken the first essential step in an experiment by omitting from their chemical structure a special catalyst which at one point in the procession of metabolic chemical events is essential to its continuance. At this point there is arrest, and intermediate products come to light (Hopkins, 1913, p. 218).

There was a serious experimental difficulty, however; as J. B. S. Haldane put it, ". . . alcaptonuric men are not available by the dozen for research work" (Haldane, 1937, p. 2). This difficulty, as well as the limited knowledge regarding enzymes and metabolic pathways, were not conducive to the study of the biochemical aspects of genetics. Nor did the research during 1910–1935 on the relation of tyrosinase to the inheritance of coat color in animals (Onslow, Wright) or on the genetic control of the pigmentation of flowers (Wheldale, Scott-Moncrieff) encourage many others to enter this field. A significant change came during the late 1930s, as a consequence of the work of Beadle and Ephrussi on the inheritance of eye color in Drosophila; we noted on p. 255 that these studies led Beadle and Tatum to turn to the mold Neurospora.

Among the first evidences of the power of the Beadle-Tatum technique to illuminate metabolic pathways were their studies (with David Bonner) on the biosynthesis of tryptophan in Neurospora. In contrast to mammals, which require tryptophan in their diet, Neurospora can make this amino acid from sucrose as a source of carbon and ammonia as a source of nitrogen. The mutant strains of Neurospora found among the survivors after irradiation with X-rays included several that had lost the ability to make tryptophan. There had been indications, from earlier work by Paul Fildes and by Esmond Snell, that the compounds anthranilic acid and indole (see formulas) might be intermediates in the microbial formation of tryptophan. When these compounds were tested, it

Anthranilic acid

Indole

was found that one of the strains of "tryptophanless" Neurospora could use either anthranilic acid or indole in place of tryptophan for growth; another strain grew on indole but not on anthranilic acid. Moreover, the latter strain produced anthranilic acid in amounts large enough to permit its isolation and chemical characterization. Tatum, Bonner, and Beadle (1944) concluded therefore that anthranilic acid is an intermediate in the biosynthesis of tryptophan in Neurospora.

It was evident that, in the strain unable to use anthranilic acid for tryptophan synthesis, one of the enzymic steps in that conversion had been blocked as a consequence of mutation. In terms of its metabolic consequences, such a block was equivalent to the inhibition of that step by a chemical reagent. By the late 1930s several examples of such chemical effects were known; we have seen that the inhibition of fermentation by iodoacetate or fluoride led the accumulation of intermediates whose identification contributed to the formulation of the Embden-Meyerhof pathway. Also, the discovery of the action of sulfanilamide (see formula) by

H$_2$N⬡SO$_2$NH$_2$ H$_2$N⬡COOH

Sulfanilamide   **p—Aminobenzoic acid**

Gerhard Domagk led to the concept developed by Fildes that such antibacterial agents interfere with the metabolic utilization of "essential metabolites." In particular, his associate Donald Woods offered experimental evidence for the view that

> . . . the enzyme reaction involved in the further utilization of p-aminobenzoic acid is subject to competitive inhibition by sulphanilamide, and that this inhibition is due to a structural relationship between sulphanilamide and p-aminobenzoic acid (which is the substrate of the enzyme reaction in question) (Woods, 1940, p. 88).

A year after Woods's discovery, p-aminobenzoic acid was identified as a naturally occurring metabolite, and in 1945 it was shown to be a constituent of the vitamin folic acid. It should be added that the idea of the competitive inhibition of an enzyme by a structural analogue of its substrate had been foreshadowed during the 1920s by the work of Juda Quastel on the inhibition of succinic dehydrogenase by malonate (see p. 381). Not until after the work of Fildes and Woods, however, did biochemists undertake the systematic study of such "antimetabolites." The intensive search for new chemotherapeutic agents like sulfanilamide produced many inhibitors of bacterial growth, and the enzyme-catalyzed reactions blocked by some of them were identified as steps in important metabolic pathways; for a summary of the early developments, see Woolley (1952).

Shortly after the initial reports on the biosynthesis of tryptophan, further evidence of the value of the Neurospora mutants for the study of intermediary metabolism came from the work of Adrian Srb and Norman Horowitz (1944) on several "arginineless" strains. Three groups of mutants were found: (a) those that grew upon the addition of arginine, citrulline, or ornithine; (b) those that used arginine or citrulline for growth, but not ornithine; (c) those that grew only in the presence of arginine. No mutants appeared that could use arginine and ornithine, but not citrulline. Since Neurospora was found to contain arginase, the Krebs ornithine cycle clearly is operative in this organism (see accompanying scheme); the urea formed by the action of arginase is decomposed to

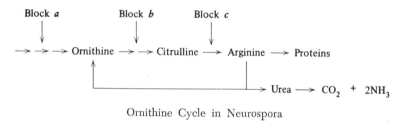

Ornithine Cycle in Neurospora

CO_2 and NH_3 by urease, present in microorganisms, but not in animal tissues. Furthermore, in view of our earlier mention of the use of the isotope technique in the study of the biosynthesis of sterols, it should be added that among the many other Neurospora mutants isolated during the course of Tatum's early studies was one that required acetate for growth. When grown on ^{14}C-labeled acetate, this strain was found (in 1950) to produce ergosterol essentially without dilution of the isotope, showing that acetate is a relatively direct precursor of the carbon atoms of the sterol skeleton. These early examples of the use of microbial mutants to study metabolic pathways were multiplied many times over in the years that followed. Moreover, after Tatum had described mutants of *Escherichia coli* in 1945, a valuable technique for the rapid isolation of such metabolically deficient strains was developed in 1948 by Bernard Davis and by Joshua Lederberg and Norton Zinder; this organism (along with others) then provided a rich harvest of data on intermediary metabolism and its genetic control. The details of the many important discoveries that were made may be found in textbooks such as that of Wagner and Mitchell (1964).

By the mid-1950s a large number of metabolic pathways had been traced in a variety of biological organisms, and many multistep processes were found to share common intermediates (*e.g.*, acetyl-coenzyme A in the degradation of carbohydrates and fatty acids), thus indicating branching of the pathways. In general, processes that led to the breakdown of metabolites were seen to converge toward common intermediates on the way to end products of metabolism or to utilization for the synthesis of entirely different cell constituents. On the other hand, the synthetic processes tended to diverge from common intermediates, with the occurrence of enzyme-catalyzed reactions that channel such intermediates into a particular pathway; an example is the formation of mevalonic acid as a precursor of the isoprenoid units of the steroids and terpenes. Furthermore, "alternative" pathways for the metabolic conversion of cell constituents had been discovered; one of the first to be identified was the "pentose phosphate pathway" for the oxidative breakdown of glucose-6-phosphate, as a counterpart of the Embden-Meyerhof scheme. For those who sought the "unity of biochemistry," there were numerous examples of the occurrence of the same (or very similar) pathways in very different biological forms; we have just noted the operation, in the mold Neurospora, of the Krebs ornithine cycle, first identified in mammalian liver. There was also much evidence of biochemical diversity, even among closely related organisms; the differences in the chemical structure of the alkaloids elaborated by plants or of the antibiotics made by microorganisms provided examples of variations in individual metabolic pathways. For the biochemist, these separate pathways offered a happy hunting-ground (see Dagley and Nicholson, 1970). Through the use of presumed precursors labeled with one or more isotopes, of microbial mutants blocked at a step, of cell extracts (as in the study of anaerobic glycolysis), or of subcellular elements such as mitochondria (as in the study of oxidative phosphorylation) there emerged during the 1950s a bewildering array of interconnected metabolic pathways. Moreover, when particular enzymes were identified as participants in a multistep process, efforts were made to isolate them, and to discern from their properties as catalytic proteins how they cooperate with the other enzymic components in a pathway. The resulting proliferation of new enzymes (or of names assigned to apparently new enzymic activities) led to listings that rivaled an urban telephone directory.

This embarrassment of riches pointed up a limitation of the chemical dissection of integrated biological systems that was clearly recognized by the investigators who were engaged in this enterprise during the first half of this century. As Krebs (1942) stated with regard to the ornithine cycle (see p. 444), the metabolic pathways represented hypotheses; repeated test by the various experimental methods that we have considered in this chapter strengthened the validity of some hypotheses, and disproved others. Furthermore, despite the concern of some biologically minded philosophers and some philosophically minded biologists whether hypotheses of this kind could adequately describe the physiological functions of intact biological systems, the cumulation of biochemical knowledge made it possible to study such functions from new and fruitful points of view, and to provide useful guides to practice in medicine and agriculture. The principal objection to the chemical dissection of biological systems has revolved about a question that has reappeared in one form or another since 1800—can the functions of biological organisms be explained in terms of the properties of their chemical constituents? Those who have insisted on either the logical or practical impossibility of achieving such explanation have emphasized that living things are organized systems in which the whole has properties that cannot be inferred from those of its parts. The historical background of this point of view merits attention in relation to the study of intermediary metabolism, and may suggest some reasons for its persistence.

The Whole and Its Parts

Throughout the interplay of biology and chemistry since 1800, efforts to explain physiological processes on the basis of chemical analyses met the objection that such studies told nothing about the "organization" of the individual parts of a biological system into a whole having the properties associated with life. In this respect, recent criticisms of the biological inferences drawn from the Watson-Crick model of DNA are in the tradition of Bichat's objection to Lavoisier's experiments on animal respiration:

> Physics, chemistry, etc. touch each other, because the same laws govern their phenomena; but an immense distance separates them from the science of organized bodies, because there is an enormous difference between these laws and the laws of life (Bichat, 1805, p. 84).

Many centuries earlier, Aristotle had criticized speculations about physiological functions in terms of the mechanical action of the water, air, fire, and earth of which biological organisms were thought to be composed (*De Partibus Animalium,* 640[b]). For Aristotle, and for successive generations of biologists, the whole is not only more than the sum of its parts ("the formal nature is of greater importance than the material nature"), but the proper method of studying a physiological process is to determine its purpose. For example,

> In dealing with respiration, we must show that it takes place for such or such a final object, and we must also show that this or that part of the process is necessitated by this or that stage of it (*ibid.*, 642ª).

We cannot do justice to the history of this tradition, which opposed organicism to atomism, and teleology to mechanism, except to note its respectable age, and to mention some highlights of its more recent development.

As we saw at the beginning of this chapter, one of the contributions to physiology of the chemistry developed by Lavoisier and his contemporaries was the clearer definition of the problem of animal nutrition. The mid-eighteenth century picture of living things as Cartesian machines with vibrating fibers, advocated by *philosophes* such as Julien Offray de La Mettrie in his *l'Homme Machine*, and for which Jacques Vaucanson's mechanical automata were popular models, gave way to theories that invoked the operation of some kind of "vital principle" or "vital force." Among these theories was that of Johann Christian Reil (see p. 69); in the words of his biographer, Reil considered that in biological organization

> . . . the laws of chemistry are not annulled, but rather acquire here their highest, truest meaning . . . . Organic chemistry is the continued dissection of the organ, an anatomy that reaches farther than the knife. This is the deeper sense, the real spirit of chemical physiology, when it does not remain in dark laboratories and narrow-mindedly deals with analyses, when it ventures instead into the light of life, and takes up its more significant and more general phenomena (Steffens, 1815, p. 21).

Reil's view that the "vital force" is a property of living matter arising from its organization was reflected in the writings of many chemists during the first decades of the nineteenth century. As we have mentioned before, the idea was adopted by Berzelius, who had considerable influence on his contemporaries. On the other hand, Chevreul did not

> . . . share the opinion of scientists who claim to explain the mysteries of organization by means of one or more forces that they call vital. When the adherents of this view tell me that living beings exhibit phenomena different from those of inanimate bodies, that nothing in in-

organic nature can give us an idea of human reason, I approve their words, I reflect, and I admire! But, when by means of one or more vital forces they claim to explain physiological phenomena, I find that their mode of philosophizing is just like the one I have condemned among scientists who advocate the contrary opinion: because they take a phenomenon without first having examined whether it is simple or complex, and they ask whether this phenomenon can be explained by affinities, by electric or magnetic forces, etc. And, because at present, as I have said, it is impossible to answer this question affirmatively on the basis of clear proof, they therefore conclude that the phenomenon depends on one or more forces that are peculiar to living bodies . . . Are the nature and effects of physical and chemical forces sufficiently well known at present to allow us reasonably to affirm that a particular phenomenon of living bodies is absolutely independent of them? Certainly not (Chevreul, 1824, pp. 241–242).

Some years later, the botanist Schleiden wrote:

In the present state of our natural sciences, only ignorance and indolence of spirit are the defenders of a vital force that can do and explain everything, and of which nobody can tell where it resides, how it acts, or what laws it obeys. The savage, who calls a locomotive a wild animal, is not more ignorant than the scientist who talks about vital force in the organism (Schleiden, 1845, Vol. I, p. 60).

The "mysteries of biological organization" constantly deepened during the early nineteenth century, as chemical analysis continued to reveal new complexity in the matter of which living things are composed. The "living molecules" (a term derived from *molécules organiques*, introduced by Buffon in 1749) had acquired a gelatinous or albuminoid character, and could not be fitted into the chemistry based on Dalton's atomic theory. Raspail was not alone in insisting that:

The organic molecule actually results from a chemical combination of known inorganic elements; but this kind of combination is such that from it alone there arises a new class of phenomena, and constitutes a separate realm. The foundations of the chemical theory of organized beings should not be looked for in the inorganic realm, but in organization itself (Raspail, 1833, p. 77).

For this reason, he advocated the use of the microscope in the study of living matter; during his brief career as a scientist (see Weiner, 1968),

Raspail pioneered in the application of chemistry to the microscopic examination of plant tissues.

The emergence of microscopy as a respectable biological technique after 1830 provided an important new approach to the problem of the organization of living matter; a decisive factor in this development was the improvement of the compound achromatic microscope, notably by Amici. Among the early fruits of the new technique was the rediscovery of the cell nucleus by Purkyně, Brown, and Mirbel, and the subsequent formulation of the cell theory by Schleiden and Schwann at the end of the 1830s. During this period, Felix Dujardin (1835) concluded from his observations on protozoa that these organisms do not have organs resembling those in higher animals, especially the numerous stomachs attributed to them by Ehrenberg, but that they are constituted of a glutinous material (*sarcode*) having the properties of organized living matter, with some chemical similarity to albumen. In the next decade, sarcode was renamed protoplasm by Purkyně and Mohl, and by 1861 the work of Virchow, Schultze, and Brücke had defined cells as independent units whose protoplasm is associated with their vital processes. In particular, Brücke wrote:

> I call cells elementary organisms as we now call elements those substances which thus far have not been decomposed chemically. Just as the unresolvability of these [chemical substances] is not yet proven, so we cannot dismiss the possibility that cells consist of other smaller organisms, which have the same relation to them as the cells have to the entire organism; but at present there is no basis for this assumption (Brücke, 1861, p. 381).

Brücke was similarly cautious about the apparent homogeneity of protoplasm, and noted:

> We cannot imagine a living, growing cell with a homogeneous nucleus and a homogeneous membrane, and which contains a simple solution of albumen, since we do not find in this protein those phenomena which we call the life process. We must therefore ascribe to the living cell, besides the molecular structure of its constituent organic compounds, a more complex structure of another order, which we call organization (*ibid.*, p. 386).

This anticipation of the supramolecular organization of cell constituents was not to be realized fully, however, until almost a century later, when the electron microscope enlarged the scope of cytological observation.

On 8 November 1868, Thomas Henry Huxley delivered in Edinburgh a Sunday lay sermon entitled *On the Physical Basis of Life* (see Geison, 1969). In it, he called attention to

> . . . the striking uniformity of material composition in living matter . . . . [A]ll the forms of protoplasm which have yet been examined contain the four elements, carbon, hydrogen, oxygen, and nitrogen, in very complex union, and . . . they behave similarly towards several re-agents. To this complex combination, the nature of which has never been determined with exactness, the name of Protein has been applied . . . . Enough, has, perhaps, been said to prove the existence of a general uniformity in the character of the protoplasm, or physical basis of life, in whatever group of living beings it may be studied. But it will be understood that this general uniformity by no means excludes any amount of special modifications of the fundamental substance (Huxley, 1870, pp. 17–19).

For Huxley, "all vital action may . . . be said to be the result of the molecular forces of the protoplasm which displays it" (*ibid.*, p. 26) and, in characteristic style, he asserted that

> . . . as surely as every future grows out of past and present, so will the physiology of the future gradually extend the realm of matter and law until it is coextensive with knowledge, with feeling, and with action. The consciousness of this great truth weighs like a nightmare, I believe, upon many of the best minds of these days. They watch what they conceive to be the progress of materialism, in such fear and powerless anger as a savage feels, when, during an eclipse, the great shadow creeps over the face of the sun. The advancing tide of matter threatens to drown their souls; the tightening grasp of law impedes their freedom; they are alarmed lest man's moral nature be debased by the increase of his wisdom (*ibid.*, p. 31).

During the hundred years after Huxley's lecture, the issues he raised, both scientific and philosophical, have reappeared in successive generations, as modified by the cumulation of chemical and biological knowledge. The persistence of the debate is indicated by the frequency with which the title Huxley gave his lecture was later used by others: Hardy (1906), Wilson (1923), and Bernal (1951) are a few examples.

It should be recalled that Huxley's lecture came before the remarkable development of cytology and the emergence of ideas about intracellular metabolism during the period 1870–1900. The apparent homogeneity and uniformity of protoplasm gave way to increasing complexity, and there was a proliferation of hypothetical subcellular particles, both as units of metabolic activity and of heredity (see p. 197). The evidence for the existence of some of these particles was better than for others, but as noted by Yves Delage (1895) in his valuable critique, there was uncertainty about all of them. In his view,

> Living protoplasm may be considered as a very complex chemical substance, essentially composed of albuminous materials, some mixed with each other, others separated in distinct aggregates whose structural parts are laid out in a precise arrangement (*ibid.*, pp. 748–749).

He also envisaged the possibility that ". . . all individualized parts of the cell are bounded by a membrane" (*ibid.*, p. 752), thus suggesting an intracellular anatomy not made evident until the advent of electron microscopy.

Physiological chemists deplored this preoccupation with the microscopically visible elements of cells, as well as the speculations about their intracellular function. Hofmeister reflected a growing conviction that one should proceed

> . . . not from the visible structures of the cell or of protoplasm, but from their activity, and thus study how the cell or protoplasm must be built in order to make this activity possible (Hofmeister, 1901, p. 9).

We have seen in this chapter that in 1901 the experimental knowledge about intermediary metabolism was essentially limited to the role of the liver in effecting such processes as the breakdown and synthesis of glycogen, or the synthesis of urea. By that time, however, Buchner's preparation of zymase had strengthened the opinion that all these processes are effected by enzymes, and this view was warmly espoused by Hofmeister. Furthermore, in order to explain the ordered conversion of metabolites, he suggested the intracellular localization of the enzymes because of their ability, as colloids, to form membranes; the interest in physicochemical studies on colloidal systems was lively, and a few years earlier Otto Bütschli (1892) had described artificial emulsions as models of protoplasm. Clearly, Hofmeister's view of the "chemical organization of the cell" reflected the aspirations, rather than the achievements, of physiological chemists at the turn of the century:

> If the morphologist attempts to elucidate the structure of protoplasm in the smallest detail, and the biochemist seeks to determine the chem-

ical activity of the same protoplasm with his seemingly cruder yet more penetrating methods, one deals here with two different aspects of the same thing. For the one, the goal is as detailed a description of the structure of protoplasm as possible, for the other it is the description of the entirety of the protoplasmic processes by means of a connected chain of chemical and physical formulas. It will be the difficult, but not thankless, task of the future to bring together these two widely different conceptions (Hofmeister, 1901, p. 29).

The advances made during the succeeding three decades, though considerable, were limited by the uncertainty about the nature of enzymes and their relation to proteins, as well as by the conviction that many metabolic processes (*e.g.*, intracellular oxidation, urea synthesis) are indissolubly linked to the integrity of "living protoplasm." In the face of these difficulties, there was a widespread view among biologists that the efforts of the biochemists were misdirected.

Among English physiologists, J. S. Haldane was an unflagging critic of the biochemical approach. His quantitative studies on the regulation of human breathing, together with those of Christian Bohr and of Joseph Barcroft, had provided striking evidence for the dictum that "the constancy of the internal environment is a requirement for free independent life" (Bernard, 1878–1879, Vol. I, p. 113). Haldane began from the philosophical position that the fact of life is elementary, a view reiterated by Niels Bohr (1933), and that efforts to analyze living organisms into physical and chemical mechanisms were futile. For example, in 1908, Haldane stated:

If it is a fundamental axiom that an organism actively asserts or maintains a specific structure and specific activities, it is clear that nutrition itself is only a constant state of reproduction: for the material of the cell is constantly changing. Not only is there constant molecular change, but the living cells are constantly being cast off and reproduced. It is only a step from this to the reproduction of lost parts which occurs so readily among lower organisms; and a not much greater step to the development of a complete organism from a single one of the constituent cells of an embryo in its early stages. In all these facts we have simply manifestations of the fundamental characters of the living organism. The reproduction of the parent organism from a single one of its constituent cells separated from the body seems to me only another such manifestation. Heredity, or, as it is sometimes metaphorically expressed, organic memory, is for Biology an axiom and not a problem . . . . Those who aim at physico-chemical explanations of

life are simply running their heads at a stone wall, and can only expect
sore heads as a consequence . . . . [I]n Physiology, and Biology generally, we are dealing with phenomena which, so far as our present
knowledge goes, not only differ in complexity, but differ in kind from
physical and chemical phenomena; and . . . the fundamental working
hypotheses of Physiology must differ correspondingly from those of
Physics and Chemistry.

That a meeting-point between Biology and Physical Science may at
some time be found, there is no reason for doubting. But we may con-
fidently predict that when that meeting-point is found, and one of the
two sciences is swallowed up, that one will not be Biology (Haldane,
1908, pp. 555–556).

Many years later, in acknowledging the work biochemists had done since
1908, Haldane restated his position and added:

When we discover . . . the existence of an intraprotoplasmic enzyme
or other substance on which life depends, we are at the same time
faced by the question how this particular substance is present at the
right time and place, and reacts to the right amount to fulfil its normal
functions. It is always, therefore, to the conception of life as a whole
that we are driven forwards (Haldane, 1931, p. 79).

Haldane's repeated criticisms of the biochemical approach were firmly
countered by Hopkins (see Needham, 1962), for whom the life of the
cell ". . . *is the expression of a particular dynamic equilibrium which
obtains in a polyphasic system*" (Hopkins, 1913, p. 220):

The task of the biochemist wishing to get to the heart of his problem
is exceptional in that he must study systems in which the organization
of chemical events counts for more, and is carried far beyond, such
simpler co-ordinations as may be found in non-living systems. He
would be over-bold were he to claim at present that such high organization can depend alone upon adjusted concentrations and ordered
structural distribution among specialized colloidal catalysts, but he is
justified, I think, in feeling sure that such factors contribute to that
organization in a significant sense. The biochemist, when he aims at
describing living systems in his own language, comes in contact with
philosophical thought. Current philosophy is busy in emphasizing the
truism that the properties of the Whole do not merely summarize but
emerge from the properties of its parts, and some exponents hold *a
priori* that biochemical data can throw no real light on the nature of

an organism which, in its very essence, is a unit. The biologist has long studied living organisms as wholes and will continue to do so with ever-increasing interest. But these studies can tell us nothing of the nature of the "physical basis of life", which no form of philosophy can ignore. It is for chemistry and physics to replace the vague concept "protoplasm"—a pure abstraction—by something more real and descriptive. I know of nothing which has shown that current efforts to this end do not deal with realities. It is only necessary for the biochemist to remember that his data gain their full significance only when he can relate them with the activities of the organism as a whole. He should be bold in experiment but cautious in his claims. His may not be the last word in the description of life, but without his help the last word will never be said (Hopkins, 1931, pp. 19–20).

The thirty years that separated the lectures of Hopkins and Hofmeister had produced, as we have seen, considerable biochemical knowledge, but for the organicists the later lecture was no less a statement of misplaced aspirations than the earlier one. During the succeeding thirty years, however, the elucidation of metabolic pathways and the chemical study of their component enzymes offered new biochemical approaches to the problems of the integration and regulation of biological activities. To see the recent development in better perspective, a brief summary of earlier views on the control of metabolism is necessary.

The Integration of Biochemical Processes. For the experimental physiologists of the nineteenth century, whether inclined toward mechanism or vitalism, the nervous system constituted the regulatory apparatus of the animal organism. The early studies of Charles Bell (in 1811), and of François Magendie, Pierre Flourens, and Johannes Müller during the 1820s laid the groundwork for the concept of the integrative action of the nervous system, enunciated by Charles Sherrington at the end of the century. We have already seen that around 1870 Claude Bernard emphasized the role of the nervous system in the control of the glycogenic function of the liver. By that time, however, there had accumulated a sizable body of evidence indicating that ductless or endocrine (Greek *endon,* within + *krinein,* to separate) glands such as the thyroid secrete into the blood chemical substances that affect body functions (see Rolleston, 1936). Of particular importance were clinical studies such as those of Thomas Addison (in 1855) on the consequences of adrenal insufficiency and of William Gull (in 1879) on the myxedema arising from a deficiency in thyroid function. Soon afterward, "internal secretions" (the term used by Bernard for the release of glucose from the liver) derived from those glands were seen as an additional mode of metabolic regulation. A pioneer in this development was Charles Édouard Brown-Séquard:

> We recognize that each tissue and, more generally, each cell of the organism secretes . . . special products or ferments into the blood and which thereby influence all the other cells thus integrated with each other by a mechanism other than the nervous system (Brown-Séquard and d'Arsonval, 1891, p. 267).

By 1905 Ernest Starling had termed these agents "chemical messengers" or "hormones," and during the succeeding half-century many were isolated in crystalline form, characterized in terms of their chemical structure, and synthesized in the laboratory. None of them turned out to be "ferments," although before 1930 the hormones were often termed catalysts, and grouped with the vitamins in this respect, reflecting the uncertainty about the nature of enzymes. During the 1930s the male and female sex hormones, as well as those of the adrenal cortex, were shown to be steroids; after the introduction of the isotope technique, their metabolic formation from cholesterol was largely elucidated. Except for a few amino acid derivatives, such as epinephrine and thyroxine, the other hormones turned out to be proteins (*e.g.,* insulin) or peptides, whose structure was determined after World War II with the help of the new methods of protein chemistry. Furthermore, the release of the characteristic hormones of the adrenal cortex, the thyroid, and the gonads was found to be under the control of "tropic" hormones (*e.g.,* adrenocorticotropic hormone, ACTH) produced by the anterior pituitary. When the activity of this gland was recognized to be under the control of the adjacent part of the brain (the hypothalamus), it became evident that an intricate and coordinated network involving both the nervous system and the endocrine organs is operative in the regulation of metabolism in the mammalian organism.

At each stage in the unravelling of the pattern of metabolic pathways, new efforts were made to correlate the physiological action of a hormone with its effect on particular enzymes or enzyme systems. We mentioned on p. 418 the important work of Carl and Gerty Cori on the effect of insulin and epinephrine on the carbohydrate metabolism of the intact rat. Each subsequent advance in the elucidation of the metabolic pathway linking glycogen and glucose was followed by extensive, but largely indecisive, attempts to pinpoint the site of the action of these two hormones. As the use of the isotope technique revealed the details of other pathways, similar studies were made on the effect of various hormones on the steps in these pathways. Most of these efforts only served to indicate that, in addition to the complexity of the biochemical mechanisms for the interconversion of metabolites, the mammalian organism has complex multistep mechanisms for regulating the rates of these interconversions. The difficulty was made all the greater by the multiplicity of physiological consequences of the administration of a hormone. A striking example is thyroxine, which stimulates the basal metabolic rate of human subjects, promotes the growth of immature mammals, and has a dramatic effect on amphibian metamorphosis. Indeed, the last of these is among the most sensitive biological assays for thyroid hormone activity; the phenomenon was discovered in 1912 by Gudernatsch, who fed tadpoles thyroid extract and observed precocious metamorphosis. During the 1920s the multiple physiological effects were attributed to the action of thyroxine in making tissues more responsive to nervous stimulation, but after the emergence of enzyme chemistry such ideas were gradually replaced by the conviction that the hormone interacts directly with a key rate-limiting enzyme operative in several metabolic processes. Among the results of efforts to discern the nature of this key enzyme was the discovery, during the 1950s, that thyroxine "uncouples" the link between electron transport and the generation of ATP in liver mitochondria, and that it promotes protein synthesis (as measured by the incorporation of labeled amino acids) by various fortified subcellular systems. These findings narrowed the area of search, but much still remains to be learned about the enzymic mechanisms of oxidative phosphorylation and of protein biosynthesis, so that the mode of action of thyroxine is still uncertain. Similar uncertainty applies to the other known hormones. The regulation they exhibit has been ascribed, in some cases, to the control of the biosynthesis of key enzymes, by actions on the intracellular apparatus involving DNA, RNA, and the protein catalysts concerned with the specific assembly of amino acid units into proteins; in other cases, an action on the cell membrane has been postulated, thus assigning to these hormones a role in controlling the rate of influx of essential metabolites. As in the investigation of other difficult biochemical problems, the finding of a new compound (in this case, 3',5'-cyclic AMP, discovered by Sutherland in 1957) has provided in recent years a new focus of attention, since this substance affects the action of a variety of enzymes, and the rate of its formation from ATP by a membrane-bound enzyme (adenyl cyclase) is influenced by several hormones. Whatever the outcome of current efforts to correlate the physiological action of individual hormones with their effect on the adenyl cyclase of target cells, it is clear that the chemical mechanisms involved in the integration of metabolic pathways are accessible to experimental study, but that many difficulties still remain to be overcome.

In the face of these advances and uncertainties, it is of interest to recall that, before the explosive growth of biochemistry through the study of enzymes as chemical entities and the use of isotopes and bacterial mutants to determine metabolic pathways, William Bate Hardy wrote:

> The integrating mechanism within the cell is the little cherub which sits up aloft conducting the orchestra. We know he is there, but his features are veiled for he is the Theoria—the mystery of life itself!
>
> Have I exaggerated the mystery? If any one doubt, let him consider the hepatic cell. There is no evidence of specialization in the mammalian liver—indeed the evidence is definitely against it. Any or every cell seems capable of synthesizing glycogen from sugar or from lactic acid, of solving the chemical conundrum: how to pass directly from carbohydrates to fats and back or proteins to fats, of dealing with metallic poisons, of controlling the chemical cycle of hemoglobin, of synthesizing uric acid, so on and so on. Has the biologist any picture, even of the vaguest kind, of how so diverse a chemical factory can operate in a fluid mass, say 10^{-8} cubic millimetres in volume?
>
> Some day it will be necessary to return to the standpoint of the 'eighties, and to realise that there is some master process, some integrating principle now vaguely but truly spoken of as "structure" which subdues a galaxy of catalysts to its purpose, and deserves the title of "living" (Hardy, 1936, p. 899).

Since Hardy wrote these lines in 1931, electron microscopy has revealed the striking compartmentalization in the intracellular structure of liver cells, and the isotope technique has shown the chemical components of this structure to be in a dynamic state. For the biochemist, therefore, the "integrating principle" (whatever Hardy meant by the term) has become the property of an intricate chemical assembly whose parts not only are associated with each other in specific arrangements, but also are undergoing continuous breakdown and regeneration. Furthermore, he considers that the steady state essential for survival of the cell in a changing environment is maintained by virtue of the interconnected network of enzyme-catalyzed metabolic pathways, and of the mechanisms that control their rates. With the growth of knowledge about the three-dimensional structure of proteins and nucleic acids, and about their interaction with each other and with other cell components, it has seemed reasonable to hope that a study of the detailed properties of the parts would reveal the factors which determine the structural and functional integrity of the whole. This hope has guided recent efforts to describe the "wholeness" of living systems in chemical terms; in particular, significant new results have come from the study of enzyme systems, especially in bacteria.

Before these recent developments, several investigators (notably Jacques Monod and Ernest Gale) had provided clear evidence for variation and adaptation in the enzymic constitution of bacteria. As was stated by Marjory Stephenson:

> Biochemically, bacterial cells are the most plastic of living material, even as compared with other micro-organisms which have developed a more regulated hereditary mechanism. Higher animals and plants have in addition elaborated machinery for maintaining a constant environment, any severe disturbance in which is reflected in immediate pathological symptoms. The bacterial cell, by reason of its small size and consequent relatively large surface, cannot develop by maintaining a constant chemical environment, but reacts by adapting its enzyme systems so as to survive and grow in changing conditions. It is immensely tolerant of experimental meddling and offers material for the study of processes of growth, variation and development of enzymes without parallel in any other biological material. Other living forms have long passed their experimental stage and have evolved relatively satisfactory stable systems; biochemically speaking the stages of their evolution are unknown to us. But with bacteria constant evolutionary changes occur under our eyes and can be controlled and imitated in the laboratory. Bacterial studies pay the highest dividends on biochemical investigation (Stephenson, 1949, pp. 311–312).

Indeed, studies on the regulation of metabolic pathways in bacteria (especially the colon bacillus *Escherichia coli*) have been exceedingly fruitful.

The definition of individual pathways in bacterial metabolism, through the combined use of the isotope technique, microbial mutants, and enzyme chemistry, also provided clues to the possible intracellular mechanisms of the control of these pathways. An important factor in the appreciation of the significance of these clues was the development, during World War II, of electronic control devices for military purposes, and the application of the theory underlying this technology to the consideration of the biological problem of homeostasis. Thus in 1956 Edwin Umbarger reported that, in *Escherichia coli*, isoleucine specifically inhibits the first enzyme-catalyzed reaction in the sequence of steps leading to its biosynthesis from threonine; his paper began as follows:

Recent developments in automation have led to the use in industry of machines capable of performing operations that have been compared with certain types of human activity. In the internally regulated machine, as in the living organism, processes are controlled by feedback loops that prevent any one phase of the process from being carried to a catastrophic extreme. The consequences of such feedback control can be observed at all levels of organization in a living animal —for example, proliferation of cells to form a definite structure, the maintenance of muscle tone, and such homeostatic mechanisms as temperature regulation and the maintenance of a relatively constant blood sugar level (Umbarger, 1956, p. 848).

In the same year, Richard Yates and Arthur Pardee reported another example of such "negative feedback" in the inhibition, by cytidine triphosphate, of the first committed step in the pathway leading to its biosynthesis in *E. coli*. Thereafter, some earlier instances of enzyme inhibition were seen to fall into the same category, and new ones were found. Furthermore, chemical studies on the isolated enzymes provided evidence for the view that such regulatory phenomena are a consequence of the specific interaction of the effector with the catalytic protein or with one of the subunits of an oligomeric enzyme complex (Atkinson, 1965). To this feedback inhibition of an early enzyme in a metabolic pathway, Henry Vogel, Boris Magasanik, and others added in 1957 another regulatory mechanism, in which the product of a microbial biosynthetic sequence causes a "repression" of the cellular apparatus that manufactures the enzyme proteins for the pathway; later work (notably by Bruce Ames) provided strong evidence for this kind of intracellular regulation. Moreover, other specific regulatory mechanisms were identified, involving the interconversion of a more active and a less active form of an enzyme through its chemical modification. Enough has perhaps been said here to indicate the current trend of considering the problem of metabolic regulation in terms of the detailed chemistry of catalytic proteins; a fuller account may be found in biochemical treatises. It remains to be seen to what extent the results of studies on the regulation of microbial enzymes can be applied to the organized enzyme assemblies of animal cells, and whether the conclusions that are drawn can be correlated with the physiological control mechanisms in the living animal. That the activity of some liver enzymes can fluctuate in response to excess metabolite or to adrenal cortical hormones, in a manner comparable to the induction of enzymic activity in bacterial cultures, was demonstrated by Eugene Knox and others during the 1950s, but much remains to be learned about the integration of the biochemical functions of the liver cell. If one considers the multiplicity of interconnected metabolic pathways, and the various known possibilities of their control by hormones and metabolites, the task is truly formidable.

These difficulties, which at present seem to be insurmountable, have given encouragement to those who have continued in the tradition of Bichat and Haldane, and have believed that living organisms follow special laws, different in principle from those of chemistry and physics. This view has assumed various forms during this century, and its adherents have shifted their ground in the face of biochemical advances. Around 1900 a leading spokesman of vitalism was the embryologist Hans Driesch who, in 1889, had shown that half of a sea-urchin egg could give rise to a whole larva. He developed a theory of "dynamic teleology" (Driesch, 1908), with the antimechanist argument that machines cannot do what the developing embryo (or other living things) can do; this position was opposed by Wilhelm Roux and by Jacques Loeb, who called attention to the ability of physicochemical systems to do what a machine cannot do. Between the two World Wars, the vitalist position was espoused by Hans Spemann, one of the two dominant figures in embryology at that time, whereas the other (Ross Harrison) stated in 1936:

This quality of "wholeness" in the parts of the organism, particularly the embryo, has led to much speculation and even a system of philosophy. It is the capital problem of embryology to find the physical-chemical basis for it (Harrison, 1945, p. 281).

After 1930 the philosophical ideas of Driesch and Spemann largely ceased to have currency among experimental biologists, principally owing to the successes achieved in the chemical explanation of some biological phenomena. The organismic point of view continues to be upheld, however, with emphasis on the integration of intracellular constituents and of cells in hierarchally ordered assemblies (see Weiss, 1962). Although some leading biologists deplore the current preoccupation with the chemistry of individual cell components, and stress the importance of studying living organisms as wholes, they generally acknowledge the biological significance of such chemical work and use its fruits in their own researches.

Since World War II, doubts about the value of the detailed chemical description of biological processes have come from other sectors of the scientific community. In 1949 Max Delbrück, a theoretical physicist who had turned to the study of bacterial viruses, wrote:

Biology is a very interesting field to enter for anyone, by the vastness of its structure and the extraordinary variety of strange facts it has collected, but to the physicist it is also a depressing subject, because, insofar as physical explanations of seemingly physical phenomena go, like excitation, or chromosome movements, or replication, the analysis seems to have stalled around in a semidescriptive manner without progressing towards a radical physical explanation. He may be told that the only real access of atomic physics to biology is through biochemistry. Listening to the story of modern biochemistry he might become persuaded that the cell is a sack full of enzymes acting on substrates converting them through various intermediate stages either into cell substance or into waste products. The enzymes must be situated in their proper strategic positions to perform their duties in a well regulated fashion. They in turn must be synthesized and must be brought into position by manoevers which are not yet understood, but which, at first sight at least, do not necessarily seem to differ in nature from the rest of biochemistry. Indeed, the vista of the biochemist is one with an infinite horizon. And yet, this program of explaining the simple through the complex smacks suspiciously of the program of explaining atoms in terms of complex mechanical models. It looks sane until the paradoxes crop up and come into sharper focus. In biology we are not yet at the point where we are presented with clear paradoxes and this will not happen until the analysis of the behaviour of living cells has been carried into far greater detail. This analysis should be done on the living cell's own terms and the theories should be formulated without fear of contradicting molecular physics. I believe that it is in this direction that physicists will show the greatest zeal and will create a new intellectual approach to biology which would lend meaning to the ill-used term biophysics (Delbrück, 1949, p. 190).

We have recalled on p. 251 some of the important experimental achievements of Delbrück and of other investigators (for example, Seymour Benzer) who entered genetics from physics. By 1968, however, "no paradoxes had come into focus, no 'other laws of physics' had turned up" and the successes after 1950 "depended on the appearance on the molecular biological scene of many other highly intelligent and gifted experimentalists, not a few of whom had recourse to the recently despised methods of biochemistry" (Stent, 1968, p. 394).

In recent years the "antireductionist" thesis has been restated by the noted theoretical physicist Eugene Wigner (1969), and Michael Polanyi has emphasized anew the irreducibility of biology to chemistry and physics:

The recognition of certain basic impossibilities has laid the foundations of some major principles of physics and chemistry; similarly, recognition of the impossibility of understanding living things in terms of physics and chemistry, far from setting limits to our understanding of life, will guide it in the right direction. And even if the demonstration of this impossibility should prove of no great advantage in the pursuit of discovery, such a demonstration would help to draw a truer image of life and man than that given by the present basic concepts of biology (Polanyi, 1968, p. 1312).

The issue raised by Polanyi has not been decided, but, since his position may be "of no great advantage in the pursuit of discovery," it is not likely to be adopted by working biologists and chemists engaged in that pursuit.

Historically, the question whether the functions of a living organism can be explained in terms of the properties of its components has been related to the use of physical models drawn from contemporary technology. For those who believed that such explanation is possible, models of this kind have served to illustrate the applicability to biological systems of the physical principles underlying the operation of the man-made devices. In the eighteenth century the living machinery was thought to resemble a mechanical clockwork in the function of its parts; in the nineteenth, a combustion engine in the production of work and heat or the electric telegraph in the operation of the nervous system; in the twentieth, an electronic computer in the storage, transmission, and expression of genetic information or a servomechanism in the control of metabolic processes. The utility of all these models as aids to biological thought was limited by contemporary chemical knowledge. Although each of them highlighted a physical principle applicable to the understanding of biological function, and contributed new words to biological language, none of them, however sophisticated in its time, could accommodate the discovery of new details of the chemical structure and reactivity of cell constituents. The subtlety of chemical specificity in the interactions of enzymes with their substrates and with other molecules that influence their catalytic action, and in the interaction of individual nucleic acids and proteins with one another and with other biochemical compounds, has indicated a highly ordered complexity greater than that of man-made technological devices. Moreover, the biological application of the theoretical principles underlying the operation of these devices (*e.g.*, information theory) has been limited by the available knowledge of the details of the specifically ordered chemical heterogeneity of living systems.

We have seen that, although the unraveling of the chemical complexity of biological systems has already produced impressive results, this seemingly tedious effort has gone only a small part of the way. At each stage in the advance of biochemistry, voices have been raised to question the value of this effort, either for the advancement of biology, or for the dignity of man, or for both. Impatient minds have sought to short-circuit the need for more detailed chemical and biological knowledge through flights of the imagination, many of which stimulated fruitful discovery. In the test of these speculations, the appearance of new chemical compounds and the emergence of new techniques, often from unexpected sources, usually proved decisive. At the present stage of the interplay of biology and chemistry, attention is focused on the specific cooperative interactions of macromolecules with one another and with smaller molecules to form organized functional assemblies. What new chemical insights, accidental observations, or new techniques will lead to the next stage is a matter of conjecture, but the achievements of the past half-century suggest that an essential component of future effort must be ever more refined study of the structure and reactivity of cell constituents. As Hopkins stated in 1931, the chemical knowledge so obtained will gain its full significance only when it is related with the activities of the organism as a whole, and one of the main objectives of present-day experimental biology is to narrow the conceptual gap that still separates it from chemistry. In such biological studies, new insights, new observations, and new techniques in the investigation of whole cells and organisms, and of their interaction with one another and with their environment, will continue to be decisive. Whatever the outcome of such future efforts in the interplay of biology and chemistry, they seem likely to contribute more to the understanding of the phenomena of life, including human life, than mystery writings that stress the uniqueness of the biological order and the impossibility of fully describing the behavior of living things in the language of chemistry.

We conclude with the words of a great biologist, spoken in 1922. After summarizing the knowledge provided by cytology, genetics, and chemistry about the parts of the living cell, Edmund Wilson asked:

How shall we put it together again? It is here that we first fairly face the real problem of the physical basis of life; and here lies the unsolved riddle. We try to disguise our ignorance concerning this problem with learned phrases. We are forever conjuring with the word "organization" as a name for the integrating and unifying principle in the vital processes; but which one of us is really able to translate this word into intelligible language? We say pedantically—and no doubt correctly—

that the orderly operation of the cell results from a dynamic equilibrium in a polyphasic colloidal system. In our mechanistic treatment of the problem we commonly assume this operation to be somehow traceable to an original pattern or configuration of material particles in the system, as is the case with a machine. Most certainly conceptions of this type have given us an indispensable working method—it is the method which almost alone is responsible for the progress of modern biology—but the plain fact remains that there are still some of the most striking phenomena of life of which it has thus far failed to give us the most rudimentary understanding . . . . We are ready with the time-honored replies: It is the "organism as a whole"; it is a "property of the system as such"; it is "organization." These words, like those of Goldsmith's country parson, are "of learned length and thundering sound." Once more, in the plain speech of everyday life, their meaning is: *We do not know* . . . . Shall we then join hands with the neo-vitalists in referring the unifying and regulatory principle to the operation of an unknown power, a directive force, an archaeus, an entelechy or a soul? Yes, if we are ready to abandon the problem and have done with it once for all. No, a thousand times, if we hope really to advance our understanding of the living organism. To say *ignoramus* does not mean that we must also say *ignorabimus* (Wilson, 1923, pp. 284–286).

The half-century since these words were spoken has provided much understanding of the chemical basis of "some of the most striking phenomena of life," and has justified the conviction that the continued interplay of biology and chemistry will bring better understanding in the future.

References

(For abbreviations of serial publications, see p. 552)

ABDERHALDEN, E. AND KOMM, E. (1924). Über die Anhydridstruktur der Proteine. Z. *physiol. Chem.* **139**, 181–204.

ABELOUS, E. AND GÉRARD, E. (1899). Sur la coexistence d'une diastase réductrice et d'une diastase oxydante dans les organes animaux. *Comp. Rend.* **129**, 1023–1025.

AHRENS, F. B. (1902). Das Gärungsproblem. *Sammlung chemischer und chemisch-technischer Vorträge* **7**, 445–495.

ALLEN, R. S. AND MURLIN, J. R. (1925). Biuret-free insulin. *Am. J. Physiol.* **75**, 131–139.

ALLEN, W. AND PEPYS, W. H. (1808). On the changes produced in atmospheric air, and oxygen gas, by respiration. *Phil. Trans.* **98**, 249–281.

ALTMANN, R. (1889). Ueber Nucleinsäuren. *Arch. Anat. Physiol.*, pp. 524–536.

ALTMANN, R. (1890). *Die Elementarorganismen und ihre Beziehung zu den Zellen,* Leipzig: Veit.

ALTMANN, R. (1892). Über Kernstruktur und Netzstruktur. *Arch. Anat. Physiol.*, pp. 223–230.

ANSCHÜTZ, R. (1929). *August Kekulé,* Berlin: Verlag Chemie.

ANSON, M. L. AND MIRSKY, A. E. (1930). Hemoglobin, the heme pigments, and cellular respiration. *Physiol. Revs.* **10**, 506–546.

ARMSTRONG, E. F. AND ARMSTRONG, H. E. (1904). Studies on enzyme action. III. The influence of the products of change on the rate of change conditioned by sucro-clastic enzymes. *Proc. Roy. Soc.* **73**, 516–526.

ARMSTRONG, H. E. (1895). The nature of chemical change and the conditions which determine it. *J. Chem. Soc.* **67**, 1122–1172.

ARMSTRONG, H. E. (1927). Poor common salt! *Nature* **120**, 478.

ARNON, D. I. (1955). The chloroplast as a complete photosynthetic unit. *Science* **122**, 9–16.

ARNON, D. I. (1967). Photosynthetic activity of isolated chloroplasts. *Physiol. Revs.* **47**, 317–358.

ARTHUS, M. (1896). *Nature des Enzymes,* Paris: Jouve.

ASTBURY, W. T. (1943). X-Rays and the stoichiometry of proteins. *Adv. Enzymol.* **3**, 63–108.

ASTBURY, W. T. (1947a). On the structure of biological fibres and the problem of muscle. *Proc. Roy. Soc.* **B134**, 303–328.

ASTBURY, W. T. (1947b). X-Ray studies of nucleic acids. *Symp. Soc. Exp. Biol.* **1**, 66–76.

ASTBURY, W. T. AND BELL, F. O. (1938). X-Ray studies of thymonucleic acid. *Nature* **141**, 747–748.

ASTIER, C. (1813). Expériences faites sur le sirop et le sucre de raisin. *Ann. Chim.* **87**, 271–285.

ATKINSON, D. E. (1965). Biological feedback control at the molecular level. *Science* **150**, 851–857.

AUERBACH, C. (1967). The chemical production of mutations. *Science* **158**, 1141–1147.

AUERBACH, C., ROBSON, J. M. AND CARR, J. G. (1947). The chemical production of mutations. *Science* **105**, 243–247.

AULIE, R. P. (1970). Boussingault and the nitrogen cycle. *Proc. Am. Phil. Soc.* **114**, 435–479.

AVERY, O. T., MAC LEOD, C. M. AND MC CARTY, M. (1944). Studies on the chemical nature of the substance inducing transformation of pneumococcal types. *Journal of Experimental Medicine* **79**, 137–158.

BACH, A. (1911). Zur Kenntnis der Reduktionsfermente. I. Mitteilung. Über das Schardinger-enzym (Perhydridase). *Biochem. Z.* **31**, 443–449.

BAEYER, A. (1870). Über die Wasserentziehung und ihre Bedeutung für das Pflanzenleben und die Gärung. *Ber. chem. Ges.* **3**, 63–75.

BAEYER, A. AND VILLIGER, V. (1900). Benzoylwasserstoffsuperoxyd und die Oxydation des Benzaldehyds an der Luft. *Ber. chem. Ges.* **33**, 1569–1585.

BAKER, J. R. (1955). The cell-theory: a restatement, history, and critique. Part V. The multiplication of nuclei. *Quarterly Journal of Microscopical Science* **96**, 449–481.

BALL, E. G. (1942). Oxidative mechanisms in animal tissues. In *A Symposium on Respiratory Enzymes*, pp. 16–32, Madison: University of Wisconsin Press.

BALTZER, F. (1967). *Theodor Boveri: Life and Work of a Great Biologist* (translation by Dorothea Rudnick), Berkeley and Los Angeles: University of California Press.

BANTING, F. G. AND BEST, C. H. (1922). The internal secretion of the pancreas. *Journal of Laboratory and Clinical Medicine* **7**, 251–266.

BARCROFT, J. (1914). *The Respiratory Function of the Blood*, Cambridge: University Press.

BARCROFT. J. AND HALDANE, J. S. (1902). A method of estimating the oxygen and carbonic acid in small quantities of blood. *J. Physiol.* **28**, 232–240.

BARENDRECHT, H. P. (1904). Enzymwirkung. *Z. physik. Chem.* **49**, 456–482.

BÁRON, J. AND POLANYI, M. (1913). Über die Anwendigkeit des zweiten Hauptsatzes der Thermodynamik auf Vorgänge im tierischen Organismus. *Biochem. Z.* **53**, 1–20.

BARRON, E. S. G. AND HARROP, G. A. (1928). Studies on blood cell metabolism. II. The effect of methylene blue and other dyes upon the glycolysis and lactic acid formation of mammalian and avian erythrocytes. *J. Biol. Chem.* **79**, 65–87.

BARTH, M. (1878). Zur Kenntnis des Invertins. *Ber. chem. Ges.* **11**, 474–482.

BATESON, W. (1909). *Mendel's Principles of Heredity*, Cambridge: University Press.

BATESON, W. (1913). *Problems of Genetics*, New Haven: Yale University Press.

BATESON, W. AND SAUNDERS, E. R. (1902). The facts of heredity in the light of Mendel's discovery. *Reports to the Evolution Committee of the Royal Society of London* 1, 125–160.

BATTELLI, F. AND STERN, L. (1909). Die akzessorische Atmung in den Tiergeweben. *Biochem. Z.* 21, 488–509.

BATTELLI, F. AND STERN, L. (1910). Die Katalase. *Erg. Physiol.* 10, 531–597.

BATTELLI, F. AND STERN, L. (1911). Die Oxydation der Bernsteinsäure durch Tiergewebe. *Biochem. Z.* 30, 172–194.

BATTELLI, F. AND STERN, L. (1912). Die Oxydationsfermente. *Erg. Physiol.* 12, 96–268.

BATTELLI, F. AND STERN, L. (1914). Einfluss der mechanischen Zerstörung der Zellstruktur auf die verschiedenen Oxydationsprozesse der Tiergewebe. *Biochem. Z.* 67, 443–471.

BATTELLI, F. AND STERN, L. (1920). Nature des ferments oxydants et des ferments réducteurs. *Comptes Rendus de la Société de Biologie* 83, 1544–1545.

BAUMANN, E. AND KOSSEL, A. (1895). Zur Erinnerung an Felix Hoppe-Seyler. *Z. physiol. Chem.* 21, I–LXI.

BAWDEN, F. C., PIRIE, N. W., BERNAL, J. D. AND FANKUCHEN, I. (1936). Liquid crystalline substances from virus-infected plants. *Nature* 138, 1051–1052.

BAYLISS, W. M. (1908). *The Nature of Enzyme Action*, London: Longmans, Green.

BAYLISS, W. M. (1924). *Principles of General Physiology* (fourth edition), London: Longmans, Green.

BEACH, E. F. (1961). Beccari of Bologna. The discoverer of vegetable protein. *J. Hist. Med.* 16, 354–373.

BEADLE, G. W. (1945a). Biochemical genetics. *Chem. Revs.* 37, 15–96.

BEADLE, G. W. (1945b). The genetic control of biochemical reactions. *Harvey Lectures* 40, 179–194.

BEADLE, G. W. (1963). *Genetics and Modern Biology*, Philadelphia: American Philosophical Society.

BEADLE, G. W. (1969). Genes, chemistry, and the nature of man. In *Biology and the Physical Sciences*, pp. 1–13 (S. Devons, ed.), New York: Columbia University Press.

BEADLE, G. W. AND EPHRUSSI, B. (1936). The differentiation of eye pigments in Drosophila as studied by transplantation. *Genetics* 21, 225–247.

BEADLE, G. W. AND TATUM, E. L. (1941). Genetic control of biochemical reactions in Neurospora. *Proc. Natl. Acad. Sci.* 27, 499–506.

BEAUMONT, W. (1833). *Experiments and Observations on the Gastric Juice and the Physiology of Digestion*, Plattsburgh: Allen.

BÉCHAMP, A. (1883). *Les Microzymas*, Paris: Baillière.

BEER, J. J. (1959). *The Emergence of the German Dye Industry*, Urbana: University of Illinois Press.

BEGUIN, J. (1624). *Les Elemens de Chymie* etc., Troisiesme Edition, Paris: Le Maistre.

BEIJERINCK, M. W. (1917). The enzyme theory of heredity. *Proceedings of the Royal Academy of Sciences, Amsterdam* 19, 1275–1289.

BELITSER, V. A. AND TSIBAKOVA, E. T. (1939). [The mechanism of phosphorylation associated with respiration.] *Biokhimiya* 4, 516–535.

BENCE JONES, H. (1867). *Lectures on some of the Applications of Chemistry and Mechanics to Pathology and Therapeutics,* London: Churchill.

BÉRARD, J. E. (1817). Essai sur l'analyse des substances animales. *Ann. Chim.* 2e Sér. 5, 290–298.

BERGMANN, M. (1938). The structure of proteins in relation to biological problems. *Chem. Revs.* 22, 423–435.

BERGMANN, M. AND ZERVAS, L. (1932). Über ein allgemeines Verfahren der Peptid-Synthese. *Ber chem. Ges.* 65, 1192–1201.

BERMAN, A. (1963). Conflict and anomaly in the scientific orientation of French pharmacy 1800–1873. *Bull. Hist. Med.* 37, 440–462.

BERNAL, J. D. (1951). *The Physical Basis of Life,* London: Routledge and Kegan Paul.

BERNAL, J. D. (1963). William Thomas Astbury 1898–1961. *Biog. Mem. F. R. S.* 9, 1–35.

BERNAL, J. D. AND CROWFOOT, D. (1934). X-Ray photographs of crystalline pepsin. *Nature* 133, 794.

BERNARD, C. (1848). De l'origine du sucre dans l'économie animale. *Archives Générales de Médecine* 18, 303–319.

BERNARD, C. (1853). *Nouvelle Fonction du Foie Considéré comme Organe Producteur de Matière Sucrée chez l'Homme et les Animaux,* Paris: Baillière.

BERNARD, C. (1855a). *Leçons de Physiologie Expérimentale Appliquée a la Médecine,* Paris. Baillière.

BERNARD, C. (1855b). Sur le mécanisme de la formation du sucre dans le foie. *Comp. Rend.* 41, 461–469.

BERNARD, C. (1857a). *Leçons sur les Effets des Substances Toxiques et Médicamenteuses,* Paris: Baillière.

BERNARD, C. (1857b). Sur le mécanisme physiologique de la formation du sucre dans le foie. *Comp. Rend.* 44, 578–586.

BERNARD, C. (1859). *Leçons sur les Propriétés Physiologiques et les Alterations Pathologiques des Liquides de l'Organisme,* Paris: Baillière.

BERNARD, C. (1865). *Introduction à l'Étude de la Médecine Expérimentale,* Paris: Baillière.

BERNARD, C. (1872). *De la Physiologie Générale,* Paris: Hachette.

BERNARD, C. (1876). *Leçons sur la Chaleur Animale,* Paris: Balliére.

BERNARD, C. (1877a). *Leçons sur le Diabète et la Glycogenèse Animale,* Paris: Baillère.

BERNARD, C. (1877b). Critique expérimentale sur le mécanisme de la formation du sucre dans le foie. *Comp. Rend.* 85, 519–525.

BERNARD, C. (1878–1879). *Leçons sur les Phénomènes de la Vie Communs aux Animaux et aux Végétaux,* Paris: Baillière.

BERNARD, C. (1927). *An Introduction to the Study of Experimental Medicine* (translated by H. C. Greene), New York: Macmillan.

BERRY, A. J. (1960). *Henry Cavendish,* London: Hutchinson.

BERT, P. (1870). *Leçons sur la Physiologie Comparée de la Respiration,* Paris: Baillière.

BERTHELOT, M. (1859). Remarques sur la fermentation alcoolique de la levure de bière. *Comp. Rend.* **48**, 691–692.

BERTHELOT, M. (1860a). *Chimie Organique Fondée sur la Synthèse,* Paris: Mallet-Bachelier.

BERTHELOT, M. (1860b). Sur la fermentation glucosique du sucre de canne. *Comp. Rend.* **50**, 980–984.

BERTHELOT, M. (1875). Sur les principes généraux de la thermochimie. *Ann. Chim.* 5e Sér. **4**, 5–131.

BERTHELOT, M. (1890). *La Révolution Chimique: Lavoisier,* Paris: Alcan.

BERTRAND, G. (1894). Recherches sur le latex de l'arbre à laque du Tonkin. *Bulletin de la Société Chimique de Paris* 3e Sér. **11**, 718–721.

BERZELIUS, J. J. (1813). *A View of the Progress and Present State of Animal Chemistry* (translated by G. Brunnmark), London: Hatchard, Johnson, and Boosey.

BERZELIUS, J. J. (1836). Einige Ideen über bei der Bildung organischer Verbindungen in der lebenden Natur wirksame, aber bisher nicht bemerkte Kraft. *Jahres-Ber.* **15**, 237–245.

BERZELIUS, J. J. (1839). Weingährung. *Jahres-Ber.* **18**, 400–403.

BERZELIUS, J. J. (1843). Thierchemie. *Jahres-Ber.* **22**, 535–538.

BERZELIUS, J. J. (1916). *Jac. Berzelius Bref* (edited by H. G. Söderbaum; V. Correspondence between Berzelius and Mulder, 1834–1847), Uppsala: Almquist and Wiksells.

BERZELIUS, J. J. (1934). *Autobiographical Notes* (published by Royal Swedish Academy of Sciences through H. G. Söderbaum; translated by O. Larsell), Baltimore: Williams and Wilkins Co.

BICHAT, X. (1801). *Anatomie Générale Appliquée à la Physiologie et à la Médecine,* Paris: Brosson, Gabon et Cie.

BICHAT, X. (1805). *Recherches Physiologiques sur la Vie et la Mort* (third edition), Paris: Marchant.

BIDDER, F. AND SCHMIDT, C. (1852). *Die Verdauungssaefte und der Stoffwechsel,* Mitau and Leipzig: Reyher.

BING, F. C. (1971). The history of the word 'metabolism.' *J. Hist. Med.* **26**, 158–180.

BISCHOFF, T. L. W. (1847). Theorie der Befruchtung und über die Rolle, welche die Spermatozoiden dabei spielen. *Arch. Anat. Physiol.,* pp. 422–442.

BISCHOFF, T. L. W. (1853). *Der Harnstoff als Maass des Stoffwechsels,* Giessen: Ricker.

BISCHOFF, T. L. W. AND VOIT, C. (1860). *Die Gesetze der Ernährung des Fleischfressers durch Neue Untersuchungen Festgestellt,* Leipzig: Winter.

BITTING, A. W. (1937). *Appertizing or the Art of Canning: Its History and Development,* San Francisco: The Trade Pressroom.

BLOCH, K. (1965). The biological synthesis of cholesterol. *Science* **150**, 19–28.

BLOCH, K. AND SCHOENHEIMER, R. (1941). The biological precursors of creatine. *J. Biol. Chem.* **138**, 167–194.

BODLÄNDER, G. (1895). Moritz Traube. *Ber. chem. Ges.* **28**, 1085–1108.

BODLÄNDER, G. (1899). Ueber langsame Verbrennung. *Sammlung chemischer und chemisch-technischer Vorträge* **3**, 385–488.

BOERHAAVE, H. (1735). *Elements of Chemistry* (translated by T. Dallowe), London: Pemberton, Clarke, Millar, and Gray.

BOHR, N. (1933). Light and Life. *Nature* 131, 421–423, 457–459.

BOIVIN, A. (1947). Directed mutation in colon bacilli, by an inducing principle of desoxyribonucleic nature: its meaning for the general biochemistry of heredity. *Cold Spr. Harb. Symp.* 12, 7–17.

BONNER, D. M. (1946). Biochemical mutations in Neurospora. *Cold Spr. Harb. Symp.* 11, 14–24.

BORSOOK, H. AND KEIGHLEY, G. L. (1935). The "continuing" metabolism of nitrogen in animals. *Proc. Roy. Soc.* B118, 488–521.

BORSOOK, H. AND SCHOTT, H. F. (1931). The role of the enzyme in the succinate-enzyme-fumarate equilibrium. *J. Biol. Chem.* 92, 535–557.

BOUCHARDAT, A. AND SANDRAS, C. M. (1842). Recherches sur la digestion. *Annales des Sciences Naturelles* 18, 225–241.

BOURQUELOT, E. (1896). *Les Ferments Solubles (Diastases—Enzymes)*, Paris: Société Editions Scientifiques.

BOUSSINGAULT, J. B. (1839). Analyses comparées des aliments consommés et des produits rendus par une vache laitière; recherches entreprises dans le but d'examiner si les animaux herbivores empruntent de l'azote a l'atmosphère. *Ann. Chim.* 2e Sér. 71, 113–127.

BOUTRON-CHARLARD, A. F. AND FRÉMY, E. (1841). Recherches sur la fermentation lactique. *Ann. Chim.* 3e Sér. 2, 257–274.

BOVERI, T. (1904). *Ergebnisse über die Konstitution der Chromatischen Substanz des Zellkerns*, Jena: Fischer.

BOYLE, R. (1680). *The Sceptical Chymist* (second edition), London: Hall.

BRADLEY, H. C. (1913). The problem of enzyme synthesis. I. Lipase and fat of animal tissues. *J. Biol. Chem.* 13, 407–418.

BRAGG, W. L. (1967). Reminiscences of fifty years' research. *Proceedings of the Royal Institution of Great Britain* 41, 92–100.

BRAGG, W. L., KENDREW, J. C., AND PERUTZ, M. F. (1950). Polypeptide chain configurations in crystalline proteins. *Proc. Roy. Soc.* A203, 321–357.

BRAND, E., SAIDEL, L. J., GOLDWATER, W. H., KASSELL, B., AND RYAN, F. J. (1945). The empirical formula of beta-lactoglobulin. *J. Am. Chem. Soc.* 67, 1524–1531.

BRAUNSTEIN, A. E. AND KRITZMANN, M. G. (1937). Über den Ab- und Aufbau von Aminosäuren durch Umaminierung. *Enzymologia* 2, 129–146.

BRIDGES, C. B. (1935). Salivary chromosome maps. *Journal of Heredity* 26, 60–64.

BRIGGS, G. E. AND HALDANE, J. B. S. (1925). A note on the kinetics of enzyme action. *Biochem. J.* 19, 388–339.

BRODIE, B. C. (1812). Further experiments and observations on the action of poisons on the animal system. *Phil. Trans.* 102, 205–227.

BRODIE, T. G. (1910). Some new forms of apparatus for the analysis of the gases of the blood by the chemical method. *J. Physiol.* 39, 391–396.

BROOKE, J. H. (1968). Wöhler's urea, and its vital force?—A verdict from the chemists. *Ambix* 15, 84–114.

BROWN, S. W. (1966). Heterochromatin. *Science* 151, 417–425.

BROWN-SÉQUARD, C. AND D'ARSONVAL, J. (1891). Additions à une note sur l'injection des extraits liquides de divers organes, comme méthode thérapeutique. *Comptes Rendus de la Société de Biologie* (9) 3, 265–268.

BROWNE, C. A. (1944). A source book of agricultural chemistry. *Chronica Botanica* 8, 1–290.

BRÜCKE, E. V. (1861). Die Elementarorganismen. *Sitzungsberichte der Akademie der Wissenschaften Wien, Mathematisch-wissenschaftliche Classe* 44 (2), 381–406.

BUCHNER, E. (1897). Alkoholische Gährung ohne Hefezellen (Vorläufige Mittheilung). *Ber. chem. Ges.* 30, 117–124.

BUCHNER, E. (1903). In *Die Zymasegärung* (by E. Buchner, H. Buchner, and M. Hahn), Munich: Oldenbourg.

BUCHNER, E. AND MEISENHEIMER, J. (1904). Die chemische Vorgänge bei der alkoholischen Gärung. *Ber. chem. Ges.* 37, 417–428.

BUCHNER, E. AND RAPP, R. (1898). Alkoholische Gärung ohne Hefezellen. *Ber. chem. Ges.* 31, 212–213.

BÜCHER, T. (1947). Über ein phosphatübertragendes Gärungsferment. *Biochimica et Biophysica Acta* 1, 292–314.

BÜTSCHLI, O. (1892). *Untersuchungen über Mikroskopische Schäume und das Protoplasma. Versuche und Beobachtungen zur Lösung der Frage nach den Physikalischen Bedingungen der Lebenserscheinungen,* Leipzig: Engelmann.

BUNGE, G. (1885). Ueber die Assimilation des Eisens. *Z. physiol. Chem.* 9, 49–59.

BUNGE, G. (1902). *Text-book of Physiological and Pathological Chemistry* (translated from fourth German edition by F. A. Starling and E. H. Starling), Philadelphia: Blakiston.

BURIAN, R. (1906). Chemie der Spermatozoen II. *Erg. Physiol.* 5, 768–846.

BURK, D. (1929). The free energy of glycogen-lactic acid breakdown in muscle. *Proc. Roy. Soc.* B104, 153–170.

BURKE, J. G. (1966). *Origins of the Science of Crystals,* Berkeley: University of California Press.

BUTTERFIELD, H. (1931). *The Whig Interpretation of History,* London: Bell.

CAGNIARD-LATOUR, C. (1838). Mémoire sur la fermentation vineuse. *Ann. Chim.* 2e Sér. 68, 206–223.

CALVIN, M. (1956). The photosynthetic carbon cycle. *J. Chem. Soc.,* pp. 1895–1915.

CANNON, W. B. (1932). *The Wisdom of the Body,* New York: Norton.

CARLSON, E. A. (1971). An unacknowledged founding of molecular biology: H. J. Muller's contributions to gene theory, 1910–1936. *J. Hist. Biol.* 4, 149–170.

CAROTHERS, W. H. (1931). Polymerization. *Chem. Revs.* 8, 353–426.

CASPERSSON, T. (1936). Über den chemischen Aufbau der Strukturen des Zellkerns. *Skand. Arch. Physiol.* 73, Suppl. 8.

CATHCART, E. P. (1909). The influence of carbohydrates and fats on protein metabolism. *J. Physiol.* 39, 311–330.

CATHCART, E. P. (1912). *The Physiology of Protein Metabolism,* London: Longmans, Green.

CHARGAFF, E. (1950). Chemical specificity of nucleic acids and mechanism of their enzymatic degradation. *Experientia* 6, 201–209.

CHARGAFF, E. (1968). What really is DNA? Remarks on the changing aspects of a scientific concept. *Progress in Nucleic Acid Research and Molecular Biology* 8, 297–333.

CHEVREUL, M.-E. (1824). *Considérations sur l'Analyse Organique et sur ses Applications*, Paris: Levraut.

CHEVREUL, M.-E. (1847). Rapports de l'agriculture avec les autres connaissances humaines. *Journal des Savants,* pp. 577–591.

CHIBNALL, A. C. (1939). *Protein Metabolism in the Plant,* New Haven: Yale University Press.

CHIBNALL, A. C. (1942). Amino-acid analysis and the structure of proteins. *Proc. Roy. Soc.* B131, 136–160.

CHIBNALL, A. C. (1966). The road to Cambridge. *Ann. Rev. Biochem.* 35, 1–22.

CHIBNALL, A. C. (1967). The Armstrongs and the polypeptide chain in 1909. *Chemistry and Industry,* p. 1289.

CHICK, H. AND MARTIN, C. J. (1912). On the "heat coagulation" of proteins. Part IV. The conditions controlling the agglutination of proteins already acted on by hot water. *J. Physiol.* 45, 261–295.

CHICK, H., HUME, M., AND MACFARLANE, M. (1971). *War on Disease: A History of the Lister Institute,* London: Andre Deutsch.

CHITTENDEN, R. H. (1930). *The Development of Physiological Chemistry in the United States,* New York: Chemical Catalog Co.

CHODAT, R. AND BACH, A. (1903). Untersuchungen über die Rolle der Peroxyde in der Chemie der lebenden Zelle. V. Zerlegung der sogennanten Oxydasen in Oxygenasen und Peroxydasen. *Ber chem. Ges.* 36, 606–609.

CLARK, R. (1968). *J. B. S., The Life of J. B. S. Haldane,* London: Hodder and Stoughton.

CLARK, W. M. (1923). Studies on oxidation-reduction. I. Introduction. *Public Health Reports (U.S.)* 38, 443–455.

CLARK, W. M. (1925). Recent studies on reversible oxidation-reduction in organic systems. *Chem. Revs.* 2, 127–178.

CLARK, W. M. (1960). *Oxidation-Reduction Potentials of Organic Systems,* Baltimore: Williams and Wilkins Co.

CLARKE, H. T. (1955). The Journal of Biological Chemistry. *J. Biol. Chem.* 216, 449–454.

CLAUDE, A. (1946). Fractionation of mammalian liver cells by differential centrifugation. *Journal of Experimental Medicine* 84, 51–89.

COCHIN, D. (1880). Recherches du ferment alcoholique soluble. *Ann. Chim.* 5e Sér. 21, 430–432.

COHEN, E. (1912). *Jacobus Henricus van't Hoff: Sein Leben und Wirken,* Leipzig: Akademische Verlagsgesellschaft.

COHEN, P. P. AND HAYANO, M. (1946). Urea synthesis by liver homogenates. *J. Biol. Chem.* 166, 251–259.

COHEN, S. S. AND STANLEY, W. M. (1942). The molecular size and shape of the nucleic acid of tobacco mosaic virus. *J. Biol. Chem.* 144, 589–598.

COHN, E. J. (1925). The physical chemistry of proteins. *Physiol. Revs.* 5, 349–437.

COHNHEIM, O. (1901). Die Umwandlung des Eiweiss durch die Darmwand. *Z. physiol. Chem.* 33, 451–465.

COLEMAN, W. (1965). Cell, nucleus, and inheritance. *Proc. Am. Phil. Soc.* 109, 124–158.

COLEY, N. G. (1967). The Animal Chemistry Club: assistant society to the Royal Society. *Notes and Records of the Royal Society of London* 22, 173–185.

COLEY, N. G. (1971). Animal chemists and urinary calculi. *Ambix* 18, 69–93.

COLIN, J. J. (1825). Mémoire sur la fermentation du sucre. *Ann. Chim.* 2e Sér. 28, 128–142.

COLLIE, J. N. (1893). The production of naphthalene derivatives from dehydracetic acid. *J. Chem. Soc.* 63, 329–337.

COLLIE, J. N. (1907). Derivatives of the multiple keten group. *J. Chem. Soc.* 91, 1806–1813.

COMTE, A. (1838). *Cours de Philosophie Positive,* Paris: Bachelier.

CONANT, J. B. (1970). Theodore William Richards and the periodic table. *Science* 168, 425–428.

CONANT, J. B. AND FIESER, L. F. (1925). Methemoglobin. *J. Biol. Chem.* 62, 595–622.

CONNSTEIN, W. AND LÜDECKE, K. (1919). Über Glyceringewinnung durch Gärung. *Ber. chem. Ges.* 52, 1385–1391.

CONSDEN, R., GORDON, A. H., AND MARTIN, A. J. P. (1944). Quantitative analysis of proteins: a partition chromatographic method using paper. *Biochem. J.* 38, 224–232.

COOKSON, G. H. AND RIMINGTON, C. (1954). Porphobilinogen. *Biochem. J.* 57, 476–484.

COREY, R. B. AND PAULING, L. (1953). Fundamental dimensions of polypeptide chains. *Proc. Roy. Soc.* B141, 10–20.

CORI, C. F. (1969). The call of science. *Ann. Rev. Biochem.* 38, 1–20.

CORI, C. F. AND CORI, G. T. (1929). Glycogen formation in the liver from *d*- and *l*-lactic acid. *J. Biol. Chem.* 81, 389–403.

CORI, C. F. AND CORI, G. T. (1936). Mechanism of formation of hexosemonophosphate in muscle and isolation of a new phosphate ester. *Proceedings of the Society for Experimental Biology and Medicine* 34, 702–705.

CORI, G. T., CORI, C. F., AND SCHMIDT, G. (1939). The role of glucose-1-phosphate in the formation of blood sugar and synthesis of glycogen in the liver. *J. Biol. Chem.* 129, 629–639.

CORI, G. T., SLEIN, M. W., AND CORI, C. F. (1945). Isolation and crystallization of *d*-glyceraldehyde-3-phosphate dehydrogenase from rabbit muscle. *J. Biol. Chem.* 159, 565–566.

CORNER, G. W. (1964). *A History of the Rockefeller Institute, 1901–1953,* New York: Rockefeller Institute Press.

CORRAN, H. S., GREEN, D. E., AND STRAUB, F. B. (1939). On the catalytic function of heart flavoprotein. *Biochem. J.* 33, 793–801.

CORRENS, C. (1900). G. Mendel's Regel über das Verhalten der Nachkommenschaft der Rassenbastarde. *Ber. bot. Ges.* 18, 158–168.

CORYELL, C. D. (1940). The proposed terms "exergonic" and "endergonic" for thermodynamics. *Science* 92, 380.

COSTA, A. B. (1962). *Michel Eugene Chevreul: Pioneer of Organic Chemistry*, Madison: State Historical Society of Wisconsin.

CRANDALL, D. I. AND GURIN, S. (1949). Studies of acetoacetate formation with labeled carbon. I. Experiments with pyruvate, acetate, and fatty acids in washed liver homogenates. *J. Biol. Chem.* 181, 829–843.

CRICK, F. H. C. (1958). On protein synthesis. *Symp. Soc. Exp. Biol.* 12, 138–163.

CRICK, F. H. C. AND WATSON, J. D. (1954). The complementary structure of deoxyribonucleic acid. *Proc. Roy. Soc.* A223, 80–96.

CROFT HILL, A. (1898). Reversible zymohydrolysis. *J. Chem. Soc.* 73, 634–658.

CROSLAND, M. (1967). *The Society of Arcueil*, London: Heinemann.

CULOTTA, C. A. (1970). On the color of the blood from Lavoisier to Hoppe-Seyler, 1777–1864: a theoretical dilemma. *Episteme* 4, 219–233.

CUVIER, G. (1828). *Rapport Historique sur les Progrès des Sciences Naturelles depuis 1789, et sur leur État Actuel* (second edition), Paris: Verdière and Ladrange.

DAGLEY, S. AND NICHOLSON, D. E. (1970). *An Introduction to Metabolic Pathways*, New York: Wiley.

DAGOGNET, F. (1967). *Méthodes et Doctrines dans l'Oeuvre de Pasteur*, Paris: Presses Universitaires de France.

DAKIN, H. D. (1909). The mode of oxidation in the animal organism of phenyl derivatives of fatty acids. Part V. Studies on the fate of phenylvaleric acid and its derivatives. *J. Biol. Chem.* 6, 221–233.

DAKIN, H. D. (1912). *Oxidations and Reductions in the Animal Body*, London: Longmans Green.

DAKIN, H. D. AND DUDLEY, H. W. (1913). Glyoxalase. III. The distribution of the enzyme and its relation to the pancreas. *J. Biol. Chem.* 15, 463–474.

DALE, H. H. (1963). Otto Loewi. *Erg. Physiol.* 52, 1–19.

DARLINGTON, C. D. (1947). Nucleic acid and the chromosomes. *Symp. Soc. Exp. Biol.* 1, 252–269.

DAVIDSON, J. N. (1947). The distribution of nucleic acids in tissues. *Symp. Soc. Exp. Biol.* 1, 77–85.

DEBUS, A. G. (1965). *The English Paracelsians*, London: Oldbourne.

DELAGE, Y. (1895). *La Structure du Protoplasma, et les Théories de l'Hérédité et les Grands Problèmes de la Biologie*, Paris: Reinwald.

DELAUNAY, A. (1962). *L'Institut Pasteur de ses Origines à Aujourd'hui*, Paris: France-Empire.

DELBRÜCK, M. (1941). A theory of autocatalytic synthesis of polypeptides. *Cold Spr. Harb. Symp.* 9, 122–126.

DELBRÜCK, M. (1949). A physicist looks at biology. *Trans. Conn. Acad.* 38, 173–190.

DELBRÜCK, M. AND SCHROHE, A. (1904). *Hefe, Gärung, und Fäulnis*, Berlin: Parey.

DEMEREC, M. (1935). Role of genes in evolution. *Am. Natur.* 69, 125–138.

DIXON, M. (1929). Oxidation mechanisms in animal tissues. *Biol. Revs.* 4, 352–397.

DIXON, M. AND ELLIOTT, K. A. C. (1929). The effect of cyanide on the respiration of animal tissues. *Biochem. J.* 23, 812–830.

DIXON, M. AND THURLOW, S. (1925). Studies on xanthine oxidase. VI. A cell oxidation system independent of iron. *Biochem. J.* 19, 672–675.

DIXON, M. AND ZERFAS, L. G. (1940). The role of coenzymes in dehydrogenase systems. *Biochem. J.* 34, 371–391.

DOBZHANSKY, T. (1936). Position effects on genes. *Biol. Revs.* 11, 364–382.

DODGE, B. O. (1939). Some problems in the genetics of the fungi. *Science* 90, 379–385.

DOKE. T. (1969). Establishment of biochemistry in Japan. *Japanese Studies in the History of Science* 8, 145–153.

DOLE, M. (1941). *The Glass Electrode,* New York: Wiley.

DRABKIN, D. L. (1958). *Thudichum, Chemist of the Brain,* Philadelphia: University of Pennsylvania Press.

DRECHSEL, E. (1889). Zur Kenntnis der Spaltungsprodukte des Caseins. *J. prakt. Chem.* 174, 425–429.

DRECHSEL, E. (1890). Ueber die Bildung von Harnstoff aus Eiweiss. *Ber. chem. Ges.* 23, 3096–3102.

DRIESCH, H. (1908). *The Science and Philosophy of the Organism,* London: Black.

DRUMMOND, J. C. AND FUNK, C. (1914). The chemical investigation of the phosphotungstate precipitate from rice-polishings. *J. Physiol.* 8, 598–615.

DRURY, A. N. (1914). The validity of the microchemical test for the oxygen place in tissues. *Proc. Roy. Soc.* B88, 166–176.

DU BOIS-REYMOND, E. (1912). *Reden* (second edition), Leipzig: Veit.

DU BOIS-REYMOND, E. (1927). *Zwei Grosse Naturforscher des 19. Jahrhunderts. Ein Briefwechsel zwischen Emil duBois-Reymond und Karl Ludwig,* Leipzig: Barth.

DUBOS, R. J. (1950). *Louis Pasteur: Free Lance of Science,* Boston: Little Brown.

DUBRUNFAUT, A. P. (1846). Note sur quelques phénomènes rotatoires et sur quelques propriétés des sucres. *Ann. Chim.* 3e Sér. 18, 99–108.

DUCLAUX, E. (1883). *Microbiologie,* Paris: Masson.

DUCLAUX, E. (1896). *Pasteur: Histoire d'un Esprit,* Sceaux: Charaire.

DUCLAUX, E. (1898). Lois générales de l'action des diastases. *Annales de l'Institut Pasteur,* Paris 12, 96–127.

DUHEM, P. (1915). *La Science Allemande,* Paris: Hermann.

DUHEM, P. (1916), *La Chimie est-elle une Science Française?,* Paris: Hermann.

DUJARDIN, F. (1835). Recherches sur les organismes inférieurs. *Annales des Sciences Naturelles* 4, 343–376.

DUMAS, J. B. (1841). Leçon sur la statique chimique des êtres organisés. *Annales des Sciences Naturelles* 16, 33–61.

DUMAS, J. B. (1844). *Essai de Statique Chimique des Êtres Organisés* (third edition), Paris: Fortin, Masson.

DUMAS, J. B. (1856). [no title] *Comp. Rend.* 43, 548–550.

DUMAS, J. B. (1874). Recherches sur la fermentation alcoolique. *Ann. Chim.* 5e Sér. 3, 57–108.

DUNN, L. C. (1966). *A Short History of Genetics,* New York: McGraw-Hill.

DU VIGNEAUD, V. (1952). *A Trail of Research,* Ithaca: Cornell University Press.

EDELMAN, J. (1956). The formation of oligosaccharides by enzymic transglycosylation. *Adv. Enzymol.* 17, 189–232.

EDSALL, J. T. (1955). Edwin J. Cohn. *Erg. Physiol.* 48, 23–48.

EDSON, N. L. AND LELOIR, L. F. (1936). Ketogenesis-antiketogenesis. V. Metabolism of ketone bodies. *Biochem. J.* 30, 2319–2332.

EGGLETON, P. AND EGGLETON, G. P. (1927). The physiological significance of "phosphagen." *J. Physiol.* 63, 155–161.

EHRLICH, P. (1885). *Das Sauerstoff-Bedürfnis des Organismus. Eine Farbenanalytische Studie,* Berlin: Hirschwald.

ELLINGER, A. (1904). Die Entstehung der Kynurensäure. *Z. physiol. Chem.* 43, 325–337.

ELLINGER, P. AND KOSCHARA, W. (1934). The lyochromes: a new group of animal pigments. *Nature* 133, 553–556.

ELLIS, E. L. AND DELBRÜCK, M. (1939). The growth of bacteriophage. *J. Gen. Physiol.* 22, 365–384.

ELVEHJEM, C. A., MADDEN, R. J., STRONG, F. M., AND WOOLLEY, D. W. (1937). Relation of nicotinic acid and nicotinic acid amide to canine black tongue. *J. Am. Chem. Soc.* 59, 1767–1768.

EMBDEN, G. AND LAQUER, F. (1914). Über die Chemie des Lactacidogens. I. Mitteilung. Isolierungsversuche. *Z. physiol. Chem.* 93, 94–123.

EMBDEN, G. AND LAQUER, F. (1921). Über die Chemie des Lactacidogens. III. *Z. physiol. Chem.* 113, 1–9.

EMBDEN, G. AND OPPENHEIMER, M. (1912). Über den Abbau der Brenztraubensäure im Tierkörper. *Biochem. Z.* 45, 186–206.

ENGELHARDT, V. A. (1932). Die Beziehungen zwischen Atmung und Pyrophosphatumsatz in Vogelerythrocyten. *Biochem. Z.* 251, 343–368.

ENGELHARDT, V. A. (1942). Enzymatic and mechanical properties of muscle proteins. *Yale J. Biol. and Med.* 15, 21–38.

ENGELHARDT, V. A. AND LYUBIMOVA, M. N. (1939). Myosine and adenosinetriphosphatase. *Nature* 144, 668–669.

ERLENMEYER, E. AND SCHÖFFER, A. (1859). Ein experimentellkritischer Beitrag zur Kenntnis der Eiweisskörper. *Zeitschrift für Chemie* 2, 315–343.

EULER, H. V. (1936). Die Cozymase. *Erg. Physiol.* 38, 1–30.

EULER, H. V. (1938). Bedeutung der Wirkstoffe (Ergone), Enzyme und Hilfsstoffe im Zellenleben. *Ergebnisse der Vitamin und Hormonforschung* 1, 159–190.

EULER, H. V. AND MYRBÄCK, K. (1923). Gärungs-co-Enzym (Co-Zymase) der Hefe. I. *Z. physiol. Chem.* 131, 179–203.

EULER, H. V., MYRBÄCK, K., AND NILSSON, R. (1928). Neuere Forschungen über den enzymatischen Kohlehydratabbau (I). Die Mutation als einleitende Reaktion des Glucose-Abbaues und das daran beteiligte Enzymsystem. *Erg. Physiol.* 26, 531–567.

EULNER, H. H. (1970). *Die Entwicklung der Medizinischen Spezialfächer an den Universitäten des Deutschen Sprachgebietes*, Stuttgart: Enke.

EWALD, P. P. (1962). *Fifty Years of X-Ray Diffraction*, Utrecht: International Union of Crystallography.

FABRONI, G. (1799). D'un mémoirie du cit. Fabroni sur les fermentations etc. (by Fourcroy). *Ann. Chim.* 31, 299–327.

FARADAY, M. (1839). *Experimental Researches in Electricity*, London: Taylor and Francis.

FAYET, J. (1960). *La Révolution Française et la Science*, Paris: Rivière.

FENTON, H. J. H. (1894). Oxidation of tartaric acid in presence of iron. *J. Chem. Soc.* 65, 899–910.

FEULGEN, R. (1914). Über die "Kohlenhydratgruppe" in der echten Nukleinsäure. *Z. physiol. Chem.* 92, 154–158.

FEULGEN, R. (1923). *Chemie und Physiologie der Nukleinstoffe*, Berlin: Bornträger.

FEULGEN, R. AND ROSSENBECK, H. (1924). Mikroskopisch-chemischer Nachweis einer Nukleinsäure vom Typus Thymusnukleinsäure und die darauf beruhende elektive Färbung von Zellkernen in mikroskopischen Präparaten. *Z. physiol. Chem.* 135, 203–248.

FICK, A. (1882). *Mechanische Arbeit und Wärmeentwicklung bei der Muskelthätigkeit*, Leipzig: Brockhaus.

FISCHER, E. (1894). Einfluss der Konfiguration auf die Wirkung der Enzyme. *Ber. chem. Ges.* 27, 2985–2993.

FISCHER, E. (1898). Bedeutung der Stereochemie für die Physiologie. *Z. physiol. Chem.* 26, 60–87.

FISCHER, E. (1899). Ueber die Spaltung einiger racemischer Amidosäuren in die optisch-activen Componenten. *Ber. chem. Ges.* 32, 2451–2471.

FISCHER, E. (1906). Untersuchungen über Aminosäuren, Polypeptide, und Proteine. *Ber. chem. Ges.* 39, 530–610.

FISCHER, E. (1907a). Synthese von Polypeptide XVII. *Ber. chem. Ges.* 40, 1754–1767.

FISCHER, E. (1907b). Synthetical chemistry in its relation to biology. *J. Chem. Soc.* 91, 1749–1765.

FISCHER, E. (1913). Synthese von Depsiden, Flechtenstoffen und Gerbstoffen. *Ber. chem. Ges.* 46, 3253–3289.

FISCHER, E. (1923). *Untersuchungen über Aminosäuren, Polypeptide und Proteine II, 1907–1919*, Berlin: Springer.

FISCHER, E. (1924). Die Kaiser-Wilhelm-Institute und der Zusammenhang von organischer Chemie und Biologie. In *Untersuchungen aus Verschiedenen Gebieten*, pp. 796–809 (M. Bergmann, ed.), Berlin: Springer.

FISCHER, E. AND FOURNEAU, E. (1901). Ueber einige Derivate des Glycocolls. *Ber. chem. Ges.* 34, 2868–2877.

FISCHER, E. AND FREUDENBERG, K. (1913). Über das Tannin und die Synthese ähnlicher Stoffe. III. Hochmolekulare Verbindungen. *Ber. chem. Ges.* 46, 1116–1138.

FISCHER, E. AND THIERFELDER, H. (1894). Verhalten der verschiedenen Zucker gegen reine Hefen. *Ber. chem. Ges.* 27, 2031–2037.

FISCHER, H. O. L. (1960). Fifty years "Synthetiker" in the service of biochemistry. *Ann. Rev. Biochem.* 29, 1–14.

FISCHER, H. O. L. AND BAER, E. (1932). Über die 3-Glycerinaldehyd-phosphorsäure. *Ber. chem. Ges.* 65, 337–345.

FISKE, C. H. AND SUBBAROW, Y. (1927). The nature of the "inorganic phosphate" in involuntary muscle. *Science* 65, 401–403.

FISKE, C. H. AND SUBBAROW, Y. (1929). Phosphorus compounds of muscle and liver. *Science* 70, 381–382.

FLEISCH, A. (1924). Some oxidation processes of normal and cancer tissue. *Biochem. J.* 18, 294–311.

FLEMMING, W. (1882). *Zellsubstanz, Kern und Zellteilung,* Leipzig: Vogel.

FLETCHER, W. M. AND HOPKINS, F. G. (1907). Lactic acid in amphibian muscle. *J. Physiol.* 35, 247–309.

FLORKIN, M. (1960). *Naissance et Déviation de la Théorie Cellulaire dans l'Oeuvre de Théodore Schwann,* Paris: Hermann.

FLORY, P. J. (1953). *Principles of Polymer Chemistry,* Ithaca: Cornell University Press.

FOLIN, O. (1905). A theory of protein metabolism. *Am. J. Physiol.* 13, 117–138.

FORBES, R. J. (1954). Chemical, culinary and cosmetic arts. In *A History of Technology,* Vol. I, pp. 238–298 (C. Singer, E. J. Holmyard, and A. R. Hall, eds.), Oxford: Clarendon Press.

FORSTER, M. O. (1920). Emil Fischer memorial lecture. *J. Chem. Soc.* 117, 1157–1201.

FOSTER, M. (1877). *A Text Book of Physiology,* London: Macmillan.

FOURCROY, A. F. (1789a). Extrait d'un mémoire ayant pour titre, Recherches pour servir a l'histoire du gaz azote ou de la mofette, comme principe des matières animales. *Ann. Chim.* 1, 40–46.

FOURCROY, A. F. (1789b). Mémoire sur l'existence de la matière albumineuse dans les végétaux. *Ann. Chim.* 3, 252–262.

FOURCROY, A. F. (1801). *Système des Connaissances Chimiques,* Paris: Baudouin.

FOURCROY, A. F. (1806). *Philosophie Chimique* (third edition), Paris: Tourneisen.

FOX, R. (1971). *The Caloric Theory of Gases from Lavoisier to Regnault,* London: Oxford University Press.

FOX, S. W. (1945). Terminal amino acids in peptides and proteins. *Adv. Protein Chem.* 2, 155–177.

FRANK, F. C. (1936). [no title] *Nature* 138, 242.

FREUDENBERG, K. (1966). Emil Fischer and his contributions to carbohydrate chemistry. *Adv. Carb. Chem.* 21, 1–38.

FREUNDLICH, M. M. (1963). Origin of the electron microscope. *Science* 142, 185–188.

FRIEDMANN, E. (1913). Zur Kenntnis des Abbaues der Karbonsäuren im Tierkörper. XVII. Mitteilung. Über die Bildung von Acetessigsäure aus Essigsäure bei der Leberdurchblutung. *Biochem. Z.* 55, 436–442.

FRUTON, J. S. (1950). The role of proteolytic enzymes in the biosynthesis of peptide bonds. *Yale J. Biol. and Med.* 22, 263–271.

FRUTON, J. S. (1951). The place of biochemistry in the university. *Yale J. Biol. and Med.* 23, 305–310.

FRUTON, J. S. (1957). Enzymic hydrolysis and synthesis of peptide bonds. *Harvey Lectures* 51, 64–87.

FRUTON, J. S. AND BERGMANN, M. (1938). The specificity of pepsin action. *Science* 87, 557.

FÜRTH, O. V. (1903). Die chemische Zustandsänderungen des Muskels. *Erg. Physiol.* 2, 574–611.

FURBERG, S. (1952). On the structure of nucleic acids. *Acta Chemica Scandinavica* 6, 634–640.

GAMOW, G. (1954). Possible relation between deoxyribonucleic acid and protein structures. *Nature* 173, 318.

GARROD, A. E. (1902). The incidence of alkaptonuria: a study in chemical individuality. *Lancet* 2, 1616–1620.

GARROD, A. E. (1909). *Inborn Errors of Metabolism,* London: Frowde, Hodder, and Stoughton.

GARROD, A. E. AND HELE, T. S. (1905). The uniformity of the homogentisic acid excretion in alkaptonuria. *J. Physiol.* 33, 198–205.

GAUTIER, A. (1869). *Les Fermentations,* Paris: Savy.

GAY-LUSSAC, J. L. (1810). Extrait d'un mémoire sur la fermentation. *Ann. Chim.* 76, 245–259.

GEISON, G. L. (1969). The protoplasmic theory of life and the vitalist-mechanist debate. *Isis* 60, 273–292.

GERHARDT, C. (1856). *Traité de Chimie Organique,* Paris: Firmin Didot.

GIBBS, J. W. (1906). *The Scientific Papers of J. Willard Gibbs,* London: Longmans, Green.

GILLESPIE, L. J. AND LIU, T. H. (1931). The reputed dehydrogenation of hydroquinone by palladium black. *J. Am. Chem. Soc.* 53, 3969–3972.

GINSBURG, A. AND STADTMAN, E. R. (1970). Multienzyme systems. *Ann. Rev. Biochem.* 39, 429–472.

GOLDSCHMIDT, R. (1916). Genetic factors and enzyme reaction. *Science* 43, 98–100.

GOLDSCHMIDT, R. (1938). The theory of the gene. *Scientific Monthly* 46, 268–273.

GOPPELSROEDER, F. (1901). *Capillaranalyse,* Basel: Birkhäuser.

GORDON, A. H., MARTIN, A. J. P., AND SYNGE, R. L. M. (1941). A study of the partial acid hydrolysis of some proteins, with special reference to the mode of linkage of the basic amino acids. *Biochem. J.* 35, 1369–1387.

GORTNER, R. A. (1929). *Outlines of Biochemistry,* New York: Wiley.

GRAHAM, T. (1876). *Chemical and Physical Researches,* Edinburgh: University Press.

GRAY, R. D. (1952). *Goethe the Alchemist,* Cambridge: University Press.

GREEN, A. A. AND CORI, G. T. (1943). Crystalline muscle phosphorylase I. Preparation, properties and molecular weight. *J. Biol. Chem.* 151, 21–29.

GREEN, D. E. (1951). The cyclophorase complex of enzymes. *Biol. Revs.* 26, 410–455.

GREEN, D. E., DEWAN, J. G., AND LELOIR, L. F. (1937a). The β-hydroxybutyric dehydrogenase of animal tissues. *Biochem. J.* 31, 934–949.

GREEN, D. E., NEEDHAM, D. M., AND DEWAN, J. G. (1937b). Dismutations and oxidoreductions. *Biochem. J.* 31, 2327–2352.

GREEN, J. R. (1898). The alcohol-producing enzyme of yeast. *Annals of Botany* 12, 491–497.

GREEN, J. R. (1899). *The Soluble Ferments and Fermentation,* Cambridge: University Press.

GRIFFITH, F. (1928). The significance of pneumococcal types. *Journal of Hygiene* 27, 113–159.

GRIMAUX, E. (1886). Les substances colloidales et la coagulation. *Soc. Chim. Paris,* pp. 89–107.

GRMEK, M. D. (1968). First steps in Claude Bernard's discovery of the glycogenic function of the liver. *J. Hist. Biol.* 1, 141–154.

GRÜNHUT, L. (1896). Die Einführung der Reinhefe in die Gärungsgewerbe. *Sammlung chemischer und chemisch-technischer Vorträge* 1, 393–452.

GÜNTHER, H. (1921). Über den Muskelfarbstoff. *Virchows Archiv für pathologische Anatomie und Physiologie und für klinische Medizin* 230, 146–178.

GUERLAC, H. (1957). Joseph Black and fixed air: a bicentenary retrospective with some new or little known material. *Isis* 48, 124–151, 433–456.

GUERLAC, H. (1961). *Lavoisier—The Crucial Year,* Ithaca: Cornell University Press.

GULICK, A. (1939). Growth and reproduction as problems in the synthesis of protein molecules. *Growth* 3, 241–260.

GULLAND, J. M. (1938). Nucleic acids. *J. Chem. Soc.,* pp. 1722–1734.

GULLAND, J. M. (1947a). The structures of nucleic acids. *Symp. Soc. Exp. Biol.* 1, 1–14.

GULLAND, J. M. (1947b). The structures of nucleic acids. *Cold Spr. Harb. Symp.* 12, 95–103.

HAAS, E. (1938). Isolierung eines neuen gelben Ferments. *Biochem. Z.* 298, 378–390.

HABER, F. (1901). Bemerkungen über Elektrodenpotentiale. *Zeitschrift für Elektrochemie* 7, 1042–1053.

HADELICH, W. (1862). Ueber die Bestandtheile des Guaiakharzes. *J. prakt. Chem.* 87, 321–343.

HAGEDOORN, A. L. (1911). *Autokatalytical Substances the Determinants for the Inheritable Characters,* Leipzig: Engelmann.

HAHN, A. (1931). Zur Thermodynamik des Erholungsvorganges im Muskel. *Z. Biol.* 91, 444–448.

HALDANE, J. B. S. (1937). The biochemistry of the individual. In *Perspectives in Biochemistry,* pp. 1–10 (J. Needham and D. E. Green, eds.), Cambridge: University Press.

HALDANE, J. B. S. (1942). *New Paths in Genetics,* New York: Harper.

HALDANE, J. B. S. (1945). A physicist looks at genetics. *Nature* 155, 375–376.

HALDANE, J. B. S. (1966). An autobiography in brief. *Persp. Biol. and Med.* 9, 476–481.

HALDANE, J. S. (1895). The relation of the action of carbonic oxide to oxygen tension. *J. Physiol.* 18, 201–217.

HALDANE, J. S. (1908). The relation of physiology to physics and chemistry. *Nature* 78, 553–556.

HALDANE, J. S. (1917). *Organism and Environment as Illustrated by the Physiology of Breathing*, New Haven: Yale University Press.

HALDANE, J. S. (1931). *The Philosophical Basis of Biology*, London: Hodder and Stoughton.

HALDANE, J. S. AND SMITH, J. L. (1896). The oxygen tension of arterial blood. *J. Physiol.* **20**, 497–520.

HALL, T. S. (1969). *Ideas of Life and Matter*, Chicago: University Press.

HALLÉ, J. N. (1791). Essai de théorie sur l'animalisation et l'assimilation des alimens. *Ann. Chim.* **11**, 158–174.

HAMILTON, W. C. (1970). The revolution in crystallography. *Science* **169**, 133–141.

HAMMARSTEN, E. (1924). Zur Kenntnis der biologischen Bedeutung der Nucleinsäure-verbindungen. *Biochem. Z.* **144**, 383–469.

HAMMARSTEN, O. (1885). Studien über Mucin und mucinähnliche Substanzen. *Pflügers Arch.* **36**, 373–456.

HAMMARSTEN, O. (1894). Zur Kenntnis der Nucleoproteide. *Z. physiol. Chem.* **19**, 19–37.

HANES, C. S. (1940). The breakdown and synthesis of starch by an enzyme system from pea seeds. *Proc. Roy. Soc.* **B128**, 421–450; **B129**, 174–208.

HARDEN, A. (1903). Ueber alkoholische Gährung mit Hefe-Pressstoff (Buchner's Zymase) bei Gegenwart von Blutserum. *Ber. chem. Ges.* **36**, 715–716.

HARDEN, A. (1923). *Alcoholic Fermentation* (third edition), London: Longmans, Green.

HARDEN, A. AND MACLEAN, H. (1911a). On the alleged presence of an alcoholic enzyme in animal tissues and organs. *J. Physiol.* **42**, 64–92.

HARDEN, A. AND MACLEAN, H. (1911b). The oxidation of isolated animal tissues. *J. Physiol.* **43**, 34–45.

HARDEN, A. AND YOUNG, W. J. (1906). The alcoholic ferment of yeast-juice. *Proc. Roy. Soc.* **B77**, 405–520.

HARDY, W. B. (1899). Structure of cell protoplasm. *J. Physiol.* **24**, 158–210.

HARDY, W. B. (1900). A preliminary investigation of the conditions which determine the stability of irreversible hydrosols. *Proc. Roy. Soc.* **66**, 110–125.

HARDY, W. B. (1903). Colloidal solution, the globulin system. *J. Physiol.* **29**, xxvi–xxix.

HARDY, W. B. (1906). The physical basis of life. *Science Progress* **1**, 177–205.

HARDY, W. B. (1936). To remind. A biological essay. In *Collected Scientific Papers*, pp. 896–917, Cambridge: University Press.

HARRIES, C. (1917). Eduard Buchner. *Ber. chem. Ges.* **50**, 1843–1876.

HARRIS, H. (1963). *Garrod's Inborn Errors of Metabolism*, London: Oxford University Press.

HARRISON, R. G. (1945). Relations of symmetry in the developing embryo. *Trans. Conn. Acad.* **36**, 277–330.

HARVEY, E. N. (1957). *A History of Luminescence from the Earliest Times Until 1900*, Philadelphia: American Philosophical Society.

HASSENFRATZ, J. H. (1791). Mémoire sur la combinaison de l'oxigène avec le carbone et l'hydrogène du sang, sur la dissolution de l'oxigène dans le sang, et sur la manière dont le calorique se dégage. *Ann. Chim.* **9**, 261–274.

HASSID, W. Z. (1969). Biosynthesis of oligosaccharides and polysaccharides in plants. *Science* 165, 137–144.

HASTINGS, A. B. (1970). A biochemist's anabasis. *Ann. Rev. Biochem.* 39, 1–24.

HAYES, W. (1968). *The Genetics of Bacteria and Their Viruses* (second edition), New York: Wiley.

HEFFTER, A. (1908). Gibt es reduzierende Fermente im Tierköper? *Arch. exp. Path. Pharm.* Suppl. pp. 253–260.

HEHRE, E. J. (1951). Enzymic synthesis of polysaccharides. *Adv. Enzymol.* 11, 297–337.

HEIDENHAIN, M. (1894). Neue Untersuchungen über die Centralkörper und ihre Beziehungen zum Kern und Zellenprotoplasma. *Arch. mikr. Anat.* 43, 423–758.

HEINEMANN, K. (1939). Zur Geschichte der Entdeckung der roten Blutkörperchen. *Janus* 43, 1–42.

HELMHOLTZ, H. (1845). Ueber den Stoffverbrauch bei der Muskelaction. *Arch. Anat. Physiol.*, pp. 72–83.

HELMHOLTZ, H. (1881). On the modern development of Faraday's conception of electricity. *J. Chem. Soc.* 39, 277–304.

HELMHOLTZ, H. (1882). *Wissenschaftliche Abhandlungen*, Leipzig: Barth.

HELMONT, J. B. VAN (1662). *Oriatrike Or, Physick Refined* (translation by John Chandler), London: Lodowick Lloyd.

HENDERSON, L. J. (1928). *Blood, a Study in General Physiology*, New Haven: Yale University Press.

HENRI, V. (1903). *Lois Générales de l'Action des Diastases*, Paris: Hermann.

HERBST, R. M. AND ENGEL, L. L. (1934). A reaction between α-ketonic acids and α-amino acids. *J. Biol. Chem.* 107, 505–512.

HERING, E. (1888). Zur Theorie der Vorgänge in der lebendigen Substanz. *Lotos* 37, 35–70.

HERMANN, L. (1867). *Untersuchungen über den Stoffwechsel der Muskeln*, Berlin: Hirschwald.

HERSHEY, A. D. AND CHASE, M. (1952). Independent functions of viral protein and nucleic acid in growth of bacteriophage. *J. Gen. Physiol.* 36, 39–56.

HERTWIG, O. (1885). Das Problem der Befruchtung und der Isotropie des Eies, eine Theorie der Vererbung. *Jenaische Zeitschrift für Medizin und Naturwissenschaft* 18, 276–318.

HESS, E. L. (1970). Origins of molecular biology. *Science* 168, 644–669.

HESS, G. H. (1840). Thermochemische Untersuchungen. *Ann. Phys.* 50, 385–404.

HEVESY, G. (1923). The absorption and translocation of lead by plants. A contribution to the application of the method of radioactive indicators in the investigation of the change of substance in plants. *Biochem. J.* 17, 439–445.

HEVESY, G. (1948). Historical sketch of the biological application of tracer elements. *Cold Spr. Harb. Symp.* 13, 129–150.

HILL, A. V. (1912). The heat-production of surviving amphibian muscles during rest, activity, and rigor. *J. Physiol.* 44, 466–513.

HILL, A. V. (1913). The combination of haemoglobin with oxygen and with carbon monoxide. I. *Biochem. J.* 7, 471–480.

HILL, A. V. (1932). The revolution in muscle physiology. *Physiol. Revs.* 12, 56–67.

HILL, A. V. (1950). A challenge to biochemists. *Biochemica et Biophysica Acta* 4, 4–11.

HILL, A. V. (1956). Why biophysics? *Science* 124, 1233–1237.

HILL, A. V. (1959). The heat production of muscle and nerve, 1848–1914. *Ann. Rev. Physiol.* 21, 1–18.

HILL, R. (1926). The chemical nature of haemochromogen and its carbon monoxide compound. *Proc. Roy. Soc.* B100, 419–430.

HILL, R. (1965). The biochemists' green mansions: the photosynthetic electron transport chain in plants. *Essays in Biochemistry* 1, 121–152.

HIRSCH, R. (1950). The invention of printing and the diffusion of alchemical and chemical knowledge. *Chymia* 3, 115–141.

HIS, W. (1874). *Unsere Körperform und das Physiologische Problem ihrer Entstehung,* Leipzig: Vogel.

HLASIWETZ, H. AND HABERMANN, J. (1871). Ueber die Proteinstoffe. *Ann.* 159, 304–333.

HLASIWETZ, H. AND HABERMANN, J. (1873). Ueber die Proteinstoffe. *Ann.* 169, 150–166.

HODGKIN, D. C. (1965). The X-ray analysis of complicated molecules. *Science* 150, 979–988.

HODGKIN, D. C. AND RILEY, D. P. (1968). Some ancient history of protein X-ray analysis. In *Structural Chemistry and Molecular Biology,* pp. 15–28 (A. Rich and N. Davidson, eds.), San Francisco: Freeman.

HOESCH, K. (1921). *Emil Fischer,* Berlin: Verlag Chemie.

HOFF, J. H. VAN'T (1898). Über der zunehmende Bedeutung der anorganischen Chemie. *Z. anorg. Chem.* 18, 1–13.

HOFMANN, A. W. (1870). Zur Erinnerung an Gustav Magnus. *Bur. chem. Ges.* 3, 993–1101.

HOFMANN, A. W. (1888). *Aus Justus Liebig's und Friedrich Wöhler's Briefwechsel,* Braunschweig: Vieweg.

HOFMEISTER, F. (1882). Zur Lehre vom Pepton. V. Das Verhalten des Peptons in der Magenschleimhaut. *Z. physiol. Chem.* 6, 69–73.

HOFMEISTER, F. (1894). Ueber Methylierung im Thierkörper. *Arch. exp. Path. Pharm.* 33, 198–215.

HOFMEISTER, F. (1900). Willy Kühne. *Ber. chem. Ges.* 33, 3875–3880.

HOFMEISTER, F. (1901). *Die Chemische Organisation der Zelle,* Braunschweig: Vieweg.

HOFMEISTER, F. (1902). Über Bau und Gruppierung der Eiweisskörper. *Erg. Physiol.* 1, 759–802.

HOGEBOOM, G. H., SCHNEIDER, W. C., AND PALADE, G. E. (1948). Cytochemical studies of mammalian tissues. I. Isolation of intact mitochondria; some biochemical properties of mitochondria and submicroscopic particulate material. *J. Biol. Chem.* 172, 619–635.

HOGNESS, D. S., COHN, M., AND MONOD, J. (1955). Studies on the induced synthesis of β-galactosidase in *Escherichia coli:* the kinetics and mechanism of sulfur incorporation. *Biochimica et Biophysica Acta* 16, 99–116.

HOGNESS, T. R. (1942). The flavoproteins. In *A Symposium on Respiratory Enzymes,* pp. 134–148, Madison: University of Wisconsin Press.

HOLIDAY, E. R. (1930). The characteristic absorption of ultraviolet radiation by certain purines. *Biochem. J.* 24, 619–625.

HOLLAENDER, A. (1961). Review of *Studies on Quantitative Radiation Biology* (by K. G. Zimmer). *Science* 134, 1233.

HOLMES, F. L. (1963). Elementary analysis and the origins of physiological chemistry. *Isis* 54, 50–81.

HOLMES, F. L. (1964). Introduction to facsimile edition of J. Liebig, *Animal Chemistry* (Cambridge, Mass., 1842), pp. vii–cxvi, New York: Johnson Reprint Corp.

HOLMES, F. L. (1972). *Claude Bernard and Animal Chemistry,* Cambridge, Mass.: Harvard University Press [in press].

HOOYKAAS, R. (1958). The concepts of "individual" and "species" in chemistry. *Centaurus* 5, 307–322.

HOPKINS, F. G. (1906). The analyst and the medical man. *Analyst* 31, 385–404.

HOPKINS, F. G. (1912). Feeding experiments illustrating the importance of accessory factors in normal dietaries. *J. Physiol.* 44, 425–460.

HOPKINS, F. G. (1913). The dynamic side of biochemistry. *Nature* 92, 213–223.

HOPKINS, F. G. (1921a). On an autoxidizable constituent of the cell. *Biochem. J.* 15, 286–305.

HOPKINS, F. G. (1921b). The chemical dynamics of muscle. *Johns Hopkins Hospital Bulletin* 32, 359–367.

HOPKINS, F. G. (1926). On current views concerning the mechanisms of biological oxidation. *Skand. Arch. Physiol.* 49, 33–59.

HOPKINS, F. G. (1929). On glutathione: a reinvestigation. *J. Biol. Chem.* 84, 269–320.

HOPKINS, F. G. (1930a). Denaturation of proteins by urea and related substances. *Nature* 126, 328–330; 383–384.

HOPKINS, F. G. (1930b). The earlier history of vitamin research. In *Les Prix Nobel en 1929,* pp. 1–12, Stockholm: Nobelstiftelsen.

HOPKINS, F. G. (1931). *The Problems of Specificity in Biochemical Catalysis,* London: Oxford University Press.

HOPKINS, F. G. AND DIXON, M. (1922). On glutathione. II. A thermostable oxidation-reduction system. *J. Biol. Chem.* 54, 527–563.

HOPPE-SEYLER, F. (1866). Beiträge zur Kenntnis der Constitution des Blutes. 1. Ueber die Oxydation im lebenden Blute. *Med. chem. Unt.,* pp. 133–140.

HOPPE-SEYLER, F. (1871). Ueber die chemische Zusammensetzung des Eiters. *Med. chem. Unt.,* pp. 486–501.

HOPPE-SEYLER, F. (1873). Ueber den Ort der Zersetzung von Eiweiss- und andern Nährstoffen im thierischen Organismus. *Pflügers Arch.* 7, 399–417.

HOPPE-SEYLER, F. (1876). Ueber die Processe der Gährungen und ihre Beziehung zum Leben des Organismus. *Pflügers Arch.* 12, 1–17.

HOPPE-SEYLER, F. (1877a). Vorwort. *Z. physiol. Chem.* 1, I–III.

HOPPE-SEYLER, F. (1877b). *Physiologische Chemie,* Berlin: Hirschwald.

HOPPE-SEYLER, F. (1878). Ueber Gährungsprozesse. *Z. physiol. Chem.* 2, 1–28.

HOPPE-SEYLER, F. (1884). *Über die Entwicklung der physiologischen Chemie und ihre Bedeutung für die Medicin,* Strassburg: Trübner.

HORECKER, B. L. (1962). Interdependent pathways of carbohydrate metabolism. *Harvey Lectures* 57, 35–61.

HOROWITZ, N. H. (1948). The one-gene one-enzyme hypothesis. *Genetics* 33, 612–613.

HOTCHKISS, R. D. (1965). Oswald T. Avery 1877–1955. *Genetics* 51, 1–10.

HOTCHKISS, R. D. (1970). Review of *The Coming of the Golden Age* (by G. S. Stent). *Science* 169, 664–666.

HOUSSAY, B. A. (1952). The discovery of pancreatic diabetes—the role of Oscar Minkowski. *Diabetes* 1, 112–116.

HÜFNER, G. (1884). Ueber das Oxyhämoglobin des Pferdes. *Z. physiol. Chem.* 8, 358–365.

HÜNEFELD, F. L. (1840). *Der Chemismus in der thierischen Organisation,* Leipzig: Brockhaus.

HUME, E. D. (1923). *Béchamp or Pasteur? A Lost Chapter in the History of Biology,* Chicago: Covici-McGee.

HUNTER, A. (1912). On urocanic acid. *J. Biol. Chem.* 11, 537–545.

HUNTER, A. (1928). *Creatine and Creatinine,* London: Longmans, Green.

HUNTER, J. (1837). *The Works of John Hunter, Vol. I, Lectures on the Principles of Surgery* (edited by J. F. Palmer), London: Longmans.

HUXLEY, H. E. (1971). The structural basis of muscular contraction. *Proc. Roy. Soc.* B178, 131–149.

HUXLEY, T. H. (1870). *On the Physical Basis of Life* (third edition), New Haven, Conn.: Chatfield.

INGENKAMP, C. (1886). Die geschichtliche Entwicklung unserer Kenntnis von Fäulnis und Gärung. *Zeitschrift für klinische Medizin* 10, 59–107.

INGRAM, V. M. (1956). A specific chemical difference between the globins of normal human and sickle-cell anaemia hemoglobin. *Nature* 178, 792–794.

IRONMONGER, E. E. (1958). The Royal Institution and the teaching of chemistry in the nineteenth century. *Nature* 182, 80–83.

JACOB, F. AND MONOD, J. (1961). Genetic regulatory mechanisms in the synthesis of proteins. *Journal of Molecular Biology* 3, 318–356.

JAGER, L. DE (1890). Erklärungsversuch ueber die Wirkungsart der ungeformten Fermente. *Virchows Archiv für pathologische Anatomie und Physiologie und für klinische Medizin* 121, 182–187.

JAQUET, A. (1892). Ueber die Bedingungen der Oxydationsvorgänge in den Geweben. *Arch. exp. Path. Pharm.* 29, 386–396.

JEVONS, F. R. (1962). Boerhaave's biochemistry. *Medical History* 6, 343–362.

JOHANNSEN, W. (1909). *Elemente der Exakten Erblichkeitslehre,* Jena: Fischer.

JONES, M. E. (1953). Albrecht Kossel, a biographical sketch. *Yale J. Biol. and Med.* 26, 80–97.

JONES, M. E., SPECTOR, L., AND LIPMANN, F. (1955). Carbamyl phosphate, the carbamyl donor in enzymatic citrulline synthesis. *J. Am. Chem. Soc.* 77, 819–820.

References

JONES, W. (1920a). *Nucleic Acids* (second edition), London: Longmans, Green.

JONES, W. (1920b). The action of boiled pancreas extract on yeast nucleic acid. *Am. J. Physiol.* 52, 203–207.

JORPES, J. E. (1966). *Jac. Berzelius: His Life and Work*, Stockholm: Almquist and Wiksell.

KAHANE, E. (1968). La vie et l'oeuvre scientifique de Charles Gerhardt. *Bulletin de la Société Chimique de France*, pp. 4733–4742.

KAHLBAUM, G. W. A. (1904). *Justus von Liebig und Friedrich Mohr in ihren Briefen von 1834–1870*, Leipzig: Barth.

KALCKAR, H. M. (1937). Phosphorylation in kidney tissue. *Enzymologia* 2, 47–52.

KALCKAR, H. M. (1941). The nature of energetic coupling in biological syntheses. *Chem. Revs.* 28, 71–178.

KALCKAR, H. M. (1942). The enzymatic action of myokinase. *J. Biol. Chem.* 143, 299–300.

KALCKAR, H. M. (1969). *Biological Phosphorylation: Development of Concepts*, Englewood Cliffs, N.J.: Prentice-Hall.

KAMEN, M. D. (1963). The early history of carbon-14. *J. Chem. Ed.* 40, 234–242.

KAPLAN, N. O. AND KENNEDY, E. P. (1966). *Current Aspects of Biochemical Energetics*, New York: Academic Press.

KAPOOR, S. C. (1969). The origins of Laurent's organic classification. *Isis* 60, 477–527.

KARGON, R. H. (1966), *Atomism in England from Hariot to Newton*, Oxford: Clarendon Press.

KASTLE, J. H. AND LOEVENHART, A. S. (1900). Concerning lipase, the fat-splitting enzyme, and the reversibility of its action. *American Chemical Journal* 24, 491–525.

KAUFFMAN, G. B. (1966). *Alfred Werner, Founder of Coordination Chemistry*, Berlin: Springer.

KEILIN, D. (1925). On cytochrome, a respiratory pigment, common to animals, yeast, and higher plants. *Proc. Roy. Soc.* B98, 312–339.

KEILIN, D. (1927). Influence of carbon monoxide and light on indophenol oxidase of yeast cells. *Nature* 119, 670–671

KEILIN, D. (1929). Cytochrome and respiratory enzymes. *Proc. Roy. Soc.* B104, 206–252.

KEILIN, D. (1930). Cytochrome and intracellular oxidase. *Proc. Roy. Soc.* B106, 418–444.

KEILIN, D. (1966). *The History of Cell Respiration and Cytochrome*, Cambridge: University Press.

KEILIN, D. AND HARTREE, E. F. (1939). Cytochrome and cytochrome oxidase. *Proc. Roy. Soc.* B127, 167–191.

KEKULÉ, A. (1858). Ueber die Constitution und die Metamorphosen der chemischen Verbindungen. *Ann.* 106, 129–159.

KEKULÉ, A. (1861). *Lehrbuch der Organischen Chemie*, Erlangen: Enke.

KEKULÉ, A. (1867). On some points of chemical philosophy. *The Laboratory* 1, 303–306.

KENDREW, J. C. (1967). Review of *Phage and the Origins of Molecular Biology* (J. Cairns, G. S. Stent, J. D. Watson, eds.). *Scientific American* **216** (March), 141–144.

KENDREW, J. C., BODO, G., DINTZIS, H. M., PARRISH, R. G., AND WYCKOFF, H. (1958). A three-dimensional model of the myoglobin molecule obtained by X-ray analysis. *Nature* **181**, 662–666.

KENDREW, J. C., DICKERSON, R. E., STRANBERG, B. E., HART, R. G, DAVIES, D. R., PHILLIPS, D. C., AND SHORE, V. C. (1960). Structure of myoglobin. A three-dimensional Fourier synthesis at 2 Å resolution. *Nature* **185**, 422–427.

KENDREW, J. C. *et al.* (1968). *Report of the Working Group on Molecular Biology*, London: H. M. S. O. (Cmnd. 3675).

KENNEDY, E. P. AND LEHNINGER, A. L. (1949). Oxidation of fatty acids and tricarboxylic acid cycle intermediates by isolated rat liver mitochondria. *J. Biol. Chem.* **179**, 957–972.

KENT, A. (1950). *An Eighteenth Century Lectureship in Chemistry*, Glasgow: Jackson.

KINGZETT, C. T. (1878). *Animal Chemistry*, London: Longmans, Green.

KLEIBER, M. (1961). *The Fire of Life*, New York: Wiley.

KLUG, A. (1968). Rosalind Franklin and the discovery of the structure of DNA. *Nature* **219**, 808–810, 843–844.

KLUYVER, A. J. AND DONKER, H. J. L. (1925). The catalytic transference of hydrogen as the basis of the chemistry of dissimilation processes. *Proceedings of the Royal Academy of Sciences, Amsterdam* **28**, 605–618.

KLUYVER, A. J. AND DONKER, H. J. L. (1926). Die Einheit in der Biochemie. *Chemie der Zelle und Gewebe* **13**, 134–190.

KNIERIEM, W. V. (1874). Beiträge zur Kenntnis der Bildung des Harnstoffs im thierischen Organismus. *Z. Biol.* **10**, 263–294.

KNIGHT, D. M. (1967). *Atoms and Elements*, London: Hutchinson.

KNOOP, F. (1904). *Der Abbau aromatischer Fettsäuren im Tierkörper*, Freiburg: Kuttruff.

KNOOP, F. (1910). Über den physiologischen Abbau der Säuren und die Synthese einer Aminosäure im Tierkörper. *Z. physiol. Chem.* **67**, 489–520.

KNOOP, F. AND MARTIUS, C. (1936). Über die Bildung der Citronensäure. *Z. physiol. Chem.* **242**, 204.

KNOOP, F. AND OESTERLIN, H. (1925). Über die natürliche Synthese der Aminosäuren und ihre experimentelle Reproduktion. *Z. physiol. Chem.* **148**, 294–315.

KNOX, W. E. (1958). Sir Archibald Garrod's "Inborn Errors of Metabolism." *American Journal of Human Genetics* **10**, 3–32, 95–124, 249–267, 385–397.

KÖLLIKER, A. (1885). Die Bedeutung der Zellkerne für die Vorgänge der Vererbung. *Zeitschrift für wissenschaftliche Zoologie* **42**, 1–46.

KOHLER, R. (1971). The background to Eduard Buchner's discovery of cell-free fermentation. *J. Hist. Biol.* **4**, 35–61.

KOLBE, H. (1877). Zeichen der Zeit. *J. prakt. Chem.* **123**, 473–477.

KOPP, H. (1847). *Geschichte der Chemie*, Braunschweig: Vieweg.

KORNBERG, A. (1960). Biologic synthesis of deoxyribonucleic acid. *Science* **131**, 1503–1508.

KORNBERG, H. L. (1968). H. A. Krebs: a pathway in metabolism. *Biochemical Society Symposia* **27**, 3–9.

KOSSEL, A. (1879). Über das Nuklein der Hefe. *Z. physiol. Chem.* **3**, 284–291.

KOSSEL, A. (1881). *Untersuchungen über die Nucleine und ihre Spaltungsprodukte,* Strassburg: Trübner.

KOSSEL, A. (1884). Über ein peptonartigen Bestandteil des Zellkerns. *Z. physiol. Chem.* **8**, 511–515.

KOSSEL, A. AND DAKIN, H. D. (1904). Über die Arginase. *Z. physiol. Chem.* **41**, 321–331; **42**, 181–185.

KOSSEL, A. AND NEUMANN, A. (1896). Über Nukleinsäure und Thyminsäure. *Z. physiol. Chem.* **22**, 74–82.

KOSTYCHEV, S. (1926). Über die Nichtexistenz einiger Fermente. *Z. physiol. Chem.* **154**, 262–275.

KREBS, H. A. (1935a). Metabolism of amino-acids. III. Deamination of amino-acids. *Biochem. J.* **29**, 1620–1644.

KREBS, H. A. (1935b). Metabolism of amino-acids. IV. The synthesis of glutamine from glutamic acid and ammonia, and the enzymic hydrolysis of glutamine in animal tissues. *Biochem. J.* **29**, 1951–1969.

KREBS, H. A. (1942). Urea formation in mammalian liver. *Biochem. J.* **36**, 758–767.

KREBS, H. A. (1943). The intermediary stages in the biological oxidation of carbohydrate. *Adv. Enzymol.* **3**, 191–252.

KREBS, H. A. (1947). Cyclic processes in living matter. *Enzymologia* **12**, 88–100.

KREBS, H. A. (1970). The history of the tricarboxylic acid cycle. *Persp. Biol. and Med.* **14**, 154–170.

KREBS, H. A. AND EGGLESTON, L. V. (1940). The oxidation of pyruvate in pigeon breast muscle. *Biochem. J.* **34**, 442–459.

KREBS, H. A. AND HENSELEIT, K. (1932). Untersuchungen über die Harnstoffbildung im Tierkörper. *Z. physiol. Chem.* **210**, 33–66.

KREBS, H. A. AND JOHNSON, W. A. (1937). The role of citric acid in intermediate metabolism in animal tissues. *Enzymologia* **4**, 148–156.

KREBS, H. A. AND KORNBERG, H. L. (1957). A survey of the energy transformations in living matter. *Erg. Physiol.* **49**, 212–298.

KŘÍŽENECKÝ, J. (1965). *Fundamenta Genetica,* Brno: Moravian Museum.

KÜHLING, O. (1899). Ueber die Reduction des Toalloxazins. *Ber. chem. Ges.* **32**, 1650–1653.

KÜHNE, W. (1866). *Lehrbuch der physiologischen Chemie,* Leipzig: Engelmann.

KÜHNE, W. (1878). Erfahrungen und Bemerkungen über Enzyme und Fermente. *Untersuchungen aus dem physiologischen Institut Heidelberg* **1**, 291–324.

KÜTZING, F. (1837). Mikroskopische Untersuchungen über die Hefe und Essigmutter, nebst mehreren anderen dazu gehörigen vegetabilischen Gebilden. *J. Prakt. Chem.* **11**, 385–409.

KUHN, R. (1935). Sur les flavines. *Bull. Soc. Chim. Biol.* **17**, 905–926.

KUHN, R., RUDY, H., AND WEYGAND, F. (1936). Synthese der Lactoflavin-5′-phosphorsäure. *Ber. chem. Ges.* **69**, 1543–1547.

KUHN, T. S. (1971). The relations between history and history of science. *Daedalus* **100**, 271–304.

KUNCKEL, J. (1716). *Laboratorium Chymicum*, Hamburg: Heyl.

KURZER, F. AND SANDERSON, P. M. (1956). Urea in the history of organic chemistry. *J. Chem. Ed.* **33**, 452–459.

LANDSTEINER, K. (1945). *The Specificity of Serological Reactions* (second edition), Cambridge: Harvard University Press.

LANGMUIR, I. (1939). The structure of proteins. *Proceedings of the Physical Society* **51**, 592–612.

LANKESTER, E. R. (1871). Ueber das Vorkommen von Haemoglobin in den Muskeln der Mollusken und die Verbreitung desselben in den lebendigen Organismen. *Pflügers Arch.* **4**, 315–320.

LANKESTER, E. R. (1872). A contribution to the knowledge of haemoglobin. *Proc. Roy. Soc.* **21**, 70–81.

LARDER, D. F. (1967). Historical aspects of the tetrahedron in chemistry. *J. Chem. Ed.* **44**, 661–666.

LARDY, H. A. AND FERGUSON, S. M. (1969). Oxidative phosphorylation in mintochondria. *Ann. Rev. Biochem.* **38**, 991–1034.

LASKOWSKI, N. (1846). Ueber die Proteintheorie. *Ann.* **58**, 129–166.

LAURENT, A. (1837). Suite de recherches diverses de chimie organique. *Ann. Chim.* 2e Sér. **66**, 314–335.

LAVOISIER, A. L. (1799). *Elements of Chemistry* (fourth edition, translated by R. Kerr), Edinburgh: Creech.

LAVOISIER, A. L. (1862). *Oeuvres de Lavoisier,* Paris: Imprimerie Impériale.

LAWES, J. B. AND GILBERT, J. H. (1866). On the sources of the fat of the animal body. *Phil. Mag.* (4) **32**, 439–451.

LAYTON, D. (1965). The original observations of Brownian motion. *J. Chem. Ed.* **42**, 367–368.

LEATHES, J. B. (1906). *Problems in Animal Metabolism*, Philadelphia: Blakiston.

LEBEDEV, A. (1912). Über den Mechanismus der alkoholischen Gärung. *Biochem. Z.* **46**, 483–489.

LECANU, L. R. (1838). Études chimiques sur le sang humain. *Ann. Chim.* 2e Sér. **67**, 54–70.

LEETE, E. (1965). Biosynthesis of alkaloids. *Science* **147**, 1000–1006.

LEHMANN, C. G. (1855). *Physiological Chemistry* (translated from second edition by G. E. Day; edited by R. E. Rogers), Philadelphia: Blanchard and Lea.

LEHMANN, J. (1930). Zur Kenntnis biologischer Oxydations-Reduktionspotentiale. Messungen im System: Succinat-Fumarat-Succinodehydrogenase. *Skand. Arch. Physiol.* **58**, 173–312.

LEHNARTZ, E. (1933). Die chemische Vorgänge bei der Muskelkontraktion. *Erg. Physiol.* **35**, 874–966.

LEHNINGER, A. L. (1951). Phosphorylation coupled to oxidation of dihydrodiphospho-pyridine nucleotide. *J. Biol. Chem.* **190**, 345–359.

LELOIR, L. F. (1964). Nucleoside diphosphate sugars and saccharide synthesis. *Biochem. J.* **91**, 1–8.

LEMAY, P. (1949). Desormes et Clément decouvrent et expliquent la catalyse. *Chymia* **2**, 45–49.

LEVENE, P. A. (1903. On the chemistry of the chromatin substance of the nerve-cell. *Journal of Medical Research* **10**, 204–211.

LEVENE, P. A. (1909). Über die Hefenucleinsäure. *Biochem. Z.* **17**, 120–131.

LEVENE, P. A. (1921). On the structure of thymus nucleic acid and on its possible bearing on the structure of plant nucleic acid. *J. Biol. Chem.* **48**, 119–125.

LEVENE, P. A. AND BASS, L. W. (1931). *Nucleic Acids,* New York: Chemical Catalogue Co.

LEVENE, P. A. AND JACOBS, W. A. (1908, 1909). Ueber die Inosinsäure. *Ber chem. Ges.* **41**, 2703–2707; **42**, 335–338, 1198–1203.

LEVENE, P. A. AND JACOBS, W. A. (1912). On the structure of thymus nucleic acid. *J. Biol. Chem.* **12**, 411–420.

LEVENE, P. A. AND RAYMOND, A. L. (1928). Hexose diphosphate. *J. Biol. Chem.* **80**, 633–638.

LEVENE, P. A. AND TIPSON, R. S. (1935). The ring structure of thymidine. *J. Biol. Chem.* **109**, 623–630.

LEVY, L. (1889). Ueber Farbstoffe in den Muskeln. *Z. physiol. Chem.* **13**, 309–325.

LIBAVIUS, A. (1964). *Die Alchemie des Andreas Libavius* (reprint of 1597 edition), Weinheim: Verlag Chemie.

LIEBEN, F. (1935). *Geschichte der Physiologischen Chemie,* Leipzig and Vienna: Deuticke.

LIEBIG, G. (1850). Ueber die Respiration der Muskeln. *Arch. Anat. Physiol.,* pp. 393–416.

LIEBIG, J. (1834). Über einige Stickstoffverbindungen. *Ann.* **10**, 1–47.

LIEBIG, J. (1839). Ueber die Erscheinung der Gährung, Fäulnis und Verwesung und ihre Ursachen. *Ann.* **30**, 250–287.

LIEBIG, J. (1841). Ueber die stickstoffhaltigen Nahrungsmittel des Pflanzenreiches. *Ann.* **39**, 129–169.

LIEBIG, J. (1842). *Die Organische Chemie in ihrer Anwendung auf Physiologie und Pathologie,* Braunschweig: Vieweg.

LIEBIG, J. (1843). *Die Thier-chemie oder die Organische Chemie in ihrer Anwendung auf Physiologie und Pathologie,* Braunschweig: Vieweg.

LIEBIG, J. (1846a). Ueber den Schwefelgehalt des stickstoffhaltigen Bestandtheils der Erbsen. *Ann.* **57**, 131–133.

LIEBIG, J. (1846b). *Die Thierchemie oder die Organische Chemie in ihrer Anwendung auf Physiologie und Pathologie* (third edition), Braunschweig: Vieweg.

LIEBIG, J. (1847a). *Researches on the Chemistry of Food* (translated by W. Gregory), London: Taylor and Walton.

LIEBIG, J. (1847b). Ueber die Bestandteile der Flüssigkeiten des Fleisches. *Ann.* **62**, 257–369.

LIEBIG, J. (1870). Ueber die Gährung und die Quelle der Muskelkraft. *Ann.* **153**, 1–47, 137–228.

LILLIE, R. S. (1924). Reactivity of the cell. In *General Cytology*, pp. 167–233 (E. V. Cowdry, ed.), Chicago: University Press.

LINDEBOOM, G. A. (1968). *Herman Boerhaave*, London: Methuen.

LINDERSTRØM-LANG, K. (1939). S. P. L. Sørensen. *Compt. Rend. Carlsberg* 23, I–XXI.

LINDERSTRØM-LANG, K., HOTCHKISS, R. D., AND JOHANSEN, G. (1938). Peptide bonds in globular proteins. *Nature* 142, 996.

LIPMAN, T. O. (1964). Wöhler's preparation of urea and the fate of vitalism. *J. Chem. Ed.* 41, 452–458; 42, 394–397.

LIPMAN, T. O. (1967). Vitalism and reductionism in Liebig's physiological thought. *Isis* 58, 167–185.

LIPMANN, F. (1933). Über die oxydative Hemmbarkeit der Glykolyse und den Mechanismus der Pasteurschen Reaktion. *Biochem. Z.* 265, 133–140.

LIPMANN, F. (1939). An analysis of the pyruvic acid oxidation system. *Cold Spr. Harb. Symp.* 7, 248–259.

LIPMANN, F. (1941). Metabolic generation and utilization of phosphate bond energy. *Adv. Enzymol.* 1, 99–162.

LIPMANN, F. (1945). Acetylation of sulfanilamide by liver homogenates and extracts. *J. Biol. Chem.* 160, 173–190.

LIPMANN, F. (1969). Polypeptide chain elongation in protein biosynthesis. *Science* 164, 1024–1031.

LIPMANN, F. (1971). *Wanderings of a Biochemist*, New York: Wiley-Interscience.

LIPPMANN, E. O. (1906). *Abhandlungen und Vorträge zur Geschichte der Naturwissenschaften*, pp. 275–295, Leipzig: Veit.

LOEB, J. (1906). *The Dynamics of Living Matter*, New York: Columbia University Press.

LÖFFLER, W. (1920). Zur Kenntnis der Leberfunktion unter experimentell pathologischen Bedingungen. *Biochem. Z.* 112, 164–187.

LOEW, O. (1896). *The Energy of Living Protoplasm*, London: Kegan Paul, Trench, Trübner and Co.

LOEWI, O. (1898). Ueber das "harnstoffbildende" Ferment der Leber. *Z. physiol. Chem.* 25, 511–522.

LOEWI, O. (1902). Ueber Eiweisssynthese im Tierkörper. *Arch. exp. Path. Pharm.* 48, 303–330.

LOHMANN, K. (1929). Über die Pyrophosphatfraktion im Muskel. *Naturwissenschaften* 17, 624–625.

LOHMANN, K. (1934). Über die enzymatische Aufspaltung der Kreatinphosphorsäure; zugleich ein Beitrag zum Chemismus der Muskelkontraktion. *Biochem. Z.* 271, 264–277.

LOHMANN, K. (1935). Konstitution der Adenylpyrophosphorsäure und Adenosindiphosphorsäure. *Biochem. Z.* 282, 120–123.

LOHMANN, K. AND SCHUSTER, P. (1937). Untersuchungen über die Cocarboxylase. *Biochem. Z.* 294, 188–214.

LOVEJOY, A. O. (1936). *The Great Chain of Being*, Cambridge, Mass.: Harvard University Press.

LUDWIG, K. (1852). *Lehrbuch der Physiologie des Menschen*, Heidelberg: Winter.

LUND, E. W. (1965). Guldberg and Waage and the law of mass action. *J. Chem. Ed.* 42, 548–550.

LUNDSGAARD, E. (1930). Untersuchungen über Muskelkontraktion ohne Milchsäurebildung. *Biochem. Z.* 217, 162–177.

LUNIN, N. (1881). Ueber die Bedeutung der anorganischen Salze für die Ernährung des Thieres. *Z. physiol. Chem.* 5, 31–39.

LUSK, G. (1910). The fate of the amino acids in the organism. *J. Am. Chem. Soc.* 32, 671–680.

LYLE, R. E. AND LYLE, G. G. (1964). A brief history of polarimetry. *J. Chem. Ed.* 41, 308–313.

MACALPINE, I., HUNTER, R., AND RIMINGTON, C. (1966). Porphyria in the royal houses of Stuart, Hanover, and Prussia. *British Medical Journal* (1) pp. 65–71.

MC COLLUM, E. V. (1957). *A History of Nutrition,* Boston: Houghton Mifflin.

MC CORMMACH, R. (1969). Henry Cavendish: A study of rational empiricism in eighteenth-century natural philosophy. *Isis* 60, 293–306.

MAC FADYEN, A., MORRIS, G. H., AND ROWLAND, S. (1900). On expressed yeast-cell plasma (Buchner's "zymase"). *Proc. Roy. Soc.* 67, 250–266.

MC GUCKEN, W. (1969). *Nineteenth-century Spectroscopy,* Baltimore: Johns Hopkins Press.

MC KIE, D. (1944). Wöhler's "synthetic" urea and the rejection of vitalism: a chemical legend. *Nature* 153, 608–610.

MC KIE, D. (1952). *Antoine Lavoisier,* London: Constable.

MC KIE, D. AND HEATHCOTE, N. H. DE V. (1935). *The Discovery of Specific and Latent Heats,* London: Arnold.

MAC MUNN, C. A. (1886). Researches on myohaematin and the histohaematins. *Phil. Trans.* 177, 267–298.

MAC MUNN, C. A. (1887). Further observations on myohaematin and the histohaematins. *J. Physiol.* 8, 51–65.

MAC MUNN, C. A. (1890). Über das Myohämatin. *Z. physiol. Chem.* 14, 328–329.

MAC MUNN, C. A. (1914). *Spectrum Analysis Applied to Biology and Medicine,* London: Longmans, Green.

MACQUER, P. J. (1777). *A Dictionary of Chemistry* (second English edition, translated by J. Keir), London: Cadell and Elmsly.

MADDEN, S. C. AND WHIPPLE, G. H. (1940). Plasma proteins: their source, production and utilization. *Physiol. Revs.* 20, 194–217.

MAGENDIE, F. (1816–1817). *Précis Élémentaire de Physiologie,* Paris: Méquignon-Marvis.

MAGNUS, G. (1837). Ueber die im Blute enthaltenen Gase, Sauerstoff, Stickstoff, und Kohlensäure. *Ann. Phys.* 40, 583–606.

MAITLAND, P. (1950). Biogenetic origin of the pyrrole pigments. *Quarterly Reviews, Chemical Society* 4, 45–68.

MALFATTI, H. (1891). Beiträge zur Kenntnis der Nucleine. *Z. physiol. Chem.* 16, 68–86

MANASSEIN, M. (1897). Zur Frage von der alkoholischen Gärung. *Ber. chem. Ges.* 30, 3061–3062.

MANCHOT, W. (1902). Ueber Peroxydbildung beim Eisen. *Ann.* 325, 10–124.

MANDELSTAM, J. (1958). Turnover of protein in growing and non-growing populations of *Escherichia coli. Biochem. J.* 69, 110–119.

MANI, N. (1956). Das Werk von Friedrich Tiedemann und Leopold Gmelin: "Die Verdauung nach Versuchen", und seine Bedeutung für die Entwicklung der Ernährungslehre in der ersten Hälfte des 19. Jahrhunderts. *Gesnerus* 13, 190–214.

MANI, N. (1967). *Die Historischen Grundlagen der Leberforschung. II. Teil. Die Geschichte der Leberforschung von Galen bis Claude Bernard,* Basle: Schwabe.

MANN, G. (1906). *Chemistry of the Proteids,* London: Macmillan.

MANN, T. (1964). David Keilin 1887–1963. *Biog. Mem. F. R. S.* 10, 183–205.

MARCHAND, R. F. (1838). Fortgesetzte Versuche über die Bildung des Harnstoffes im thierischen Körper. *J. prakt. Chem.* 14, 490–497.

MARGOLIASH, E. AND SCHEJTER, A. (1966). Cytochrome c. *Adv. Protein Chem.* 21, 113–286.

MARTIN, A. J. P. (1948). Partition chromatography. *Ann. N. Y. Acad. Sci.* 49, 249–264.

MARTIN, A. J. P. AND SYNGE, R. L. M. (1945). Analytical chemistry of the proteins. *Adv. Protein. Chem.* 2, 1–83.

MATHEWS, A. P. (1924). Some general aspects of the chemistry of cells. In *General Cytology* (E. V. Cowdry, ed.), Chicago: University Press.

MATHEWS, A. P. AND WALKER, S. (1909). The action of cyanides and nitriles on the spontaneous oxidation of cystein. *J. Biol. Chem.* 6, 29–37.

MATHIAS, P. (1959). *The Brewing Industry in England 1700–1830,* Cambridge: University Press.

MATTEUCCI, C. (1856). Recherches sur les phénomènes physiques et chimiques de contraction musculaire. *Ann. Chim.* 3e Sér. 47, 129–153.

MAYER, A. (1882). *Die Lehre von den Chemischen Fermenten oder Enzymologie,* Heidelberg: Winter.

MAYER, R. (1893). *Die Mechanik der Wärme* (third edition, J. J. Weyrauch, ed.), Stuttgart: Cotta.

MEDAWAR, P. B. (1963). The 'M.R.C.' fifty years ago and now. *Nature* 200, 1039–1042.

MEISTER, A. (1956). Metabolism of glutamine. *Physiol. Revs.* 36, 103–127.

MELDRUM, A. N. (1930). *The Eighteenth Century Revolution in Science—The First Phase,* Calcutta: Longmans, Green.

MENDEL, G. (1865). Versuche über Pflanzen-hybriden. *Verhandlungen des Naturforschenden Vereines in Brünn* 4, 3–47.

MENDEL, L. B. AND JACKSON, H. C. (1898). On the excretion of kynurenic acid. *Am. J. Physiol.* 2, 1–28.

MENDELSOHN, E. (1963). Cell theory and the development of general physiology. *Archives Internationales d'Histoire des Sciences* 6, 419–429.

MENDELSOHN, E. (1964). *Heat and Life,* Cambridge: Harvard University Press.

MERING, J. V. (1877). Zur Glycogenbildung in der Leber. *Pflügers Arch.* 14, 274–284.

MERTON, R. K. (1961). Singletons and multiples in scientific discovery. *Proc. Am. Phil. Soc.* 105, 470–486.

MESELSON, M. AND STAHL, F. W. (1958). The replication of DNA in *Escherichia coli. Proc. Natl. Acad. Sci.* 44, 671–682.

METZGER, H. (1918). *La Genèse de la Science des Cristaux,* Paris: Alcan.

METZGER, H. (1930). *Newton, Stahl et la Doctrine Chimique,* Paris: Alcan.

MEYER, K. H. (1928). Neue Wege in der organischen Strukturlehre und in der Erforschung hochpolymeren Verbindungen. *Z. angew. Chem.* 41, 935–946.

MEYER, K. H. (1943). The chemistry of glycogen. *Adv. Enzymol.* 3, 109–135.

MEYERHOF, O. (1918). Über das Vorkommen des Coferments der alkoholischen Hefegärung im Muskelgewebe und seine mutmassliche Bedeutung im Atmungsmechanismus. *Z. physiol. Chem.* 101, 165–175.

MEYERHOF, O. (1925). Über den Zusammenhang der Spaltungsvorgänge mit der Atmung in der Zelle. *Ber. chem. Ges.* 58, 991–1001.

MEYERHOF, O. (1926a). Über die enzymatische Milchsäurebildung im Muskelextrakt. *Biochem. Z.* 178, 395–418, 462–490.

MEYERHOF, O. (1926b). Thermodynamik des Lebensprozesses. In *Handbuch der Physik,* Vol. 11, pp. 238–271 (H. Geiger and K. Scheel, eds.), Berlin: Springer.

MEYERHOF, O. (1930). *Die Chemische Vorgänge im Muskel,* Berlin: Springer.

MEYERHOF, O. AND KIESSLING, W. (1935). Über den Hauptweg der Milchsäurebildung in der Muskulatur. *Biochem. Z.* 283, 83–113.

MEYERHOF, O. AND LOHMANN, K. (1931). Über die Energetik der anaeroben Phosphagensynthese ("Kreatinphosphorsäure") im Muskelextrakt. *Naturwissenschaften* 19, 575–576.

MEYERHOF, O., OHLMEYER, P. AND MÖHLE, W. (1938). Über die Koppelung zwischen Oxydoreduktion und Phosphatveresterung bei der anaeroben Kohlenhydratspaltung. *Biochem. Z.* 297, 90–133.

MIALHE, L. (1856). *Chimie appliquée à la Physiologie et à la Thérapeutique,* Paris: Masson.

MICHAELIS, L. (1900). Die vitale Färbung, eine Darstellungsmethode der Zellgranula. *Arch. mikr. Anat.* 55, 558–575.

MICHAELIS, L. (1935). Semiquinones, the intermediate steps of reversible oxidation-reduction. *Chem. Revs.* 16, 243–286.

MICHAELIS, L. AND MENTEN, M. L. (1913). Zur Kinetik der Invertinwirkung. *Biochem. Z.* 49, 333–369.

MIESCHER, F. (1871a). Ueber die chemische Zusammensetzung des Eiters. *Med. chem. Unt.,* pp. 441–460.

MIESCHER, F. (1871b). Die Kerngebilde im Dotter des Hühnereies. *Med. chem. Unt.,* pp. 502–509.

MIESCHER, F. (1879). *Die Histochemischen und Physiologischen Arbeiten,* Leipzig: Vogel.

MILAS, N. A. (1932). Auto-oxidation. *Chem. Revs.* 10, 295–364.

MILNE EDWARDS, H. (1862). *Leçons sur la Physiologie et l'Anatomie Comparée de l'Homme et des Animaux,* Vol. 7, Paris: Masson.

MINCHIN, E. A. (1916). The evolution of the cell. *Am. Natur.* **50**, 5–39.

MIRSKY, A. E. (1951). Some chemical aspects of the cell nucleus. In *Genetics in the Twentieth Century*, pp. 127–153 (L. C. Dunn, ed.), New York: Macmillan.

MIRSKY, A. E. AND PAULING, L. (1936). On the structure of native, denatured, and coagulated proteins. *Proc. Natl. Acad. Sci.* **22**, 439–447.

MITSCHERLICH, E. (1821). Sur la relation qui existe entre la forme crystalline et les proportions chimiques. II. Mémoire sur les arseniates et les phosphates. *Ann. Chim.* 2e Sér. **19**, 350–419.

MITSCHERLICH, E. (1834). Ueber die Aetherbildung. *Ann. Phys.* **31**, 273–282.

MITTASCH, A. (1940). *Julius Robert Mayers Kausalbegriff*, Berlin: Springer.

MOKRUSHIN, S. G. (1962). Thomas Graham and the definition of colloids. *Nature* **195**, 861.

MONNIER, M. (1942). Federico Battelli (1867–1941). *Rivista di Biologia* **33**, 267–280.

MOORE, B. AND WHITLEY, E. (1909). The properties and classification of the oxidizing enzymes, and analogies between enzymic activity and the effects of immune bodies and complement. *Biochem. J.* **4**, 136–167.

MORGAN, T. H. (1910). Sex limited inheritance in Drosophila. *Science* **32**, 120–122.

MORGAN, T. H. (1928). *The Theory of the Gene* (second edition), New Haven: Yale University Press.

MÜLLER, J. (1843). *Elements of Physiology* (translated by W. Baly; arranged from second London edition by J. Bell), Philadelphia: Lea and Blanchard.

MULDER, G. J. (1838). Zusammensetzung von Fibrin, Albumin, Leimzucker, Leucin u. s. w. *Ann.* **28**, 73–82.

MULDER, G. J. (1839). Ueber die Zusammensetzung einiger thierischen Substanzen. *J. prakt. Chem.* **16**, 129–152.

MULDER, G. J. (1844–1851). *Versuch einer allgemeinen physiologischen Chemie*, Braunschweig: Vieweg.

MULLER, H. J. (1922). Variation due to change in the individual gene. *Am. Natur.* **56**, 32–50.

MULLER, H. J. (1927). Artificial transmutation of the gene. *Science* **66**, 84–87.

MULLER, H. J. (1929). The gene as the basis of life. *Proceedings of the International Congress of Plant Sciences* **1**, 897–921.

MULLER, H. J. (1935a). Physics in the attack on the fundamental problems of genetics. *Scientific Monthly* **44**, 210–214.

MULLER, H. J. (1935b). On the dimensions of chromosomes and genes in dipteran salivary glands. *Am. Natur.* **69**, 405–411.

MULLER, H. J. (1946). A physicist stands amazed at genetics. *Journal of Heredity* **37**, 90–92.

MULLER, H. J. (1947). The gene. *Proc. Roy. Soc.* **B134**, 1–37.

MULTHAUF, R. P. (1954). John of Rupescissa and the origin of medical chemistry. *Isis* **45**, 359–367.

MULTHAUF, R. P. (1955). J. B. Van Helmont's reformation of the Galenic doctrine of digestion. *Bull. Hist. Med.* **29**, 154–163.

MURNAGHAN, J. H. AND TALALAY, P. (1967). John Jacob Abel and the crystallization of insulin. *Persp. Biol. and Med.* **10**, 334–380.

NÄGELI, C. (1879). Theorie von Gärung. *Abhandlungen der königlichen Akademie der Wissenschaften, Munich* 13(2), 77–205.

NÄGELI, C. AND LOEW, O. (1878). Ueber die chemische Zusammensetzung der Hefe. *Ann.* 193, 322–348.

NAQUET, A. (1867). *Principes de Chimie Fondés sur les Théories Modernes* (second edition). Paris: Savy.

NASSE, O. (1869). Beiträge zur Physiologie der contractilen Substanz. *Pflügers Arch.* 2, 97–121.

NATTA, G. (1965). Macromolecular chemistry. *Science* 147, 261–272.

NEEDHAM, D. M. (1930). A quantitative study of succinic acid in muscle. III. Glutamic and aspartic acids as precursors. *Biochem. J.* 24, 208–227.

NEEDHAM, D. M. (1971). *Machina Carnis: The Biochemistry of Muscular Contraction in its Historical Development,* Cambridge: University Press.

NEEDHAM, D. M. AND PILLAI, R. K. (1937). The coupling of oxido-reductions and dismutations with esterification of phosphate in muscle. *Biochem. J.* 31, 1837–1851.

NEEDHAM, J. (1962). Frederick Gowland Hopkins. *Persp. Biol. and Med.* 6, 2–46.

NEEDHAM, J. AND BALDWIN, E. (1949). *Hopkins and Biochemistry 1861–1947,* Cambridge: Heffer.

NEGELEIN, E. AND BRÖMEL, H. (1939a). Protein der d-Aminosäureoxydase. *Biochem. Z.* 300, 225–239.

NEGELEIN, E. AND BRÖMEL, H. (1939b). R-Diphosphoglycerinsäure, ihre Isolierung und Eigenschaften. *Biochem. Z.* 303, 132–144.

NEGELEIN, E. AND WULFF, H. (1937). Diphosphopyridinproteid: Alkohol, Acetaldehyd. *Biochem. Z.* 293, 351–389.

NENCKI, M. (1872). Die Wasserentziehung im Thierkörper. *Ber. chem. Ges.* 5, 890–893.

NENCKI, M. (1878). Ueber den chemischen Mechanismus der Fäulniss. *J. prakt. Chem.* 17, 105–124.

NENCKI, M. (1885). Ueber das Parahämoglobin. *Arch. exp. Path. Pharm.* 20, 332–343.

NENCKI, M. (1904). *Marceli Nencki Opera Omnia,* Braunschweig: Vieweg.

NENCKI, M. AND SIEBER, N. (1882). Untersuchungen über die physiologische Oxydation. *J. prakt. Chem.* 26, 1–41.

NEUBAUER, O. AND FALTA, W. (1904). Über das Schicksal einiger aromatischer Säuren bei der Alkaptonurie. *Z. physiol. Chem.* 42, 81–101.

NEUBAUER, O. AND FROMHERZ, K. (1911). Über den Abbau der Aminosäuren bei der Hefegärung. *Z. physiol. Chem.* 70, 326–350.

NEUBERG, C. AND KERB, J. (1913). Über zuckerfreie Hefegärungen. XII. Über die Vorgänge bei der Hefegärung. *Biochem. Z.* 53, 406–419.

NEUBERG, C. AND REINFURTH, E. (1919). Weitere Untersuchungen über die korrelative bildung von Acetaldehyd und Glycerin bei der Zuckerspaltung und neue Beiträge zur Theorie der alkoholischen Gärung. *Ber. chem. Ges.* 52, 1677–1703.

NEUMEISTER, R. (1897). Bemerkungen zu Eduard Buchner's Mittheilungen über "Zymase." *Ber. chem. Ges.* 30, 2963–2966.

NEWTON, I. (1630). *Opticks: or, a Treatise of the Reflections, Refractions, Inflections and Colours of Light* (fourth edition), London: Innys.

NIER, A. O. (1955). Determination of isotopic masses and abundances by mass spectrometry. *Science* 121, 737–744.

NORD, F. F. (1940). Facts and interpretations in the mechanism of alcoholic fermentation. *Chem. Revs.* 26, 423–472.

NORTHROP, J. H. (1930). Crystalline pepsin. I. Isolation and tests of purity. *J. Gen. Physiol.* 13, 739–766.

NORTHROP, J. H. (1935). The chemistry of pepsin and trypsin. *Biol. Revs.* 10, 263–282.

NORTHROP, J. H. (1951). Growth and phage production of lysogenic B. megatherium. *J. Gen. Physiol.* 34, 715–735.

OCHOA, S. (1941). "Coupling" of phosphorylation with oxidation of pyruvic acid in brain. *J. Biol. Chem.* 138, 751–773.

OCHOA, S. (1943). Efficiency of aerobic phosphorylation in cell-free heart extracts. *J. Biol. Chem.* 151, 493–505.

OERTMANN, E. (1887). Ueber den Stoffwechsel entbluteter Frösche. *Pflügers Arch.* 15, 381–398.

OGSTON, A. G. AND SMITHIES, O. (1948). Some thermodynamic and kinetic aspects of metabolic phosphorylation. *Physiol. Revs.* 28, 283–303.

OHLE, H. (1931). Die Chemie der Monosaccharide und der Glykolyse. *Erg. Physiol.* 33, 558–701.

OLBY, R. (1970). Francis Crick, DNA, and the central dogma. *Daedalus* 99, 938–987.

OLBY, R. (1971). Schrödinger's problem: what is life? *J. Hist. Biol.* 4, 119–148.

OLMSTED, J. M. D. (1944). *François Magendie*, New York: Schuman.

OPPENHEIMER, C. (1900). *Die Fermente und ihre Wirkungen*, Leipzig: Vogel.

OPPENHEIMER, C. (1926). *Die Fermente und ihre Wirkungen* (fifth edition), Leipzig: Thieme.

OSANN, G. (1853). Ueber eine Modification des Wasserstoffs. *J. prakt. Chem.* 58, 385–391.

OSBORNE, T. B. (1892). Crystallised vegetable proteids. *American Chemical Journal* 14, 662–689.

OSBORNE, T. B. AND MENDEL, L. B. (1914). Amino-acids in nutrition and growth. *J. Biol. Chem.* 17, 325–349.

OSTERHOUT, W. J. V. (1930). Biographical memoir of Jacques Loeb 1859–1924. *Biog. Mem. Natl. Acad. Sci.* 13, 318–401.

OSTWALD, W. (1900). Über Oxydationen mittels freien Sauerstoffs. *Z. physik. Chem.* 34, 248–252.

O'SULLIVAN, C. AND TOMPSON, F. W. (1890). Invertase: a contribution to the history of an enzyme or unorganized ferment. *J. Chem. Soc.* 57, 834–931.

OVERTON, E. (1901). *Studien über die Narkose*, Jena: Fischer.

PAGEL, W. (1944). The religious and philosophical aspects of Van Helmont's science and medicine. *Bull. Hist. Med.* Suppl. 2.

PAGEL, W. (1955). J. B. Van Helmont's reformation of the Galenic doctrine of digestion—and Paracelsus. *Bull. Hist. Med.* 29, 563–568.

PAGEL, W. (1956). Van Helmont's ideas on gastric digestion and the gastric acid. *Bull. Hist. Med.* 30, 524–536.

PAGEL, W. (1958). *Paracelsus,* Basel: Karger.

PAGEL, W. (1962a). *Das Medizinische Weltbild des Paracelsus,* Wiesbaden: Steiner.

PAGEL, W. (1962b). The wild spirit (Gas) of Van Helmont and Paracelsus. *Ambix* 10, 1–13.

PALADE, G. E. (1952). The fine structure of mitochondria. *Anatomical Record* 114, 427–451.

PALLADIN, W. (1909). Über das Wesen der Pflanzenatmung. *Biochem. Z.* 18, 151–206.

PALMER, W. G. (1965). *A History of the Concept of Valency to 1930,* Cambridge: University Press.

PARNAS, J. (1910). Ueber fermentative Beschleunigung der Cannizzaroschen Aldehydumlagerung durch Gewebssäfte. *Biochem. Z.* 28, 274–294.

PARNAS, J. K. (1943). Co-enzymatic reactions. *Nature* 151, 577–580.

PARNAS, J. K., OSTERN, P., AND MANN, T. (1934). Über die Verkettung der chemischen Vorgänge im Muskel. *Biochem. Z.* 272, 64–70.

PARTINGTON, J. R. (1956). The life and work of John Mayow (1641–1679). *Isis* 47, 217–230.

PARTINGTON, J. R. (1960). Joseph Black's "Lectures on the Elements of Chemistry." *Chymia* 6, 27–67.

PARTINGTON, J. R. AND MC KIE, D. (1937–1939). Historical studies on the phlogiston theory. *Ann. Sci.* 2, 361–404; 3, 1–58, 337–371; 4, 113–149.

PASTEUR, L. (1857). Mémoire sur la fermentation appelée lactique. *Comp. Rend.* 45, 913–916.

PASTEUR, L. (1858a). Mémoire sur la fermentation appelée lactique. *Ann. Chim.* 3e Sér. 52, 404–418.

PASTEUR, L. (1858b). Nouveaux faits concernant l'histoire de la fermentation alcoolique. *Comp. Rend.* 47, 1011–1013.

PASTEUR, L. (1859). Note. *Comp. Rend.* 48, 737–740.

PASTEUR, L. (1860a). Recherches sur la dissymétrie moléculaire des produits organiques naturels. *Soc. Chim. Paris,* pp. 1–48.

PASTEUR, L. (1860b). Mémoire sur la fermentation alcoolique. *Ann. Chim.* 3e Sér. 58, 323–426.

PASTEUR, L. (1860c). Note sur la fermentation alcoolique. *Comp. Rend.* 50, 1083–1084.

PASTEUR, L. (1861). Expériences et vues nouvelles sur la nature des fermentations. *Comp. Rend.* 52, 1260–1264.

PASTEUR, L. (1862). Mémoire sur les corpuscules organisés qui existent dans l'atmosphère. Examen de la doctrine des générations spontanées. *Ann. Chim.* 3e Sér. 64, 5–110.

PASTEUR, L. (1871). Note sur un mémoire de M. Liebig relatif aux fermentations. *Comp. Rend.* 73, 1419–1424.

PASTEUR, L. (1876a). Sur la fermentation de l'urine. *Comp. Rend.* 83, 5–8.

PASTEUR, L. (1876b). Réponse à M. Berthelot. *Comp. Rend.* 83, 10.

PASTEUR, L. (1876c). *Études sur la Bière,* Paris: Gauthier-Villars.

PASTEUR, L. (1878). Première reponse à M. Berthelot. *Comp. Rend.* 87, 1053–1058.

PASTEUR, L. (1879). *Examen Critique d'un Écrit Posthume de Claude Bernard sur la Fermentation,* Paris: Gauthier-Villars.

PASTEUR, L. (1886). La dissymétrie moléculaire. *Soc. chim. Paris*, pp. 24–37.

PATIN, G. (1846). *Lettres* (J. H. Reveille-Parisse, ed.), Paris: Baillière.

PATTERSON, T. S. (1937). Jean Beguin and his "Tyrocinium Chymicum." *Ann. Sci.* **2**, 243–298.

PAULI, W. (1934). The chemistry of the amino acids and proteins. *Ann. Rev. Biochem.* **3**, 111–132.

PAULING, L. AND COREY, R. B. (1950). Two hydrogen-bonded spiral configurations of the polypeptide chain. *J. Am. Chem. Soc.* **72**, 5349.

PAULING, L. AND COREY, R. B. (1953). A proposed structure for the nucleic acids. *Proc. Natl. Acad. Sci.* **39**, 84–97.

PAULING, L. AND DELBRÜCK, M. (1940). The nature of the intermolecular forces operative in biological processes. *Science* **92**, 77–79.

PAULING, L. AND NIEMANN, C. (1939). The structure of proteins. *J. Am. Chem. Soc.* **61**, 1860–1867.

PEKELHARING, C. A. (1902). Mittheilungen über Pepsin. *Z. physiol. Chem.* **35**, 8–30.

PETERS, R. (1898). Ueber Oxydations- und Reductions-ketten und den Einfluss komplexer Ionen auf ihre elektromotorische Kraft. *Z. physik. Chem.* **26**, 193–236.

PETERS, R. (1963). *Biochemical Lesions and Lethal Synthesis*, Oxford: Pergamon Press.

PETTENKOFER, M. AND VOIT, C. (1866). Untersuchungen über den Stoffverbrauch des normalen Menschen. *Z. Biol.* **2**, 459–573.

PFEFFER, W. (1872). Untersuchungen über die Proteinkörner und die Bedeutung des Asparagins beim Keimen der Samen. *Jahrb. wiss. Bot.* **8**, 429–574.

PFEFFER, W. (1881). *Pflanzenphysiologie*, Leipzig: Engelmann.

PFLÜGER, E. (1872). Ueber die Diffusion des Sauerstoffs, den Ort und die Gesetze der Oxydationsprocesse im thierischen Organismus. *Pflügers Arch.* **6**, 43–64.

PFLÜGER, E. (1875). Beiträge zur Lehre von der Respiration. I. Ueber die physiologische Verbrennung in den lebendigen Organismen. *Pflügers Arch.* **10**, 251–369, 641–644.

PFLÜGER, E. (1877). Die Physiologie und ihre Zukunft. *Pflügers Arch.* **15**, 361–365.

PFLÜGER, E. (1893). Ueber einige Gesetze des Eiweissstoffwechsels. *Pflügers Arch.* **54**, 333–419.

PFLÜGER, E. (1903). Glykogen. *Pflügers Arch.* **96**, 1–398.

PFLÜGER, E. (1910). Nachschrift. *Pflügers Arch.* **131**, 302–305.

PHILLIPS, J. P. (1966). Liebig and Kolbe, critical editors. *Chymia* **11**, 89–97.

PICCARD, J. (1874). Über Protamin, Guanin und Sarkin als Bestandteile des Lachsspermas. *Ber. chem. Ges.* **7**, 1714–1719.

PICTON, H. AND LINDER, S. E. (1892). Solution and pseudo-solution I. *J. Chem. Soc.* **61**, 148–172.

PIRIA, R. (1844). Note sur l'asparagine. *Comp. Rend.* **19**, 575–577.

PIRIE, N. W. (1940). The criteria of purity used in the study of large molecules of biological origin. *Biol. Revs.* **15**, 377–404.

PIRIE, N. W. (1966). John Burdon Sanderson Haldane 1892–1964. *Biog. Mem. F. R. S.* **12**, 219–249.

PIRIE, N. W. (1970). Retrospect on the biochemistry of plant viruses. *Biochemical Society Symposia* **30**, 43–56.

References

PLANTEFOL, L. (1968). Le genre du mot enzyme. *Comp. Rend.* **266**, 41–46.

POLANYI, M. (1968). Life's irreducible structure. *Science* **160**, 1308–1312.

PRÉVOST, J. L. AND DUMAS, J. B. (1823). Examen du sang et de son action dans les divers phénomènes de la vie. *Ann. Chim.* 2e Sér. **23**, 90–104.

PRIANISHNIKOV, D. (1904). Zur Frage der Asparaginbildung. *Ber. bot. Ges.* **22**, 35–43.

PRIANISHNIKOV, D. (1924). Asparagin und Harnstoff. *Biochem. Z.* **150**, 407–423.

PRIMEROSE, J. (1651). *Popular Errours: Or the Errours of the People in Physick* (translated by Robert Wittie), London: Bourne.

QUASTEL, J. H. (1926). Dehydrogenations produced by resting bacteria. IV. A theory of the mechanisms of oxidation and reduction *in vivo*. *Biochem. J.* **20**, 166–193.

QUASTEL, J. H. AND WHETHAM, M. D. (1924). The equilibria existing between succinic, fumaric and malic acids in the presence of resting bacteria. *Biochem. J.* **18**, 519–534.

RACKER, E. (1955). Mechanism of action and properties of pyridine nucleotide-linked enzymes. *Physiol. Revs.* **35**, 1–56.

RADZISZEWSKI, B. (1880). Ueber die Phosphorescenz der organischen und organisirten Körper. *Ann.* **203**, 305–336.

RAMSAY, W. (1918). *The Life and Letters of Joseph Black M.D.,* London: Constable.

RAPER, H. S. (1907). The condensation of acetaldehyde and its relation to the biochemical synthesis of fatty acids. *J. Chem. Soc.* **91**, 1831–1838.

RAPKINE, L. (1938). Role des groupements sulfhydrilés dans l'activité de l'oxyréductase du triosephosphate. *Comp. Rend.* **207**, 301–304.

RAPPAPORT, R. (1961). Rouelle and Stahl—The phlogistic revolution in France. *Chymia* **7**, 73–102.

RASPAIL, F. V. (1830). *Essai de Chimie Microscopique appliquée à la Physiologie,* Paris: Meilhac.

RASPAIL, F. V. (1833). *Nouveau Système de Chimie Organique,* Paris: Baillière.

RATNER, S. (1954). Urea synthesis and metabolism of arginine and citrulline. *Adv. Enzymol.* **15**, 319–387.

REED, H. S. (1949). Jan Ingenhousz, plant physiologist, with a history of the discovery of photosynthesis. *Chronica Botanica* **11**, 285–393.

REGNAULT, V AND REISET, J. (1849). Recherches chimiques sur la respiration des animaux des diverses classes. *Ann. Chim.* 3e Sér. **26**, 299–519.

REICHERT, E. T. AND BROWN, A. P. (1909). *The Differentiation and Specificity of Corresponding Proteins and other Vital Substances in Relation to Biological Classification and Organic Evolution: The Crystallography of Hemoglobins,* Washington: Carnegie Institution.

REIL, J. C. (1799). Veränderte Mischung und Form der thierischen Materie, als Krankheit oder nächste Ursache der Krankheitszufälle betrachtet. *Reil's Archiv für Physiologie* **3**, 424–461.

REINKE, J. (1883). Die Autoxydation in der lebenden Pflanzenzelle. *Botanische Zeitung* **41**, 65–76, 89–103.

REY-PAILHADE, J. DE (1888). Nouvelles recherches physiologiques sur la substance organique hydrogénant le soufre à froid. *Comp. Rend.* **107**, 43–44.

RHOADES, M. M. (1954). Chromosomes, mutations, and cytoplasm in maize. *Science* 120. 115–120.

RITTENBERG, D. AND SCHOENHEIMER, R. (1937). Deuterium as an indicator in the study of intermediary metabolism. XI. Further studies on the biological uptake of deuterium into organic substances, with special reference to fat and cholesterol formation. *J. Biol. Chem.* 121, 235–253.

RITTHAUSEN, H. (1872). *Die Eiweisskörper der Getreidearten, Hülsenfrüchte, und Ölsamen,* Bonn: Cohen.

ROBERTSON, J. M. (1972). Molecules and crystals, 1926–1970. *Helvetica Chimica Acta* 55, 119–127.

ROBIN, C. AND VERDEIL, F. (1853). *Traité de Chimie Anatomique et Physiologique Normale et Pathologique,* Paris: Baillière.

ROBINSON, R. (1917). A theory of the mechanism of the phytochemical synthesis of certain alkaloids. *J. Chem. Soc.* 111, 876–899.

ROBINSON, R. (1955). *The Structural Relations of Natural Products,* Oxford: Clarendon Press.

ROBINSON, T. (1960). Michael Tswett. *Chymia* 6, 146–161.

ROBISON, R. (1922). A new phosphoric ester produced by the action of yeast juice on hexoses. *Biochem. J.* 16, 809–824.

ROCHE, A. (1935). Le poids moléculaire des protéines. *Bull. Soc. Chim. Biol.* 17, 704–744.

ROGER, J. (1963). *Les Sciences de la Vie dan la Pensée Française du XVIIIᵉ Siècle,* Paris: Colin.

ROLLESTON, H. D. (1936). *The Endocrine Organs in Health and Disease,* Oxford: University Press.

ROSE, W. C. (1938). The nutritive significance of the amino acids. *Physiol. Revs.* 18, 109–136.

ROSE, W. C. (1969). Recollections of personalities involved in the early history of American biochemistry. *J. Chem. Ed.* 46, 759–763.

ROSS, S. (1962). *Scientist:* the story of a word. *Ann. Sci.* 18, 65–85.

ROUGHTON, F. J. W. (1970). Some recent work on the interactions of oxygen, carbon dioxide and hemoglobin. *Biochem. J.* 117, 801–812.

ROUX, E. (1898). La fermentation alcoolique et l'évolution de la microbie. *Revue Scientifique et Industrielle* (4) 10, 833–840.

RUBEN, S. (1943). Photosynthesis and phosphorylation. *J. Am. Chem. Soc.* 65, 279–282.

RUBNER, M. (1894). Die Quelle der thierischen Wärme. *Z. Biol.* 30, 73–142.

RUBNER, M.. (1913). Die Ernährungsphysiologie der Hefezelle bei der alkoholischer Gärung. *Archiv für Physiologie* Suppl. pp. 1–392.

RUDÉ, G. (1966). *The Crowd in History 1730–1848,* New York: Wiley.

RUNNSTRÖM, J. AND MICHAELIS, L. (1935). Correlation of oxidation and phosphorylation in hemolyzed blood in presence of methylene blue and pyocyanine. *J. Gen. Physiol.* 18, 717–727.

RUZICKA, L. (1959). History of the isoprene rule. *Proceedings of the Chemical Society,* pp. 341–360.

SAMBURSKY, S. (1959). *Physics of the Stoics,* London: Routledge and Kegan Paul.

SANGER, F. (1945). The free amino groups of insulin. *Biochem. J.* 39, 507–515.

SANGER, F. (1952). The arrangement of amino acids in proteins. *Adv. Protein Chem.* 7, 1–67.

SAUSSURE, T. DE (1819). Sur la décomposition de l'amidon à la température atmosphérique, par l'action de l'air et de l'eau. *Ann. Chim.* 2e Sér. 11, 379–408.

SCHILLER, J. (1967a). Wöhler, l'urée et le vitalisme. *Sudhoffs Arch.* 51, 229–243.

SCHILLER, J. (1967b). *Claude Bernard et les Problèmes Scientifiques de son Temps,* Paris: Éditions du Cèdre.

SCHLEIDEN, M. J. (1845). *Grundzüge der Wissenschaftlichen Botanik* (second edition), Leipzig: Engelmann.

SCHLENK, F. (1942). Nicotinamide nucleotide enzymes. In *A Symposium on Respiratory Enzymes,* pp. 104–133, Madison: University of Wisconsin Press.

SCHMIEDEBERG, O. (1877). Ueber das Verhältniss des Ammoniaks und der primären Monoaminbasen zur Harnstoffbildung im Thierkörper. *Arch. exp. Path. Pharm.* 8, 1–14.

SCHOEN, M. (1928). *The Problem of Fermentation* (translated by H. L. Hind), London: Chapman and Hall.

SCHÖNBEIN, C. F. (1848). Über einige chemische Wirkungen der Kartoffel. *Ann. Phys.* 75, 357–361.

SCHÖNBEIN, C. F. (1860). Fortsetzung der Beiträge zur näheren Kenntnis des Sauerstoffes. *J. prakt. Chem.* 81, 1–20.

SCHÖNBEIN, C. F. (1861). Ueber einige durch die Haarröhrchenanziehung des Papieres hervorgebrachte Trennungswirkungen. *Ann. Phys.* 114, 275–280.

SCHÖNBEIN, C. F. (1863). Ueber die katalytische Wirksamkeit organischer Materien und deren Verbreitung in der Pflanzen- und Thierwelt. *J. prakt. Chem.* 89, 323–344.

SCHOENHEIMER, R. AND RITTENBERG, D. (1935). Deuterium as an indicator in the study of intermediary metabolism. *Science* 82, 156–157.

SCHOENHEIMER, R., RATNER, S., AND RITTENBERG, D. (1939). Studies in protein metabolism. X. The metabolic activity of body proteins investigated with 1(−)-leucine containing two isotopes. *J. Biol. Chem.* 130, 703–732.

SCHOENHEIMER, R., AND RITTENBERG, D. (1938). The application of isotopes to the study of intermediary metabolism. *Science* 87, 221–226.

SCHROEDER, W. V. (1882). Ueber die Bildungsstätte des Harnstoffs. *Arch. exp. Path. Pharm.* 15, 364–402.

SCHRÖDINGER, E. (1944). *What is Life?,* Cambridge: University Press.

SCHRÖER, H. (1967). *Carl Ludwig,* Stuttgart: Wissenschaftliche Verlagsgesellschaft.

SCHRYVER, S. B. (1906). *Chemistry of the Albumens,* London: Longmans, Green.

SCHULTZ, J. (1941). The evidence of the nucleoprotein nature of the gene. *Cold Spr. Harb. Symp.* 9, 55–65.

SCHULTZEN, O. AND NENCKI, M. (1869). Die Vorstufen des Harnstoffs im Organismus. *Ber. chem. Ges.* 2, 566–571.

SCHULZ, F. N. (1903). *Die Grösse des Eiweissmoleküls,* Jena: Fischer.

SCHWANN, T. (1836). Ueber das Wesen des Verdauungsprocesses. *Arch. Anat. Physiol.*, pp. 90–138.

SCHWANN, T. (1837). Vorläufige Mitteilung, betreffend Versuche über die Weingärung und Fäulnis. *Ann. Phys.* 41, 184–193.

SCHWANN, T. (1839). *Mikroskopische Untersuchungen,* Berlin: Sander.

SCHWARZENBACH, G. (1966). Die Entwicklung der Valenzlehre und Alfred Werner. *Experientia* 22, 633–646.

SEGAL, H. L. (1959). The development of enzyme kinetics. In *The Enzymes,* Vol. I, pp. 1–48 (P. D. Boyer, H. Lardy, and K. Myrbäck, eds.), New York: Academic Press.

SÉGUIN, A. (1791). Mémoire sur l'eudométrie. *Ann. Chim.* 9, 293–303.

SHAFFER, P. A. (1923). Intermediary metabolism of carbohydrates. *Physiol. Revs.* 3, 394–437.

SHAFFER, P. A. AND RONZONI, E. (1932). Carbohydrate metabolism. *Ann. Rev. Biochem.* 1, 247–266.

SHEMIN, D. AND RITTENBERG, D. (1945). The utilization of glycine for the synthesis of a porphyrin. *J. Biol. Chem.* 159, 567–568.

SHEMIN, D. AND WITTENBERG, J. (1951). The mechanism of porphyrin formation. The role of the tricarboxylic acid cycle. *J. Biol. Chem.* 192, 315–334.

SIMMER, H. (1955). Aus den Anfängen der physiologischen Chemie in Deutschland. *Sudhoffs Arch.* 39, 216–236.

SIMMER, H. (1958). Zur Entwicklung der physiologischen Chemie. *Ciba Zeitschrift* 8, 3013–3044.

SLATOR, A. (1906). Studies in fermentation. I. The chemical dynamics of alcoholic fermentation by yeast. *J. Chem. Soc.* 89, 128–142.

SMEATON, W. A. (1962). *Fourcroy, Chemist and Revolutionary 1755–1809,* Cambridge: Heffer.

SNELL, E. E. (1945). The microbiological assay of amino acids. *Adv. Protein Chem.* 2, 85–118.

SNELL, E. E. (1958). Chemical structure in relation to biological activities of vitamin B_6. *Vitamins and Hormones* 16, 77–125.

SODDY, F. (1913). Intra-atomic charge. *Nature* 92, 399–400.

SØRENSEN, M. AND SØRENSEN, S, P. L. (1925). On the coagulation of proteins by heating. *Comp. Rend. Carlsberg* 15, 1–26.

SØRENSEN, S. P. L. (1909). Enzymstudien. II. Über die Messung und die Bedeutung der Wasserstoffionenkonzentration bei enzymatischen Prozessen. *Biochem. Z.* 21, 131–304.

SONDERHOFF, R. AND THOMAS, H. (1937). Die enzymatische Dehydrierung der Trideutero-essigsäure. *Ann.* 530, 195–213.

SORBY, H. C. (1867). On a definite method of qualitative analysis of animal and vegetable colouring-matters by means of the spectrum microscope. *Proc. Roy. Soc.* 15, 433–455.

SORBY, H. C. (1873). On comparative vegetable chromatology. *Proc. Roy. Soc.* 21, 442–483.

SOURKES, T. L. (1955). Moritz Traube, 1826–1894: His contribution to biochemistry. *J. Hist. Med.* **10**, 379–391.

SPALLANZANI, L. (1789). *Dissertations Relative to the Natural History of Animals and Vegetables* (second edition), London: Murray.

SPITZER, W. (1897). Die Bedeutung gewisser Nucleoproteide für die oxydative Leistung der Zelle. *Pflügers Arch.* **67**, 615–656.

SRB, A. M. AND HOROWITZ, N. H. (1944). The ornithine cycle in Neurospora and its genetic control. *J. Biol. Chem.* **154**, 129–139.

STADIE, W. C. (1945). The intermediary metabolism of fatty acids. *Physiol. Revs.* **25**, 395–441.

STADLER, L. J. (1954). The gene. *Science* **120**, 811–819.

STAHL, G. E. (1748). *Zymotechnia fundamentalis oder allgemeine Grund-erkänntnis der Gährungs-kunst*, Frankfurt and Leipzig: Montag.

STANLEY, W. M. (1936). Chemical studies on the virus of tobacco mosaic. VI. The isolation from diseased Turkish tobacco plants of a crystalline protein possessing the properties of tobacco-mosaic virus. *Phytopathology* **26**, 305–320.

STANLEY, W. M. (1937). Chemical studies on the virus of tobacco mosaic. VIII. The isolation of crystalline protein possessing the properties of Aucuba mosaic virus. *J. Biol. Chem.* **117**, 325–340.

STARE, F. J. AND BAUMANN, C. A. (1936). The effect of fumarate on respiration. *Proc. Roy. Soc.* **B121**, 338–357.

STAUDINGER, H. (1929). Die Chemie der hochmolekularen organischen Stoffe im Sinne der Kekuleschen Strukturlehre. *Z. angew. Chem.* **42**, 37–40, 67–73.

STAUDINGER, H. (1961). *Arbeitserinnerungen*, Heidelberg: Hüthig.

STEFFENS, H. (1815). *Johann Christian Reil*, Halle: Curt.

STENT, G. S. (1968). That was the molecular biology that was. *Science* **160**, 390–395.

STENT, G. S. (1970). DNA. *Daedalus* **99**, 909–937.

STEPHENSON, M. (1949). *Bacterial Metabolism* (third edition), London: Longmans, Green.

STERN, C. AND SHERWOOD, E. R. (1966). *The Origin of Genetics*, San Francisco: Freeman.

STERN, J. R., OCHOA, S., AND LYNEN, F. (1952). Enzymatic synthesis of citric acid. V. Reaction of acetyl coenzyme A. *J. Biol. Chem.* **198**, 313–321.

STERN, K. G. (1939). Respiratory catalysts in heart muscle. *Cold Spr. Harb. Symp.* **7**, 312–322.

STERN, K. G. AND HOLIDAY, E. R. (1934). Zur Konstitution des Photo-flavins; Versuche in der Alloxazin-reihe. *Ber. chem. Ges.* **67**, 1104–1106.

STERN, K. G. AND MELNICK, J. L. (1941). The photochemical spectrum of the Pasteur enzyme in retina. *J. Biol. Chem.* **139**, 301–323.

STOKES, G. G. (1864). On the reduction and oxidation of the colouring matter of the blood. *Proc. Roy. Soc.* **13**, 355–364.

STONEY, G. J. (1881). On the physical units of nature. *Phil. Mag.* (5) **11**, 381–390.

STRASBURGER, E. (1901). Über Befruchtung. *Botanische Zeitung* **59**, 353–358.

STRASBURGER, E. (1909). The minute structure of cells in relation to heredity. In

Darwin and Modern Science, pp. 102–111 (A. C. Seward, ed.), Cambridge: University Press.

STRICKLAND, S. P. (1971). Integration of medical research and health policies. *Science* 173, 1093–1103.

STROMINGER, J. L. (1970). Penicillin-sensitive enzymatic reactions in bacterial cell wall synthesis. *Harvey Lectures* 64, 179–213.

STRUNZ, F. (1913). *Die Vergangenheit der Naturforschung*, pp. 139–161, Jena: Diederich.

STURTEVANT, A. H. (1913). The linear association of six sex-linked factors in Drosophila, as shown by their mode of association. *Journal of Experimental Zoology* 14, 43–59.

SUMNER, J. B. (1926). The isolation and crystallization of the enzyme urease. *J. Biol. Chem.* 69, 435–441.

SUMNER, J. B. (1933). The chemical nature of enzymes. *Science* 78, 335–336.

SUTTON, W. S. (1903). The chromosomes in heredity. *Biological Bulletin* 4, 231–251.

SVEDBERG, T. (1937). The ultra-centrifuge and the study of high-molecular compounds. *Nature* 139, 1051–1062.

SYNGE, R. L. M. (1943). Partial hydrolysis products derived from proteins and their significance for protein structure. *Chem. Revs.* 32, 135–172.

SYNGE, R. L. M. (1962). Tsvet, Willstätter, and the use of adsorption for the purification of proteins. *Archives of Biochemistry and Biophysics* Suppl. 1, pp. 1–6.

SZABADVARY, F. (1964). Development of the pH concept. *J. Chem. Ed.* 41, 105–107.

SZENT-GYÖRGYI, A. (1924). Über den Mechanismus der Succin- und Paraphenylendiaminoxydation. Ein Beitrag zur Theorie der Zellatmung. *Biochem. Z.* 150, 195–210.

SZENT-GYÖRGYI, A. (1928). Observations on the function of peroxidase systems and the chemistry of the adrenal cortex. Description of a new carbohydrate derivative. *Biochem. J.* 22, 1387–1409.

SZENT-GYÖRGYI, A. (1937). *Studies on Biological Oxidation and some of its Catalysts*, Budapest and Leipzig: Eggenberg and Barth.

TAMMANN, G. (1895). Zur Wirkung ungeformten Fermente. *Z. physik. Chem.* 18, 426–442.

TATUM, E. L., BONNER, D., AND BEADLE, G. W. (1944). Anthranilic acid and the biosynthesis of indole and tryptophan in Neurospora. *Archives of Biochemistry* 3, 477–478.

TAYLOR, F. S. (1953). The idea of the quintessence. In *Science Medicine, and History*, Vol. I, pp. 247–265 (E. A. Underwood, ed.), London: Oxford University Press.

TEICH, M. (1965a). The origins of carbohydrate chemistry in Bohemia. *Acta historiae rerum naturalium necron technicarum*. Special issue 1, pp. 85–102.

TEICH, M. (1965b). On the historical foundations of modern biochemistry. *Clio Medica* 1, 41–57.

TEMKIN, O. (1946). Materialism in French and German physiology of the early nineteenth century. *Bull. Hist. Med.* 20, 322–327.

TESKE, A. (1969). Einstein and Smoluchowski. *Sudhoffs Arch.* **53**, 292–305.

THENARD, L. J. (1803). Mémoire sur la fermentation vineuse. *Ann. Chim.* **46**, 294–320.

THEORELL, H. (1935). Das gelbe Oxydationsferment. *Biochem. Z.* **278**, 263–290.

THEORELL, H, (1936). Keilin's cytochrome c and the respiratory mechanism of Warburg and Christian. *Nature* **138**, 687.

THEORELL, H. (1947). Heme-linked groups and mode of action of some hemoproteins. *Adv. Enzymol.* **7**, 265–303.

THOMSON, T. (1843). *Chemistry of Animal Bodies*, Edinburgh: Black.

THUDICHUM, J. L. W. (1872). *A Manual of Chemical Physiology*, London: Longmans, Green, Reader, and Dyer.

THUDICHUM, J. L. W. (1881a). On the colouring matters of the rods and the cells of the choroid coat of the retina. *Annals of Chemical Medicine* **2**, 64–76.

THUDICHUM, J. L. W. (1881b). On modern text-books as impediments to the progress of animal chemistry. *Annals of Chemical Medicine* **2**, 183–189.

THUNBERG, T. (1911). Untersuchungen über autoxydable Substanzen und autoxydable Systeme von physiologischem Interesse. *Skand. Arch. Physiol.* **24**, 90–96.

THUNBERG, T. (1920). Zur Kenntnis des intermediären Stoffwechsels und der dabei wirksamen Enzyme. *Skand. Arch. Physiol.* **40**, 1–91.

THUNBERG, T. (1926). Das Reduktions-Oxydationspotential eines Gemisch von Succinat-Fumarat. *Skand. Arch. Physiol.* **46**, 339–340.

THUNBERG, T. (1937). Biologische Aktivierung, Übertragung und endgültige Oxydation des Wasserstoffs. *Erg. Physiol.* **39**, 76–116.

TIEDEMANN, F. AND GMELIN, L. (1827). *Die Verdauung nach Versuchen*, Heidelberg: Groos.

TILDEN, W. A. (1884). On the decomposition of terpenes by heat. *J. Chem. Soc.* **45**, 410–420.

TRAUBE, M. (1858). Zur Theorie der Gährungen und Verwesungserscheinungen, wie der Fermentwirkungen überhaupt. *Ann. Phys.* **103**, 331–344.

TRAUBE, M. (1874). Ueber das Verhalten der Alkoholhefe in sauerstoffgasfreien Medien. *Ber. chem. Ges.* **7**, 872–887.

TRAUBE, M. (1878). Die chemische Theorie der Fermentwirkungen und der Chemismus der Respiration. *Ber. chem. Ges.* **11**, 1984–1992.

TRAUBE, M. (1899). *Gesammelte Abhandlungen*, Berlin: Mayer and Müller.

TRENEER, A. (1963). *The Mercurial Chemist: A Life of Sir Humphry Davy*, London: Methuen.

TROLAND, L. T. (1917). Biological enigmas and the theory of enzyme action. *Am. Natur.* **51**, 321–350.

TSCHERMAK, E. (1900). Ueber künstliche Kreuzung bei Pisum sativum. *Ber. bot. Ges.* **18**, 232–239.

TSWETT, A. (1906). Physikalisch-chemische Studien über das Chlorophyll. Die Adsorptionen. *Ber. bot. Ges.* **24**, 316–232.

TURPIN, P. J. F. (1838). Mémoire sur la cause et les effets de la fermentation alcoolique et aceteuse. *Comp. Rend.* **7**, 369–402.

TUTTON, A. E. H. (1924). *The Natural History of Crystals,* London: Kegan Paul, Trench, Trubner.

TYRRELL, H. J. V. (1964). The origin and present status of Fick's diffusion law. *J. Chem. Ed.* 41, 397–400.

UMBARGER, H. E. (1956). Evidence for a negative-feedback mechanism in the biosynthesis of isoleucine. *Science* 123, 848.

UNNA, P. G. (1911). Die Reduktionsorte und Sauerstofforte des tierischen Gewebes. *Arch. mikr. Anat.* 78, 1–21.

UREY, H. C. (1934). Deuterium and its compounds in relation to biology. *Cold. Spr. Harb. Symp.* 2, 47–56.

VALLERY-RADOT, R. (1900). *La Vie de Pasteur,* Paris: Hachette.

VAN NEIL, C. B. (1931). On the morphology and physiology of the purple and green sulphur bacteria. *Archiv für Mikrobiologie* 3, 1–112.

VAN NIEL, C. B. (1949). The "Delft school" and the rise of general microbiology. *Bacteriological Reviews* 13, 161–174.

VAN SLYKE, D. D. AND JACOBS, W. A. (1944). Phoebus Aaron Theodor Levene. *Biog. Mem. Natl. Acad. Sci.* 23, 75–126.

VAN SLYKE, D. D. AND MEYER, G. M. (1913). The fate of protein digestion products in the body. III. The absorption of amino-acids from the blood by the tissues. *J. Biol. Chem.* 16, 197–212.

VERNON, H. M. (1911). The indophenol oxidase of mammalian and avian tissues. *J. Physiol.* 43, 96–108.

VERNON, H. M. (1912). Die Abhängigkeit der Oxydasewirkung von Lipoiden. *Biochem. Z.* 47, 374–395.

VERWORN, M. (1903). *Die Biogenhypothese,* Jena: Fischer.

VICKERY, H. B. (1931). Biographical memoir of Thomas Burr Osborne 1859–1929. *Biog. Mem. Natl. Acad. Sci.* 14, 261–304.

VICKERY, H. B. (1946). The early years of the Kjeldahl method to determine nitrogen. *Yale J. Biol. and Med.* 18, 473–516.

VICKERY, H. B. (1950). The origin of the word protein. *Yale J. Biol. and Med.* 22, 387–393.

VICKERY, H. B. AND OSBORNE, T. B. (1928). A review of hypotheses of the structure of proteins. *Physiol. Revs.* 8, 393–446.

VICKERY, H. B. AND SCHMIDT, C. L. A. (1931). The history of the discovery of the amino acids. *Chem. Revs.* 9, 169–318.

VOIT, C. (1870). Ueber die Entwicklung der Lehre von der Quelle der Muskelkraft und einiger Theile der Ernährung seit 25 Jahren. *Z. Biol.* 6, 303–401.

VOIT, C. (1891). Ueber die Glykogenbildung nach Aufnahme verschiedener Zuckerarten. *Z. Biol.* 28, 245–292.

VOLHARD, J. (1909). *Justus von Liebig,* Leipzig: Barth.

VRIES, H. DE (1900). Das Spaltungsgesetz der Bastarde. Vorläufige Mittheilung. *Ber. bot. Ges.* 18, 83–90.

VRIES, H. DE (1910). *Intracellular Pangenesis* (translated by C. S. Gager), Chicago: Open Court.

WADA, M. (1930). Über Citrullin, eine neue Aminosäure in Presssaft der Wassermelone, Citrullus vulgaris shrad. *Biochem. Z.* **224**, 420–429.

WADDINGTON, C. H. (1939). The physicochemical nature of the chromosome and the gene. *Am. Natur.* **73**, 300–314.

WADDINGTON, C. H. (1969). Some European contributions to the prehistory of molecular biology. *Nature* **221**, 318–321.

WAGNER, R. P. AND MITCHELL, H. K. (1964). *Genetics and Metabolism*, New York: Wiley.

WAITE, A. E. (1888). *Lives of Alchemystical Philosophers*, London: Redway.

WAKSMAN, S. A. (1953). *Sergei N. Winogradsky: His Life and Work*, New Brunswick: Rutgers University Press.

WALDEN, P. (1928). Die Bedeutung der Wöhlerschen Harnstoffsynthese. *Naturwissenschaften* **16**, 835–849.

WALDEYER, W. (1888). Ueber Karyokinese und ihre Bezeihung zu den Befruchtungsvorgängen. *Arch. mikr. Anat.* **32**, 1–122.

WALDSCHMIDT-LEITZ, E. (1933). The chemical nature of enzymes. *Science* **78**, 189–190.

WALLACH, O. (1901). *Briefwechsel zwischen J. Berzelius und F. Wöhler*, Leipzig: Engelmann.

WARBURG, O. (1911). Über Beeinflussung der Sauerstoffatmung. *Z. physiol. Chem.* **70**, 413–432.

WARBURG, O. (1913). Über sauerstoffatmende Körnchen aus Leberzellen und über Sauerstoffatmung in Berkfeld-Filtraten wässeriger Leberextrakte. *Pflügers Arch.* **154**, 599–617.

WARBURG, O. (1914a). Beiträge zur Physiologie der Zelle, insbesondere über die Oxydationsgeschwindigkeit in Zellen. *Erg. Physiol.* **14**, 253–337.

WARBURG, O. (1914b). Über die Rolle des Eisens in der Atmung des Seeigeleies nebst Bemerkungen über einige durch Eisen beschleunigte Oxydationen. *Z. physiol. Chem.* **92**, 231–256.

WARBURG, O. (1923). Über die Grundlagen der Wielandschen Atmungstheorie. *Biochem. Z,* **142**, 518–523.

WARBURG, O. (1924). Über Eisen, den sauerstoffübertragenden Bestandteil des Atmungsferments. *Biochem. Z.* **152**, 479–494.

WARBURG, O. (1925a). Über Eisen, den sauerstoffübertragenden Bestandteil des Atmungsferments. *Ber. chem. Ges.* **58**, 1001–1011.

WARBURG, O. (1925b). Bemerkung zu einer Arbeit von M. Dixon und S. Thurlow sowie zu einer Arbeit von G. Ahlgren. *Biochem. Z.* **163**, 252.

WARBURG, O. (1926). *Über den Stoffwechsel der Tumoren*, Berlin: Springer.

WARBURG, O. (1927). Über die Wirkung von Kohlenoxyd und Stickoxyd auf Atmung und Gärung. *Biochem. Z.* **189**, 354–380.

WARBURG, O. (1928). *Über die Katalytische Wirkung der Lebendigen Substanz*, Berlin: Springer.

WARBURG, O. (1929). Atmungsferment und Oxydasen. *Biochem. Z.* **214**, 1–3.

WARBURG, O. (1932). Das sauerstoffübertragende Ferment der Atmung. *Z. angew. Chem.* **45**, 1–6.

WARBURG, O. (1938). Chemische Konstitution von Fermenten. *Ergebnisse der Enzymforschung* **7**, 210–245.

WARBURG, O. AND CHRISTIAN, W. (1931). Über Aktivierung der Robisonschen Hexose-mono-phosphorsäure in roten Blutzellen und die Gewinnung aktivierender Fermentlösungen. *Biochem. Z.* **242**, 206–227.

WARBURG, O. AND CHRISTIAN, W. (1932). Über ein neues Oxydationsferment und sein Absorptionsspektrum. *Biochem. Z.* **254**, 438–458.

WARBURG, O. AND CHRISTIAN, W. (1933). Über das gelbe Ferment und seine Wirkungen. *Biochem. Z.* **266**, 377–411.

WARBURG, O. AND CHRISTIAN, W. (1935). Co-Fermentproblem. *Biochem. Z.* **275**, 464.

WARBURG, O. AND CHRISTIAN, W. (1936). Pyridin, der wasserstoffübertragende Bestandteil von Gärungsfermenten (Pyridin-Nucleotide). *Biochem. Z.* **287**, 291–328.

WARBURG, O. AND CHRISTIAN, W. (1938). Bemerkung über gelbe Fermente. *Biochem. Z.* **298**, 368–377.

WARBURG, O. AND CHRISTIAN, W. (1939). Isolierung und Kristallisation des Proteins des oxydierenden Gärungsferments. *Biochem. Z.* **303**, 40–68.

WARBURG, O., CHRISTIAN, W., AND GRIESE, A. (1935a). Die Wirkungsgruppe des Co-Ferments aus roten Blutzellen. *Biochem. Z.* **279**, 143–144.

WARBURG, O., CHRISTIAN, W., AND GRIESE, A. (1935b). Wasserstoffübertragendes Co-Ferment, seine Zusammensetzung und Wirkungsweise. *Biochem. Z.* **282**, 157–205.

WARBURG, O., KUBOWITZ, F., AND CHRISTIAN, W. (1930a). Kohlenhydratverbrennung durch Methämoglobin. *Biochem. Z.* **221**, 494–497.

WARBURG, O., KUBOWITZ, F., AND CHRISTIAN, W. (1930b). Über die katalytische Wirkung von Methylenblau in lebenden Zellen. *Biochem. Z.* **227**, 245–271.

WARBURG, O. AND NEGELEIN, E. (1929). Über das Absorptionsspektrum des Atmungsferments. *Biochem. Z.* **214**, 64–100.

WASTENEYS, H. AND BORSOOK, H. (1930). The enzymatic synthesis of protein. *Physiol. Revs.* **10**, 110–145.

WATSON, J. D. (1968). *The Double Helix*, New York: Atheneum.

WATSON, J. D. AND CRICK, F. H. C. (1953a). A structure for deoxyribose nucleic acid. *Nature* **171**, 737–738.

WATSON, J. D. AND CRICK, F. H. C. (1953b). Genetical implications of the structure of deoxyribonucleic acid. *Nature* **171**, 964–967.

WEAVER, W. (1970). Molecular biology: origin of the term. *Science* **170**, 581–582.

WEINER, D. B. (1968). *Raspail, Scientist and Reformer,* New York: Columbia University Press.

WEINHOUSE, S., MEDES, G., AND FLOYD, N. F. (1944). Fatty acid metabolism. The mechanism of ketone body synthesis from fatty acids, with isotopic carbon as tracer. *J. Biol. Chem.* **155**, 143–151.

WEISMANN, A. (1883). *Ueber die Vererbung,* Jena: Fischer.

WEISMANN, A. (1893). *The Germ-Plasm: A Theory of Heredity* (translated by W. N. Parker and H. Rönnfeldt), New York: Scribner's.

WEISS, P. (1962). From cell to molecule. In *The Molecular Control of Cellular Activity,* pp. 1–72 (J. M. Allen, ed.), New York: McGraw-Hill.

WERKMAN, C. H. AND WOOD, H. G. (1942). Heterotrophic assimilation of carbon dioxide. *Adv. Enzymol.* 2, 139–182.

WERNER, E. A. (1923). *The Chemistry of Urea*, London: Longmans, Green.

WEYL, T. (1877). Beiträge zur Kenntnis thierischer und pflanzlicher Eiweisskörper. Z. *physiol. Chem.* 1, 72–100.

WHEELER, L. P. (1951). *Josiah Willard Gibbs*, New Haven: Yale University Press.

WHELDALE, M. (1911). On the direct guaiacum reaction given by plant extracts. *Proc. Roy. Soc.* B84, 121–124.

WHITE, J. H. (1932). *The History of the Phlogiston Theory*, London: Arnold.

WICHMANN, A. (1899). Ueber die Krystallform der Albumine. Z. *physiol. Chem.* 27, 575–593.

WIELAND, H. (1913). Über den Mechanismus der Oxydationsvorgänge. *Ber. chem. Ges.* 46, 3327–3342.

WIELAND, H. (1922a). Über den Verlauf der Oxydationsvorgänge. *Ber. chem. Ges.* 55, 3639–3648.

WIELAND, H. (1922b). Über den Mechanismus der Oxydationsvorgänge. *Erg. Physiol.* 20, 477–518.

WIGNER, E. P. (1969). Are we machines? *Proc. Am. Phil. Soc.* 113, 95–101.

WILKIE, J. S. (1961). Carl Nägeli and the fine structure of living matter. *Nature* 190, 1145–1150.

WILLIAMS, L. P. (1965). *Michael Faraday*, London: Chapman and Hall.

WILLIAMS, R. J. (1943). The chemistry and biochemistry of pantothenic acid. *Adv. Enzymol.* 3, 253–287.

WILLIS, T. (1684). *Dr. Willis' Practice of Physick*, etc. (translated by S. Pordage), London: Dring, Harper and Leigh.

WILLSTÄTTER, R. (1922). Über Isolierung von Enzyme. *Ber. chem. Ges.* 55, 3601–3623.

WILLSTÄTTER, R. (1926). Über Sauerstoff-übertragung in der lebenden Zelle. *Ber. chem. Ges.* 59, 1871–1876.

WILLSTÄTTER, R. (1927). *Problems and Methods in Enzyme Research*, Ithaca: Cornell University Press.

WILLSTÄTTER, R. (1949). *Aus meinem Leben* (edited by A. Stoll), Weinheim: Verlag Chemie.

WILSON, E. B. (1895). *An Atlas of the Fertilization and Karyokinesis of the Ovum*, New York: Macmillan.

WILSON, E. B. (1896). *The Cell in Development and Inheritance*, New York: Macmillan.

WILSON, E. B. (1923). The physical basis of life. *Science* 57, 277–286.

WILSON, E. B. (1925). *The Cell in Development and Heredity* (third edition), New York: Macmillan.

WINOGRADSKY, S. (1887). Ueber Schwefelbacterien. *Botanische Zeitung* 45, 606–610.

WÖHLER, F. (1842). Ueber die im lebenden Organismus vor sich gehende Umwandlung der Benzoësäure in Hippursäure. *Ann. Phys.* 56, 638–641.

WÖHLER, F. AND FRERICHS, F. (1848). Ueber die Veränderungen, welche namentlich organische Stoffe bei ihrem Uebergang in den Harn erlangen. *Ann.* 65, 335–349.

WOHL, A. (1907). Die neuere Ansichten über den chemischen Verlauf der Gärung. *Biochem. Z.* **5**, 45–64.

WOLKOW, M. AND BAUMANN, E. (1891). Ueber das Wesen der Alkaptonurie. *Z. physiol. Chem.* **15**, 228–285.

WOOD, H. G. AND WERKMAN, C. H. (1936). The utilization of CO_2 in the dissimilation of glycerol by the propionic acid bacteria. *Biochem. J.* **30**, 48–53.

WOODS, D. D. (1940). The relation of *p*-aminobenzoic acid to the mechanism of the action of sulphanilamide. *British Journal of Experimental Pathology* **21**, 74–89.

WOODS, D. D. (1953). The integration of research on the nutrition and metabolism of micro-organisms. *Journal of General Microbiology* **9**, 151–173.

WOODWARD, R. B. AND BLOCH, K. (1953). The cyclization of squalene in cholesterol synthesis. *J. Am. Chem. Soc.* **75**, 2023.

WOOLLEY, D. W. (1952). *A Study of Antimetabolites,* New York: Wiley.

WORM-MÜLLER, J. (1874). Zur Kenntnis der Nucleine. *Pflügers Arch.* **8**, 190–194.

WORTMANN, J. (1880). Ueber die Beziehungen der intramolekularen zur normalen Athmung der Pflanze. *Botanische Zeitung* **38**, 25–27.

WRIGHT, S. (1917). Color inheritance in mammals. *Journal of Heredity* **8**, 224–235.

WRIGHT, S. (1945). Physiological aspects of genetics. *Annual Review of Physiology* **7**, 75–106.

WRÓBLEWSKI, A. (1901). Ueber den Buchner'schen Hefepresssaft. *J. prakt. Chem.* **172**, 1–70.

WURTZ, A. (1880). *Traité de Chimie Biologique,* Paris: Masson.

WYATT, H. V. (1972). When does information become knowledge? *Nature* **235**, 86–89.

YOUNG, F. G. (1937). Claude Bernard and the theory of the glycogenic function of the liver. *Ann. Sci.* **2**, 47–83.

ZACHARIAS, E. (1881). Ueber die chemische Beschaffenheit des Zellkerns. *Botanische Zeitung* **39**, 169–176.

ZEILE, K. (1935). Über Cytochrom c. *Z. physiol. Chem.* **236**, 212–215.

ZEKERT, O. (1931). *Carl Wilhelm Scheele,* Mittenwald: Nemayer.

ZEKERT, O. (1963). *Carl Wilhelm Scheele,* Stuttgart: Wissenschaftliche Verlagsgesellschaft.

ZILVA, S. S. (1928). The antiscorbutic fraction of lemon juice VII. *Biochem. J.* **22**, 779–785.

ZINOFFSKY, O. (1886). Ueber die Grösse des Hämoglobinmoleküls. *Z. physiol. Chem.* **10**, 16–34.

Abbreviations of Serial Publications

Adv. Carb. Chem.	Advances in Carbohydrate Chemistry
Adv. Enzymol.	Advances in Enzymology
Adv. Protein Chem.	Advances in Protein Chemistry
Am. J. Physiol.	American Journal of Physiology
Am. Natur.	American Naturalist
Ann.	Annalen der Pharmacie (1832–1839)
	Annalen der Chemie und Pharmacie (1840–1873)
	Justus Liebigs Annalen der Chemie (1873–)
Ann. Chim.	Annales de Chimie (1789–1815)
	Annales de Chimie et de Physique (1816–1914)
Ann. N. Y. Acad. Sci.	Annals of the New York Academy of Sciences
Ann. Phys.	Poggendorf's Annalen der Physik und Chemie
Ann. Rev. Biochem.	Annual Review of Biochemistry
Ann. Sci.	Annals of Science
Arch. Anat. Physiol.	Müllers Archiv für Anatomie, Physiologie, und wissenschaftliche Medizin
Arch. exp. Path. Pharm.	Archiv für experimentelle Pathologie und Pharmakologie
Arch. mikr. Anat.	Archiv für mikroskopische Anatomie
Ber. bot. Ges.	Berichte der deutschen botanischen Gesellschaft
Ber. chem. Ges.	Berichte der deutschen chemischen Gesellschaft
Biochem. J.	Biochemical Journal
Biochem. Z.	Biochemische Zeitschrift
Biog. Mem. F. R. S.	Biographical Memoirs of Fellows of the Royal Society
Biog. Mem. Natl. Acad. Sci.	Biographical Memoirs of the National Academy of Sciences
Biol. Revs.	Biological Reviews
Bull. Hist. Med.	Bulletin of the History of Medicine
Bull. Soc. Chim. Biol.	Bulletin de la Société Chimie Biologique
Chem. Revs.	Chemical Reviews
Cold Spr. Harb. Symp.	Cold Spring Harbor Symposia on Quantitative Biology
Comp. Rend.	Comptes Rendus Hebdomadaires des Séances de l'Académie des Sciences, Paris
Comp. Rend. Carlsberg	Comptes Rendus des Travaux du Laboratoire Carlsberg
Erg. Physiol.	Ergebnisse der Physiologie
Jahrb. wiss. Bot.	Jahrbücher für wissenschaftliche Botanik
Jahres-Ber.	Jahres-Bericht über die Fortschritte der Chemie
J. Am. Chem. Soc.	Journal of the American Chemical Society
J. Biol. Chem.	Journal of Biological Chemistry
J. Chem. Ed.	Journal of Chemical Education
J. Chem. Soc.	Journal of the Chemical Society, London
J. Gen. Physiol.	Journal of General Physiology
J. Hist. Biol.	Journal of the History of Biology

J. Hist. Med.	Journal of the History of Medicine and Allied Sciences
J. Physiol.	Journal of Physiology
J. prakt. Chem.	Journal für praktische Chemie
Med. chem. Unt.	Medicinisch-chemische Untersuchungen (Tübingen)
Persp. Biol. and Med.	Perspectives in Biology and Medicine
Pflügers Arch.	Pflügers Archiv für die gesamte Physiologie des Menschen und der Tiere
Phil. Mag.	London, Edinburgh, Dublin Philosophical Magazine and Journal of Science
Phil. Trans.	Philosophical Transactions of the Royal Society of London
Physiol. Revs.	Physiological Reviews
Proc. Am. Phil. Soc.	Proceedings of the American Philosophical Society
Proc. Natl. Acad. Sci.	Proceedings of the National Academy of Sciences (U. S.)
Proc. Roy. Soc.	Proceedings of the Royal Society of London
Skand. Arch. Physiol.	Skandinavisches Archiv für Physiologie
Soc. chem. Paris	Société Chimique de Paris, Leçons de Chimie
Sudhoffs Arch.	Sudhoffs Archiv für Geschichte der Medizin und der Naturwissenschaften
Symp. Soc. Exp. Biol.	Symposia of the Society of Experimental Biology
Trans. Conn. Acad.	Transactions of the Connecticut Academy of Arts and Sciences
Yale J. Biol. and Med.	Yale Journal of Biology and Medicine
Z. angew. Chem.	Zeitschrift für angewandte Chemie
Z. anorg. Chem.	Zeitschrift für anorganische Chemie
Z. Biol.	Zeitschrift für Biologie
Z. physik. Chem.	Zeitschrift für physikalische Chemie
Z. physiol. Chem.	Zeitschrift für physiologische Chemie

Index of Names

Abbe, Ernst Karl [1840-1905], 195
Abderhalden, Emil [1877-1950], 152, 153, 155, 415, 432, 505
Abel, John Jacob [1857-1938], 171, 535
Abelous, Jacques Émile [1864-1940], 317, 505
Adair, Gilbert Smithson [1896-], 136, 137, 286
Adamkiewicz, Albert [1850-1921], 119
Addison, Thomas [1793-1860], 493
Agricola, Georgius [1494-1555], 28
Agrippa, Heinrich Cornelius [1486-1535], 28
Ahrens, Felix Benjamin [1863-1910], 22, 505
Albertus Magnus [c. 1200-1280], 24, 26
Alexandrov, Nicholas Alexandrovich [1858-1935], 133
Allen, Richard Sweetnam [1896-], 171, 505
Allen, William [1770-1843], 268, 505
Alloway, James Lionel [1900-1954], 247
Altmann, Richard [1852-1900], 190, 191, 197, 389-391, 505
Ambrosioni, Felice [1790-1843], 406
Ames, Bruce Nathan [1928-], 498
Amici, Giovanni Battista [1786-1863], 488
Ampère, André Marie [1775-1836], 41, 105
Anderson, John F. [1873-1958], 125
Anderson, Thomas Foxen [1911-], 219, 252
Andrews, Thomas [1813-1885], 273
Andronicos of Rhodes [fl. c. 100 B.C.], 23
Anschütz, Richard [1852-1937], 104, 505
Anson, Mortimer Louis [1901-1968], 146, 160, 173, 311, 312, 505

Appert, François Nicolas [1750-1841], 42
Aquinas, Thomas [c. 1225-1274], 24
Araki, Torasaburo [1866-1942], 207, 248
Aristotle [384-322 B.C.], 23, 24, 26-28, 397, 485
Armstrong, Edward Frankland [1878-1945], 80, 505, 512
Armstrong, Henry Edward [1848-1937], 80, 130, 141, 505, 512
Arnon, Daniel Israel [1910-], 465, 468, 505
Arreguin, Barbarin [1917-], 473
Arrhenius, Svante August [1859-1927], 18, 82, 141, 231, 325
d'Arsonval, Jacques Arsène [1851-1940], 494, 511
Arthus, Nicolas Maurice [1862-1945], 77, 505
Aschoff, Karl Albert Ludwig [1866-1942], 458
Ascoli, Alberto [1877-1957], 192
Astbury, William Thomas [1898-1961], 14, 161, 163, 164, 220-223, 505, 506, 508
Astier, Charles Bernard [1771-1837], 43, 506
Aston, Francis William [1877-1945], 458
Atkinson, Daniel Edward [1921-], 498, 506
Auerbach, Charlotte [1899-], 249, 506
Auhagen, Ernst [1904-], 359, 369
Aulie, Richard Paul [1926-], 403, 506
Averroës [1126-1198], 23
Avery, Oswald Theodore [1877-1955], 12, 246-249, 252, 253, 260, 506, 525
Avicenna [980-1037], 23
Avogadro, Amedeo [1776-1856], 17, 41, 105, 138, 139

Bach, Aleksei Nikolaevich [1857-1946],

555

Northern Michigan University

3 1854 001 178 410

EZNO
QP511 F78
Molecules and life; historical essays on